Neglected Tropical Diseases and Conditions of the Nervous System

Marina Bentivoglio • Esper A. Cavalheiro
Krister Kristensson • Nilesh B. Patel
Editors

Neglected Tropical Diseases and Conditions of the Nervous System

Springer

Editors
Marina Bentivoglio
Department of Neurological
and Movement Sciences
University of Verona
Verona, Italy

Esper A. Cavalheiro
Department of Neurology and Neurosurgery
Escola Paulista de Medicina/Universidade
Federal de São Paulo (EPM/UNIFESP)
São Paulo, Brazil

Krister Kristensson
Department of Neuroscience
Karolinska Institute
Stockholm, Sweden

Nilesh B. Patel
Department of Physiology
School of Medicine
College of Health Science
University of Nairobi
Nairobi, Kenya

ISBN 978-1-4614-8099-0 ISBN 978-1-4614-8100-3 (eBook)
DOI 10.1007/978-1-4614-8100-3
Springer New York Heidelberg Dordrecht London

Library of Congress Control Number: 2014935159

Printed on acid-free paper

Springer is part of Springer Science+Business Media (www.springer.com)

Preface

This volume on the "*Neglected Tropical Diseases and Conditions of the Nervous System*" arose from a symposium on neglected tropical diseases held at the 8th International Brain Research Organization (IBRO) World Congress of Neuroscience, Florence, Italy, in 2011. This was only the second time a symposium on this topic had been held at a major neuroscience conference, the first being at the 7th IBRO World Congress, Melbourne, Australia, in 2007. Neglected tropical diseases are a group of chronic infections affecting hundreds of millions of people in impoverished areas of the world, and these diseases have been historically neglected by the public health agenda of wealthy countries. In the world regions in which neglected tropical diseases are prevalent, there are also "neglected conditions" which have a negative impact on health.

Besides the general neglect of research on neglected diseases and conditions (NTDC), they are even more neglected in the basic and clinical neuroscience community. Yet, NTDC cause profound changes in brain function. Research on NTDC can not only open new avenues for therapeutic interventions but also provide novel models for the study of brain function.

This book addresses such issues to graduate students and postdoctoral fellows in different fields of biomedicine, including travel medicine, residents in different clinical disciplines, neurologists, physicians, and medical personnel, targeting primarily a neuroscience readership to increase the awareness of NTDC and how neuroscience can play a role in combating NTDC and in better management of the patients.

The decline, in high-income countries, of infectious diseases of the nervous system among the general population, due to discoveries of effective treatments and vaccinations, was followed by a general decline in research and funding on these diseases. Moreover, as the European colonial period ended, research on tropical diseases, needed for the colonization, became marginalized. As repeatedly emphasized in editorials and articles, pharmaceutical companies do not have the financial incentive to commit to NTDC the vast resources needed to produce drugs, because the return on their investments is not sufficient. The regions of the world where NTDC cause a heavy toll on lives and standard of living have neither the financial nor scientific resources to undertake sustained research in NTDC. It is a Catch 22 situation: those who need the most do not have the resources and those who have the resources do not have the need.

In this book, we begin with an overview of nervous system involvement in NTDC, followed by the spread of neglected tropical diseases by migration and modern travel to other parts of the world, and by the discussion of the stigma attached to these diseases. The current limited state of knowledge of consequences for the central nervous system of major parasitic infections is discussed: nematodes, amoebae, schistosomes, tape worms, and trypanosomes. Then bacterial infections (leprosy), and the viral infections causing rabies, Japanese encephalitis and dengue fever. While the focus is on infectious diseases, neglected conditions of the nervous system, with severe effects on the health and economy of low-incomes countries, are covered: khat addiction, konzo, dangers in food preservation, and nervous system complications due to snake bite envenoming. Finally, long-term sequels of the infections are discussed: epilepsy, cognitive dysfunction, and behavioral disorders, even after successful treatment following malaria and HIV infection. While these are not classified as neglected diseases, the study of epilepsy and cognitive and behavioral impairments following cerebral malaria and HIV infection is a neglected area of investigation.

We greatly appreciate the work the authors of the various chapters and we thank them for their, at times solitary, effort in the study of NTDC. Many of them work in countries plagued by NTDC and are therefore well familiar with the needs of research on these issues. IBRO should also be acknowledged, not only for organizing the conferences mentioned above but also for the extensive involvement in organizing schools and courses in NTDC-afflicted countries, activities in which the editors of this volume have had long-term experience. We are also grateful for the support by the NIH-Fogarty program on "BRAIN Disorders in the Developing World". Regional neuroscience associations and activities should also be acknowledged, and, in particular, the Society of Neuroscientists of Africa (SONA) and the Latin-American Summer School on Epilepsy (LASSE) for their efforts in bringing NTDC out from neglect. Finally, we would like to thank Springer for stimulating this publication and the editorial team for their support.

Verona, Italy — Marina Bentivoglio
São Paulo, Brazil — Esper A. Cavalheiro
Stockholm, Sweden — Krister Kristensson
Nairobi, Kenya — Nilesh B. Patel

Contents

Part I The "Global World Village" and Neglected Diseases

Overview of Neglected Tropical Diseases and Conditions of the Nervous System: Past, Present and Perspectives 3
Marina Bentivoglio, Esper A. Cavalheiro, Krister Kristensson, and Nilesh B. Patel

Travel, Migration and Neglected Tropical Diseases 21
Zeno Bisoffi, Dora Buonfrate, and Andrea Angheben

Stigma in Neurological Diseases in the Tropics 45
Earnest N. Tabah, Faustin Yepnjio, and Alfred K. Njamnshi

Part II Neglected Infectious Diseases of the Nervous System

Nematode Infections: Neurological Involvement and Neurobiology 67
Pewpan M. Intapan, Wanchai Maleewong, and Yukifumi Nawa

Amoebiasis: Neurological Involvement and Neurobiology 93
Ruqaiyyah Siddiqui and Naveed Ahmed Khan

Neuroschistosomiasis 111
Teresa Cristina de Abreu Ferrari

Neurocysticercosis: Neurology and Neurobiology 127
Arturo Carpio and Agnès Fleury

Onchocerciasis: Neurological Involvement 147
Alfred K. Njamnshi, Anne-Cécile Zoung-Kanyi Bissek, and Daniel Etya'ale

Human African Trypanosomiasis: A Highly Neglected Neurological Disease 165
Alain Buguet, Ghislain Mpanzou, and Marina Bentivoglio

Neurobiology of African Trypanosomiasis 183
Willias Masocha, Krister Kristensson, and Martin E. Rottenberg

Chagas Disease: Neurology and Neurobiology 201
José Rodrigues Coura

Neurodegeneration in Leprosy: Insights from Model Systems and Patients 217
Toshihiro Masaki and Anura Rambukkana

Rabies: Neurology 233
Alan C. Jackson

Rabies: Neurobiology 251
Monique Lafon

Japanese Encephalitis and Dengue Virus Infection: Neurology and Neurobiology 269
Usha K. Misra and Jayantee Kalita

Part III Neglected Toxic Disorders of the Nervous System

Mental Health Problems Associated with the Use and Abuse of Khat (*Catha edulis*) 293
Michael Odenwald

Neurobiology of Khat (*Catha edulis* Forsk) 307
Nilesh B. Patel

Konzo: Neurology of a Permanent and Non-progressive Motor Neuron Disorder Associated with Food (Cassava) Toxicity 321
Daniel Okitundu, Dieudonné Mumba, and Desiré Tshala-Katumbay

Food Preservation, Snake Venoms and Stroke in the Tropics 335
Albert K. Akpalu

Part IV Sequels and Consequences

Infectious Causes of Epilepsy? 355
Charles R. Newton and Ryan G. Wagner

Cognitive Impairment and Behavioural Disturbances Following Malaria or HIV Infection in Childhood 369
Michael Kihara, Amina Abubakar, and Charles R.J.C. Newton

Index 391

Part I
The “Global World Village” and Neglected Diseases

Overview of Neglected Tropical Diseases and Conditions of the Nervous System: Past, Present and Perspectives

Marina Bentivoglio, Esper A. Cavalheiro, Krister Kristensson, and Nilesh B. Patel

Abstract Neglected tropical diseases consist of a group of infectious diseases with the unifying features of being poverty-related, prevalent in tropical and subtropical regions, and absent from the public health agenda in wealthy countries. In the same world regions, there are "neglected conditions" which also have a negative impact on health. The nervous system involvement in neglected tropical diseases and conditions is an especially neglected field of investigation. Yet, neurological impairments in neglected tropical diseases and conditions constitute together a major category of invalidating disorders, which cause profound changes in nervous system function, often associated with severe sequels or late-onset disturbances. Here are highlighted (1) infectious diseases of the nervous system that nowadays, with the advances of vaccines and antibiotic therapy, are becoming forgotten (2), those belonging to still ravaging neglected tropical diseases, and (3) knowledge perspectives. It is time for clinical and basic neuroscience to be at the forefront and take up the fight against neglected tropical diseases and conditions, and provide means for better management of the patients.

M. Bentivoglio (✉)
Department of Neurological and Movement Sciences, School of Medicine,
University of Verona, Strada Le Grazie 8, 37134 Verona, Italy
e mail: marina.bentivoglio@univr.it; marina.bentivoglio@gmail.com

E.A. Cavalheiro
Department of Neurology and Neurosurgery, Escola Paulista de Medicina/Universidade,
Federal de São Paulo (EPM/UNIFESP), São Paulo, Brazil
e-mail: esper.cavalheiro@unifesp.br

K. Kristensson
Department of Neuroscience, Karolinska Institutet, Stockholm, Sweden
e-mail: krister.kristensson@ki.se

N.B. Patel
Department of Physiology, School of Medicine, College of Health Science,
University of Nairobi, P. O. Box 30197-00100, Nairobi, Kenya
e-mail: npatel@uonbi.ac.ke

M. Bentivoglio et al. (eds.), *Neglected Tropical Diseases and Conditions of the Nervous System*,
DOI 10.1007/978-1-4614-8100-3_1,

Keywords Brain disorders • Central nervous system • Peripheral nervous system • Autonomic nervous system • Parasites • Bacteria • Viruses • Nematodes • Epilepsy • Cognition • Toxins

1 The Neglected Nervous System in the Neglected Tropical Diseases and Conditions

The "neglected tropical diseases" (NTD) are a group of infectious diseases caused by worms, protozoa, bacteria, viruses and fungi, with high prevalence in tropical and subtropical regions, where the majority of the middle- and low-income countries are located. Of relevance, the unifying features of NTD, even when considering slightly different definitions (see Bisoffi et al. 2014), are that they are poverty-related diseases (both a cause and an effect of poverty in a vicious circle), and have been neglected by the public health agenda in high-income countries. Sufferings of the patients often go beyond the malaise of the diseases because the stigma attached to many of them leads to avoidance and neglect by the society (Tabah et al. 2014). The distribution of two or more NTD frequently overlaps in endemic countries.

The "brand name" neglected tropical diseases was introduced in 2005, when HIV/AIDS, tuberculosis and malaria became the "big three" infectious diseases (Molyneux et al. 2005). Initiatives of organizations and partnerships for disease surveillance and control, advocacy, research funding, search for novel drugs and diagnostic tools were launched at the beginning of the twenty-first century and are increasing (Hotez et al. 2007; Utzinger et al. 2012). However, the availability of treatment for many of the NTD still represents a burning issue for which public-private partnership can be very effective. The burden of NTD in terms of incidence and prevalence is still incredibly high, and has been estimated to cause every year more than half a million deaths and a loss of 57 million disability-adjusted life years (Hotez et al. 2006).

In addition to the infectious diseases that are now under the NTD umbrella, in the same world regions there are several "neglected conditions", such as khat addiction (Odenwald 2014; Patel 2014), exposure to neurotoxins (including snake bite envenoming), dangers in food preparations (Okitundu et al. 2014; Akpalu 2014). Taken together, all these represent "neglected tropical diseases and conditions" (NTDC).

Though mainly prevalent in the tropics, neglected infections, including those that affect the brain such as toxoplasmosis, are documented in the territory of the Arctic (Hotez 2010), in Oceania (Kline et al. 2013), in pockets of poverty in the USA (Hotez 2008), and Europe (Hotez and Gurwith 2011). In addition, migration and the increase in travels and in mass tourism make NTDC an increasing problem in wealthy countries (Bisoffi et al. 2014).

Importantly, many of the NTDC affect the nervous system. A search on PubMed on December 1st, 2013 using the search words "neglected tropical diseases" results in 1,020 hits; adding "brain" results in 11 hits, and adding "nervous system" in 13 hits. Most of the articles on the involvement of the nervous system in NTDC have not been published in neuroscience journals. It is therefore important to bring the neurology and neurobiology of NTDC out of neglect, and to increase, in the basic and clinical neuroscience community, awareness of the relevant problems and challenges.

2 Past

Infections of the nervous system have been a major plague to humans from time immemorial. A historical account is beyond the scopes of this chapter, but notes are here presented on neurosyphilis, poliomyelitis, and lethargic encephalitis, just to mention three of the many diseases that nowadays, with passing of the time, are becoming forgotten.

2.1 Neurosyphilis

Syphilis is named after the legend of the shepherd Syphilus, struck by the disease, recounted by a poem published by Girolamo Fracastoro (1483–1553). In a book published in 1546. Fracastoro, who pioneered the germ theory of disease, introduced the concept of contagion writing that "contagion is an infection that passes from one thing to another" (Lechevalier and Solotorovsky 1965). Syphilis is an example of a travel medicine disease due to the search for new global trading routes, since it is suspected to have been brought to Europe by the returning crew of Christopher Columbus (1451–1506) from America in 1493 (Lechevalier and Solotorovsky 1965; Sherman 2006). Syphilis swept over Europe and the rest of the world, its spread also aided by troops and invasions.

In return, the so-called "Columbian exchange" brought about the complex biological and ecological consequences (see, for example, Crosby 1972). The New World population, immunologically naïve and with a low population density, was exposed to Old World pathogens, such as the causative agents of smallpox (Fig. 1), yellow fever, tuberculosis, cholera, typhoid and bubonic plague (Sherman 2006).

Treponema pallidum was identified in 1905 as causative agent of syphilis by Fritz Schaudinn (1871–1906) and Erich Hoffmann (1868–1959). Following infection, neurosyphilis can appear as late-onset form typically 4–25 years later. The semeiology of spinal cord disease, tabes dorsalis, included, among other signs and symptoms, the characteristic Babinski sign, named after the neurologist Joseph Babinski (1857–1932). Working at Hôpital de la Pitié in Paris, Babinski described, in patients affected by tabes dorsalis, the sign named after him, which has become in neurology a cardinal sign of the pyramidal syndrome.

Another late-onset form of neurosyphilis was general paresis with dementia, which made many victims including several famous artists, philosophers and politicians. Interestingly, in 1917 malaria fever therapy became a treatment of choice of neurosyphilis after the description by Julius Wagner-Jauregg (1857–1940). To be awarded the Nobel Prize in Physiology or Medicine in 1927, he had to wait the retirement of the psychiatry expert of the Nobel Committee, in whose view

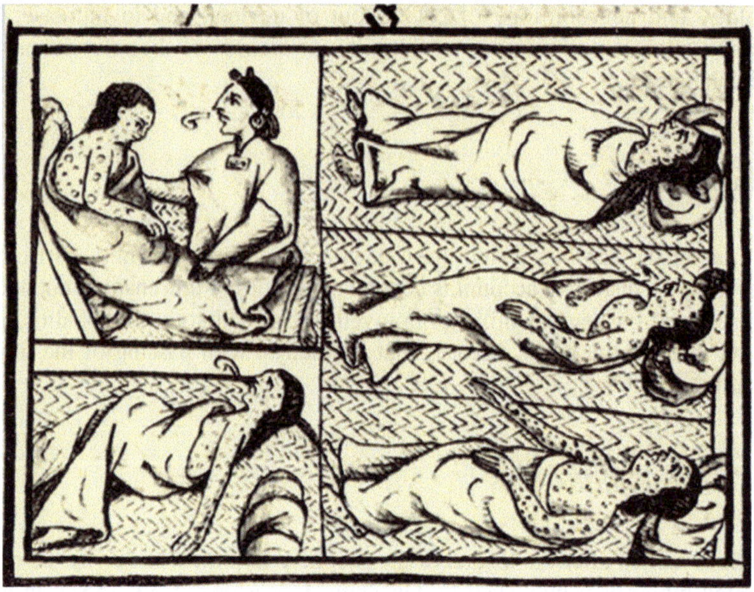

Fig. 1 Smallpox contagion in Mexico from Old World incomers, as illustrated in Bernardino de Sahagún, *Historia general de las cosas de Nueva España* (General History of the Things of New Spain, circa 1575–1580). The text says "*… an epidemic broke out, a sickness of pustules. It began in Tepeilhuitl. Large bumps spread on people; some were entirely covered… The pustules that covered people caused great desolation; very many people died of them, and many just starved to death; starvation reigned, and no one took care of others any longer*" [reproduced with permission from Lockhart (1993)]

Wagner-Jauregg's treatment based on injecting malaria was criminal (Whitrow 1990). The citation of the Prize (http://www.nobelprize.org/nobel_prizes/medicine/laureates/1927/press.html) included the following:

> How then did Wagner-Jauregg proceed to heal the unfortunate victims of this terrible disease? There is a saying «one must expel evil with evil» that might aptly have been coined as a motto for his treatment of paralysis. He healed the mental patients by infecting them with another disease—malaria.

Malaria fever treatment of general paresis was still described in detail in Lord Brain's classical textbook "Diseases of the Nervous System" as late as 1962. The treatment was given by transmission of parasites from a patient with malaria to patients suffering from general paresis or by the bite of mosquitoes infected with the relatively benign *Plasmodium vivax*, which in England could be obtained through the Ministry of Health.

The peculiar wisdom to contract malaria before being exposed to the risk of syphilis was already well known by sailors. Syphilis is, thus, an example from the past of how a microbe can control the growth of another microbe. Neurosyphilis also highlights the potentials of microbes to cause severe neuropsychiatric disorders and dementia that appear with a very late onset after the initial infection. The mechanisms of these late-onset forms of the disease, however, have not been

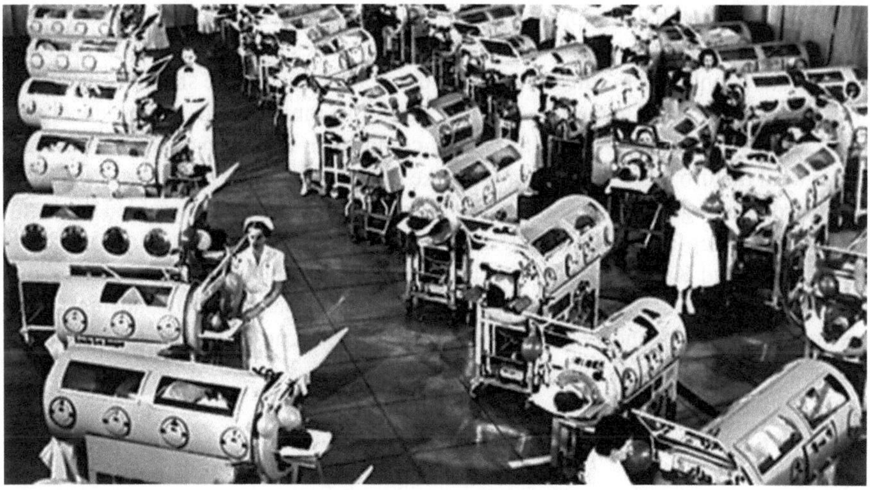

Fig. 2 Patients with "infantile paralysis" during an epidemic of poliomyelitis that filled up the iron lung ward at Rancho Los Amigos Hospital in California (USA), circa 1953. Source: Food and Drug Administration, public domain

clarified because, with the advent of efficient prevention and antibiotic treatment, neurosyphilis is no longer a major public health problem, and lost its position as major neurological or neuropsychiatric disease.

2.2 *Poliomyelitis*

This disease is typical of the twentieth century (Fig. 2): sizable epidemics appeared at the beginning of the century and were close to elimination at the end of the century. Before Karl Landsteiner (1868–1943) discovered in 1909, in cooperation with Erwin Popper (1879–1955), its causative pathogen, poliovirus, the paralytic aspects of the disease were considered to be its characteristic feature and the disease was denominated "infantile paralysis".

With the introduction of mass vaccinations, this severe, disabling, disease has almost completely disappeared. This illustrates that results obtained from experimental basic research, i.e. isolation of the virus in cell culture, can be translated into the elimination of a severe nervous system disorder. However, the elimination of polio left a number of interesting problems concerning its pathogenetic mechanisms unsolved (Nathanson 2008).

2.3 *Lethargic Encephalitis*

This disease is also called von Economo disease after Constantin von Economo (1876–1931; a disciple of Wagner-Jauregg), who presented his studies on this new nosological entity in 1929.

Lethargic encephalitis appeared in 1916 and disappeared as mysteriously as it had appeared 10 years later after sweeping three times around the world infecting millions of people and killing one-third of them. It caused severe sleep disturbances: most patients were lethargic, others were insomniac, and some developed narcolepsy-like sleep episodes. Survivors could, with a late onset, develop severe behavioral disturbances, or a Parkinson-like disorder, or both.

Neuropathological studies of victims of lethargic encephalitis heralded the localization of sleep-wakefulness regulating areas in the brain (Triarhou 2006; Bentivoglio and Kristensson 2007). The pathogen was never isolated, but the disease highlights that we should remember that infectious diseases causing severe mental disturbances and neurodegenerative disorders can suddenly become a real threat and be spread rapidly over the world.

3 Present

With discoveries and progress of vaccines and antibiotic therapy, the infectious diseases of the nervous system have, during the past 50–60 years, fallen more or less out of favor in the neurological community, and are currently handled mostly by health authorities and infectious disease clinics. Since the early 1970s, the rapidly growing field of neuroscience has focused on other important areas of clinical and basic research. Thus, with the exception of HIV infection, infectious diseases of the nervous system have become a relatively neglected area of research.

As mentioned previously, the NTD do not include "the big three" infections (malaria, tuberculosis, HIV-AIDS). It is, however, frequently forgotten that important sequels may follow or may appear as late-onset disorders caused by cerebral malaria and HIV infections in infancy, namely epilepsy and cognitive disturbances (Newton and Wagner 2014; Kihara et al. 2014). These also represent an area of investigation up till now neglected in neuroscience. Importantly, the occurrence of seizures increases misunderstandings, stigmatization, social isolation caused by NTD, though epilepsy can be often managed by treatment with anti-epileptic drugs (Scorza et al. 2013).

Infections with worms affect annually hundreds of millions of people worldwide and some are often accompanied by complications of the central nervous system (CNS), e.g. cysticercosis (Fig. 3), causing neurocysticercosis (Carpio 2014). Others spread more rarely to the CNS, e.g. schistosomiasis (Fig. 4) (Ferrari 2014), onchocerciasis (Njamnshi and Zoung-Kanyi Bissek 2014) and other nematode infections (Intapan et al. 2014). As these infections are very common and causative pathogens are widely spread, the CNS is under a constant threat for millions of people. For example, over 230 million people are estimated to require treatment for schistosomiasis yearly (WHO 2012), and the blood flukes of the genus *Schistosoma* can cause neuroschistosomiasis in a proportion of cases (Ferrari 2014).

Among protozoa targeting the nervous system, infections with *Toxoplasma gondii* are common in high-income countries and are not included in the NTD group,

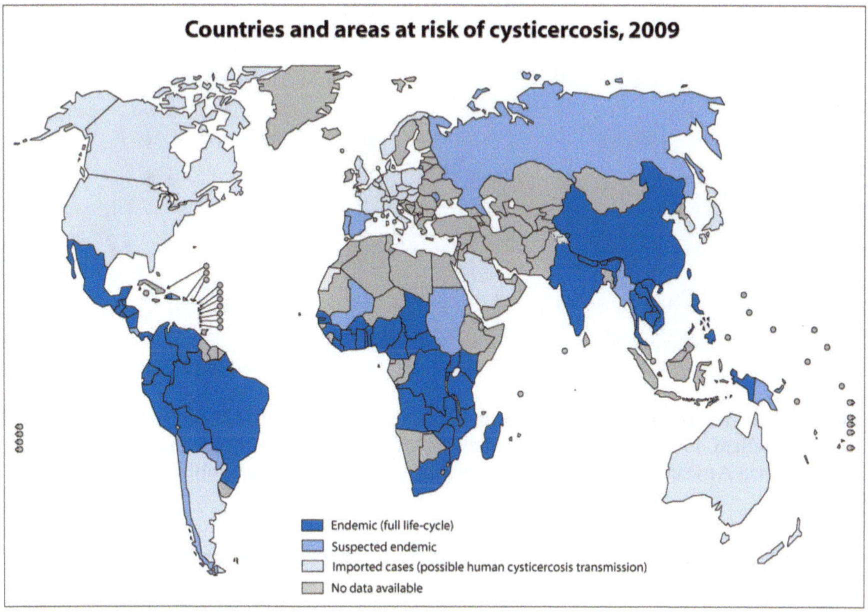

Fig. 3 Countries and areas at risk of cysticercosis in 2009 (reproduced by the permission from WHO)

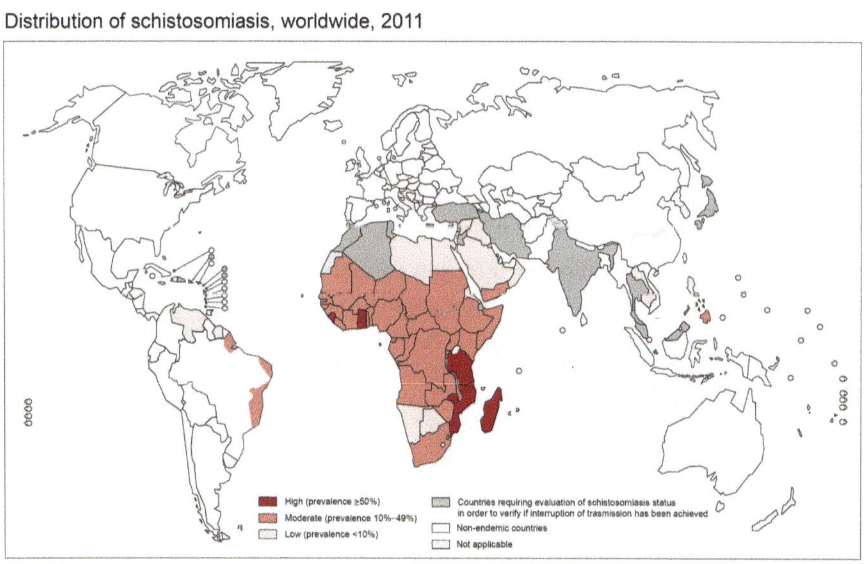

Fig. 4 WHO map on prevalence of schistosomiasis in 2011 (reproduced by the permission from WHO)

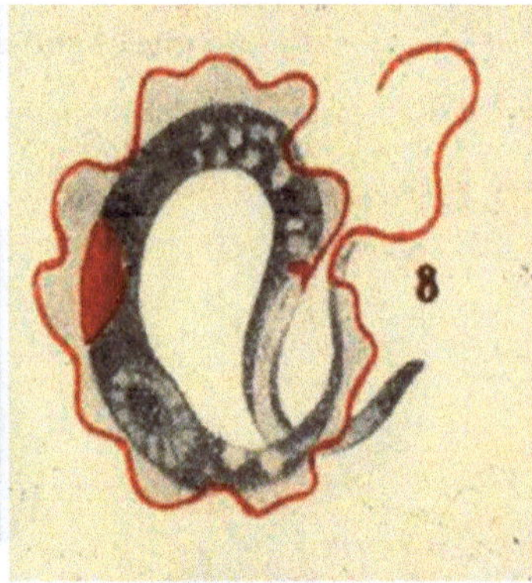

Fig. 5 *Right*: The adobe of the First Sleeping Sickness Commission in Entebbe, Uganda, in 1903 (from Castellani 1960). *Left*: historical drawing of *Trypanosoma brucei*, the protozoan which causes human African trypanosomiasis or sleeping sickness [from Bertarelli and Figuier (1915)]

while infections by *Trypanosoma (T.) brucei* in Africa and *T. cruzi* in Latin America represent severe NTD.

T. brucei, the causative agent of human African trypanosomiasis or sleeping sickness, is an example of a parasite that has predilection for the brain (Buguet et al. 2014; Masocha et al. 2014). This parasite caused large epidemics in sub-Saharan Africa at the turn of the twentieth century (Fig. 5). Two-thirds of the population in Uganda died during the 1896–1906 epidemic. African trypanosomiasis, which also causes infections of livestock (nagana), probably severely impeded the pre-colonial social and economic development of sub-Saharan Africa. With surveillance and better control of the *T. brucei* vector, the tsetse fly, the prevalence of human African trypanosomiasis rapidly declined in the second part of the twentieth century. However, the disease then re-emerged as control measures were neglected due to political unrest and civil wars, and was reported to affect >300,000 persons in 1998. Continued control measures have now brought the number of reported cases down to below 10,000 (WHO 2013), although this is probably an underestimate.

Chagas disease is the human infection caused by *T. cruzi* (Coura 2014). The infection, which affects primarily rural populations, has serious consequences for public health in Latin America, where it is estimated to have killed 10,000 people in 2008 (WHO 2012). It is an infection probably as old as mankind, as *T. cruzi* DNA was found in mummified bodies of 9,000 years ago. Carlos Chagas (1879–1933) in Brazil, in 1909, not only described the disease but also discovered the new *Trypanosoma* species he would prove to be the pathogen, and identified an insect vector, the triatomines (Fig. 6). Charles Darwin (1809–1899) is described to have been attacked by a triatomine bug during his expedition in 1835, and developed much later symptoms compatible with Chagas disease (Adler 1997). Triatomines are called "kissing bugs" because they bite near the lips and the eyes (the so-called kiss of death).

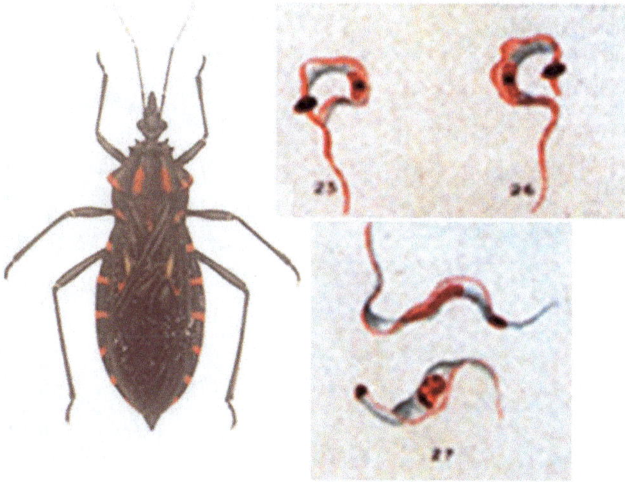

Fig. 6 Drawings from the first description, by Carlos Chagas of the vector (the triatomine bug *Panstrongylus megistus* is here depicted) and causative parasite, *Trypanosoma cruzi*, of Chagas disease or American trypanosomiasis [reproduced from Chagas (1909); kindly provided by JR Coura]

About 30 % of chronically infected people develop after a very long clinical latency the cardiac and/or gastrointestinal forms of Chagas disease, with progressive denervation of viscera which makes of this disease a major disorder of the autonomic nervous system. The pathogenesis of neurodegenerative events in Chagas disease is still enigmatic. Spread of *T. cruzi* to the brain, with reactivation of the acute form of Chagas disease, is of increasing concern in immunosuppressed individuals, including HIV-AIDS patients.

Leprosy can serve as an example of a NTD caused by bacteria. This chronic infection attacks the peripheral nervous system (Masaki and Rambukkana 2014), and is a severe social stigmatizing condition due to the accompanying conspicuous damages to the skin and extremities that are the result of sensory disturbances.

Disfigurement caused by leprosy has fired imagination in all cultures (Fig. 7). There are accounts of leprosy throughout recorded history, in ancient civilizations in Egypt (leprosy was probably described in an Egyptian papyrus written around 1550 B.C.), China, and India (where the earliest certain account of leprosy was reported between 600 and 400 B.C.). Leprosy may also represent a travel medicine disease, believed to have been brought from India to Greece by the soldiers of Alexander the Great in the fourth century B.C. (Sherman 2006). Segregation and even harsh punishment of lepers (up to burning them alive), the use of a leper clapper or bell to make them manifest and isolate lepers are well documented not only in Europe in the Middle Ages, but also elsewhere and throughout centuries. The renowned parable of the "healing-miracle" of Jesus encountering the ten lepers on

Fig. 7 Nigerian mask (*idiok ehpo* mask, Ibibio, Nigeria) representing the disfigurement caused by leprosy to indicate moral ugliness. Photo credit: Charles Davis (reproduced with permission)

Fig. 8 Illustration of the Gospel parable "The Cleansing of the Ten Lepers", or "The Grateful Samaritan" (in addition to the healing the Samaritan shares with the other nine lepers, he has become a believer in Jesus), from the "*Codex Aureus*" (the early medieval "Golden Code" that collected the Gospels) (available in the public domain)

the road to Galilee (Gospel of Luke 17:11–19) presented leprosy in the context of faith and salvation (Fig. 8).

Medical salvation of leprosy had, however, to wait for many centuries. The causative pathogen, *Mycobacterium leprae*, was discovered in 1873 by Gerhard Armauer Hansen (1841–1912) through his work in the Norwegian seaport Bergen. Hansen was sued and lost his post at the hospital when trying to prove his discovery by infecting a patient suffering from the anesthetic form of leprosy (Jay 2000). However, his efforts lead to a marked decline of the disease in Norway, and represented an obvious turning point in the fight against leprosy.

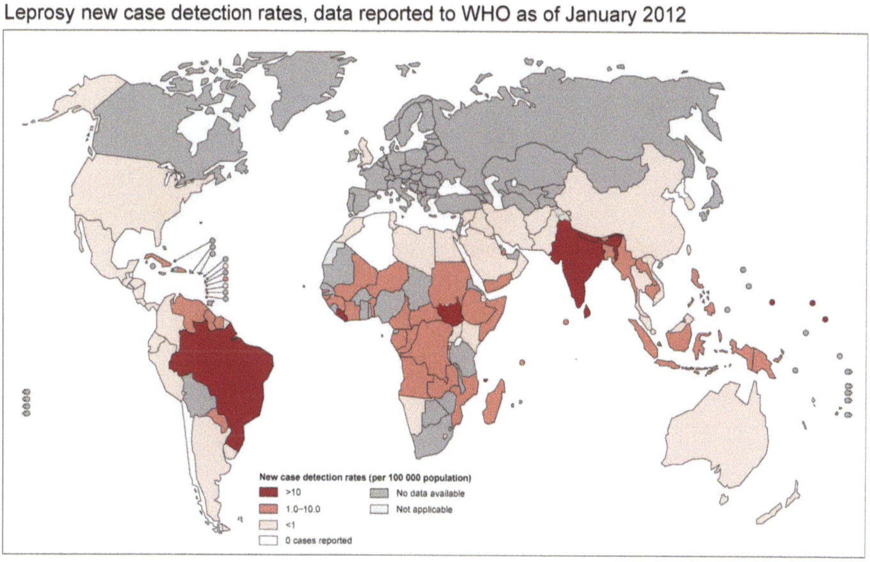

Fig. 9 WHO map of detection rates of leprosy in 2012 (reproduced by the permission from WHO)

Mycobacterium leprae colonizes Schwann cells in peripheral nerves, and replicates very slowly to cause disease after a very long incubation time, mostly 5 years; symptoms can take even 20 years to appear. Through early diagnosis and multidrug treatments, the global burden of leprosy has dropped. However, this infection is still in the process of being eliminated, but is no longer a priority as public health concern, and has therefore become a NTD. Although an estimated four million leprosy patients have been cured over the last 10 years, the official figures show that about 182,000 persons are affected in endemic pockets in India, Indonesia, Brazil and sub-Saharan countries (WHO 2012). About 219,000 new cases of this chronic disease have been reported in 2011 (WHO 2012) (Fig. 9).

Neglected viral infections of the nervous system can be exemplified by rabies (Jackson 2014; Lafon 2014). Rabies ("fury" in Latin) is another disease recorded throughout history. This terrifying disease was documented by Mesopotamian codes of law on bites by rabid dogs (Fales 2010), in early writings in China, in famous Greek and Roman writings. The development of the first rabies virus vaccine from the spinal cord of rabid rabbits by Louis Pasteur (1822–1895) is a milestone in the history of medicine. The discovery by Adelchi Negri (1876–1912) in 1903 of the characteristic intraneuronal inclusions named Negri bodies after him (Fig. 10) provided the means for a rapid histopathological diagnosis in rabid dogs, formerly requiring quarantine to verify whether the animal was rabid (Kristensson et al. 1996; Bentivoglio et al. 2011). Of note, Negri bodies are the first intraneuronal inclusions ever described, a topic which is now at the forefront of neuroscience for

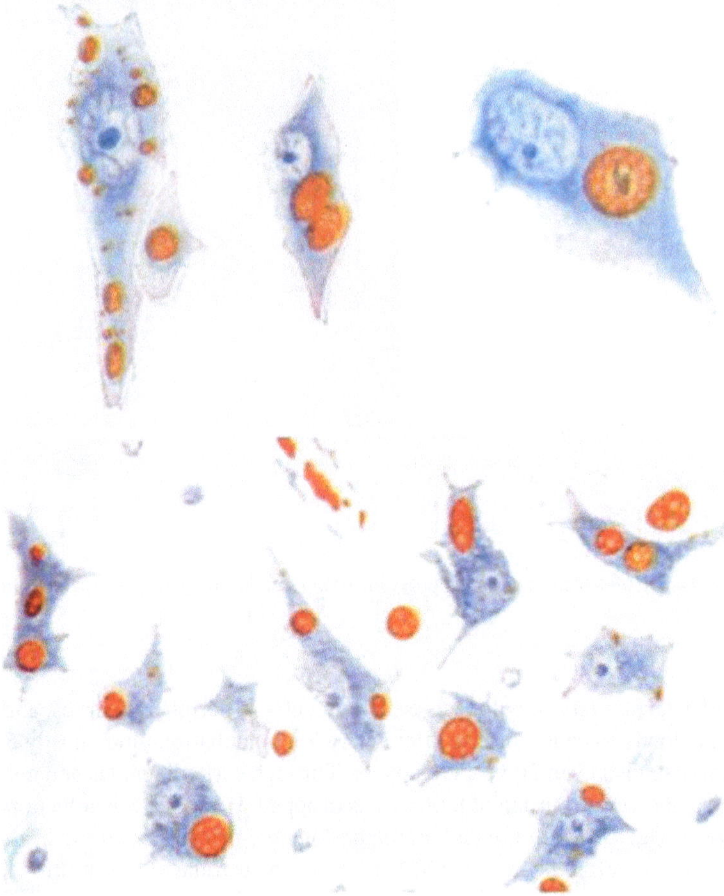

Fig. 10 Negri bodies (intraneuronal inclusion bodies pathognomic of rabies) in the hippocampal Ammon horn of a rabid dog; Mann staining [reproduced from Fermi (1951)]

the investigation of intraneuronal inclusions (due to protein misfolding or other causes) in neurological non-infectious disorders.

Rabies is an example of a virus that evades the immune response in the nervous system and does not kill neurons. However, the infection, in its furious and paralytic forms, elicits dramatic and progressive functional disturbances leading to spread of the virus and death of the infected host. Rabies occurs nowadays in more than 150 countries and territories and kills about 55,000 persons every year mostly in Asia and Africa (WHO 2012) (Fig. 11). The principal source of infection is canine rabies, and national control programs should engage in prevention and control activities, including animal control strategies, in local communities (Meslin and Briggs 2013).

The mosquito-borne Japanese encephalitis virus is another example of a neurotropic virus (Misra and Kalita 2014), with an increasing geographical spread. The

Distribution of risk levels for humans contracting, rabies, worldwide, 2009

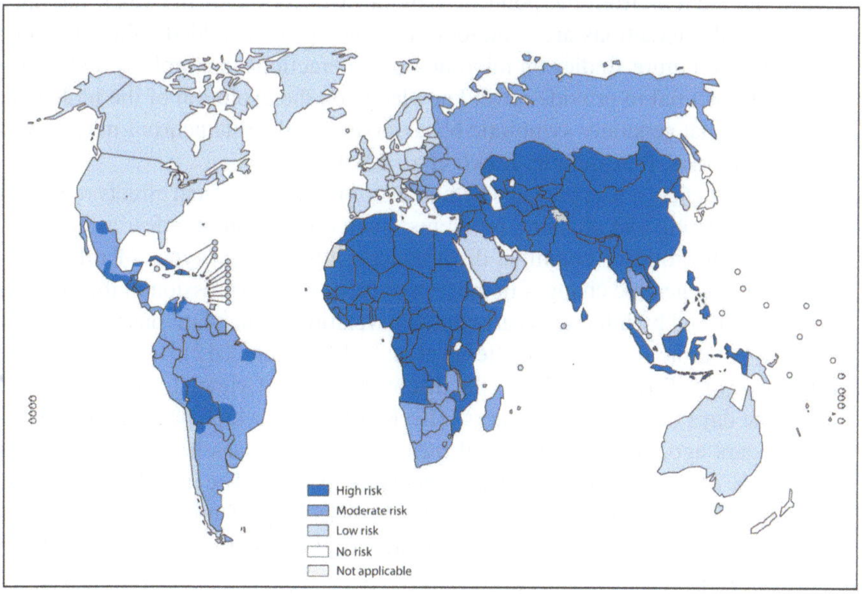

Fig. 11 WHO map on the risk for humans of contracting rabies in the world (reproduced by the permission from WHO)

disease started from Japan in 1920s to become prevalent in China by the 1950s and in Indonesia and India by the 1990s. About 50,000 people are infected each year and 15,000 of them die (Griffin 2010). In survivors, signs of Parkinsonism of still unclear pathogenetic mechanisms are common.

4 Perspectives

Why is it important that neuroscience should contribute to meet the challenges and problems of NTDC, how can this be implemented and what could be the impact of neuroscience in this field? Only a few examples are here highlighted.

(a) _Neural-immune interactions_: Studies of anti-microbial or immune response molecules involved in combating an invading pathogen that does not harm directly neuronal cells and networks may lead to a better understanding of mechanisms that regulate the balance between neuroprotective and neurodegenerative signaling during disease. Such studies may also lead to discoveries of new host-derived molecules of potential therapeutic use. Search for such molecules will be crucial for the increasing problem of resistance to antibiotics (Schwegmann and Brombacher 2008).

(b) *Microbe-host symbiotic behavior*: Although it is often useful to discuss microbe-host interactions from the point of view of favor to one or the other, the most successful interactions are symbiotic, i.e., they provide added values for both organisms. Future studies on microbe-host interactions in simple systems have a great potential to provide novel knowledge on the evolution of the molecular interplay that generates symbiotic behavior and on the role neurons may play in immunity (see, for example, Kawli et al. 2010).

(c) *Late-onset effects*: Mechanisms behind late-onset neurological effects represent very interesting and puzzling problems. Such effects can manifest as epilepsy, and/or behavioral or cognitive disturbances, after several types of infections. Potential epigenetic changes in the repertoire of gene expression in the nervous system caused by infections early in life in relation to such late-onset effects are now a hot topic for research (Bierne et al. 2012; Lubin 2012).

(d) *Pathways to discoveries*: From the discovery point of view, it should be noted that the days of "Microbe Hunters" (De Kruif 1926), which peaked about 100–150 years ago, are not over. Viral metagenomic analyses have estimated that only 1 % of the viral diversity has been explored so far. The introduction of new techniques may lead to a revolution in the discovery of infectious agents and neuroscience will be much needed to uncover their interactions and effects on the nervous system (Mokili et al. 2012).

(e) *Microbes for nervous system investigations*: Finally, in a historical perspective, microbes should be viewed not only as causes of nervous system diseases, but also as tools to unravel nervous system organization and functions. For example (1) the blood-brain barrier was detected following attempts to treat trypanosome infections with trypan dyes (see Bentivoglio and Kristensson 2014; Ribatti et al. 2006; Wainwright 2010). (2) As mentioned above, sleep-wakefulness-regulating areas in the brain were discovered in neuropathological studies of victims of lethargic encephalitis. (3) The existence of retrograde axonal transport was predicted from studies of neurotropic viruses (Kristensson and Olsson 1973).

In recent years, neuroscience has used extensively microbes for studies on the nervous system. For instance, viruses are used as tracers to visualize connectivities in neuronal circuits and as vectors for gene delivery; nematodes have been used for basic studies on microbe-host interactions, and new and conserved molecular mechanisms for mammals have thereby been found (Kawli et al. 2010).

5 Fighting Nervous System Neglect in Neglected Tropical Diseases and Conditions

It is time for the community of basic and clinical neuroscientists to increase the engagement in the endeavor of combating NTDC. This important group of disorders is a neglected area of research in the neuroscience community.

It has been emphasized that malaria, tuberculosis, HIV/AIDS were still considered neglected diseases in articles published in 2002 in major biomedical journals

(Lancet, EMBO Reports) (see Utzinger et al. 2012). The sad "upgrading" of these diseases to the ranking of "the big three" soon after shows that advocacy, and especially research and engagement of the scientific community, urging also stakeholders, can overcome neglect and indifference.

Taken together, NTDC are currently the cause of some of the most prevalent and disabling nervous system disorders worldwide. In the "battle-field" of microbe-host interactions in the nervous system, antimicrobial therapies are not enough to combat the culprits. Means have to be devised to protect the nervous tissue from life-threatening and invalidating, permanent or late-onset dysfunctions.

It is time to reintroduce into the neuroscience community some of the fascinating, intriguing challenges in understanding nervous system functioning in health and disease posed by NTDC. It is time for the neuroscience community to provide knowledge and means for better diagnosis, treatment and follow-up of patients affected by these neurological disorders. It is time for the neuroscience global world village to strengthen scientific and research capacity on these issues in the local communities of disease-endemic countries.

References

Adler J (1997) The dueling diagnoses of Darwin. JAMA 277:1275

Akpalu AK (2014) Food preservation, snake venoms and stroke in the tropics. In: Bentivoglio M, Cavalheiro EA, Kristensson K, Patel N (eds) Neglected tropical diseases and conditions of the nervous system. Springer, New York

Bentivoglio M, Kristensson K (2007) Neural-immune interactions in disorders of sleep-wakefulness organization. Trends Neurosci 30:645–652

Bentivoglio M, Kristensson K (2014) Tryps and trips: cell trafficking across the 100 year-old blood-brain barrier. Trends Neurosci, in press

Bentivoglio M, Mariotti R, Bertini G (2011) Neuroinflammation and brain infections: historical context and current perspectives. Brain Res Rev 66:152–179

Bertarelli E, Figuier L (1915) Conosci te stesso. Nozioni di fisiologia. Treves, Milan

Bierne H, Hamon M, Cossart P (2012) Epigenetics and bacterial infections. Cold Spring Harb Perspect Med. doi:10.1101/cshperspect.a010272

Bisoffi Z, Buonfrate D, Angheben A (2014) Travel, migration and neglected tropical diseases. In: Bentivoglio M, Cavalheiro EA, Kristensson K, Patel N (eds) Neglected tropical diseases and conditions of the nervous system. Springer, New York

Brain L (1962) Syphilis of the nervous system. In: Brain L (ed) Diseases of the nervous system, 6th edn. Oxford University Press, London

Buguet A, Mpanzou G, Bentivoglio M (2014) Human African trypanosomiasis: a highly neglected neurological disease. In: Bentivoglio M, Cavalheiro EA, Kristensson K, Patel N (eds) Neglected tropical diseases and conditions of the nervous system. Springer, New York

Carpio A (2014) Neurocysticercosis: neurology and neurobiology. In: Bentivoglio M, Cavalheiro EA, Kristensson K, Patel N (eds) Neglected tropical diseases and conditions of the nervous system. Springer, New York

Castellani A (1960) Microbes, men and monarchs. A doctor's life in many lands. Victor Gollancz, London

Chagas C (1909) Nova Tripanozomiaze Humana. Estudo sobre a morfolojia e o ciclo evolutivo do *Schizotrypanum cruzi* n. gen.sp. ajente etiolojico de uma nova entidade mórbida do homem. Mem Inst Oswaldo Cruz 1:159–218

Coura JR (2014) Chagas disease: neurology and neurobiology. In: Bentivoglio M, Cavalheiro EA, Kristensson K, Patel N (eds) Neglected tropical diseases and conditions of the nervous system. Springer, New York
Crosby AW (1972) The Columbian exchange: biological and cultural consequences of 1492. Greenwood, Westport, CT
De Kruif P (1926) Microbe hunters. Harcourt, Brace, New York
Fales FM (2010) Mesopotamia. Handb Clin Neurol 95:15–27
Fermi C (1951) La rabbia. Istituto Sieroterapico e Vaccinogeno Toscano "SCLAVO", Siena
Ferrari TCA (2014) Neuroschistosomiasis. In: Bentivoglio M, Cavalheiro EA, Kristensson K, Patel N (eds) Neglected tropical diseases and conditions of the nervous system. Springer, New York
Griffin D (2010) Emergence and re-emergence of viral diseases of the central nervous system. Prog Neurobiol 91:95–101
Hotez PJ (2008) Neglected infections of poverty in the United States of America. PLoS Negl Trop Dis 2:e256
Hotez PJ (2010) Neglected infections of poverty among the indigenous peoples of the Arctic. PLoS Negl Trop Dis 4:e606
Hotez PJ, Gurwith M (2011) Europe's neglected infections of poverty. Int J Infect Dis 15:e611–e619
Hotez PJ, Molyneux DH, Ferwick A, Ottersen E, Ehrlich Sachs S, Sachs JD (2006) Incorporating a rapid-impact package for neglected tropical diseases with programs for HIV/AIDS, tuberculosis, and malaria. PLoS Med 3:e102
Hotez PJ, Molyneux DH, Fenwick A, Kumaresan J, Sachs SE, Sachs JD, Savioli L (2007) Control of neglected tropical diseases. N Engl J Med 357:1018–1027
Intapan PM, Maleewong W, Nawa Y (2014) Nematode infections: neurological involvement and neurobiology. In: Bentivoglio M, Cavalheiro EA, Kristensson K, Patel N (eds) Neglected tropical diseases and conditions of the nervous system. Springer, New York
Jackson AC (2014) Rabies: neurology. In: Bentivoglio M, Cavalheiro EA, Kristensson K, Patel N (eds) Neglected tropical diseases and conditions of the nervous system. Springer, New York
Jay V (2000) The legacy of Armauer Hansen. Arch Pathol Lab Med 124:496–497
Kawli T, He F, Tan M-W (2010) It takes nerves to fight infections: insights on neuro-immune interactions from *C. elegans*. Dis Model Mech 3:721–731
Kihara M, Abubakar A, Newton CRJC (2014) Cognitive impairment and behavioural disturbances following malaria or HIV infection in childhood. In: Bentivoglio M, Cavalheiro EA, Kristensson K, Patel N (eds) Neglected tropical diseases and conditions of the nervous system. Springer, New York
Kline K, McCarthy JS, Pearson M, Loukas A, Hotez PJ (2013) Neglected tropical diseases of Oceania: review of their prevalence, distribution, and opportunities for control. PLoS Negl Trop Dis 7:e1755
Kristensson K, Olsson Y (1973) Diffusion pathways and retrograde axonal transport of protein tracers in peripheral nerves. Prog Neurobiol 1:87–109
Kristensson K, Dastur DK, Manghani DK, Tsiang H, Bentivoglio M (1996) Rabies. Interactions between neurons and viruses. A review of the history Negri inclusion bodies. Neuropathol Appl Neurobiol 22:179–187
Lafon M (2014) Rabies: neurobiology. In: Bentivoglio M, Cavalheiro EA, Kristensson K, Patel N (eds) Neglected tropical diseases and conditions of the nervous system. Springer, New York
Lechevalier HA, Solotorovsky M (1965) Three centuries of microbiology. McGraw-Hill Inc., New York
Lockhart J, ed., tr. (1993) We people here: Nahuatl accounts of the conquest of Mexico. University of California Press, Berkeley.
Lubin FD (2012) Epileptogenesis: can the science of epigenetics give us answers? Epilepsy Curr 12:105–110
Masaki T, Rambukkana A (2014) Neurodegeneration in leprosy: insights from model systems and patients. In: Bentivoglio M, Cavalheiro EA, Kristensson K, Patel N (eds) Neglected tropical diseases and conditions of the nervous system. Springer, New York

Masocha W, Kristensson K, Rottenberg ME (2014) Neurobiology of African trypanosomiasis. In: Bentivoglio M, Cavalheiro EA, Kristensson K, Patel N (eds) Neglected tropical diseases and conditions of the nervous system. Springer, New York

Meslin FX, Briggs DJ (2013) Eliminating canine rabies, the principal source of human infection: what will it take? Antiviral Res 98:291–296

Misra UK, Kalita J (2014) Japanese encephalitis and dengue virus infection: neurology and neurobiology. In: Bentivoglio M, Cavalheiro EA, Kristensson K, Patel N (eds) Neglected tropical diseases and conditions of the nervous system. Springer, New York

Mokili JL, Rohwer F, Dutilh BE (2012) Metagenomics and future perspectives in virus discovery. Curr Opin Virol 2:63–77

Molyneux DH, Hotez PJ, Fenwick A (2005) "Rapid impact interventions": how a policy of integrated control for Africa's neglected tropical diseases could benefit the poor. PLoS Med 2(11):e336

Nathanson N (2008) The pathogenesis of poliomyelitis: what we don't know. Adv Virus Res 71:1–50

Newton CR, Wagner RG (2014) Infectious causes of epilepsy? In: Bentivoglio M, Cavalheiro EA, Kristensson K, Patel N (eds) Neglected tropical diseases and conditions of the nervous system. Springer, New York

Njamnshi AK, Bissek AC, Etya'ale D (2014) Onchocerciasis: neurological involvement. In: Bentivoglio M, Cavalheiro EA, Kristensson K, Patel N (eds) Neglected tropical diseases and conditions of the nervous system. Springer, New York

Odenwald M (2014) Mental health problems associated with the use and abuse of khat (*Catha edulis*). In: Bentivoglio M, Cavalheiro EA, Kristensson K, Patel N (eds) Neglected tropical diseases and conditions of the nervous system. Springer, New York

Okitundu D, Mumba D, Tshala-Katumbay D (2014) Konzo: neurology of a permanent and non-progressive motor neuron disorder associated with food (cassava) toxicity. In: Bentivoglio M, Cavalheiro EA, Kristensson K, Patel N (eds) Neglected tropical diseases and conditions of the nervous system. Springer, New York

Patel NB (2014) Neurobiology of khat (*Catha edulis* Forsk). In: Bentivoglio M, Cavalheiro EA, Kristensson K, Patel N (eds) Neglected tropical diseases and conditions of the nervous system. Springer, New York

Ribatti D, Nico B, Crivellato E, Artico M (2006) Development of the blood-brain barrier: a historical point of view. Anat Rec B New Anat 289:3–8

Schwegmann A, Brombacher F (2008) Host-directed drug targeting of factors hijacked by pathogens. Sci Signal 1(29):re8. doi:10.1126/scisignal.129re8, 1-7

Scorza FA, Yacubian EM, Calderazzo L, Scorza CA, de Albuquerque M, Cavalheiro EA (2013) Training and workforce: an expert panel presents a new approach to epilepsy in the tropics. Clinics 68:127–128

Sherman IW (2006) The power of plagues. ASM, Washington, DC

Tabah EN, Yepnjio FN, Njamnshi AK (2014) Stigma in neurological diseases in the tropics. In: Bentivoglio M, Cavalheiro EA, Kristensson K, Patel N (eds) Neglected tropical diseases and conditions of the nervous system. Springer, New York

Triarhou LC (2006) The percipient observations of Constantin von Economo on encephalitis lethargica and sleep disruption and their lasting impact on contemporary sleep research. Brain Res Bull 69:244–258

Utzinger J, Becker SL, Knopp S, Blum J, Neumayr AL, Keiser JJ, Hatz CF (2012) Neglected tropical diseases: diagnosis, clinical management, treatment and control. Swiss Med Wkly 142:w13727

Wainwright M (2010) Dyes, trypanosomiasis and DNA: a historical and critical review. Biotech Histochem 85:341–354. doi:10.3109/10520290903297528

Whitrow M (1990) Wagner-Jauregg and fever therapy. Med Hist 34:294–310

WHO (2012, 2013) website: http://www.who.int/topics/tropical_diseases

Travel, Migration and Neglected Tropical Diseases

Zeno Bisoffi, Dora Buonfrate, and Andrea Angheben

Abstract There are slightly different definitions of "neglected" tropical diseases. However, the main concept is that they are diseases related to poverty, absent from the global public health agenda. The exponential growth of travel and migration worldwide, which has occurred in the past few decades, has caused an increased circulation of tropical diseases outside the countries of origin. Actually, neglected tropical diseases of migrants and travellers could be considered as a mirror of the huge public health problems they cause in the affected countries, which roughly, though not perfectly, correspond to the tropical world. In Western countries, health care providers are often unfamiliar with these infections, potentially causing a delay in the diagnosis, and therefore exposing many patients to short-term or, more often, long-term, severe consequences. In this chapter we describe some neglected tropical diseases of particular interest in relation to travel and migration, because of either their frequency, or their severity, or both.

Keywords Poverty-related diseases • Migrant health • Dengue • Filariasis • Loiasis • Strongyloidiasis

1 Background

With the exponential growth of travel and migration worldwide that has occurred in the past few decades, tropical diseases have been increasingly observed outside the countries of origin (Norman et al. 2010). Among them, most of the so-called

Z. Bisoffi (✉) • D. Buonfrate • A. Angheben
Centre for Tropical Diseases (CTD), Sacro Cuore – Don Calabria Hospital,
Via Sempreboni 5, 37024 Negrar (Verona), Italy
e-mail: zeno.bisoffi@sacrocuore.it; dora.buonfrate@sacrocuore.it;
andrea.angheben@sacrocuore.it

M. Bentivoglio et al. (eds.), *Neglected Tropical Diseases and Conditions of the Nervous System*,
DOI 10.1007/978-1-4614-8100-3_2, © Springer Science+Business Media New York 2014

"neglected tropical diseases" (NTD) have been diagnosed with variable frequency in Western countries. Clearly, their burden outside the endemic areas is comparatively limited. NTD of migrants and travellers could be considered as a mirror of the huge public health problems they cause in the affected countries, which roughly, though not perfectly, correspond to the tropical world. Nevertheless, NTD in migrants and travellers are also a problem in itself, especially because the lack of familiarity of Western doctors with most of these infections makes their recognition problematic (Buonfrate et al. 2012). This potentially exposes many patients to the short- or, more often, long-term severe consequences of a missed diagnosis.

2 Neglected Tropical Diseases: What Are They?

These are a group of several diseases caused by protozoa, worms, bacteria and viruses which share as common feature a high prevalence in the tropics. However, not all "tropical" diseases are comprised among NTD. WHO (WHO 2007) provides the following definition: "Chronically endemic and epidemic-prone tropical diseases, which have a very significant negative impact on the lives of poor populations [and] remain critically neglected in the global public health agenda". A slightly different definition is provided by the journal *PLoS Neglected Tropical Diseases (PLoSNTD)*: "The NTDs are defined as a group of poverty-promoting chronic infectious diseases, which primarily occur in rural areas and poor urban areas of low-income and middle-income countries. They are poverty-promoting because of their impact on child health and development, pregnancy, and worker productivity, as well as their stigmatizing features".

Combining the two definitions helps understanding some of the key features of NTD. The crucial issue is the "neglect". Just to give an example, malaria, which is probably the most classical of all tropical diseases, is not in the NTD list as it is considered to receive comparatively more attention and funding for research and control, and the same applies to HIV-AIDS and tuberculosis (TB). The latter diseases are also called "the big three" to indicate the major burden they represent in global health, but also a comparatively much higher volume of resources they receive from the international community (e.g. the Global Fund, the Gates Foundation, and other major funding sources and initiatives), as opposed, precisely, to the NTD. Institutions and research groups specifically involved in NTD argue that the burden caused by these diseases is out of proportion with the volume of funding devoted to their control, also considering that many NTD are chronic infections which, besides mortality (50,000 deaths estimated annually), cause a higher number of disability-adjusted life years (DALY) lost than malaria or TB. NTD also have a great economic impact due to the productive time lost and, in some cases, to the abandon of large agricultural areas where these diseases are more frequently transmitted.

However, there is no universal agreement on the list of NTD, either. The main characteristics of the diseases considered NTD by WHO are summarized in Table 1.

Table 1 Summary of the main neglected tropical diseases, NTD (those more relevant to travel and migration are underlined)

NTD	Description	Burden, availability of treatment
<u>Dengue</u>	***Flavivirus*** (RNA virus). Mosquito borne (genus *Aedes*). Fever, bone pain, haemorrhagic complications, shock	Fifty to 100 million cases yearly worldwide. >20,000 deaths. No treatment, vaccine under trial
<u>Rabies</u>	***Lyssavirus*** (RNA virus). Transmitted by bite of infected dog or other mammal. Causes fatal encephalitis	Estimated 55,000 deaths per year. Vaccine preventable (ore and postexposure)
Trachoma	***Chlamydia trachomatis*** (bacterium). Direct transmission, linked to poor hygiene	Six millions blinds worldwide. Easily curable with azithromycin single dose (international donation)
Buruli ulcer	***Mycobacterium ulcerans*** (bacterium). Devastating skin ulcers, bone lesions	Underestimated, probably tens of thousands cases in at least 36 countries. Anti-mycobacterials, surgery
Endemic treponematosis	***Treponema pertenue, T. endemicum, T. carateum*** (bacteria). Cutaneous and mucosal lesions, often devastating	Thousands of cases but rapidly declining. Penicillin, other antibiotics. Targeted for elimination
Leprosy	***Mycobacterium leprae*** (bacterium). Skin and mucosal lesions, peripheral nerve involvement. Stigmatizing	About 200,000 new cases in 2010. Multi drug treatment regimens for years
<u>Chagas disease</u>	***Trypanosoma cruzi*** (protozoan). Transmitted by faeces of blood-sucking infected bugs, or more rarely by oral route, transfusion or transplant. Chronic cardiac and gastrointestinal complications	Estimated prev. 10 million cases with 12,500 yearly deaths in Latin Americans. Treatment with benznidazole or nifurtimox, more effective in early stage, toxic
<u>Human African Trypanosomiasis</u>	***Trypanosoma brucei*** (protozoan). Transmitted by *Glossina* ("tse tse") flies. *T.b. gambiense* causes sleeping sickness, chronic, fatal if untreated	Declining. Estimated 7,000 new cases in 2010. Treatment improved in recent years with combination eflornithin-nifurtimox
<u>Leishmaniasis</u>	Genus ***Leishmania*** (protozoan), several species. Transmitted by phlebotomus flies. Cutaneous, muco-cutaneous and visceral forms, the latter fatal if untreated	About 2 million new cases/year, ¼ visceral. Present also in temperate climates. Oral treatment (miltefosine) available in recent years
<u>Cysticercosis</u>	***Taenia solium*** (worm, tenia of the pig). Can cause epilepsy and other neurological symptoms	Fifty to 100 million persons infected with *T. solium*. Treatment with albendazole or praziquantel, not always effective
Dracunculosis	***Dracunculus medinensis*** (worm). Transmitted via drinking water contaminated with a small shellfish, intermediate host. Causes abscess-like skin lesions	Close to eradication, elimination achieved in most countries

(continued)

Table 1 (continued)

NTD	Description	Burden, availability of treatment
Echinococcosis	***Echinococcus granulosus*** (worm, tenia of dog and sheep). Hydatid cysts in liver, lungs or other organs (rarely brain)	About 200,000 new cases per year. Treatment with oral albendazole or local injection of hypertonic fluid or surgery
Filariasis	***Wuchereria bancrofti, Brugia malayi, Loa loa*** and other (worms). The former two, transmitted by mosquitoes, cause the nocturnal, lymphatic form (lymphedema, elephantiasis), while *Loa loa*, transmitted by *Chrysops* fly, causes the diurnal form (migrant oedema, other manifestations)	About 120 million cases of lymphatic form, and 12–13 million cases of loiasis. Treatment with ivermectin, albendazole, DEC. Declining thanks to mass treatment (donation programmes) and vector control
Onchocerciasis	***Onchocerca volvulus*** (worm). Transmitted by *Simulium* fly. Skin lesions, generalized itching. Ocular lesions cause blindness	About 500,000 blind people in the world. Declining thanks to mass treatment with ivermectin (donation programme)
Schistosomiasis	Genus ***Schistosoma*** (worm), several species. Acquired bathing in freshwater harboring snails acting as intermediate hosts. Infective larvae penetrate intact skin. Cause of several distinct clinical syndromes according to species and organ involved	Prevalence of about 200 million cases, and at least 40,000 deaths per year (underestimated). Declining due to mass treatment with praziquantel
Soil transmitted helminths	Comprise different nematode worms, classically: ***Ascaris lumbricoides*** and ***Trichuris trichiura*** (acquired via ingestion of food contaminated with soil), ***Ancylostoma duodenale*** and ***Necator americanus*** (acquired via penetration of larvae through intact skin). ***Strongyloides stercoralis*** (acquired as the latter two) recently added to WHO list among "other neglected tropical conditions". Variety of gastrointestinal symptoms, failure to thrive, severe anemia, respiratory problems	Estimated over a billion people infected. Mass treatment programme ongoing with mebendazole, albendazole and other drugs, but poorly effective against *Strongyloides*

Eleven diseases or groups of diseases (underlined in Table 1) are of particular interest in relation to travel and migration, because of either their frequency, or their severity, or both. We will briefly mention hereafter four of them, while the other six will be treated more in depth in the following paragraphs.

2.1 Dengue, Leishmaniasis, Onchocerciasis

Dengue is currently considered a pandemic disease. It is estimated to cause 50–100 million cases and >20,000 deaths annually worldwide. It is caused by a Flavivirus and transmitted by the bite of a mosquito of the genus *Aedes*, such *Aedes aegypti* or *Aedes albopictus*. The latter is increasingly common in temperate climates and autochthonous cases of dengue have already been reported in the USA and Western Europe, including some local epidemics. A mild febrile disease (dengue fever) may become a fatal threat (dengue haemorrhagic fever and dengue shock syndrome) in case of secondary infection with a different viral serotype. Therefore, particularly at risk are frequent travellers (especially to Asia and Latin America) and the so-called VFR ("visiting friends and relatives") travellers, a definition which includes immigrants who briefly visit their country of origin. Moreover, travellers and migrants entering a non endemic country during the acute phase of infection may be the source of transmission to the local population in the presence of a suitable vector.

No etiologic treatment is available. Vaccines are under study, but not yet at the stage of a Phase III clinical trial. Dengue is not primarily a disease of the nervous system, but several case reports of dengue-associated encephalopathy have been published (Kanade and Shah 2011), as reviewed in Misra and Kalita (2014).

Leishmaniasis is not a single disease but a group of diseases caused by different species of protozoa of the genus *Leishmania*, and transmitted by sandflies in all continents, including the Mediterranean areas of Europe. Two main forms of the infection are known, cutaneous (including muco-cutaneous) and visceral. The latter is a life-threatening disease affecting the bone marrow and the reticulo-endothelial system, closely simulating a haematologic malignancy: fever, huge splenomegaly and secondary pancytopenia. Central nervous system (CNS) involvement is uncommon, but neurological signs and symptoms (both central and peripheral) have been reported in the disseminated form of the disease, which usually affects immune-suppressed patients, especially in HIV co-infection (Walker et al. 2006).

Leishmaniasis is not among the most common travel-related infections in travellers, though cases (of both the main presentations) are regularly reported in European travellers (Odolini et al. 2012).

Echinococcosis in humans is caused by the tapeworms *E. granulosus* ("cystic echinococcosis") and *E. multilocularis* ("alveolar echinococcosis"). Human infection occurs accidentally by ingestion of *Echinococcus* eggs excreted with the faeces of the definitive host (dogs, foxes) (McManus et al. 2012). The cysts mostly develop in the liver and/or in the lungs (>90 % of cases), but other sites can be involved. For instance, localization in the spinal cord and brain can cause neurological symptoms, such as seizures or paralysis (Brunetti and White 2012). Another tapeworm which can rarely infest humans is *Taenia multiceps*, which causes coenurosis (Ing et al. 1998). The cysts usually settle in the CNS, eye, muscles or subcutaneous tissue. The signs and symptoms caused by the CNS localization of coenurosis resemble those of neurocysticercosis (see Carpio and Fleury 2014), from which coenurosis is indistinguishable at computerized tomography/magnetic resonance imaging. The only method to differentiate cestode cysts is surgical removal and subsequent microscopic examination.

Transmission of *E. multilocularis* is limited to the northern hemisphere, while *Echinococcus granulosus* is widely diffused, especially in parts of Eurasia, South America and north east Africa (McManus et al. 2012). The risk of echinococcosis in travellers to specific countries is difficult to define clearly due to the broad diffusion of the disease.

Onchocerciasis as a cause of neurological involvement is discussed in Njamnshi et al. (2014). The most common symptoms of this infection are cutaneous (generalized pruritus with a variety of skin lesions). The ocular involvement is characterized by corneal lesions. Onchocerciasis has been reported as the most common filarial infection affecting travelers and immigrants treated in the clinics belonging to the Geosentinel network (Lipner et al. 2007). Although the disease occurs much more frequently, as expected, in immigrants and long-term expatriates than in tourists, some cases have also been observed in tourists after short exposure.

3 Rabies

France had been declared rabies-free since 2001, and Cracotte was a pet dog that had always been living in that country. However, when on February 2008 it presented behavioural changes and bit two people, rabies was suspected and then confirmed. Subsequent investigations were conducted, in order to identify the possible index case. Cracotte's owners had a second dog, Youpee, which had died in January after showing symptoms compatible with rabies. Presumably, Youpee had acquired the infection from a dog named Gamin, the probable index case, illegally introduced into France from Morocco. Both Youpee and Gamin had been incinerated and not tested, although they had symptoms compatible with rabies. After the diagnosis of rabies in Cracotte, the health and veterinary authorities traced people and animals that might have been exposed, and offered the prophylaxis. Neither human nor other animal cases in relation to this event were reported (Team FMI 2008).

Rabies is invariably fatal. It is therefore essential to focus on prevention, which includes behavioral measures (i.e., avoiding potentially infected animals) and pre-exposure immunization. Post-exposure prophylaxis (PEP) is also available, and should be administered as soon as possible in case of exposure to rabies virus.

Table 2 WHO guidelines for rabies post-exposure prophylaxis

Category	Exposure to rabid animal[a]	Treatment
I	Touching or feeding; licked unbroken skin	None
II	Nibbled uncovered skin; minor scratches or abrasions without bleeding	Local treatment of the wound and immediate vaccination
III	One or more transdermal bites or scratches; licked broken skin; contamination of mucous membrane with saliva from licks; any degree of exposure to potentially rabid bats	Local treatment of the wound, immediate vaccination, and administration of rabies immunoglobulin

[a]Exposure to a confirmed or suspected rabid animal or to an animal unavailable for testing

According to WHO standards, the vaccine schedule is in three doses on days 0, 7, 28 (WHO, http://www.who.int/mediacentre/factsheets/fs099/en/). When a subject gets in contact with a suspect rabid animal, the measures to be taken depend on the type of contact (Table 2), the likelihood of the animal being rabid (including its clinical features) and its availability for observation and laboratory testing. Immediate cleaning of the wound with water and soap or other available substances (detergent, povidone iodine) has proven effective in increasing survival rates.

If a previously vaccinated person is bitten by a rabid animal, only two booster doses with vaccine on days 0 and 3 are indicated as PEP. In case the exposed person had not been vaccinated, the vaccine doses should be four: on days 0, 3, 7, 14. In case of category III exposure (Table 2), also immunoglobulins should be administered as soon as possible, injecting them in part around the wound, and in part intramuscularly, distant from the anatomic site of vaccine injection.

An estimated 10–16 million people undergo rabies PEP worldwide every year (Both et al. 2012). Death of a patient who has received PEP is rare, and usually results from inadequate prophylactic regimen, that may occur. Unfortunately, many subjects who get exposed in Africa and Asia receive inadequate PEP because of low availability and high cost of immunoglobulins. This is an additional reason to promote the vaccination in travellers, who still do not seem to be sufficiently covered by vaccination (Gautret and Parola 2012). The case reported above (Team FMI 2008) reminds us that diseases are not strictly confined within the borders of a country because people and animals migrate and travel.

Although rabies is ubiquitous, the highest incidence is in Africa and Asia, where most of human deaths caused by this virus occur (about 95 %). WHO estimates 55,000 human deaths worldwide every year, but this is probably an underestimate because of misdiagnosis and lack of efficient surveillance system in endemic countries (Both et al. 2012). According to a recent review (Gautret and Parola 2012), 48 confirmed cases of rabies have been reported in tourists/migrants returned to Europe and the USA from 1990 to May 2012, and this is also probably an underestimate. People travelling to rural areas or areas heavily populated with stray dogs in those countries are at highest risk.

Dogs remain the main source of rabies, but monkeys are also a common source in developing countries, probably because travelers very often are not aware of the risk related to a monkey bite.

3.1 When to Suspect Rabies

- History of animal bite/scratch is usually a key point. Although symptoms usually develop from 1 to 3 months from the virus inoculation, it should be kept in mind that a longer incubation period (>1 year) is also possible.
- Compatible clinical signs (see Jackson 2014). Unfortunately, when symptoms are present the disease is fatal.

3.1.1 Prevention in Travelers

Avoid contacts with wild animals. Vaccination, as described above.

4 Chagas Disease

S.A.Z.F., an Ecuadorian woman, migrated to Italy in 1999. In 2002, when she was 46 years old, migraine with aura started and she underwent cerebral computed tomography (CT) which only revealed a right parietal angioma. In 2003 and 2005 she was hospitalized in a neurological ward for severe migraine attacks accompanied by visual aura and motor symptoms. The last attack was characterized by right hemisyndrome and right homonymous lateral hemianopsia. Cerebral magnetic resonance imaging (MRI) revealed ischemic lesions of the Ammon's horn, basal ganglia and left cerebral peduncle. Thromboembolia was suspected and hypokinetic dilatative myocardiopathy with left apical ventricular thrombus was diagnosed by echocardiography. In the work-up, serology for Chagas disease was done with positive result. The cardiac ejection fraction was not severely impaired (40–45 %) but an apical wall akynesia and inferior-basal hypokynesia were present. At the anamnesis the patient referred a lipothymia with loss of consciousness in 2000. A 24 h electrocardiogram recording showed premature repetitive atrial and ventricular beats. The patient was treated with oral anticoagulants and the ventricular thrombus disappeared. After some months, only the homonymous lateral hemianopsia persisted, with marked hypostenia without motor symptoms. PCR for Trypanosoma cruzi in the peripheral blood gave negative results. Even in the absence of strong evidence in the literature, she was treated with benznidazole (5 mg/kg) for 60 days in 2005. In the follow-up until 2011 the patient continued to have positive Trypanosoma cruzi serology (repeated every 2 years) with progression of the myocardiopathy and is now candidate to heart transplant.

Chagas disease (CD) or American trypanosomiasis, is a complex zoonosis, endemic in continental Latin America, caused by the hemoflagellate *Trypanosoma cruzi* (Fig. 1) (see Coura 2014). Transmission to humans is primarily vectorial, mediated by various species of triatomine insects (blood-sucking bugs), but other non-vectorial mechanisms of transmission are possible (blood transfusion, cells/tissues/organs transplant, vertical transmission (from mother to infant) and with

Fig. 1 Detection of a trypomastigote of *Trypanosoma cruzi* in blood by the Quantitative Buffy Coat (QBC) technique (photo by Maria Gobbo, CTD)

food contaminated by vector dejections) (Prata 2001). According to WHO estimates, about ten million people are affected by this plague (WHO, http://www.who.int/mediacentre/factsheets/fs340/en/index.html) and about one-fifth of the population is considered at risk in 21 Latin American countries (Argentina, Belize, the Bolivarian Republic of Venezuela, Brazil, Chile, Colombia, Costa Rica, Ecuador, El Salvador, French Guyana, Guatemala, Guyana, Honduras, Mexico, Nicaragua, Panama, Paraguay, Peru, the Plurinational State of Bolivia, Suriname and Uruguay).

At present, CD ranks first among parasitological diseases for impact on health and social systems in endemic areas (Hotez et al. 2008) due to its effects on the productivity of people of working age and to disability and mortality. In the last 20 years, many factors contributed to a dramatic change in the epidemiological profile of CD, and these include implementation of different control initiatives in Latin America, sharp rise in international travels and migration, urbanization and internal migration in endemic countries among others (Schmunis and Yadon 2010). In the context of non-endemic countries, Europe is not spared, and the majority of cases are recorded in Spain and Italy (Guerri-Guttenberg et al. 2009).

CD is a neglected tropical disease with particular clinical features. The acute form is frequently asymptomatic or pauci-symptomatic (more than 95 % of cases), therefore contributing to underdiagnosis, and evolves into a chronic form, without any clinical manifestation, which can last for decades. In 20–30 % of cases, however, through a complex interaction between parasite and immune system, patients develop lesions in target organs (mainly myocardiopathy, dysrhythmias, megaviscera, cerebrovascular accidents). In this phase the haemoprotozoan is rarely found in the blood.

Disease expression is especially severe in immune-deficient patients such as organ transplant recipients, patient under immune-suppression or HIV-seropositive patients.

The CNS is involved in the rare symptomatic severe acute forms of CD, in the reactivation during HIV-linked immune deficiency (encephalitis) or during the

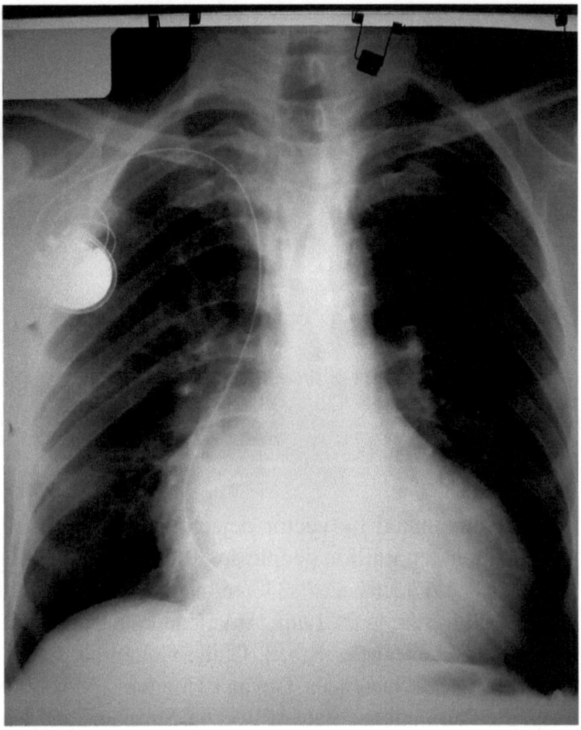

Fig. 2 Chest X-rays of a Chagasic patient with cardiomegaly

chronic phase (ischemic stroke), the latter being the commonest (see Coura 2014). Several epidemiological studies have indeed shown a correlation between ischemic stroke and CD. Chronic cardiac form of CD may be characterized by cardiomegaly (Fig. 2), heart failure, mural thrombosis, apical aneurism or arrhythmias, all risk factors for stroke. However, stroke may be rarely the first signal of CD in asymptomatic patients without heart systolic dysfunction (Carod-Artal and Gascon 2010; Paixao et al. 2009).

4.1 When to Suspect Chagas Disease?

- Accurate evaluation of epidemiological risk for CD such as Latin American origin; to be born from Latin American mother (to be considered also in adoptees); to have been transfused in Latin America; to be transplant-recipient from Latin American donor; travel to Latin America with a lifestyle at risk of infection (adventurous travel at risk for contact with vectors, drinking non-pasteurized fruit or sugar juices).
- CD can be suspected in symptomatic people with epidemiological risk, but it has been demonstrated that it is cost-effective to test for CD in asymptomatic conditions only for pregnant Latin American women (Sicuri et al. 2011);

- Travellers from Latin America with fever (associated or not with oedemas, malaise, increased liver enzymes, lactic dehydrogenase elevation), presence of Romaña sign (unilateral painless periorbital swelling and conjunctivitis);
- Newborn form Latin American mother with general signs of infection (beginning from low APGAR score at birth);
- Myocardiopathy or arrhythmia in a patient with epidemiological risk for CD (see above);
- Digestive disorders (especially megaviscera in the digestive tract, achalasia-like disturbances, colopathy) in a patient with epidemiological risk for CD (see above);
- Fever-associated or not to myocarditis/encephalitis/panniculitis in immune-suppressed or HIV-seropositive patients with epidemiological risk for CD (see above).

4.1.1 Treatment

Treatment of CD is limited to two drugs: nifurtimox and benznidazole. The latter, better tolerated, is considered the first choice. Currently, only a few methodologically rigorous trials have been conducted to study the efficacy of CD treatment (Marin-Neto et al. 2009). The complex history of the disease, unavailability of cure biomarkers, the need of prolonged follow-up to record clinical outcomes (such as myocardiopathy evolution, arrhythmias) contribute to limit the possibility to collect evidences with randomized controlled clinical trials.

4.1.2 Prevention in Travellers

No vaccine is available. Triatomine bites (generally at nighttime) should be prevented by sleeping under pyrethroid-impregnated bed nets avoiding lodges not adequately insulated or stuccoed. Moreover, drinking fresh, non-pasteurized fruit or sugar cane juices should be avoided.

5 African Trypanosomiasis

A 54-year-old nun who had lived for 30 years in the Central African Republic consulted the Centre for Tropical Diseases (CTD) in Negrar, Verona (Italy) for recurrent fever, headache, insomnia, and fatigue. At presentation, she was afebrile, with a huge spleen and marked pancytopenia. A quantitative buffy coat (QBC) for malaria was negative, and she was discharged. The working diagnosis was Hyper reactive Malarial Splenomegaly (HMS). Three days later she returned with fever. A QBC test turned out to be negative for malaria but showed the unexpected finding of trypanosomes. Similar forms were then found in peripheral blood smears. Serology results for T. brucei was positive. Results of CSF examination were normal (Bisoffi et al. 2005).

Human African trypanosomiasis (HAT) is discussed extensively in Buguet et al. (2014). *Trypanosoma brucei (T.b.) rhodesiense*, transmitted predominantly by *Glossina morsitans* which feeds on wild animals in savannah areas in eastern and southern Africa, is a zoonosis, and humans occasionally visiting affected areas can become accidental hosts (Brun et al. 2010). Humans are, instead, the only substantial reservoir of *T. b. gambiense*, which is mainly transmitted by *Glossina palpalis*, in areas of vegetation near rivers and cultivated fields. HAT *gambiense* is usually subacute/chronic, lasting months to years. HAT *rhodesiense* has a more acute and aggressive course, clinically similar in the first phase to malaria (diagnosis is often made while examining a blood film for malaria); if left untreated, death occurs within weeks or months. Both forms of HAT are characterized by two distinct phases: the early or haemolymphatic stage and the late or meningoencephalitic stage with trypanosome invasion of the CNS of patients surviving the early stage (Urech et al. 2011).

Although *T. b. gambiense* infection accounts for more than 90 % of all reported HAT cases worldwide, *T. b. rhodesiense* has been the cause of most imported cases (Blum et al. 2012), often detected as clusters by surveillance networks (Jelinek et al. 2002). From 2000 to 2010, 94 HAT cases have been diagnosed in non-endemic countries (Simarro et al. 2012), 56 (72 %) due to *T. b. rhodesiense*; among these, the vast majority (61/68) were tourists, and 82 % were diagnosed in the first stage. Most had visited game parks of Tanzania and Kenya, namely Serengeti and more recently Masai Mara (Clerinx et al. 2012; Wolf et al. 2012). No tourists were instead recorded among 26 cases of HAT *gambiense* observed in the decade 2000–2010. Ten of these were expatriates (for business or mission), 14 were migrants or refugees, and one was a sailor; only 6 (33 %) were diagnosed in the first stage of disease, witnessing the difficulty to suspect/diagnose HAT *gambiense* before CNS invasion. One case of HAT *gambiense* and three cases of HAT *rhodesiense* cases died during treatment, therefore the case fatality rate was high (4 %) in both cohorts.

5.1 When to Suspect HAT

- HAT *rhodesiense* must be included in the differential diagnosis of any febrile traveler returning from areas at risk. Repeated thick films (same as for malaria) may be necessary.
- A trypanosomal “chancre” at the site of the tsetse fly bite is seen frequently in *T. b. rhodesiense* infection, while is missing in more than half of the cases of *T. b. gambiense* infection.
- *T. b. gambiense* infection at first stage is missed in most cases. Suspicion should be raised when an immigrant or expatriate coming from an endemic country presents with recurrent fever, is negative for malaria and has splenomegaly. Sensitivity of direct diagnosis (blood films) is much lower than with *T. b. rhodesiense* infection, therefore repeated examinations with more sensitive techniques, including serology, are usually necessary.

5.1.1 Treatment

Currently, for HAT *rhodesiense* suramin is the drug of choice in the first stage and melarsoprol (both are highly toxic drugs) in the second stage, while for HAT *gambiense* pentamidine is used in the first stage, and a combination of nifurtimox and eflornithine in the second stage (Priotto et al. 2009).

5.1.2 Prevention in Travelers

No vaccine is available. Bites (occurring during daytime) can be prevented by wearing thick and light-colored clothing, better if impregnated with permethrin. The application of a skin repellent is also useful (Sholdt et al. 1989).

6 Cysticercosis

B.S., Nepali girl, was adopted at the age of 3 years and 6 months by an Italian family. At the age of 4 years, she had a seizure lasting about 10 min before spontaneous resolution, associated to vomit, post-critical sleeping and headache in the left head side. An electro-encephalogram (EEG) showed the presence of slow anomalies in the left parietal region and a brain computed tomography found: "On the left parietal region, hypodense finger shaped lesion not univocally interpretable (oedema underlying inflammatory or expansive lesion?)". The next day she had a second critical episode, followed by prolonged weakness of the right upper limb. MRI showed a circular lesion with sharp margins in the left parietal region, surrounded by mild oedema and with ring enhancement after injection of paramagnetic contrast medium. In the suspicion of neurocysticercosis, serology was performed with positive result with western blot and negative result with ELISA technique. Stool tests with enrichment were negative for ova and parasites. Antiepileptic drugs were started and then B.S. was treated with albendazole plus dexamethasone. After 1 year, B.S. was in good health conditions and did not show any neurological symptom; brain MRI was normal.

Human cysticercosis is a disease caused by the larval stage of a tapeworm called *Taenia solium*. Incidentally, other *Taenia* species (e.g., *multiceps*, *serialis*, *brauni*, *taeniaeformis*, *crassiceps*) can infect humans mimicking cysticercosis, with various localizations in the body (including soft tissues, eyes, CNS) (CDC).

As described in Carpio and Fleury (2014), humans are the definitive host of *Taenia solium* and become infected after ingesting undercooked pork containing cysticerci (cystic larval form of the tapeworm). Taeniasis develops when the cysts hatch in the stomach attaching to the intestinal wall and growing to the adult worm which produces eggs released in the stools. Pork ingests eggs or fragments of the tapeworm (proglottids) and completes the life cycle developing tissue cysticerci. Besides pigs, also humans can occasionally be intermediate hosts, ingesting contaminated food or as a result of autoinfestation (reverse passage of proglottids to stomach).

Cysticerci that reach the CNS cause neurocysticercosis, which is one of the most common parasitic infections of the human nervous system. Infected people may remain asymptomatic for a long time and the disease usually gives rise to symptoms when an inflammatory reaction appears around the cysts for spontaneous degeneration or anti-helmintic treatment.

Actually, the majority of low-income countries are endemic for cysticercosis, including sub-Saharan African countries (except countries with low pork consumption, mainly Muslim countries), India and Indian sub-continent, Latin American countries. China, South-East Asia and Indonesia (mostly Papua) are also included in the list.

6.1 When to Suspect Cysticercosis?

- Epidemiological risk factors: to harbor *Taenia solium*, to live in contact with people harboring *Taenia solium* in their intestine, to live in an endemic country;
- Neurocysticercosis is suspected in presence of neurological symptoms (including chronic headaches, seizures, blindness, dementia and generally symptoms related to CNS space-occupying lesions such as motor disturbances, hydrocephalus) in patients with epidemiological risk factor;
- Epilepsy (especially late-onset, but also juvenile) in patients with epidemiological risk is recognized as the commonest manifestation of neurocysticercosis (WHO 2011).
- Presence of cysts, calcifications also with oedema in brain imaging in a patient with epidemiological risk factor.
- Cysticercosis can be suspected in patients with epidemiological risk factor and subcutaneous/muscular cysts or nummular calcifications in a soft tissue radiography (normally forearms or thighs).

6.1.1 Treatment

Treatment of neurocysticercosis is primarily based on control of neurological manifestations: anti-epileptic drugs, steroidal/diuretic treatment for cerebral oedema. Depending on the stage of the cysts in the CNS, etiologic treatment with albendazole and/or praziquantel plus steroids can be associated. Some evidences suggest that anti-helmintics reduce the number of lesions and seizure reoccurrence (Abba et al. 2010).

6.1.2 Prevention in Travellers

The frequency of the disease has decreased in high-income countries owing to stricter controls at slaughterhouse services, improved hygiene and sanitation. No human vaccine is available. Proper water sanitation and safe food (not soil contaminated)

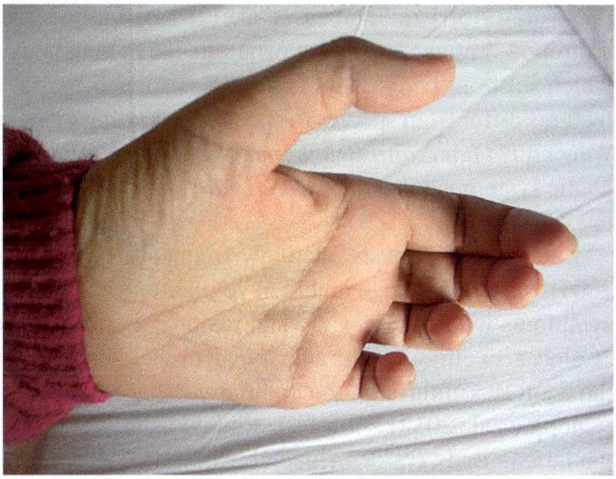

Fig. 3 Hypotrophy and sensory-motor deficit of the right forearm, and of the fourth and fifth fingers of the right hand in a patient with *Loa loa* infection (photo by Maria Gobbo, CTD)

ingestion, handwashing before meals can prevent cysticercosis. An important measure to control the spread of cysticercosis is to identify tapeworm carriers and treat them. Avoiding to eat pork meat or eating well-cooked pork meat in endemic countries can prevent tapeworm ingestion (taeniasis or intestinal tapeworm infection).

7 Filariases

In June 2008 an Italian girl presented to the CTD with spleen lesions, hypereosinophilia (eosinophil count >3,000/μl), hypotrophy and sensory-motor deficit of the right forearm, and of the fourth and fifth fingers of the right hand (Fig. 3). Neurological symptoms had appeared in May 2007. She had undergone several investigations in other hospitals, including electromyography which had revealed important suffering of the ulnar nerve, with signs of denervation. As she had been working for some months in African countries (Kenya, Mozambique, Gabon, Ethiopia), leprosy was also suspected, but then ruled out. Blood and stool resulted negative for parasites, and brain and lumbar cord MRI were negative too. In 2006 she had suffered from episodes of migrant transient oedema in both upper and lower limbs, that ceased after the development of the neurological symptoms. A filarial infection was suspected, and confirmed by examination of a peripheral blood sample, obtained at noon, through leuco-concentration method. The final diagnosis of Loa loa was obtained by examination of blood smears stained with Giemsa that showed typical microfilariae. The patient was treated with ivermectin (200 μg/kg single dose) followed by albendazole (400 mg twice a day for 28 days) and prednisone (37.5 mg/day for 7 days and then decreasing doses for 3 weeks). At follow-up, 6 and 12 months after treatment, the eosinophil count was within the normal range and the spleen lesions had disappeared. Moreover, neurological sign and symptoms gradually disappeared (Gobbi et al. 2011).

As described in Intapan et al. (2014), filariae are nematodes transmitted by arthropod vectors. The larvae penetrate the human body and develop in adult worms, which settle in body districts specific for each filariae species. Therefore, clinical features differ from species to species, according to the parasitized site. For instance, *Wuchereria bancrofti*, *Brugia malayi*, and *Brugia timori* cause damage of the lymphatic system, while *Onchocerca volvulus* is primarily associated with skin and eye involvement.

Loiasis is a filariasis relatively frequent in international travellers (Lipner et al. 2007), but has not been included yet in the WHO list of NTD despite the severity of symptoms it can cause. Classical features associated with *Loa loa* infection are the "Calabar swelling" (migrant, transient subcutaneous oedemas), and migration of the adult worm through the conjuntiva, that is clearly visible ("African eyeworm") (Knopp et al. 2012). However, this parasite can be detected also in other sites of the human body (as described above), and can exceptionally cause various central neurological manifestations, including unilateral or focal motor and/or sensory deficits, meningitis, encephalitis, or mental disturbances. On the other hand, peripheral neurological symptoms in patients with loiasis are much less known (Gobbi et al. 2011). Moreover, extremely severe neurological impairment can be observed during treatment: diethylcarbamazine and, to a lesser extent, ivermectin can induce an encephalopathy in patients harboring very high *Loa* microfilaremias. In these patients, treatment with albendazole, which slowly decreases the microfilarial load, is recommended.

Filarial nematodes are almost exclusively transmitted in tropical and subtropical areas, and they rarely affect tourists (Baaten et al. 2011; Lipner et al. 2007).

7.1 *When to Suspect* Loa loa*?*

- Accurate evaluation of epidemiological data (specific countries).
- Increased eosinophil count, otherwise unexplained.
- Unilateral, transient, peripheral oedema.

7.1.1 Prevention in Travelers

Prevention from *Loa loa* implies prevention from bites of the *Chrysops* fly. Long clothes and insect repellents are therefore needed.

8 Schistosomiasis

A group of seven young students travel to Malawi for 1 month. A couple of weeks after return, within a few days six of them present high fever with shivering and a cough. Although they are all under prophylaxis, two of them consult the CTD to exclude malaria. Blood films are negative, and the only lab finding of notice is a marked eosinophilia. Answering a specific question, they report having bathed in lake Malawi about 3 weeks before. The lab staff of CTD is requested to examine stool samples for as long as necessary; after repeated observations scanty eggs of Schistosoma mansoni are found in samples from two cases. The other four symptomatic students are then summoned and examined with identical results. The seventh student remains asymptomatic: she reports having avoided bathing as she was scared of hippos and crocodiles...

As described in Ferrari (2014), schistosomiasis is a group of diseases caused by different species of trematode worms of the genus *Schistosoma*, of which the most important affecting humans are *S. mansoni* (found in Africa and South America), *S. haematobium* (mainly found in Africa and in foci in Asia), and *S. japonicum* (South East Asia). The life cycle is complex and similar in all species, involving freshwater snails as intermediate hosts and humans as definitive host, harboring adult worm pairs inside the mesenteric or rectal veins *(S. mansoni, S. japonicum)* or the perivesical venous plexus of the bladder (*S. haematobium*). All the relevant pathological findings of the three *Schistosoma* species are related to the eggs and their migration.

8.1 Acute Schistosomiasis

Also known as Katayama fever, this is the presentation commonly found in non-immune travelers, and is mainly observed with *S. mansoni* and *S. japonicum*. When caused by the latter species, the disease can be fatal due to very severe infection. The acute phase of *S. haematobium* infection is usually milder and may be overlooked. Acute schistosomiasis is characterized by an intense immunological reaction to the deposition of the first eggs by the adult female and causes nonspecific symptoms and signs: high fever, respiratory and/or (less frequently) gastrointestinal disturbances. Eosinophilia accompanying fever and the history of exposure are clues to diagnosis. Typically, after a bath in contaminated water, virtually all exposed individuals are infected, as in the case described above. The illness has therefore often the characteristics of an outbreak (Visser et al. 1995). This form is virtually never observed in migrants due to their previous exposure since childhood.

Acute schistosomiasis is not an exceptional occurrence. In the largest prospective study published to date on fever (1,832 consecutive cases) after a stay in the tropics (Bottieau et al. 2006), collected in a referral centre in Europe (IMTA, Antwerp), acute schistosomiasis (33 cases: 1.8 %) ranked fourth among the tropical causes, after malaria, rickettsial infections, and dengue, and well before *Salmonella* enteric fever, and the ninth cause overall when cosmopolitan infections were also included. In another (retrospective) series, schistosomiasis was reported much less frequently (Wilson et al. 2007).

Missing acute schistosomiasis may expose the unaware traveler to the risk of developing the serious complications of the chronic form, although they are related to the intensity of transmission and are therefore unlikely after a single exposure.

8.2 Chronic Schistosomiasis

Due to egg migration, triggering inflammation and formation of granulomas in the affected tissues, chronic schistosomiasis can cause different clinical forms according to the *Schistosoma* species and organs involved. In the first months/years after exposure, symptoms are related to the penetration of eggs from the vessel to the nearby organ, causing abdominal pain and rectorrhagia (*S. mansoni* and *S. japonicum*) or pelvic pain, dysuria and haematuria (*S. haematobium*). The infection can then remain clinically silent for several years, before it causes severe organ damage: hepatic fibrosis and portal hypertension (*S. mansoni* and *S. japonicum*), hydronephrosis, renal failure, cancer of the bladder (*S. haematobium*) and more rarely other manifestations such as *cor pulmonale* and neurological disorders (see Ferrari 2014).

In non-endemic countries, at variance with the acute form, chronic schistosomiasis is most often seen in migrants and expatriates coming from endemic areas (Whitty et al. 2000). In countries experiencing an important migratory wave, especially from West Africa, screening for schistosomiasis is strongly recommended and should include not only the examination of stools, but also the more sensitive serology (Bierman et al. 2005). Although there is no general agreement on the need of screening asymptomatic immigrants for parasitic infections, schistosomiasis is an example of how early detection is mandatory in order to prevent late life-threatening complications.

Series of imported schistosomiasis have been published in past years by the two major networks on imported tropical and infectious diseases, TropNetEurop and Geosentinel (Grobusch et al. 2003; Jelinek 2008; Nicolls et al. 2008). Although there was not a clear distinction between the acute and chronic presentation and the diagnostic criteria were not homogeneous among the involved hospitals, it is quite clear from the reported main symptoms and signs that the acute form typically affected travellers, and the chronic form mainly expatriates and migrants, as expected.

8.3 When to Suspect Schistosomiasis

- Fever with eosinophilia in a traveler (acute form);
- Chronic abdominal pain, rectorrhagia, dysuria, haematuria, asymptomatic eosinophilia in expatriates and immigrants (chronic form before late stage complications).

8.3.1 Treatment

For all forms, praziquantel is the drug of choice, although the optimal schedule has yet to be determined and resistance is observed.

8.3.2 Prevention in Travelers

Avoid bathing in freshwater in endemic countries.

9 Strongyloidiasis

An Italian couple went on honeymoon to Southeast Asia, travelling around Malaysia, Singapore and Thailand. Along with lovely photos and souvenirs, they brought back home a series of symptoms including skin rash, itching, fever, cough and fatigue. They were admitted to a hospital in Italy, where splenomegaly and marked eosinophilia were found. They were empirically treated with antibiotics, and fever and skin rash cleared up. However, after 1 month, they still had itching, cough, weakness and eosinophilia. The couple was eventually evaluated at the CTD, where a parasitic infection was suspected. Stool culture resulted positive for larvae of Strongyloides stercoralis in both patients, and the husband had positive stool microscopy and serology, too. They were treated with ivermectin (200 μg/kg/day for 2 days, repeated after 1 month), leading to complete resolution of signs and symptoms (Angheben et al. 2011).

Soil-transmitted helminths are a group of parasites (traditionally including hookworms, *Ascaris lumbricoides* and *Trichuris trichiura*) affecting a huge number of people in countries with poor sanitation (see Intapan et al. 2014). They cause relevant morbidity (for instance, weakness, failure to thrive, anaemia), contributing to a vicious circle of marginalization, poverty, and underdevelopment.

Strongyloides stercoralis is also widely diffused and, just like other soil-transmitted helminths, can cause chronic disturbances mainly involving skin (itching, rash), intestine (diarrhoea, abdominal pain) and lungs (asthma-like symptoms). The clinical cases described above show a rarely observed presentation of strongyloidiasis: the acute phase. Strongyloidiasis commonly results in a chronic, indolent disease, characterized by mild long-lasting symptoms.

The parasite induces a peculiar "auto-infective cycle" in the human body: some larvae produced by the adult worm, settled in the intestine, are not excreted with the faeces but reinvade the organism. Therefore, the infection perpetuates over years, even in the absence of a new exposure to contaminated soil.

Moreover, *S. stercoralis* is potentially lethal: in case of immunosuppression there is an accelerated replication of larvae, which can invade all human districts, including the brain. The disseminating larvae can also cause spread of intestinal bacteria in the organism, and therefore Gram-negative sepsis and meningitis linked to severe strongyloidiasis. It is thus important to diagnose and treat the infection before it causes the invasive phase.

Despite the potential risks caused by this infection, *S. stercoralis* remains probably "the most neglected of the neglected tropical diseases" (Olsen et al. 2009), not yet considered in mass treatment campaigns which are extensively conducted for other soil-transmitted helminths (Gabrielli et al. 2011).

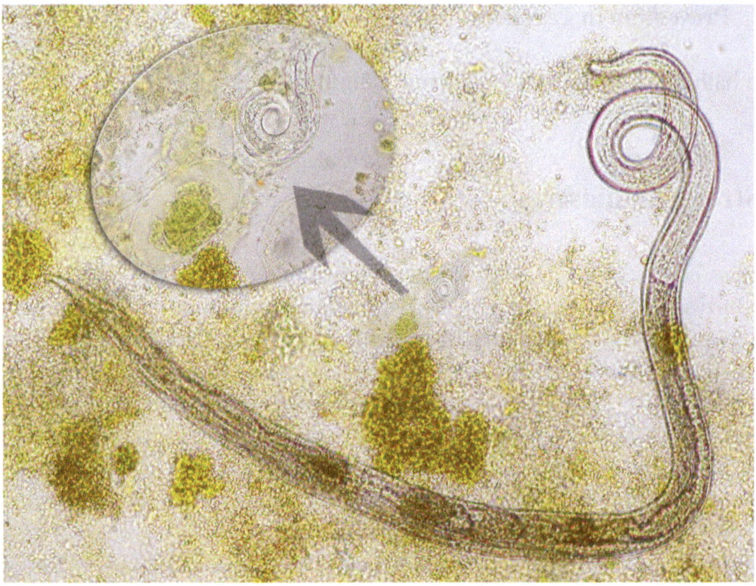

Fig. 4 Different stages of *Strongyloides stercoralis* in a sputum sample (photo by Maria Gobbo, CTD)

Recent studies based on serology (currently the most sensitive diagnostic method for strongyloidiasis) demonstrated a very high prevalence in immigrants and refugees from low-income countries. On the other hand, the risk of acquiring strongyloidiasis in travellers remains low (Buonfrate et al. 2012). Due to the auto-infective cycle, it should be also kept in mind that individuals infected during their youth may still harbor the parasite later on. This is true not only for migrants, but also for people currently living in countries where transmission has now stopped but has occurred in the past (such as elderly subjects who have always been living in Mediterranean countries).

9.1 When to Suspect Strongyloidiasis?

- Patients who have been walking barefoot in endemic areas, even decades earlier, and who present compatible, long-lasting symptoms and/or otherwise unexplained eosinophilia. For these patients serology is the best method to obtain diagnosis.
- Patients with possible exposure, history of immunosuppression (steroid therapy is frequently linked to severe strongyloidiasis!) and severely ill (abdominal obstruction, sepsis or meningitis caused by intestinal bacteria). In these cases eosinophil count is often normal, but direct examination of biological fluids (sputum—see Fig. 4, blood, cerebrospinal fluid, depending on the clinical picture) is always positive.

9.1.1 Treatment

Ivermectin has been proved to be effective in treating chronic strongyloidiasis, although the optimal dose schedule has not been clearly defined yet (Bisoffi et al. 2013).

9.1.2 Prevention in Travellers

Avoid walking barefoot on soil.

References

Abba K, Ramaratnam S, Ranganathan LN (2010) Antihelmintics for people with neurocysticercosis. Cochrane Database Syst Rev (1):CD000215

Angheben A, Mistretta M, Gobbo M, Bonafini S, Iacovazzi T, Sepe A, Gobbi F, Marocco S, Rossanese A, Bisoffi Z (2011) Acute strongyloidiasis in Italian tourists returning from Southeast Asia. J Travel Med 18:138–140

Baaten GG, Sonder GJ, van Gool T, Kint JA, van den Hoek A (2011) Travel-related schistosomiasis, strongyloidiasis, filariasis, and toxocariasis: the risk of infection and the diagnostic relevance of blood eosinophilia. BMC Infect Dis 11:84

Bierman WF, Wetsteyn JC, van Gool T (2005) Presentation and diagnosis of imported schistosomiasis: relevance of eosinophilia, microscopy for ova, and serology. J Travel Med 12:9–13

Bisoffi Z, Beltrame A, Monteiro G, Arzese A, Marocco S, Rorato G, Anselmi M, Viale P (2005) African trypanosomiasis gambiense, Italy. Emerg Infect Dis 11:1745–1747

Bisoffi Z, Buonfrate D, Montresor A, Requena-Mendez A, Muñoz J, Krolewiecki AJ, Gotuzzo E, Mena MA, Chiodini PL, Anselmi M, Moreira J, Albonico M (2013) *Strongyloides stercoralis*: a plea for action. PLoS Negl Trop Dis 7(5):e2214

Blum JA, Neumayr AL, Hatz CF (2012) Human African trypanosomiasis in endemic populations and travellers. Eur J Clin Microbiol Infect Dis 31:905–913

Both L, Banyard AC, van Dolleweerd C, Horton DL, Ma JK, Fooks AR (2012) Passive immunity in the prevention of rabies. Lancet Infect Dis 12:397–407

Bottieau E, Clerinx J, Schrooten W, Van den Enden E, Wouters R, Van Esbroeck M, Vervoort T, Demey H, Colebunders R, Van Gompel A, Van den Ende J (2006) Etiology and outcome of fever after a stay in the tropics. Arch Intern Med 166:1642–1648

Brun R, Blum J, Chappuis F, Burri C (2010) Human African trypanosomiasis. Lancet 375: 148–159

Brunetti E, White AC Jr (2012) Cestode infestations: hydatid disease and cysticercosis. Infect Dis Clin North Am 26:421–435

Buguet A, Mpanzou G, Bentivoglio M (2014) Human African trypanosomiasis: a highly neglected neurological disease. In: Bentivoglio M, Cavalheiro EA, Kristensson K, Patel N (eds) Neglected tropical diseases and conditions of the nervous system. Springer, New York

Buonfrate D, Angheben A, Gobbi F, Munoz J, Requena-Mendez A, Gotuzzo E, Mena MA, Bisoffi Z (2012) Imported strongyloidiasis: epidemiology, presentations, and treatment. Curr Infect Dis Rep 14:256–262

Carod-Artal FJ, Gascon J (2010) Chagas disease and stroke. Lancet Neurol 9:533–542

Carpio A, Fleury A (2014) Neurocysticercosis: neurology and neurobiology. In: Bentivoglio M, Cavalheiro EA, Kristensson K, Patel N (eds) Neglected tropical diseases and conditions of the nervous system. Springer, New York

CDC, http://dpd.cdc.gov/dpdx/HTML/Cysticercosis.htm

Clerinx J, Vlieghe E, Asselman V, Van de Casteele S, Maes MB, Lejon V (2012) Human African trypanosomiasis in a Belgian traveller returning from the Masai Mara area, Kenya, February 2012. Euro Surveill 17(10):pii: 20111

Coura JR (2014) Chagas disease: neurology and neurobiology. In: Bentivoglio M, Cavalheiro EA, Kristensson K, Patel NB (eds) Neglected tropical diseases and conditions of the nervous system. Springer, New York

Ferrari TCA (2014) Neuroschistosomiasis. In: Bentivoglio M, Cavalheiro EA, Kristensson K, Patel NB (eds) Neglected tropical diseases and conditions of the nervous system. Springer, New York

Gabrielli AF, Montresor A, Chitsulo L, Engels D, Savioli L (2011) Preventive chemotherapy in human helminthiasis: theoretical and operational aspects. Trans R Soc Trop Med Hyg 105: 683–693

Gautret P, Parola P (2012) Rabies pretravel vaccination. Curr Opin Infect Dis 25:500–506

Gobbi F, Boussinesq M, Mascarello M, Angheben A, Gobbo M, Rossanese A, Corachan M, Bisoffi Z (2011) Case report: Loiasis with peripheral nerve involvement and spleen lesions. Am J Trop Med Hyg 84:733–737

Grobusch MP, Muhlberger N, Jelinek T, Bisoffi Z, Corachan M, Harms G, Matteelli A, Fry G, Hatz C, Gjorup I, Schmid ML, Knobloch J, Puente S, Bronner U, Kapaun A, Clerinx J, Nielsen LN, Fleischer K, Beran J, da Cunha S, Schulze M, Myrvang B, Hellgren U (2003) Imported schistosomiasis in Europe: sentinel surveillance data from TropNetEurop. J Travel Med 10: 164–169

Guerri-Guttenberg RA, Ciannameo A, Di Girolamo C, Milei JJ (2009) Chagas disease: an emerging public health problem in Italy? Infez Med 17:5–13

Hotez PJ, Bottazzi ME, Franco-Paredes C, Ault SK, Periago MR (2008) The neglected tropical diseases of Latin America and the Caribbean: a review of disease burden and distribution and a roadmap for control and elimination. PLoS Negl Trop Dis 2:e300

Ing MB, Schantz PM, Turner JA (1998) Human coenurosis in North America: case reports and review. Clin Infect Dis 27:519–523

Intapan PM, Maleewong W, Nawa Y (2014) Nematode infections: neurological involvement and neurobiology. In: Bentivoglio M, Cavalheiro EA, Kristensson K, Patel NB (eds) Neglected tropical diseases and conditions of the nervous system. Springer, New York

Jackson AC (2014) Rabies: neurology. In: Bentivoglio M, Cavalheiro EA, Kristensson K, Patel NB (eds) Neglected tropical diseases and conditions of the nervous system. Springer, New York

Jelinek T (2008) Imported schistosomiasis in Europe: preliminary data for 2007 from TropNetEurop. Euro Surveill 13(7):pii:8038

Jelinek T, Bisoffi Z, Bonazzi L, van Thiel P, Bronner U, de Frey A, Gundersen SG, McWhinney P, Ripamonti D (2002) Cluster of African trypanosomiasis in travelers to Tanzanian national parks. Emerg Infect Dis 8:634–635

Kanade T, Shah I (2011) Dengue encephalopathy. J Vector Borne Dis 48:180–181

Knopp S, Steinmann P, Hatz C, Keiser J, Utzinger J (2012) Nematode infections: filariases. Infect Dis Clin North Am 26:359–381

Lipner EM, Law MA, Barnett E, Keystone JS, von Sonnenburg F, Loutan L, Prevots DR, Klion AD, Nutman TB (2007) Filariasis in travelers presenting to the GeoSentinel Surveillance Network. PLoS Negl Trop Dis 1:e88

Marin-Neto JA, Rassi A Jr, Avezum A Jr, Mattos AC, Rassi A, Morillo CA, Sosa-Estani S, Yusuf S (2009) The BENEFIT trial: testing the hypothesis that trypanocidal therapy is beneficial for patients with chronic Chagas heart disease. Mem Inst Oswaldo Cruz 104(Suppl 1):319–324

McManus DP, Gray DJ, Zhang W, Yang Y (2012) Diagnosis, treatment, and management of echinococcosis. BMJ 344:e3866

Misra UK, Kalita J (2014) Japanese encephalitis and dengue virus infection: neurology and neurobiology. In: Bentivoglio M, Cavalheiro EA, Kristensson K, Patel N (eds) Neglected tropical diseases and conditions of the nervous system. Springer, New York

Nicolls DJ, Weld LH, Schwartz E, Reed C, von Sonnenburg F, Freedman DO, Kozarsky PE (2008) Characteristics of schistosomiasis in travelers reported to the GeoSentinel Surveillance Network 1997-2008. Am J Trop Med Hyg 79:729–734

Njamnshi AK, Bissek AC, Etya'ale D (2014) Onchocerciasis: neurological involvement. In: Bentivoglio M, Cavalheiro EA, Kristensson K, Patel N (eds) Neglected tropical diseases and conditions of the nervous system. Springer, New York
Norman FF, Perez de Ayala A, Perez-Molina JA, Monge-Maillo B, Zamarron P, Lopez-Velez R (2010) Neglected tropical diseases outside the tropics. PLoS Negl Trop Dis 4:e762
Odolini S, Parola P, Gkrania-Klotsas E, Caumes E, Schlagenhauf P, Lopez-Velez R, Burchard GD, Santos-O'Connor F, Weld L, von Sonnenburg F, Field V, de Vries P, Jensenius M, Loutan L, Castelli F (2012) Travel-related imported infections in Europe, EuroTravNet 2009. Clin Microbiol Infect 18:468–474
Olsen A, van Lieshout L, Marti H, Polderman T, Polman K, Steinmann P, Stothard R, Thybo S, Verweij JJ, Magnussen P (2009) Strongyloidiasis – the most neglected of the neglected tropical diseases? Trans R Soc Trop Med Hyg 103:967–972
Paixao LC, Ribeiro AL, Valacio RA, Teixeira AL (2009) Chagas disease: independent risk factor for stroke. Stroke 40:3691–3694
PLoSNTD, http://www.plosntds.org/static/scope.action
Prata A (2001) Clinical and epidemiological aspects of Chagas disease. Lancet Infect Dis 1:92–100
Priotto G, Kasparian S, Mutombo W, Ngouama D, Ghorashian S, Arnold U, Ghabri S, Baudin E, Buard V, Kazadi-Kyanza S, Ilunga M, Mutangala W, Pohlig G, Schmid C, Karunakara U, Torreele E, Kande V (2009) Nifurtimox-eflornithine combination therapy for second-stage African Trypanosoma brucei gambiense trypanosomiasis: a multicentre, randomised, phase III, non-inferiority trial. Lancet 374:56–64. doi:10.1016/s0140-6736(09)61117-x
Schmunis GA, Yadon ZE (2010) Chagas disease: a Latin American health problem becoming a world health problem. Acta Trop 115:14–21
Sholdt LL, Schreck CE, Mwangelwa MI, Nondo J, Siachinji VJ (1989) Evaluations of permethrin-impregnated clothing and three topical repellent formulations of deet against tsetse flies in Zambia. Med Vet Entomol 3:153–158
Sicuri E, Munoz J, Pinazo MJ, Posada E, Sanchez J, Alonso PL, Gascon J (2011) Economic evaluation of Chagas disease screening of pregnant Latin American women and of their infants in a non endemic area. Acta Trop 118:110–117
Simarro PP, Franco JR, Cecchi G, Paone M, Diarra A, Ruiz Postigo JA, Jannin JG (2012) Human African trypanosomiasis in non-endemic countries (2000-2010). J Travel Med 19:44–53
Team FMI (2008) Identification of a rabid dog in France illegally introduced from Morocco. Euro Surveill 13(11):pii:8066
Urech K, Neumayr A, Blum J (2011) Sleeping sickness in travelers – do they really sleep? PLoS Negl Trop Dis 5:e1358
Visser LG, Polderman AM, Stuiver PC (1995) Outbreak of schistosomiasis among travelers returning from Mali, West Africa. Clin Infect Dis 20:280–285
Walker M, Kublin JG, Zunt JR (2006) Parasitic central nervous system infections in immunocompromised hosts: malaria, microsporidiosis, leishmaniasis, and African trypanosomiasis. Clin Infect Dis 42:115–125
Whitty CJ, Mabey DC, Armstrong M, Wright SG, Chiodini PL (2000) Presentation and outcome of 1107 cases of schistosomiasis from Africa diagnosed in a non-endemic country. Trans R Soc Trop Med Hyg 94:531–534
WHO (2007) Global plan to combat neglected tropical diseases 2008-2015 Geneva. WHO/CDS/NTD/2007.3
WHO (2011) Working to overcome the global impact of neglected tropical diseases – Summary. Wkly Epidemiol Rec 86:113–120
WHO, http://www.who.int/mediacentre/factsheets/fs099/en/. Accessed in Jan 2014
WHO, http://www.who.int/mediacentre/factsheets/fs340/en/index.html. Accessed in Jan 2014
Wilson ME, Weld LH, Boggild A, Keystone JS, Kain KC, von Sonnenburg F, Schwartz E (2007) Fever in returned travelers: results from the GeoSentinel Surveillance Network. Clin Infect Dis 44:1560–1568
Wolf T, Wichelhaus T, Gottig S, Kleine C, Brodt HR, Just-Nuebling G (2012) Trypanosoma brucei rhodesiense infection in a German traveller returning from the Masai Mara area, Kenya, January 2012. Euro Surveill 17(10):pii:20114

Stigma in Neurological Diseases in the Tropics

Earnest N. Tabah, Faustin Yepnjio, and Alfred K. Njamnshi

Abstract Stigma which has been defined as an "attribute that is deeply discrediting, and that reduces the bearer from a whole and usual person to a tainted, discounted, and inferior one", arises from various sources and occurs in different forms. Stigma can be internalised or anticipated by the stigmatised person who accepts perceived exclusionary views of the society or who fears enacted stigma by the society on persons with a stigmatizing conditions. Lastly, stigma may be endorsed or accepted by the society.

Stigma is associated with many neurological diseases globally, especially the neglected tropical diseases. Stigma develops within the background of rich and diversified cultural beliefs and traditions, where the population's knowledge on chronic neurological conditions is usually limited. In this context, attitudes and practices are based largely on misconceptions and myths.

Stigma has serious consequences on people affected by neglected neurological conditions and their families. It may be considered as the weakest link in the chain of disease diagnosis, treatment, prevention and eventual control or elimination. Stigma therefore constitutes a limiting factor to an acceptable quality of life for patients and the society. Various strategies have been suggested to fight stigma but this war is far from being won, although some battles have been successful.

E.N. Tabah
Swiss Tropical and Public Health Institute, The University of Basel, Basel, Switzerland

Department of Neurology, Central Hospital Yaoundé, Ministry of Public Health,
PO Box 25625, Yaoundé, Cameroon

F. Yepnjio • A.K. Njamnshi (✉)
Faculty of Medicine and Biomedical Sciences, Ministry of Public Health,
The University of Yaoundé I, Yaoundé, Cameroon

Department of Neurology, Central Hospital Yaoundé, PO Box 25265, Yaoundé, Cameroon
e-mail: aknjamnshi@yahoo.co.uk

M. Bentivoglio et al. (eds.), *Neglected Tropical Diseases and Conditions of the Nervous System*,
DOI 10.1007/978-1-4614-8100-3_3, © Springer Science+Business Media New York 2014

The brain mechanisms of stigma are largely unknown although some interesting data are now available. The growth of the young discipline of social neuroscience despite many challenges may provide leads on more effective strategies for stigma reduction in the future.

Keywords Neglected tropical diseases • Nervous system • Epilepsy • Leprosy • Onchocerciasis

1 Introduction

1.1 Stigma: From the Historical Concepts to a New Formulation

In his landmark treatise entitled *Stigma: Notes on the Management of Spoiled Identity*, the social scientist Erving Goffman (1964) noted that originally the term stigma is a Greek word that referred to a type of marking or tattooing that was cut or burned into the skin of criminals, slaves, or traitors in order to visibly identify them as blemished or morally corrupted persons. These individuals were to be avoided or shunned, particularly in public places. Goffman then defined stigma as an "attribute that is deeply discrediting, and that reduces the bearer from a whole and usual person to a tainted, discounted, and inferior one" (Goffman 1964).

Stigmatizing traits occur in three forms: (1) Overt or external deformations, such as scars, physical manifestations of leprosy, or of a physical or social disability; (2) Deviations in personal traits like behavior change in mental illness, drug addiction, homosexual tendencies, alcoholism, and criminal background; (3) "Tribal stigmas" are traits, imagined or real, of an ethnic group, a nationality or a religion that is deemed to be a deviation from the prevailing normative ethnicity, nationality or religion (Goffman 1964).

Studying the phenomenon further, Link and Phelan (2001) underscored the fact that stigma develops through a sequential process. This begins by distinguishing and labeling a trait or a human difference. The second step is to link the labeled person to undesirable characteristics referred to as negative stereotypes; the third step is, separating "them" (the labeled persons) from "us" (the normal ones). In the fourth step, the labeled persons experience loss of status and are considered inferior. Finally, the labeled persons are subjected to all forms of discrimination. The stigma process has been schematized by the The International Federation of Anti-Leprosy Associations (ILEP) (2011).

The stigma process occurs in a context of social, economic and political power that permits the identification of differences and construction of negative stereotypes, labeling, separation of labeled persons and discrimination against them (Link and Phelan 2001).

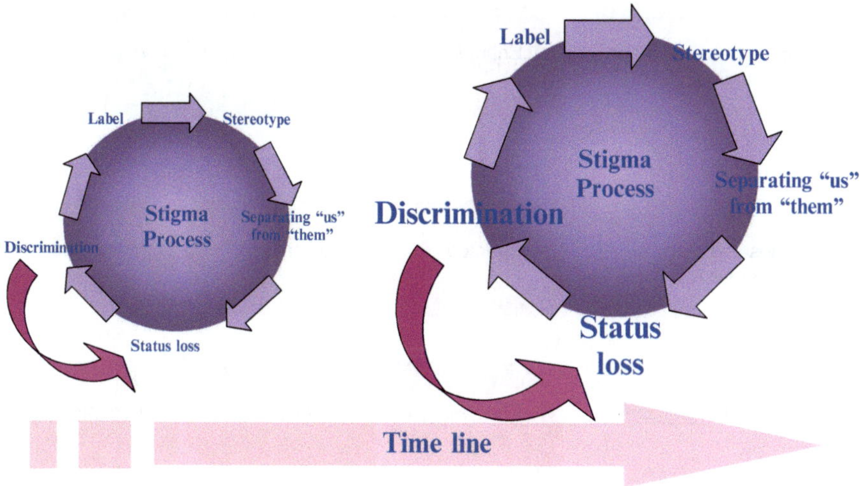

Fig. 1 The stigma process dynamics (adapted from ILEP 2011) model seen over time

As conceptualized above, the stigma process appears to be dynamic, with two of the components—status loss and discrimination—growing in relevance and scope over time. Furthermore, both of these components would appear to exert some influence one over the other in a "positive-feedback loop". This occurs as the negative stereotypes gradually become accepted by the community and internalized by the stigmatized individuals (Yanos et al. 2012). We therefore propose a "Stigma Process Dynamics" model, modified from the ILEP (2011) illustration of the stigma process, to show this dynamism (Fig. 1).

Building on Goffman's initial conceptualization, Jones et al. (1984) have identified six dimensions of stigma namely concealability, course, disruptiveness, peril, origin, and aesthetics. Concealability is the extent to which the condition can be seen by others or can be hidden; course is the severity and pattern of the condition over time, and disruptiveness is the degree of interference with the usual patterns of social interaction. The term 'aesthetic qualities' is how much the condition upsets others by way of the five senses; origin is the perceived cause and degree of responsibility of a person in contracting the illness (or condition). Finally, peril is the amount of fear and danger associated with a person's illness (Feldman and Crandall 2007; Jones et al. 1984; Quinn and Chaudoir 2009).

Goffman (1964) and Scambler and Hopkins (1990) have developed two key concepts to distinguish between "enacted" and "felt" stigma. Enacted stigma refers to acts or instances of discrimination against people with a stigmatizing condition on grounds of their perceived unacceptability or inferiority by members of the society. This could include overt discrimination in the workplace or educational institution, neglect, hostility, abuse or what is termed "fair and legitimate" discrimination, such as banning, driving or operating heavy machinery for epilepsy. Felt stigma is the anticipation or fear (on the part of persons with a stigmatizing trait) of enacted

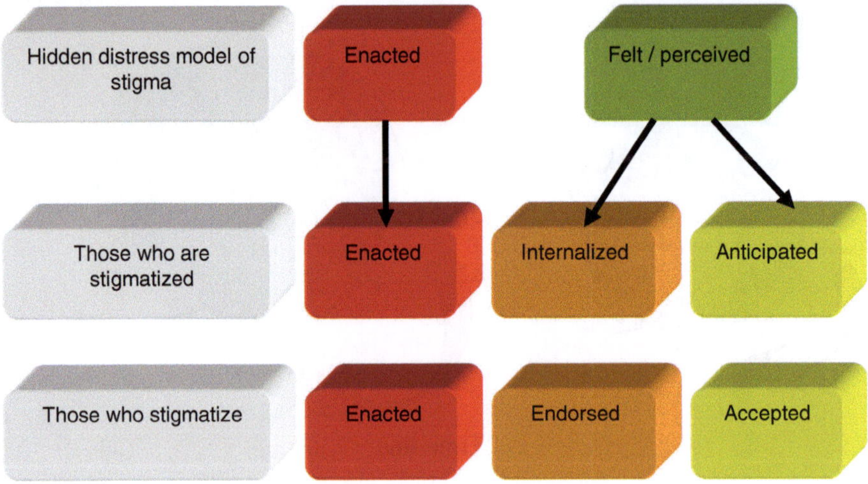

Fig. 2 Model of stigma, proposed by Weiss (2008) as extended from Scambler's "hidden distress model of stigma" (modified from Weiss 2008)

stigma, or negative reactions to the disclosure of the stigmatizing condition, and this also encompasses feelings of "difference" and shame. Felt stigma does not need to be based on personal experiences of enacted stigma, but is often built upon perceived social responses to the given stigmatizing condition, and is as debilitating as enacted stigma itself. These concepts are known as "Scambler's hidden distress model of stigma", extended by Weiss (2008) to include the concepts of "internalized, endorsed, anticipated and accepted stigma" (Fig. 2). In particular:

- Internalized stigma is a process through which a person with a stigmatized condition accepts perceived exclusionary views of the society and self-stigmatizes himself or herself.
- Anticipated stigma is fear of enacted stigma on behalf of a person with a stigmatizing condition.
- Endorsed stigma is a situation whereby some members of the society support and encourage acts of discrimination or exclusion against persons with a stigmatized condition although they do not actively engage in those acts themselves.
- Accepted stigma is the attitude of some members of the society who completely disagree with acts of discrimination or exclusion of persons with a stigmatized condition, but do nothing to stop it.

Since Goffman's seminal treatise on stigma in 1964, the number of publications on social stigma has increased sharply in recent years (from 458 articles in 2006 to 1,109 in 2011).

In 2006, Weiss and collaborators proposed a new formulation to facilitate action-oriented research on health-related stigma:

> Stigma is typically a social process, experienced or anticipated, characterized by exclusion, rejection, blame, or devaluation that results from experience or reasonable anticipation of an

Table 1 Estimated numbers of existing cases of selected NTDs which cause stigma and disablement globally (reproduced from WHO 2010)

Specific NTD	Disabilities resulting from disease	Numbers of cases/year or with permanent chronic symptoms
Buruli ulcer	Disfigurement	5,000/year
Onchocercosis/mucocutaneous leishmaniasis	Disfigurement	1.5 million/year
Onchocerciasis	Blindness; severe itching	265,000 existing cases
		746,000 existing cases
Lymphatic filariasis	Lymphoedema	15 million existing cases
	Hydrocele	25 million existing cases
Trachoma	Trichiasis	8.2 million existing cases
Yaws	Disfigurement	2.5 million (global prevalence estimate 1995)
Leprosy	Disfigurement	213,000/year
Human African trypanosomiasis	Neuropsychiatric disorders	Circa 10,000 new cases/year

adverse social judgment about a person or group. This judgment is based on an enduring feature of identity conferred by a health problem or health-related condition, and the judgment is in some essential way medically unwarranted. In addition to its application to persons or a group, the discriminatory social judgment may also be applied to the disease or designated health problem itself with repercussions in social and health policy. Other forms of stigma, which result from adverse social judgments about enduring features of identity apart from health-related conditions (e.g., race, ethnicity, sexual preferences), may also affect health; these are also matters of interest that concern questions of health-related stigma.

1.2 *Stigma and Neglected Tropical Diseases*

The concept of neglected tropical diseases (NTDs) emanated from the meetings convened in Berlin by the World Health Organization (WHO) and Deutsche Gesellschaft für Technische, now Internationale Zusammenarbeit in 2003 and 2004 (WHO 2004). Following these developments, Weiss (2008) and Hotez (2008) recently reviewed the issues of stigma in NTDs. Historically leprosy has been a major focus of stigma studies and literature. Other NTDs that generate stigma overtones include onchocerciasis, lymphatic filariasis, plague, Buruli ulcer, leishmaniasis, African trypanosomiasis and Chagas' disease. The numbers of existing cases of selected NTDs estimated by WHO (2010) and which cause stigma and disablement have increased sharply (Table 1).

In many NTDs stigma results from external deformations such as scars, physical manifestations of leprosy, or physical disabilities (Table 1). Furthermore, the impact of the meaning of the name of the disease may be as great, or even greater, source of suffering as symptoms of the disease. For example, paucibacillary leprosy may present at an early stage as a painless depigmented or anaesthetic patch.

Receiving the announcement of the diagnosis is likely to be far more troubling than these symptoms *per se* (Weiss et al. 2006; Corrigan 2007). The emotional impact of the social and cultural meanings of illness indicates another way by which stigma operates. For example, in settings where arranged marriages are a major concern of families for their children, the impact of a health problem on the ability to marry is troubling. For example, men with hydrocele suffer from embarrassment, ridicule and frustration due to their inability to perform sexual intercourse (Ahorlu et al. 2001).

1.3 Neuroscience of Stigma Behaviour

> We are, by nature, a highly affiliative species craving social contact. When social experience becomes a source of anxiety rather than a source of comfort, we have lost something fundamental – whatever we call it. (Insel 2002).

Thomas R. Insel, Editor of Biological Psychiatry, concluded that findings from various neuroscience disciplines appear to suggest that the brain has a special way of processing social behavioral information (Yizhar et al. 2011; Greimel et al. 2012; Tate et al. 2006). It would even appear that special genes determine whether some animals that do not have brains exhibit solitary or social behavior (Dreller and Page 1999). We have attempted to summarize the current data in trying to answer the following questions:

1. What happens in the brain of a person who stigmatizes another individual or a group of persons?
2. What happens in the brain of an individual who feels stigmatised or who is simply afraid of being stigmatised?

Is there any such neural network as a stigma centre in the brain? How does the brain develop social norms and are these 'programmes' fixed in time and space and if not, what modulates them? These questions are clearly very difficult to answer. The young growing discipline of behavioural or social neuroscience is going to hopefully enable us to gain a better understanding of these issues. Derks et al. (2008) have reviewed data obtained with electroencephalography, event-related potentials, or functional magnetic resonance imaging (fMRI) methods to examine neural correlates of stereotype and social identity threat. The findings that brain activation is related to the experience of being stereotyped have shed light on the cognitive processes underlying the social identity processes (Derks et al. 2008). Stereotype threat is a situational predicament in which individuals are at risk, by dint of their actions or behaviours, of confirming negative stereotypes about their group (Steele and Aronson 1995). For example, data obtained using fMRI suggest that stereotype threat affects women's mathematics performance in two ways: first, it disrupts normal recruitment of cognitive areas required for high math performance (the inferior prefrontal cortex, left inferior parietal cortex, and bilateral angular gyrus) and, second, it increases the recruitment of areas which allow for the processing and regulation of emotions (ventral anterior cingulate cortex). This implicates that women perform more poorly under stereotype threat because valuable cognitive resources are spent on emotional regulation instead of on the task at hand (Derks et al. 2008). Although this approach has limitations, it has the merit of providing some leads to the understanding of the complex phenomenon of stigma, which may in future contribute to its reduction in the targeted populations.

With regards to the second question, social neuroscience research has focused more on people who stereotype others rather than on the stigmatized individuals (see Dickter and Bartholow 2007; Ito et al. 2006). Brain imaging and electrophysiology have been applied to research on stereotyping from the perpetrator's perspective and this approach has yielded some insights into the processes that underlie prejudice and racial bias. For example, race effects have been observed in two brain areas traditionally associated with face perception, the fusiform gyrus and the posterior cingulate cortex. While these brain areas are considered to be responsible for face encoding and person knowledge respectively, evaluation and behaviour regulation appears to occur in the amygdala and anterior cingulate cortex, respectively (Ito and Bartholow 2009). This model, however, still has many unanswered questions such as the brain mechanisms of stereotype activation and regulation of stereotypic responses, and the psychological mechanisms involved.

2 Neurological Diseases Associated with Stigma and Its Determinants in the Tropics

2.1 Neurological Diseases Associated with Stigma in the Tropics

Stigma is more likely to be associated with chronic, rather than acute neurological conditions and is one of the major limitations of care provision and control of these diseases. Among chronic neurological diseases we shall discuss leprosy, epilepsy, onchocerciasis, human African trypanosomiasis (HAT) and schistosomiasis.

In the available published literature, amoebiasis, rabies, neurocysticercosis are only vaguely associated with stigma.

2.1.1 Leprosy (Hansen's Disease)

Leprosy, also known as Hansen's disease or neurodermatitis, is arguably the most extraordinary and misunderstood of diseases. Leprosy is a dreaded disease caused by *Mycobacterium leprae* akin to causative agent of tuberculosis, affecting the skin, the nerves in and close to the skin, the anterior third of the eyes, the upper respiratory tract, and the testicles (Sabin and Swift 1996). Although there is only one kind of leprosy bacillus, there are several varieties of leprosy because of the patient's immunological reaction to the infection. Three major forms have been described: tuberculoid or paucibacillary form in patients with high immune resistance, borderline form, and lepromatous form in patients with little or no immune resistance leading to progressive debilitation and gross mutilation. Patients with borderline leprosy have less skin involvement than those with lepromatous disease and may have more circumscribed skin lesions. In these patients, skin lesions are more severe than in patients with tuberculoid disease. In the spectrum of borderline leprosy, immunity may change, with patients worsening and their disease resembling lepromatous disease (downgrading reaction), or evolving toward the tuberculoid form (reversal reaction). Such shifts in the spectrum of disease may occur spontaneously or in response to drug treatment or inter-current medical conditions, such as underlying neoplasms or secondary infections (Sabin and Swift 2008).

Leprosy has been a major interest of health-related stigma studies from the outset. As cited by Hotez (2008), Berton Roueche observed that an ancient Egyptian pharaoh was known to banish people with leprosy to the edges of the Sahara desert. He coined the term *lepraphobia* to describe how, at the peak of the leprosy epidemic in Europe in the twelfth to fourteenth century, affected individuals were often subjected to their own mock funeral prior to banishment from their families and communities. In some cases, they endured torture and execution. Social constructions of leprosy are commonly guided by cultural, traditional and religious beliefs or myths about disease and illness (Wong 2004; Waxler 1981; Van den Broek et al. 1998; Nsagha et al. 2011; Opala and Boillot 1996). Biblical teachings perpetuated by missionaries associated leprosy with sin and uncleanliness, and leprosy patients came to be considered outcasts as a consequence (Edmond 2006). Leprosy-infected people are frequently considered cursed or victims of witchcraft, or blameworthy or immoral, and their disease well deserved (Nsagha et al. 2011). In many countries, treatment policies require incarceration of people affected by leprosy at various leprosaria, sometimes due to the high rates of illiteracy and misinformation about the disease (Kazeem and Adegun 2011). While people with enigmatic physical disfigurement (lepromatous form), and the distinctive ulcers consequent to untreated leprosy will face ridicule and rejection from society (Fig. 3), the diagnosis of the tuberculoid or paucibacillary form will induce fear and ultimately anticipated stigma.

2.1.2 Epilepsy

Although epilepsy is not a neglected disease, a chapter on stigma has to discuss the issue. According to the definition of the International League against Epilepsy (ILAE), epilepsy is a transient occurrence of signs and/or symptoms due to abnormal, excessive, or synchronous neuronal activity in the brain (Engel and Pedley 2008). More than 60 million people worldwide suffer from epilepsy (Ngugi et al. 2010) and it is estimated that 80 % of the burden of epilepsy worldwide is borne by resource poor countries (WHO 2004; Birbeck and Kalichi 2004; Kaiser et al. 1998; Longe and Osuntokun 1989; Osuntokun et al. 1987; Jilek-Aall and Rwiza 1992).

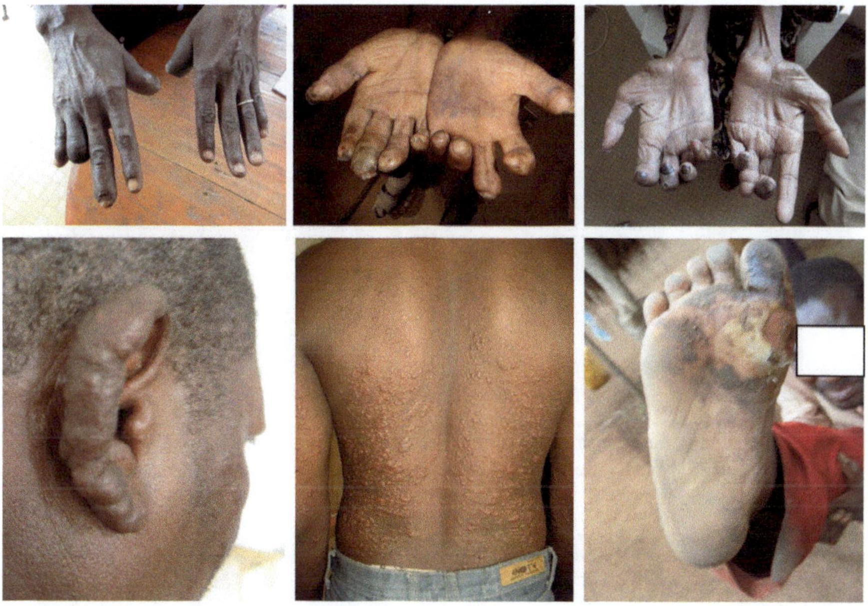

Fig. 3 Clawed and amputated fingers; muscle atrophy of the thenar and hypothenar hand regions in leprosy. Lepromatous lesions on earlobe and back in same patient, foot ulceration (Courtesy: EN Tabah). These severe disfigurements cause fear of contagion and lead to rejection of people with leprosy

Due to the treatment gap, the number of people with active epilepsy who have not accessed biomedical services or who are not on treatment or are on inadequate treatment is between 62 and 75 % in sub-Saharan Africa (Mbuba et al. 2012a; Meyer et al. 2010). Traditional belief systems attribute epilepsy to demon possession, witchcraft, and/or seek to blame the victim (Baskind and Birbeck 2005; Njamnshi et al. 2010). All these provide the ideal environment for stigma to flourish. Besides these beliefs, body disfigurement resulting from falls, burns or drug adverse effects also contribute to stigma development in epilepsy. For example, a Cameroonian suffering from sustained severe burns on the face and upper limbs, leading to the loss of her left hand. As a result of the discrimination associated with epilepsy in this part of Cameroon, she was sent on exile from the city to live far away in a farmhouse, to avoid her associating with other people (Fig. 4).

2.1.3 Onchocerciasis (River Blindness)

Onchocerciasis results from infection with the nematode *Onchocerca volvulus*, for which man is the only reservoir. Adult worms live in sub-cutaneous nodules and have a reproductive life-span of 9–11 years (Plaisier et al. 1991). The adult female worm produces microfilaria in millions, which migrate to the skin of the human host.

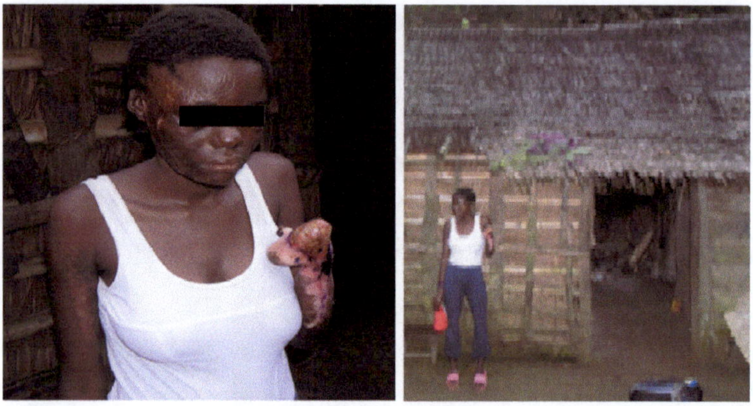

Fig. 4 Stigmatizing burn deformities in a Cameroonian epilepsy patient (courtesy: Pastor JTN Njamnshi, Cameroon Baptist Seminary, Kumba)

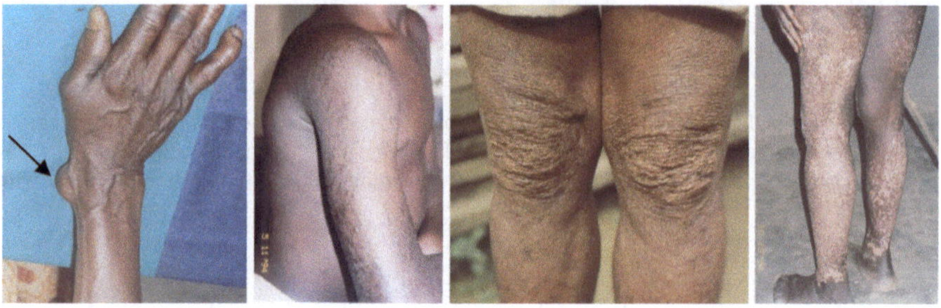

Fig. 5 Stigmatising onchocerciasis lesions: skin nodule, chronic onchodermatitis, skin atrophy and skin depigmentation in onchocerciasis (courtesy: AC Zoung-Kanyi Bissek)

The microfilaria are responsible for the clinical manifestations of the disease which include dermatitis, with concurrent pruritus; lychenified skin lesions; skin depigmentation and atrophy; and lymphadenitis, which results in the hanging groin and elephantiasis (Murdoch et al. 1993, cited in: Okoye and Onwuliri 2007). The most severe complication of onchocerciasis infection is irreversible and often stigmatizing blindness due to ocular lesions of both the anterior and the posterior chambers of the eye.

The severe itching, depigmentations as well as lichenification of the skin as a result of onchocercal infection have been a source of discrimination and stigma (Awedoba 2001). Skin nodules could also be a source of stigma depending on location (Fig. 5). Okoye and Onwuliri (2007) have found that the most worrisome consequence of onchocercal skin disease in Northeastern Nigeria included social isolation, feeling of shame and low esteem, skin blemish and marital problems. They recount the story of a 48 year-old female victim of onchocercal skin disease as follows: "*I am always afraid (anxious) that an attack of itching in a private part (buttock and groin) could occur at a public gathering; I therefore kept off; in fact, I hated myself*". In the same community, relating to stereotypes held with respect to marriage, a 26 year-old man said, "*When a lady's body has been spoilt by mbiba (popular rashes), only elderly widowers and already married men would seek her hand in marriage*". These pejorative attitudes are held across most onchocerciasis afflicted areas in sub-Saharan Africa (Awedoba 2001; Tchounkeu et al. 2012).

2.1.4 Human African Trypanosomiasis (Sleeping Sickness)

Stigma related to HAT is not clearly defined since the presenting symptoms resemble those of common conditions such as malaria in the early phase of the disease. However, in Malawi where transmission occurs around game reserves protected from human activity, individuals affected by HAT are stigmatized as deserving the condition, for having violated the band on infringing the game reserves (Chisi et al. 2011).

2.1.5 Schistosomiasis

Stigma in schistosomiasis is related to the post-micturation trickling of blood (Takoungang et al. 2004). The female genital schistosomiasis, which is an advanced form in women, is usually more stigmatizing due to symptom similarity (lower abdominal pain, bleeding after sexual intercourse) to sexually transmissible infections (Ahlberg et al. 2003). These symptoms bring about shame and guilt, resulting in concealment and delay in seeking help among young girls and women.

2.2 Measurement of Stigma

The full assessment of health-related stigma requires at least two levels of consideration that include: assessment in the community (general population as well as specific target groups void of the stigmatized condition in question) to determine enacted and felt stigma and assessment among the affected persons, to determine anticipated, internalized and experienced stigma. The impact of stigma assessment would also target the affected persons, and will seek to measure the level of participation, quality of life, self-esteem and self-efficacy (Van Brakel 2006; Rensen et al. 2010). These approaches are very important as the study of people with a stigmatized health problem provides an account of self-perceived, experienced stigma as well as their consequences. Meanwhile the study of people without the stigmatized health problem in the community clarifies the social context of stigma targeting that condition (Weiss and Ramakrishna 2006).

Different methods could be employed within each approach. The most commonly used methods include:

- Questionnaires: These are usually closed or open or interview guides, containing items that allow the collection of data on knowledge, attitudes and reported practices (KAP). This method has been widely used in the assessment of epilepsy and leprosy related stigma (Njamnshi et al. 2009a–e, 2010; Atadzhanov et al. 2010; Babikar and Abbas 2011; Van Brakel 2003).

- Qualitative methods: These are assessments based on such methods as key informant interviews, focus group discussions and observation by participants.
- Indicators: These are often used in sets. They provide separate information for each indicator, and when pooled together, they may give a profile of stigma and discrimination. They cannot however be summarised in one measure, unless they have been developed as a scale.
- Scales: These are quantitative instruments intended to give a numerical result that indicates the severity or extent of the phenomenon measured. Examples of such stigma scales have been developed and validated recently for epilepsy in Kenya (Mbuba et al. 2012a, b) and for use across various neurological conditions in the USA (Molina et al. 2013).

Several instruments have been developed for measurement of health-related stigma. The majority of the instruments have however been disease-specific and the diseases of most interest to stigma research have been: epilepsy, leprosy, mental illness and HIV/AIDS. Two major reviews have sorted out the various instruments (scales, questionnaires, indicators sets) developed and used in the assessment of general and internalized health-related stigma (Van Brakel 2006; Stevelink et al. 2012).

Attempts are being made to develop instruments that could be used to measure stigma across all stigmatizing conditions. For instance the "Explanatory Model Interview Catalogue" (EMIC) developed by Weiss et al. (1992) for assessment of leprosy stigma has been adapted and used for the assessment of stigma in epilepsy (Rafael et al. 2010), depression (Raguram et al. 1996), schizophrenia (Raguram et al. 2004), Buruli ulcer (Stienstra et al. 2002). The stigma impact scale (SIS) and the stigma experience scale (SES) have both been used in Alzheimer's dementia and Parkinson's disease (Burgener and Berger 2008). The "Internalized Stigma Mental Illness Scale" (ISMI) has been adapted and used for leprosy stigma (Rensen et al. 2010). The "Child Attitude Toward Illness Scale" (CATIS) has been used to study stigma associated with epilepsy and asthma in children (Austin and Huberty 1993) and adolescents (Heimlich et al. 2000). The most recently developed and validated scale, which seems to be the most federating, the Stigma Scale for Chronic Illnesses: 8-Item Version (SSCI-8), allows for the measurement of stigma in five neurological conditions: epilepsy, multiple sclerosis, Parkinson's disease, stroke, and amyotrophic lateral sclerosis (Molina et al. 2013). However, the SSCI-8 will still have to be tested and validated in the tropical context with a different neuroepidemiological picture from the temperate regions (Naeije et al. 2013).

It should be underscored that many instruments have been developed to assess the intensity and qualities of stigma related to neurological disorders but often these have been condition-specific. There are ongoing attempts to develop and validate more generic stigma assessment instruments that will cut across several stigmatizing neurological disorders. However, there is still much to be done, especially in the tropics, considering the huge socio-cultural variation in the expression and manifestation of health-related stigma.

2.3 Determinants of Stigma

The measurement of stigma also allows for the identification of possible determinants. In the tropics, factors associated with stigma seem to vary considerably from one culture to another and even between different communities within the same cultural setting. For instance, studies on epilepsy in different regions of Cameroon by Njamnshi et al. (2009a–e) revealed that the major determinants of negative attitudes towards people with epilepsy (PWE) included: advanced age; low level of education; the belief that epilepsy is insanity, hereditary, contagious, or caused by witchcraft. Anxiety, marital problems and social isolation were the major factors associated with epilepsy stigma in Benin (Rafael et al. 2010), while disclosure status, personal and community contagion beliefs are associated with epilepsy stigma in Zambia (Atadzhanov et al. 2010). In Mangalore, India stigmatization with respect to epilepsy was found to be related to the age and education of the respondent (Joseph et al. 2011). The factors associated with epilepsy stigma in Cambodia were related to its curability, possibility of getting married, education, fear of seizure, convulsions (tonic-clonic seizure type), memory lapses, and the ability to speak about the condition (Bhalla et al. 2012). In Bolivia, the fear linked to loss of control, the feelings of sadness and pity toward people with epilepsy, the difficulties faced by people with epilepsy (PWE) in the professional and relationship fields, the level of education and type of seizure were factors associated with epilepsy stigma (Bruno et al. 2012).

In Cameroon, enacted leprosy stigma is associated with fear of contagion, and cultural perceptions that attribute leprosy to punishment for wrong doing and witchcraft, while felt and experienced leprosy stigma is manifested by marital problems or being unable to marry, feeling of less self-esteem and shunning (Nsagha et al. 2011).

3 Consequences and Challenges of Stigma Reduction

3.1 Consequences of Stigma

Stigma, whether general or health-related, brings about numerous negative effects on its victims that may result in serious consequences.

The first consequence is the discrimination against people with a stigmatizing condition. The discrimination may be overt, for instance banishment of persons affected by leprosy as reported in India (Jacob and Franco-Paredes 2008) and some tribal communities in Cameroon (Nsagha et al. 2011). Quarantining and segregation by sex to prevent reproduction among people with leprosy was practised in India during the colonial era (Jacob and Franco-Paredes 2008).

Structural discrimination occurs, for example, when a treatment facility for a stigmatizing condition like leprosy is located in an isolated and remote neighbourhood. Another form of discrimination, which is more insidious, occurs when people with a stigmatizing condition accept and internalize the negative label. They tend to have low self-esteem or may become aggressive or may just avoid contact with "normal individuals". The outcome of this form of discrimination is usually low self-esteem, depression, participation restriction, unemployment and loss of income (Link and Phelan 2006). The fear of discrimination may lead to concealment in some individuals who become affected by a stigmatizing condition and shunning of those already labelled or stereotyped. The consequences of such an attitude are delays in seeking health care or abandonment or non compliance with ongoing treatment. This leads to delay in diagnosis and treatment and worsening of the condition. This has been noted especially in leprosy and epilepsy.

Stigma has been identified as a source of chronic stress in people with a stigmatizing condition. Stress has been attributed to experienced stigma and the fear of being stigmatized and these have led to harmful health outcomes in the victims (James et al. 1984). Link and Phelan (2006) have remarked that in addition to stress from the illness itself, stress from stigma related to a health condition can lead to worsening the clinical course of the illness and other outcomes such as the inability to lead a normal social life or work.

Stigma ultimately leads to poor quality of life in the affected persons and by extension, their families.

3.2 Approaches to Stigma Reduction

The domain of stigma reduction interventions is almost devoid of reliable evidence-based examples. For this reason, most of the literature in the area only tends to spell out broad guidelines, based on respected theories, upon which intervention programmes can be designed (Cross et al. 2011).

Based on the study of stigma and the extension of the Scambler's "hidden distress model", Weiss et al. (2006) noted that stigma reduction interventions may vary from one health condition to another, and suggested a mitigating framework indicating the focus and approach for interventions to counter the effects of stigma. According to this framework, interventions may focus on support for affected persons, changing behaviour among the stigmatizers in the general population (or specific subgroups), and dealing with the stigmatizing condition.

The guidelines by Weiss (2008) have generally been agreed upon by most investigators, but suggest that the consideration of the process of stigmatization developed by Link and Phelan (2001) would lead to more precise interventions (Cross et al. 2011).

Motivated by the perceived need for multidimensional interventions and based on the observation that programmes that included counselling (at the individual level), education and contact (at both individual and community levels) appeared to be the most promising of the many interventions, Heijnders and Van Der Meij (2006) suggested packages of strategies and interventions according to the social/ecological level under which each could be categorized (see Table 2).

Table 2 Stigma reduction strategies (adapted from Heijnders and Van Der Meij 2006)

Level	Strategies
Intrapersonal	Treatment
	Counselling
	Cognitive behavioural therapy
	Empowerment
	Group counselling
	Self help, advocacy and support groups
Interpersonal level	Care and support
	Home care teams
	Community-based rehabilitation
Organisational/institutional level	Training programmes
	Research for generating evidence
	Policies, like patient-centred and integrated approaches
Community level	Education
	Contact
	Advocacy
	Protest
Governmental/structural level	Legal and evidence-informed policy interventions
	Rights-based approaches
	Advocacy

3.3 Challenges to Stigma Reduction

The major challenge in stigma reduction is that there is no one-fit-all strategy, even for the same stigmatizing condition. Stigma is a complex issue, heavily influenced by cultural norms and beliefs. Therefore, developing any mitigating intervention adaptable to multi-cultural setting for any given stigmatizing condition remains a challenge. The need to consider several levels of interventions, each requiring multiple strategies, poses an additional challenge. These challenges are made even more difficult as the impact of stigma reduction interventions take quite long to be observed and are sometimes not very evident.

4 Conclusions and Perspectives

Dealing with stigma in neurological diseases (neglected or not) in the Tropics appears to be the weakest link in management and control or prevention as stigma has been shown to be a limiting factor to care-seeking behaviour. The young field of social neuroscience can hopefully develop, in spite of the current challenges, novel evidence for more effective, culturally sensitive and appropriate stigma reduction interventions. Success in this area may also pave the way for the removal of neglected neurological conditions from the neglected category. Otherwise, such conditions may become more neglected by this categorization, which may be discriminating in itself.

References

Ahlberg BM, Mwangi R, Poggensee G, Feldmeier H, Krantz I (2003) "Better infection than hunger". A study of illness perceptions with special focus on urinary schistosomiasis in Northern Tanzania. Afr Soc Rev 7:18–34

Ahorlu CK, Dunyo SK, Asamoah G, Simonsen PE (2001) Consequences of hydrocele and the benefits of hydrocelectomy: a qualitative study in lymphatic filariasis endemic communities on the coast of Ghana. Acta Trop 80:215–221

Atadzhanov M, Haworth A, Chomba EN, Mbewe EK, Birbeck GL (2010) Epilepsy-associated stigma in Zambia: what factors predict greater felt stigma in a highly stigmatized population? Epilepsy Behav 19:414–418. doi:10.1016/j.ybeh.2010.08.017

Austin JK, Huberty TJ (1993) Development of the child attitude toward illness scale. J Pediatr Psychol 18:467–480

Awedoba AK (2001) Help-seeking and coping strategies in onchocerciasis afflicted communities in the context of globalisation. Res Rev 17:49–62

Babikar HE, Abbas IM (2011) Knowledge, practice and attitude toward epilepsy among primary and secondary school teachers in South Gezira locality, Gezira State, Sudan. J Family Community Med 18:17–21

Baskind R, Birbeck G (2005) Epilepsy care in Zambia: a study of traditional healers. Epilepsia 46:1121–1126

Bhalla D, Chea K, Hun C, Vannareth M, Huc P et al (2012) Population-based study of epilepsy in Cambodia associated factors, measures of impact, stigma, quality of life, knowledge-attitude-practice, and treatment gap. PLoS One 7(10):e46296. doi:10.1371/journal.pone.0046296

Birbeck GL, Kalichi EM (2004) Epilepsy prevalence in rural Zambia: a door-to-door survey. Trop Med Int Health 9:92–95

Bruno E, Bartoloni A, Sofia V, Rafael F, Magnelli D, Padilla S, Quattrochi G, Bartalesi F, Segundo H, Zappia M, Preux PM, Nicoletti A (2012) Epilepsy-associated stigma in Bolivia: a community-based study among the Guarani population An International League Against Epilepsy/International Bureau for Epilepsy/World Health Organization Global Campaign Against Epilepsy Regional Project. Epilepsy Behav 25:131–136

Burgener SC, Berger B (2008) Measuring perceived stigma in persons with progressive neurological disease: Alzheimer's dementia and Parkinson's disease. Dementia 7:31–53

Chisi JE, Muula AS, Ngwira B, Kabuluzi S (2011) A retrospective study of Human African Trypanosomiasis in three Malawian districts. Tanzanian J Health Res 13:79–86

Corrigan PW (2007) How clinical diagnosis might exacerbate the stigma of mental illness. Soc Work 52:31–39

Cross HA, Heijnders M, Dalal SS, Mak S (2011) Interventions for stigma reduction – Part 1: Theoretical considerations. Reviews 22:62–70. doi:10.5463/DCID.v22i3.70

Derks B, Inzlicht M, Kang S (2008) The neuroscience of stigma and stereotype threat. Group Processes & Intergroup Relations 11:163–181. SAGE Publications, Los Angeles, London, New Delhi and Singapore. doi:10.1177/1368430207088036

Dickter CL, Bartholow BD (2007) Racial ingroup and outgroup attention biases revealed by event-related brain potentials. Soc Cogn Affect Neurosci 2:189–198

Dreller C, Page RE (1999) Genetic, developmental, and environmental determinants of honey bee foraging behaviour. In: Detrain C, Deneubourg JL, Pasteels JM (eds) Information processing in social insects. Birkhäuser, Basel, pp 187–202

Edmond R (2006) Leprosy and empire: a medical and cultural history. Cambridge University Press, Cambridge
Engel J Jr, Pedley TA (2008) Introduction: what is epilepsy? In: Engel J Jr, Pedley TA (eds) Epilepsy: a comprehensive textbook. Lippincott Williams & Wilkins–Wolters Kluwer Business, Philadelphia, pp 1–11
Feldman DB, Crandall CS (2007) Dimensions of mental illness stigma: what about mental illness causes of social rejection. J Soc Clin Psychol 26:137–154
Goffman E (1964) Stigma: notes on the management of spoiled identity. Prentice Hall, Englewood Cloffs
Greimel E, Nehrkorn B, Fink GR, Kukolja J, Kohls G, Müller K, Piefke M, Kamp-Becker I, Remschmidt H, Herpertz-Dahlmann B, Konrad K, Schulte-Rüther M (2012) Neural mechanisms of encoding social and non-social context information in autism spectrum disorder. Neuropsychologia 50:3440–3449. doi:10.1016/j.neuropsychologia.2012.09.029
Heijnders M, Van Der Meij S (2006) The fight against stigma: an overview of stigma reduction strategies and interventions. Psychol Health Med 11:353–363, http://dx.doi.org/10.1080/13548500600595327
Heimlich TE, Westbrook LE, Austin JK, Cramer JA, Devinsky O (2000) Brief report: adolescents' attitudes toward epilepsy: further validation of the Child Attitude Toward Illness Scale (CATIS). J Pediatr Psychol 25:339–345, http://dx.doi.org/10.1093/jpepsy/25.5.339
Hotez PJ (2008) Stigma: The stealth weapon of the NTD. PLoS Negl Trop Dis 2(4):e230. doi:10.1371/journal.pntd.0000230
Insel TR (2002) Social anxiety: from laboratory studies to clinical practice. Biol Psychiatry 51(1):1–3
International Federation of Anti-Leprosy Associations (ILEP) and the Netherlands Leprosy Relief (NLR) (2011) Guidelines to reduce stigma: Guide 1. ILEP & NLR, London/Amsterdam
Ito TA, Bartholow B (2009) The neural correlates of race. Trends Cogn Sci 13:524–531
Ito TA, Chiao KW, Devine PG, Lorig TS, Cacioppo JT (2006) The influence of facial feedback on race bias as measured with Implicit Association Test. Psychol Sci 17:256–261
Jacob JT, Franco-Paredes C (2008) The stigmatization of leprosy in India and its impact on future approaches to elimination and control. PLoS Negl Trop Dis 2:e113. doi:10.1371/journal.pntd.0000113.g001
James SA, LaCroix AZ, Kleinbaum DG, Strogatz DS (1984) Henryism and blood pressure differences among black men. II The role of occupational stressors. J Behav Med 7:259–275
Jilek-Aall L, Rwiza HT (1992) Prognosis of epilepsy in a rural African community: a 30-year follow-up of 164 patients in an outpatient clinic in rural Tanzania. Epilepsia 33:645–650
Jones E, Farina A, Hastorf A, Markus H, Miller D, Scott R (1984) Social stigma. The psychology of marked relationships. W.H. Freeman, New York
Joseph N, Ray A, Reshma BK, Bhat S, Herady M, Kumar A, Kiran K (2011) Assessment of quality of life, stigma associated and self management practices among patients suffering from epileptic seizures: a cross sectional study. J Neurosci Behav Health 3:91–98
Kaiser C, Asaba G, Leichsenring M, Kabagambe G (1998) High incidence of epilepsy related to Onchocerciasis in West Uganda. Epilepsy Res 30:247–251
Kazeem O, Adegun T (2011) Leprosy stigma: ironing out the creases. Lepr Rev 82:103–108
Link BG, Phelan JC (2001) Conceptualising stigma. Annu Rev Sociol 27:363–385
Link BG, Phelan JC (2006) Stigma and its public health implications. Lancet 367:328–529
Longe AC, Osuntokun BO (1989) Prevalence of neurological disorders in Udo, a rural community in southern Nigeria. Trop Geogr Med 41:36–40
Mbuba CK, Ngugi AK, Fegan G et al (2012a) Risk factors associated with the epilepsy treatment gap in Kilifi, Kenya: a cross-sectional study. Lancet Neurol 11:688–696
Mbuba CK, Abubakar A, Odermatt P, Newton CR, Carter JA (2012b) Development and validation of the Kilifi Stigma Scale for Epilepsy in Kenya. Epilepsy Behav 24(1):81–85
Meyer AC, Dua T, Ma J et al (2010) Global disparities in the epilepsy treatment gap: a systematic review. Bull World Health Organ 88:260–266
Molina Y, Choi SW, Cella D, Rao D (2013) The stigma scale for chronic illnesses 8-item version (SSCI-8): development, validation and use across neurological conditions. Int J Behav Med 20:450–460

Murdoch ME, Hay RJ, Mackenzie CD, Williams JF, Ghalib HW, Cousens S, Abiose A, Jones BR (1993) A clinical classification and grading system of the cutaneous changes in onchocerciasis. Br J Dermatol 129:260–269

Naeije G, Yepnjio FN, Bissek AC, Tabah EN, Tatah G, Fonsah JY, Fogang Y, Kuate C, Dachy B, Njamnshi AK (2013) Yield of training exchanges between Europe and Sub-Saharan Africa. Acta Neurol Belg 113:31–34

Ngugi AK, Bottomley C, Kleinschmidt I, Sander JW, Newton CR (2010) Estimation of the burden of active and life-time epilepsy: a meta-analytic approach. Epilepsia 51:883–890

Njamnshi AK, Angwafor SA, Jallon P, Muna WFT (2009a) Secondary school students' knowledge, attitudes and practice towards epilepsy in the Batibo Health District—Cameroon. Epilepsia 50:1262–1265

Njamnshi AK, Angwafor SA, Tabah EN, Muna WFT (2009b) General public knowledge, attitudes, and practices with respect to epilepsy in the Batibo Health District, Cameroon. Epilepsy Behav 14:83–88

Njamnshi AK, Tabah EN, Yepnjio FN, Angwafor SA, Dema F, Fonsah JY, Kuate CT, de Paul Djientcheu V, Angwafo F 3rd, Muna WF (2009c) General public awareness, perceptions and attitudes with respect to epilepsy in the Akwaya Health District, South West Region, Cameroon. Epilepsy Behav 15:179–185

Njamnshi AK, Yepnjio FN, Tabah EN, Dema F, Angwafor SA, Fonsah JY, Angwafo FF 3rd, Muna WF (2009d) Public awareness, perceptions and attitudes with respect to epilepsy in Ebolowa and Sangmelima—Urban Cameroon. Epilepsy Behav 14:628–633

Njamnshi AK, Yepnjio FN, Bissek AC, Tabah EN, Ongolo-Zogo P, Dema F, Angwafor SA, Fonsah JY, Lekoubou A, Angwafo FF 3rd, Jallon P, Muna WF (2009e) A survey of public knowledge, attitudes, and practices with respect to epilepsy in Badissa village, Centre Region of Cameroon. Epilepsy Behav 16:254–259

Njamnshi AK, Zoung-Kanyi Bissek AC, Yepnjio FN, Tabah EN, Angwafor AS, Kuate TC, Déma F, Fonsah YJ, Acho A, Kepeden Z-K MN, Yumo HA, Kuwoh BP, Angwafor FF III, Muna WFT (2010) A community survey of knowledge, perceptions, and practice with respect to epilepsy among traditional healers in the Batibo Health District, Cameroon. Epilepsy Behav 17:95–102

Nsagha DS, Bamgboye EA, Assob JC, Njunda AL, Kamga HL, Zoung-Kanyi Bissek AC, Tabah EN, Oyediran AB, Njamnshi AK (2011) Elimination of leprosy as a public health problem by 2000 AD: an epidemiological perspective. Pan Afr Med J 9:4.

Okoye IC, Onwuliri COE (2007) Epidemiology and psycho-social aspects of onchocercal skin disease in northeastern Nigeria. Filaria J 6:15. doi:10.1186/1475-2883-6-15

Opala J, Boillot F (1996) Leprosy among the Limba: illness and healing in the context of worldview. Soc Sci Med 42:3–19

Osuntokun BO, Adeuja AO, Nottidge VA, Bademosi O, Olumide A, Ige O, Yaria F, Bolis CL, Schoenberg BS (1987) Prevalence of the epilepsies in Nigerian Africans: a community-based study. Epilepsia 28:272–279

Plaisier AP, van Oortmarssen GJ, Remme J, Habbema JDF (1991) The reproductive lifespan of Onchocerca volvulus in West African savanna. Acta Trop 48:271–284

Quinn DM, Chaudoir SR (2009) Living with a concealable stigmatized identity: the impact of anticipated stigma, centrality, salience, and cultural stigma on psychological distress and health. J Pers Soc Psychol 97:634–651

Rafael F, Houinato D, Nabukpo P, Dubreuil CM, Tran DS, Odermatt P, Clement JP, Weiss MG, Preux PM (2010) Sociocultural and psychological features of perceived stigma reported by people with epilepsy in Benin. Epilepsia 51:1061–1068

Raguram R, Weiss MG, Channabasavanna SM, Devins GM (1996) Stigma, depression, and somatization in South India. Am J Psychiatry 153:1043–1049

Raguram R, Raghu TM, Vounatsou P, Weiss MG (2004) Schizophrenia and the cultural epidemiology of stigma in Bangalore, India. J Nerv Ment Dis 192(11):734–744

Rensen C, Bandyopadhyay S, Gopal PK, Van Brakel WH (2010) Measuring leprosy-related stigma – a pilot study to validate a toolkit of instruments. Disabil Rehabil 33:711–719, http://dx.doi.org/10.3109/09638288.2010.506942

Sabin TD, Swift TR (1996) Leprosy. In: Shakir RA, Newman PK, Poser CM (eds) Tropical neurology. Saunders, Philadelphia
Sabin TD, Swift TR (2008) Neurological complications of leprosy. In: Aminoff MJ (ed) Neurology and general medicine, 4th edn. Churchill Livingstone, Philadelphia, pp 815–825
Scambler G, Hopkins A (1990) Generating a model of epileptic stigma: the role of qualitative analysis. Soc Sci Med 30:1187–1194
Steele CM, Aronson J (1995) Stereotype threat and the intellectual test performance of African Americans. J Pers Soc Psychol 69:797–811
Stevelink SAM, Wu IC, Voorend CGN, Brakel WH (2012) The psychometric assessment of internalized stigma instruments: a systematic review. Stigma Res Action 2:100–118. doi:10.5463/SRA.v1i1.11
Stienstra Y, van der Graaf WT, Asamoa K, van der Werf TS (2002) Beliefs and attitudes toward Buruli ulcer in Ghana. Am J Trop Med Hyg 67:207–213
Takoungang I, Meli J, Fotso S, Angwafo F III, Kamajeu R, Ndumbe PM (2004) Some social determinants of urinary schistosomiasis in Northern Cameroon: implications for schistosomiasis control. Afr J Health Sci 11:111–120
Tate AJ, Fischer H, Leigh AE, Kendrick KM (2006) Behavioural and neurophysiological evidence for face identity and face emotion processing in animals. Philos Trans R Soc Lond B Biol Sci 361:2155–2172
Tchounkeu YF, Onyeneho NG, Wanji S, Kabali AT, Manianga C, Amazigo UV, Amuyunzu-Nyamongo M (2012) Changes in stigma and discrimination of onchocerciasis in Africa. Trans R Soc Trop Med Hyg 106:340–347
Van Brakel WH (2003) Measuring leprosy stigma-a preliminary review of leprosy literature. Int J Lepr 71:190–197
Van Brakel WH (2006) Measuring health-related stigma: a literature review. Psychol Health Med 11:307–334. doi:10.1080/13548500600595160
van den Broek J, O'Donoghue J, Ishengoma A, Masao H, Mbega M (1998) Evaluation of a sustained 7 year health education campaign on leprosy in Rufiji District in Tanzania. Lepr Rev 69:57–74
Waxler NE (1981) Learning to be a leper: a case study in the social construction of illness. In: Mishler EG, Amarasingham L, Hauser ST, Liem R, Osherson SD et al (eds) Social contexts of health, illness, and patient care. Cambridge University Press, Cambridge, pp 169–194
Weiss MG (2008) Stigma and the social burden of neglected tropical diseases. PLoS Negl Trop Dis 2(5):e237. doi:10.1371/journal.pntd.0000237
Weiss MG, Doongaji DR, Siddhartha S, Wypij D, Pathare S, Bhatawdekar M et al (1992) The explanatory model interview catalogue (EMIC).contribution to cross-cultural research methods from a study of leprosy and mental health. Br J Psychiatry 160:819–830
Weiss MG, Ramakrishna J (2006) Stigma interventions and research for international health. Lancet 367:536–538
Weiss MG, Ramakrishna J, Somma D (2006) Health-related stigma: rethinking concepts and interventions. Psychol Health Med 1:277–287
Wong ML (2004) Designing programmes to address stigma in leprosy: issues and challenges. Asia Pacific Disability Rehabil J 15:3–12
World Health Organization (2004) Epilepsy in the WHO Africa region, bridging the gap: the global campaign against epilepsy "out of the shadows". WHO, Geneva
World Health Organization (2010) Working to overcome the global impact of neglected tropical diseases. ©WHO/NTD 2010, http://www.who.int/neglected_diseases/2010report/en/index.html
Yanos PT, West ML, Gonzales L, Smith SM, Roe D, Lysaker PH (2012) Change in internalized stigma and social functioning among persons diagnosed with severe mental illness. Psychiatry Res 200:1032–1034
Yizhar O, Fenno LE, Prigge M, Schneider F, Davidson TJ, O'Shea DJ, Sohal VS, Goshen I, Finkelstein J, Paz JT, Stehfest K, Fudim R, Ramakrishnan C, Huguenard JR, Hegemann P, Deisseroth K (2011) Neocortical excitation/inhibition balance in information processing and social dysfunction. Nature 477:171–178

Part II
Neglected Infectious Diseases of the Nervous System

Nematode Infections: Neurological Involvement and Neurobiology

Pewpan M. Intapan, Wanchai Maleewong, and Yukifumi Nawa

Abstract Nematode parasite infections in the central nervous system (CNS) are rare and often neglected. Of the over ten species of nematode parasites known to infect CNS, the genuine neurotropic species is *Angiostrongylus cantonensis*. Other species can, as larvae, accidentally migrate into or disseminate via systemic circulation to CNS. Nematode parasites are generally prevalent in rural areas of tropical and subtropical countries, but some species like roundworm of dogs and cats (*Toxocara canis* and *T. cati*) are also common in pet animals in the high-income countries. Ease of global transport has increased the incidence of neuronematodiases among travellers and immigrants. Clinical manifestations of neuronematodiasis are similar regardless of causative species. The patients can present with symptoms of space-occupying lesions, meningitis, encephalitis or myelitis. Computed tomography and/or magnetic resonance imaging are helpful in the diagnosis. Neurological symptoms with eosinophilia, and in particular eosinophilic pleocytosis in cerebrospinal fluid (CSF), are suggestive for the helminth infections. The history of patients resident or traveling to endemic areas, eating habits or ethnic dishes, field activities, are helpful for diagnosis. Direct detection of the causative parasite is the gold standard for diagnosis, but is rarely successful. In practice, diagnosis can be made by the combination of radio-imaging and immuno-diagnosis to detect parasite-specific antibodies in serum and/or CSF, although the availability of immunodiagnostic kits and access to laboratories are limited in endemic countries. Molecular diagnostic techniques are under exploration.

P.M. Intapan • W. Maleewong
Research and Diagnostic Centre for Emerging Infectious Diseases, Faculty of Medicine, KhonKaen University, Khon Kaen, Thailand

Department of Parasitology, Faculty of Medicine, KhonKaen University, Khon Kaen, Thailand

Y. Nawa (✉)
Research Affairs, Faculty of Medicine, KhonKaen University, Khon Kaen, Thailand
e-mail: yukinawa@kku.ac.th

M. Bentivoglio et al. (eds.), *Neglected Tropical Diseases and Conditions of the Nervous System*, DOI 10.1007/978-1-4614-8100-3_4,

Keywords Parasites • Nematodes • Central nervous system • Helminths • Worms • Larvae • Eosinophilic meningitis • Loa loa • Toxocariasis

1 Introduction

A variety of nematode parasites are known to cause diseases in the central nervous system (CNS). In their classical review, Nishimura and Hung (1997) listed 11 nematodes as causative pathogens for neuronematodiasis. Recently Walker and Zunt (2005) reviewed five nematode species as the most common pathogens of CNS, namely, *Angiostrongylus cantonensis*, *Baylisascaris procyonis*, *Gnathostoma spinigerum*, *Strongyloides stercoralis* and *Toxocara* species. In relation to the recent globalization, Diaz (2008) highlighted *A. cantonensis*, *B. procyonis* and *G. spinigerum* as the emerging zoonotic pathogens causing eosinophilic meningitis. Recently Finsterer and Auer (2012) reviewed neuroparasitoses and listed five nematodes: *T. canis*, *T. cati*, *Trichinella spiralis*, *A. cantonensis* and *G. spinigerum*. Taking all those and other reviews together, over ten species of nematodes can cause CNS diseases. However, apart from *A. cantonensis*, genuine neurotropic nematodes, i.e. adult worms that parasitize the CNS or larvae that require passage through CNS for further development, is not known in humans. *A. cantonensis* larvae require passage through the arachnoid space to become adult worms in the pulmonary artery in rats. In humans, *A. cantonensis* larvae can reach the arachnoid space but do not undergo further migration or development resulting in eosinophilic meningoencephalitis (Yoshimura et al. 1980). The larvae of two animal parasitic ascarid nematodes, *Baylisascaris procyonis* and *Lagochilascaaris minor*, may have some affinity to CNS but passage is not required for their development. Likewise, all other nematode parasites can, as larvae, randomly migrate everywhere in the body and accidentally enter the CNS (e.g. *Gnathostoma* species) or disseminate by the bloodstream to be trapped in the CNS (e.g. *Toxocara* species) and cause diseases. Thus, almost exclusively, neuronematodiasis is an accidental disease caused by migration of nematoda larvae (neural larva migrans: NLM).

Despite the variety of causative species and their migration routes, clinical features of neuronematodiasis are similar with the predominance of eosinophilic meningoencephalomyelitis (Katchanov and Nawa 2010). Identification of the causative parasite species by their features is almost impossible in the clinic. Some species like *Toxocara* species and *S. stercoralis* are the cosmopolitan parasites, whereas some parasites like *A. cantonensis*, *B. procyonis* and *L. minor* show more limited areas of distribution. Although *Gnathostoma* species distributes widely in subtropical to tropical countries, (Nawa 1991; see below) neurognathostomiasis is almost exclusively found in Thailand (Katchanov et al. 2011).

Human behavior also affects the epidemiological patterns of zoonotic nematodiasis. For example, toxocariasis is globally a disease of children infected from pet dogs and cats. However, in Japan (Akao and Ohta 2007) and Korea (Kwon et al. 2006), toxocariasis is rather the disease of middle-aged men infected by ingesting raw liver of chicken (see below).

Although the definite diagnosis of neuronematodiasis can be made by the detection of worms, such a chance is extremely rare. Alternatively, diagnosis can be made based on the following three components (1) detection of parasite-specific antibodies in the sera and/or CSF of patients, (2) eosinophilia and elevation of total IgE, and (3) the history of possible contact with parasites or of ingesting food contaminated with parasites. ELISA and immunoblotting are the currently most popular methods to detect specific antibodies against causative pathogens. However, cross-reactivity due to antigenic similarities makes it difficult to identify the pathogens at the species level (Nunes et al. 1997; Fan and Su 2004). In this chapter, we review the major nematode parasites that cause NLM with the special attention on pitfalls and difficulties of diagnosis of neuronematodiases.

2 *Toxocara* Species and Other Ascarids

2.1 Toxocara *Species*

Toxocara canis and *T. cati* represent zoonotic parasites causing larva migrans in humans. They are cosmopolitan intestinal nematodes of dogs and cats, and their prevalence is high (Lee et al. 2010). When humans ingest mature (embryonated) eggs, the larvae hatch in the intestine, penetrate the intestinal wall, gain entry into the circulation either via lymphatics or mesenteric vein, pass through the liver and disseminate systemically via the bloodstream. Direct contact with infected dogs/cats, or playing at the sand pan contaminated with *Toxocara* eggs are the most common ways of infection. Thus, toxocariasis has been known as a disease of children. However, sero-epidemiological survey in Japan (Akao and Ohta 2007) and also in Korea (Kwon et al. 2006) revealed that toxocariasis is rather a disease of middle-aged men who ingested raw bovine (Yoshikawa et al. 2008) and/or chicken liver (Morimatsu et al. 2006) with the belief of its medicinal or tonic effect. In Korea, about 70 % of the patients with eosinophilia of unknown etiology were proven to be positive for anti-*T. canis* antibody (Kwon et al. 2006).

Since *Toxocara* larvae do not have any specific organ tropism, every organ or tissue can be affected by chance. Even a single larva can cause visual disturbance when lodged in retinal capillaries (ocular larva migrans; OLM). Similarly, a single larva can cause neurological and/or psychiatric symptoms when trapped in the CNS as NLM. In contrast, symptomatic visceral larva migrans (VLM), which is usually recognized as pneumonitis and/or liver injury with eosinophilia, occurs by ingestion of a large number of *Toxocara* larvae. VLM cases were frequently reported from Japan and Korea because of the custom of consuming raw liver of animals, which can harbor a large number of *Toxocara* larvae (Fig. 1) (Taira et al. 2011). One case of neurotoxocariasis after consumption of raw duck liver was reported in Germany (Hoffmeister et al. 2007).

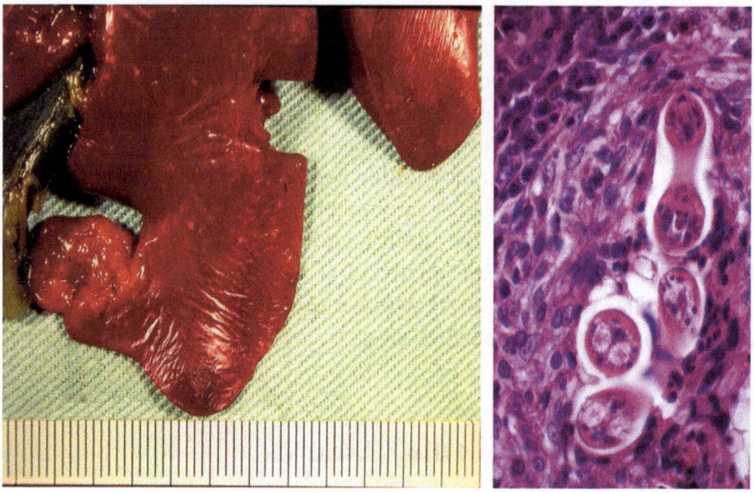

Fig. 1 Backyard chicken liver infected with *Toxocara canis*. Macroscopically numerous *white spots* are seen. Cross section revealed the presence of *Toxocara* larvae

In toxocariasis, OLM is the most common clinical manifestation presenting as a monocular retinitis/uveitis. Granuloma formation in the retina is often misdiagnosed as retinoblastoma. Recently, Shields et al. (2013) reviewed 607 cases of patients having lesions suspected of retinoblastoma and found that 22 % of them were due to toxocariasis. OLM can also occasionally be associated with optic neuritis or loss of eyesight when the parasites have migrated to the optic nerve (Komiyama et al. 1995).

Typical clinical features of toxocaral NLM are eosinophilic meningitis, encephalitis and myelitis, or a combination (Vidal et al. 2003). In cases of acute meningitis, detection of eosinophils in CSF is a strong indication of NLM. When *Toxocara* larvae are trapped in the small vessels of cortical or subcortical gray or white matter, they will cause vasculitis or microabscesses seen as single or multiple small nodular lesions (<1 cm) by radioimaging (Xinou et al. 2003). Although less common, *Toxocara* infection also causes myelitis manifested as sensory, motor, and autonomic dysfunctions, predominantly in the lower extremities (Jabbour et al. 2011). Interestingly, only 2 out of 17 patients showed a high eosinophil count in the CSF, while high blood eosinophilia was seen in six patients. Although the effects of *Toxocara* infection on mental, psychological and/or behavioral status remain unclear, its association with epilepsy has been clearly demonstrated (Quattrocchi et al. 2012). The majority of toxocariasis myelitis cases are caused by *T. canis* infection; a case of serologically confirmed *T. cati* myelitis was recently reported in Japan (Fukae et al. 2012). Taira et al. (2011) reported that *T. cati* larvae can persist and retain high infectivity in the muscles of experimentally infected chickens, posing another risk factor for humans.

2.2 Ascaris suum

Phills et al. (1972) reported that VLM, after intentional infestation of *A. suum* eggs, causes pulmonary infiltrates, asthma and eosinophilia. Outbreaks of VLM due to *A. suum* infection were also reported in southern Japan (Maruyama et al. 1996). The possibility of *Toxocara* VLM due to immunological cross-reactivity between *A. suum* and *T. canis* was pointed out by Petithory (1996) in the diagnosis of VLM by immunoblotting technique (Nunes et al. 1997). However, with the improvement of immunodiagnostic methods (Nakamura-Uchiyama et al. 2006), sporadic cases of serologically confirmed *A. suum* infection have been reported in Japan (Sakakibara et al. 2002; Izumikawa et al. 2011). In the Netherlands, Pinelli et al. (2011) demonstrated the clear-cut difference in the seroepidemiological trends of *Toxocara* and *Ascaris* among VLM/OLM-suspected patients and emphasized the importance of *A. suum* as the pathogen for VLM/OLM.

Like toxocariasis, eosinophilic meningitis/myelitis can be a clinical feature of *A. suum* larva migrans. A severe encephalopathy due to *A. suum* infection was reported (Inatomi et al. 1999). A new disease entity, *atopic myelitis,* has been established by Kira and his colleagues (Kira et al. 2001; Osoegawa et al. 2003; Isobe et al. 2012). This disease is characterized as myelitis associated with high frequency of present and/or past history of atopic disease with hyperIgEemia, high mite-specific IgEemia and high CSF levels of interleukin 9 and CCL11/eotaxin. Several patients with atopic myelitis cases have shown to be seropositive either to *A. suum* or *T. canis* (Osoegawa 2004). A spinal cord tumor-like lesion was noted in one case of *A. suum* infection (Osoegawa et al. 2001). Serial magnetic resonance imaging (MRI) in four patients with myelitis caused by either *T. canis* or *A. suum* infections revealed spinal cord swelling with or without gadolinium enhancement in three of them (Umehara et al. 2006). T2-weighted images showed high signal intensities preferentially located in both lateral and posterior columns; one patient with *T. canis* infection relapsed, which was associated with reappearance of MRI abnormalities.

Treatment of toxocariasis and ascariasis VLM is still controversial (Othman 2012). In general, antihelmintic and corticosteroid treatment was reported to yield improvement in neurologic deficits and spinal lesions. In cases of OLM, treatment is also controversial (Woodhall et al. 2012). The American Society of Ophthalmology recommends corticosteroids and not the use of antihelmintics because *Toxocara* OLM is considered to be a self-limiting disease and antihelmintic drugs often have side effects of transient liver dysfunction.

2.3 Baylisascaris procyonis

Baylisascaris procyonis is an intestinal nematode of raccoons, *Procyon lotor*. This parasite is, together with its natural final host raccoons, still prevalent in some areas of the USA (Sexsmith et al. 2009; Yeitz et al. 2009; Hernandez et al. 2013). The mode of infection and the migration route in the hosts are essentially the same as those of *Toxocara* or *Ascaris* species. However, differently from other ascarid parasites, a small but potentially devastating number of *B. procyonis* larvae (typically 5–7 %) enter the CNS leading to the debilitation or death of the intermediate hosts including humans (Gavin et al. 2005). Since the first human case of fatal eosinophilic meningoencephalitis and VLM by this parasite was reported in USA by Fox et al. (1985), up to 20 cases of NLM including recent fatal cases (Wise et al. 2005), mostly in children, have been reported in USA (Gavin et al. 2005; Murray and Kazacos 2004; Perlman et al. 2010). In addition, three cases were reported in Canada; two cases of children (Hajek et al. 2009; Haider et al. 2012) and one case of pathologically proven asymptomatic case in an elderly person (Hung et al. 2012). *Baylisascaris procyonis* infection is known also to cause OLM (Mets et al. 2003; Brasil et al. 2006; Saffra et al. 2010) or VLM (Boschetti and Kasznica 1995). There is increased risk of *B. procyonis* infection spreading out from North America to Europe (Bauer 2011) and also to Japan (Sato et al. 2003, 2005).

2.4 Lagochilascaris minor

Lagochilascaris minor is a nematode parasite of sylvatic carnivores found from Central to South America. Natural infections were reported in cats in Argentina (Romero and Led 1985) and Uruguay (Sakamoto and Cabrera 2002), bush dogs, *Speothos venaticus* (Volcán and Medrano 1991) and Costa Rican ocelots, *Felisparadalis mearnsi* (Brenes-Madrigal et al. 1972). Experimental infection in laboratory animals revealed that mice were the intermediate host in that *Lagochilascaris* larvae encysted in the skeletal muscles, while cats are the suitable definitive hosts in that the larvae developed into adult stage in the cat soft tissues around neck, nasopharyngeal region or trachea (Volcan et al. 1992; Campos et al. 1992).

Original description of this nematode was made by the morphological observation of the worms isolated from human neck soft tissue in Trinidad (Leiper 1909). A detailed human case report was first made in Suriname (Winkel and Treurniet 1956) followed by a case from Trinidad and Tobago (Draper 1963). The majority of *Lagochilascaris* infection in humans is, however, reported from Brazil, especially from Parà state (Moraes et al. 1983, 1985). A few sporadic cases have been reported from several Latin American countries, e.g., Venezuela (Volcan et al. 1982), Bolivia (Ollé-Goig et al. 1996), Mexico (Vargas-Ocampo and Alvarado-Aleman 1997; Barrera-Pérez et al. 2012), Colombia (Moncada et al. 1998), Suriname (Oostburg and Varma 1968). As described by Leiper (1909), adult worms were found in the soft tissues of the neck and also from nasal discharges of patients (Barrera-Pérez et al. 2012). Moreover, as all stages of parasites, (eggs, larvae and adults) were found in the surgical specimens of the nodular lesions of soft tissues around the neck (Barrera-Pérez et al. 2012), auto-infection seemed to occur in humans accompanied by severe clinical symptoms (Rosemberg et al. 1986). Ocular, ear and meningeal involvement was reported in one case (Aquino et al. 2008). Encephalopathy was observed in one of the fatal cases (Rosemberg et al. 1986), but generally CNS involvement is rare in *Lagochilascaris* infection.

Because of the small number of reported cases, recommendations of chemotherapy for *Lagochilascaris* infection are not established. Thiabendazole had been used for the therapy of *L. minor* infection (Oostburg 1971), but this has been gradually replaced by ivermectin (Bento et al. 1993). *In vivo* larvicidal effect of ivermectin has been shown in mice experimentally infected with this parasite (Barbosa et al. 1998).

3 Angiostrongyloidiasis

Human angiostrongyliasis is an important food-borne helminthic zoonosis, mainly caused by *Angiostrongylus cantonensis* (Prociv et al. 2000; Wang et al. 2012). *A. cantonensis* larvae are neurotropic and migrate into the arachnoid space of the human brain to cause eosinophilic meningoencephalomyelitis. Thousands of human cases have been recorded worldwide and outbreaks have been reported in endemic areas. Since the first outbreak was recorded in Wenzho in 1997 (Wang et al. 2002), many outbreaks and sporadic cases have been reported from China (Wang et al. 2008, 2012). The parasite has spread from its traditional endemic regions of the Pacific islands and Southeast Asia to the American continent including the USA, Caribbean Islands and Brazil (Wang et al. 2008, 2012). A recent molecular phylogenetic study on various *A. cantonensis* isolates from different geographic areas revealed widespread multiple lineages of this parasite (Tokiwa et al. 2012).

Human infections occur by consuming raw or semi-cooked food contaminated with the infective third stage larvae (L3) of *A. cantonensis* with the majority of infections due to eating intermediate land snails or slugs. Infection also occurs by eating animals harboring infective larvae (planarians, crustaceans, frogs, monitor lizards, wild boar, etc.) or contaminated fresh vegetables (Tsai et al. 2004).

Human angiostrongyliasis is classified into three main forms (1) eosinophilic meningitis, (2) eosinophilic encephalitis or meningoencephalitis and (3) OLM (Sawanyawisuth and Sawanyawisuth 2008; Wang et al. 2008; Diao et al. 2011). Eosinophilic meningitis is the most common form constituting of >90 % of the cases. Eosinophilic encephalitis or meningoencephalitis is less common but often leads to permanent sequelae or death in some cases (Martínez-Delgado et al. 2000; Chotmongkol and Sawanyawisuth 2002; Sawanyawisuth et al. 2009). Hearing loss, intestinal obstruction, radiculomyelitis and Bell's palsy are rare clinical manifestations of angiostrongyliasis (Sinawat et al. 2008). Postmortem examination of the brain has revealed worms and bilateral changes in the frontal and temporal lobes. Charcot-Leyden crystals, a sign of eosinophil degeneration, were observed in granulomatous meningeal lesions surrounding worms (Tangchai et al. 1967; Punyagupta et al. 1975).

High protein concentrations and high leukocyte counts in the CSF indicate a blood-CNS barrier dysfunction (Lee et al. 2006), which are mediated by several cytokines, i.e. eotaxin, urokinase-type plasminogen activator and matrix metalloprotease-9 (Chen et al. 2006; Intapan et al. 2007; Sanpool et al. 2010).

Diagnosis of human angiostrongyliasis is suggested by clinical manifestation and a coconut juice-like CSF, with eosinophils >10 % of total CSF leukocytes,

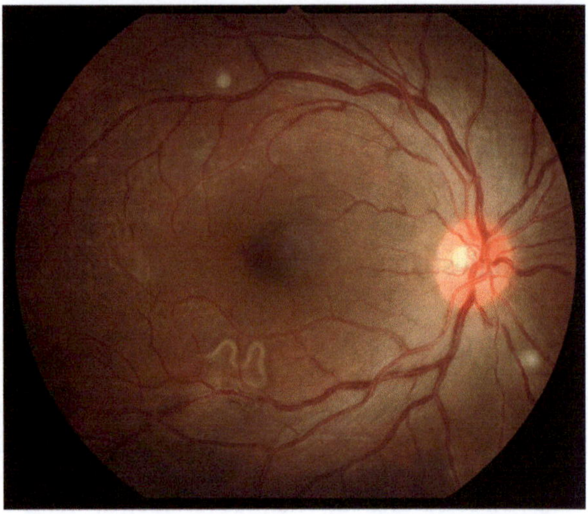

Fig. 2 Ocular angiostrongyliasis. A fundoscopic finding showing the presence of a living *A. cantonensis* larva in the vitreous chamber. (Courtesy of Dr. Patanaree Luanratanakorn)

or >10 cells/mm^3 CSF, accompanied by a history of *A. cantonensis* exposure (Chotmongkol and Sawanyawisuth 2002). A history of exposure should be within 3 months of the onset of the disease. Incubation period varies from 1 day to 6 months (Sawanyawisuth and Sawanyawisuth 2008; Kirsch et al. 2008). For ocular angiostrongyliasis, the worm can be found in any part of the eye.

The eosinophilic-meningitis shows no specific characteristics. Its duration is normally less than 1 week. Fever occurs only in about 10 % of adult patients and half of them have neck stiffness. In encephalitis patients, acute loss of consciousness leading to coma is seen, while seizures have not been reported. In elderly patients, at presentation, prolonged headache and fever are the risk factors for developing encephalitis (Sawanyawisuth et al. 2009). Ocular involvement is rare in angiostrongyliasis. Larva is usually found in the vitreous (Fig. 2) or retina of one side leading to visual impairment (Sawanyawisuth et al. 2007). Optic neuritis, necrotizing retinitis (Sawanyawisuth and Kitthaweesin 2008; Wang et al. 2006; Liu et al. 2006) and oculomotor palsy (Fig. 3) have also been recorded. The ocular form might occur without eosinophilic meningitis. The initial visual acuity is a predictor for vision outcome of this condition. No effective treatment for ocular angiostrongyliasis has been reported and the permanent damage always persists.

Although rarely done, definitive diagnosis is based on the detection of larvae or immature worms in the CSF or in the eye. Diagnosis is principally based upon neurological symptoms, CSF eosinophilia and a history of possible exposure to *A. cantonensis* larvae. CSF is cloudy, but not grossly turbid or xanthochromic. The CSF leukocyte count ranges from about 20 to 5,000 cells/mm^3 and is usually between 150 and 2,000 cells/mm^3. The CSF eosinophilia exceeds 10 % (ranging from 20 to 70 %), in most cases. The CSF protein level is usually elevated, but glucose level remains normal or only slightly reduced (Schmutzhard et al. 1988; Kuberski and Wallace 1979). Peripheral blood eosinophilia usually accompanies, but does not correlate with, the CSF eosinophilia.

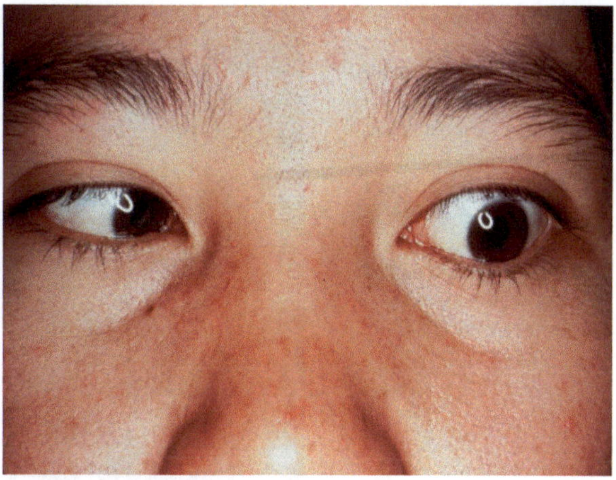

Fig. 3 Left lateral rectus muscle palsy caused by *A. cantonensis* infection. The patient has been suffering from eosinophilic meningitis. (Courtesy of Prof. Veerajit Chotemongkol)

On computed tomography (CT) or MRI imaging, *A. cantonensis* meningitis can be distinguished from other helminthic infections of the CNS by the presence of focal lesions in the brain (Weller and Liu 1993). MRI imaging of the brain of 13 *A. cantonensis* eosinophilic meningitis patients in Taiwan (Tsai et al. 2003) showed high signal intensities over the globus pallidus and cerebral peduncle, punctate areas of abnormal enhancement within the cerebral and cerebellar hemisphere on gadolinium-enhancing T1 imaging, and a hyperintense signal on T2-weighted images. MR imaging and MR spectroscopy of six cases of meningoencephalitis in Thailand (Kanpittaya et al. 2000) revealed prominence of the Virchow-Robin spaces, subcortical enhancing lesions, and abnormal high T2 signal lesions in the periventricular regions (Fig. 4, Kanpittaya et al. 2000).

Among various serological tests developed, ELISA has been widely used and still is the standard by which to evaluate new methods (Intapan et al. 2002). The intrathecal synthesis pattern of IgG1 + IgG2 and IgE antibodies is also useful for the diagnosis of eosinophilic meningoencephalitis due to *A. cantonensis* (Dorta-Contreras et al. 2003, 2005). The measurement of Th1/Th2 cytokines in the CSF by sandwich ELISA provides a suggestive evidence for the diagnosis of parasitic meningitis (Intapan et al. 2008). By immunoblotting, 29, 31 and 32-kDa antigens are potential diagnostic antigens for immunodiagnosis of human angiostrongyliasis (Nuamtanong 1996; Maleewong et al. 2001; Lai et al. 2005).

An empirical therapy of 2-week course of oral corticosteroid is effective for treatment of eosinophilic meningitis (Chotmongkol et al. 2000). Albendazole therapy was not effective as corticosteroid (Jitpimolmard et al. 2007). The combined therapy with corticosteroid and anthelminthics (Chotmongkol et al. 2004, 2006), or predonisolone and albendazole (Chotmongkol et al. 2004) was not significantly different from corticosteroids alone. Lumbar puncture is required not only for diagnosis but also for the reduction of CSF pressure (Sawanyawisuth et al. 2004a). The recommended analgesic is acetaminophen. NSAID is not superior to acetaminophen (Tsai et al. 2001) and has a risk of causing upper gastrointestinal tract bleeding if combined with corticosteroid. No effective treatment has been established for ocular angiostrongyliasis, although corticosteroid and anti-helminthics may be justified.

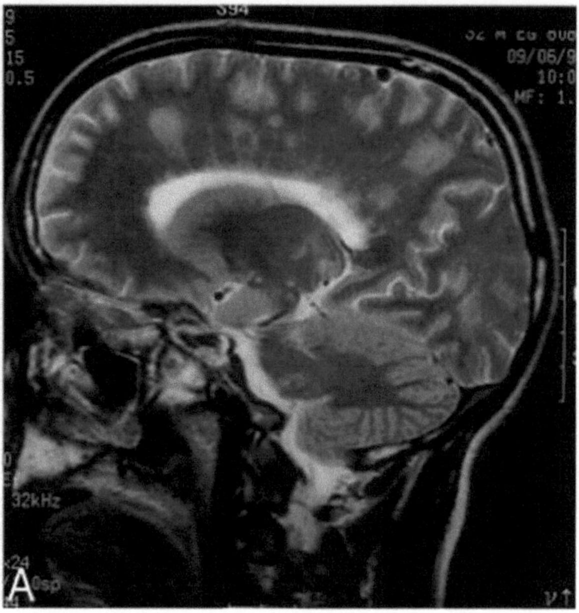

Fig. 4 MR image of eosinophilic meningoencephalitis caused by *A. cantonensis*. The patient ate raw meat of an infected monitor lizard. Sagittal T2-weighted image showing fuzzy hyperintense signal areas in the left fronto-parietal regions and centrum semiovale. (Kanpittaya et al. 2000)

4 *Gnathostoma* Species

Human gnathostomiasis is an important food-borne helminthic zoonosis caused by the spirurid nematode *Gnathostoma* spp. and is endemic in Asia and the Americas (Miyazaki 1960; Daengsvang 1981; León-Règagnon et al. 2002). The disease has been emerging among travelers returning from the endemic areas (Moore et al. 2003; Katchanov et al. 2011). Generally, gnathostomiasis is not a life-threatening disease, and death has only been recorded occasionally among neurognathostomiasis patients (Daengsvang 1981; Katchanov et al. 2011).

The most common species to cause human diseases in Asia is *G. spinigerum*. Human infections with *G. hispidum*, *G. doloresi* and *G. nipponicum* were reported only from Japan. In the American continents, *G. binucleatum* is the only proven species found in humans (Nawa 1991; León-Règagnon et al. 2002). Two outbreaks of gnathostomiasis, supposed to be due to infection with *G. spinigerum*, were recently reported in Africa (Hale et al. 2003; Herman et al. 2009).

Human infection occurs by consuming raw or inadequately cooked foods, e.g. fresh water fish, frogs, chicken, which harbors *Gnathostoma* advanced third stage larvae (aL3) (Fig. 5). In the human body, the larvae cannot sexually mature into adults and migrate around for variable periods of time causing irritation and inflammation in various organs. Migration in the subcutaneous tissue causes intermittent painful, pruritic migratory swelling (Fig. 6). Migration to visceral organs can cause various symptoms depending on the affected organs.

The larvae migrate to the CNS to cause radiculomyelitis or radiculoencephalomyelitis. Subarachnoid hemorrhage can occur, sometimes leading to death

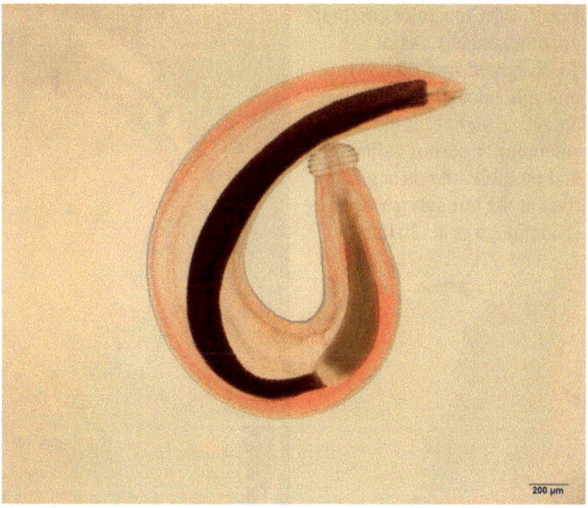

Fig. 5 Advanced third stage larva of *Gnathostoma spinigerum*. *Note* cephalic bulb armed with four rows of hooklets (scale bar = 200 µm)

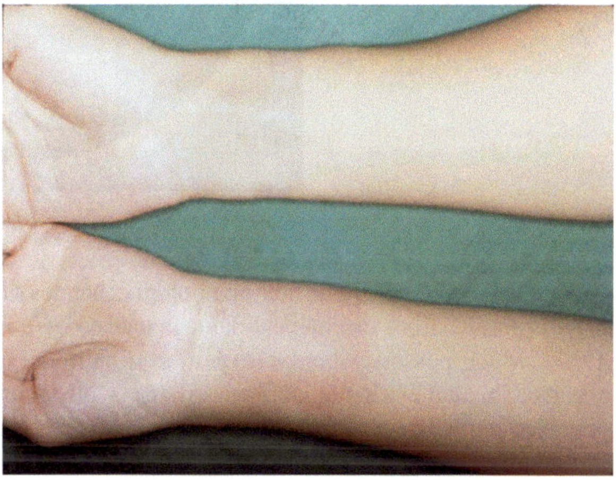

Fig. 6 Intermittent migratory swelling on the lateral forearm. A photo of a gnathastomiasis suspected case

(Chitanondh and Rosen 1967; Punyagupta et al. 1968; Boongird et al. 1977). For unknown reasons, *G. spinigerum* is up to now the only known causative agent of human neurognathostomiasis (Nawa 1991; Katchanov et al. 2011) and cases are almost exclusively found in Thailand (Katchanov and Nawa 2010). Genetic variation among geographically different populations would be of interest to examine in relationships to the pathogenicity of the nematode. *G. spinigerum* possibly penetrates the CNS by migrating along nerve roots, since the most distinctive clinical manifestation is radiculitis, characterized by excruciating nerve root pain in the extremities or trunk (Punyagupta et al. 1990). Paralysis of one or more of the extremities and urinary retention often develop, followed by involvement of the cranial nerves. Other findings include signs of myelitis and meningoencephalitis (Chitanondh and Rosen 1967; Punyagupta et al. 1968; 1990; Boongird et al. 1977; Schmutzhard et al. 1988;

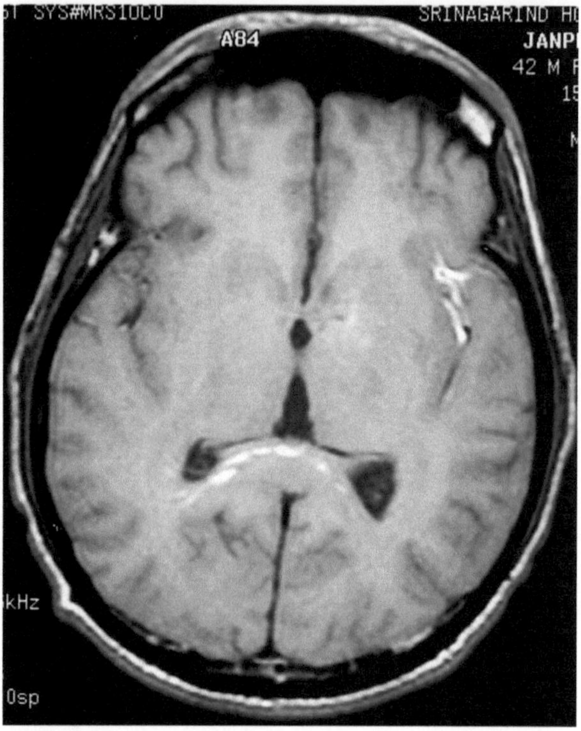

Fig. 7 MR image of cerebral gnathostomiasis. Axial T1-weighted image showing multiple linear, tiny spot of abnormal signals at the splenium of corpus callosum and possible hemorrhagic tract at the left sylvian fissure (Kanpittaya et al. 2012)

Kawamura et al. 1983). The neuropathologic changes of gnathostomiasis of the CNS arise primarily from mechanical damage caused by a migrating larva (Daengsvang 1981), which possesses hooklets on a head bulb and numerous spines on its body surface. The scattered, multiple hemorrhages with a tract-like appearance in the brain (Fig. 7, Kanpittaya et al. 2012) and spinal cord is a characteristic feature. The areas of hemorrhage differ from the sites of hypertensive hemorrhage (Punyagupta et al. 1968; Bunnag et al. 1970; Kanpittaya et al. 2012). Eosinophilic meningoencephalitis caused by *A. cantonensis* sometimes shows similar clinical manifestations and can be difficult to differentiate from those caused by *G. spinigerum.*

Definitive diagnosis of gnathostomiasis relies on identification of worms in biopsy specimen or spontaneous emergence from the skin, but the chance of finding the worms is small. Generally, diagnosis of gnathostomiasis is suspected in patients with a history of eating foods containing infective larvae and a history of intermittent migratory skin and/or subcutaneous swelling. The clinical manifestations of neurognathostomiasis are not specific and difficult to distinguish from the other tissue nematode infections, although xanthochromic or blood-containing CSF is an important diagnostic clue (Rusnak and Lucey 1993). More than half of infected patients have a CSF pleocytosis of less than 500 cells/μl (Punyagupta et al. 1990; Boongird et al. 1977). The degree of CSF eosinophilia may range from 14 to 90 % (Schmutzhard et al. 1988; Sawanyawisuth et al. 2004b). The CSF protein level is usually elevated, while the glucose level remains normal. Concomitant peripheral

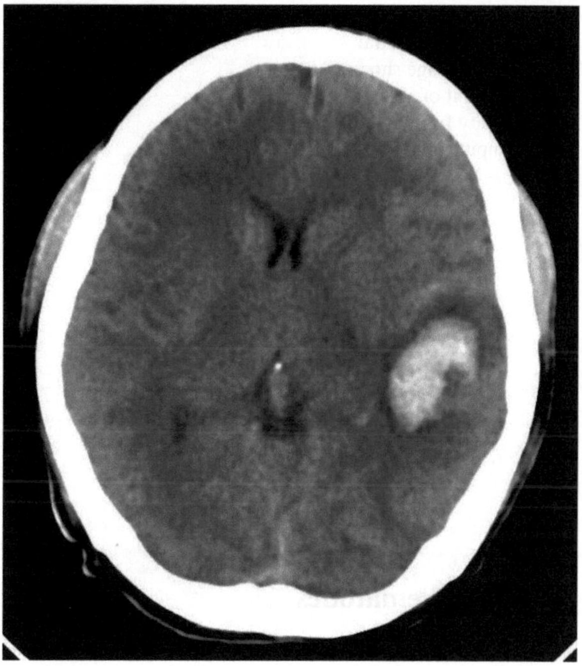

Fig. 8 CT brain image of cerebral gnathostomiasis. Axial non contrast CT image demonstrated intracerebral hemorrhage at the left temporal lobe (Kanpittaya et al. 2012)

blood eosinophilia of 10 % or higher is seen in about 55 % of patients (Punyagupta et al. 1968; Boongird et al. 1977; Schmutzhard et al. 1988). The presence of focal lesion at CT scan (Fig. 8, Kanpittaya et al. 2012) shows evidence of large intracerebral hemorrhage. MRI findings that can aid in the diagnosis have been described above, e.g. intracerebral hemorrhages scattered as multiple lesions with a tract-like appearance (Fig. 7) or myelitis patterns in the cervical, thoracic, and conus medullaris areas (Fig. 9, Kanpittaya et al. 2012). In contrast, most of angiostrongyliasis patients have normal brain CT findings, and if any, less parenchymal brain lesions.

Serological tests can be used as the supportive evidence for diagnosis. ELISA based detection of human IgG class antibody to *Gnathostoma* antigen has been widely reported (Suntharasamai et al. 1985; Dharmkrong-at et al. 1986; Diaz Camacho et al. 1998; Maleewong et al. 1988). By immunoblot, the 21 kDa and 24 kDa bands in the crude somatic third stage larvae of *G. spinigerum* antigen recognized by IgG antibody have good diagnostic value for neurognathostomiasis (Tapchaisri et al. 1991; Nopparatana et al. 1991; Wongkham et al. 2000; Anantaphruti et al. 2005; Laummaunwai et al. 2007; Intapan et al. 2010).

Randomized trials of antihelminthic therapies have been reported (Katchanov et al. 2011). Patients showed full recovery after treatment with albendazole (800 mg/day for 1 month, 800 mg/day for 3 weeks or 400 mg twice per day for 4 weeks) combined with prednisolone or dexamethasone (Górgolas et al. 2003; Elzi et al. 2004; Schmutzhard 2007). However, relapses after therapy with ivermectin and albedazole have been observed (Strady et al. 2009).

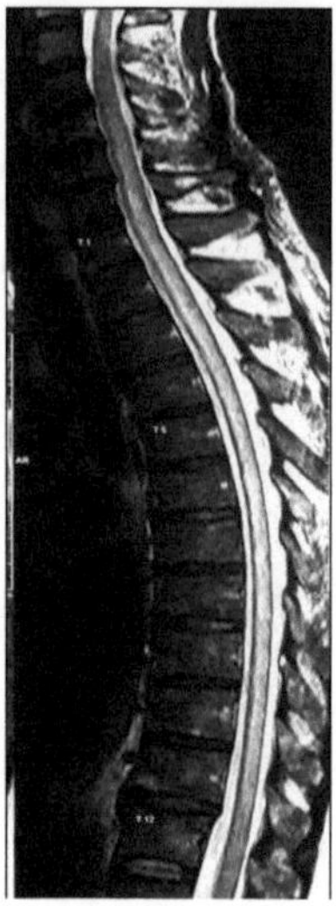

Fig. 9 MR image of spinal gnathostomiasis. Sagittal T2-weighted image showing a long segment of abnormal high intensity from C5 to T12 level (Kanpittaya et al. 2012)

5 Other Nematodes

5.1 Strongyloides stercoralis

Strongyloidiasis is primarily an infection of the upper intestinal tract caused by the nematode genus *Strongyloides*. The parasites in this genus are widely distributed from the tropical to the climate zones, and are highly species-specific. In most instances, human strongyloidiasis is caused by infection with *S. stercoralis*, but infection with other species such as *S. fullborni* has also been reported (Freedman 1991; Grove 1996).

Strongyloides species have a complicated life cycle. Infective third stage larvae (L3) reside in the humid soil or muddy water and penetrate percutaneously into the hosts. After percutaneous invasion, larvae enter circulation, reach the lungs, and emerge into the alveoli. Then, they are coughed up, swallowed and parasitize in the duodenum to the upper jejunum where they mature into adult female worms. They lay embryonated eggs from which rhabditiform first stage larvae (L1) hatch and are usually excreted in feces. In the freshwater, the L1 develop to L3 that can infect human hosts to complete their parasitic cycle. Alternatively, L1 develop into free living rhabditiform male and female adults in water. Free living male and females mate and lay eggs from which L1 emerge and take one or other path in the life cycle, i.e., either free living or parasitic. In addition, some L1 in the human intestine molt to infective L3 stage while they are descending, and L3 penetrate either through the lower intestinal mucosa or perianal skin to re-enter circulation and complete their life cycle within the same host—"auto-infection". Because of this recycling system, *Strongyloides* infection is life-long if not treated. Autoinfection is accelerated in immuno-compromised hosts resulting in a "hyperinfection syndrome" (also known as "overwhelming strongyloidiasis"). Strongyloidiasis hyperinfection syndrome was first seen in soldiers, who went to the endemic areas during the World War II and returned to their home in non-endemic areas ("War Strongyloidiasis": Grove 1980).

In strongyloidiasis hyperinfection syndrome, CNS involvement is primarily represented by meningitis or meningoencephalitis due to overwhelming co-infections with enterobacteria. In rare cases, there is aseptic meningoencephalitis, assumed to be caused by *Strongyloides* L3 stage (Foucan et al. 1997; Vishwanath et al. 1982).

As the CNS involvement in strongyloidiasis is mainly due to overwhelming co-infections with enterobacteria, both anti-parasitic and anti-bacterial treatment should be done in parallel. For *S. stercoralis*, a single dose of ivermectin 200 μg/kg for 2 weeks interval is used a standard protocol. For the cases of bacterial meningitis/menigoencephalitis, antibiotics treatment should be initiated after identification and drug resistance of bacteria. In cases in whom oral or rectal administration of drugs is impossible, or there is poor drug absorption due persistent severe diarrhea, hematochezia and vomiting, subcutaneously injectable ivermectin for veterinary use should be considered (Pacanowski et al. 2005; Takashima et al. 2008).

5.2 Trichinella *Species*

Trichinellosis is one of the major food-borne parasitic nematodiasis emerging and re-emerging in various parts of the world. Autochthonous trichinellosis occurs in 55 countries where the annual incidence of clinical trichinellosis is estimated to be 10,000 cases with a death rate of 0.2 % (Pozio 2007). In a recent literature survey of 6 international databases for 1986–2009, about 66,000 cases with 42 deaths were reported from 41 countries, the great majority being in the EU, especially in Romania during 1990–1999 (Murrell and Pozio 2011). Trichinellosis affects primarily adults without significant gender difference. Infection mainly occurs by ingesting raw or undercooked pork meat, but the meat of other domestic animals or wild game sources has also been reported.

Until recently trichinellosis was thought to be caused solely by *T. spiralis*. However, a molecular phylogenetic study identified at least 12 *Trichinella* species/genotypes (Pozio and Murrell 2006). Some of those species such as *T. pseudospiralis*, *T. papuae* and *T. murrelli* were previously considered as non-human pathogen, but confirmed human cases of infection with those species have been reported repeatedly. Thus, all except one (*T. zimbabwensis*) species/genotypes are zoonotic pathogens for humans (Pozio and Murrell 2006). Since trichinellosis outbreaks around the world are caused by different *Trichinella* species with different clinical manifestations, a more comprehensive approach is required for appropriate diagnosis and treatment (Bruschi and Murrell 2002; Dupouy-Camet and Murell 2007).

The life cycle of *Trichinella* species is maintained by predator/prey relationship of a variety of omnivorous and carnivorous mammals. When humans ingest animal meats containing infective stage of *Trichinella* larvae (muscle larvae), the larvae are released in the upper small intestine and develop into the adult stage within 48 h p.i. Then male and female adult worms mate and female worm releases newborn larvae in the gut lumen 5 days later [intestinal (enteric) phase]. The larvae immediately penetrate into lymphatic vessels, disseminate systemically via bloodstream, and penetrate into skeletal muscles where they become an infective stage and reside there for weeks to years [muscle phase (parenteral phase)].

Clinical manifestations of trichinellosis largely depend on the worm burden. In low-density infections, patients may remain asymptomatic, while infections with over a few hundred larvae may cause acute diarrhea and abdominal pain at around 2 days post-infection. However, intestinal pathology caused by adult worms and/or by penetration of newborn larvae is not serious compared with the subsequent systemic muscular damages due to migration of disseminated larvae. Although CNS is not a selective or specific target organ of this parasite, involvement can be seen in severely affected patients as part of systemic dissemination of larvae (Neghina et al. 2011). By gadolinium-enhanced MRI, multiple small subacute cortical infarcts may be observed in trichinellosis (Feydy et al. 1996).

The clinical diagnosis of trichinellosis is difficult because there are no specific signs or symptoms for trichinellosis. Epidemiological data such as eating undercooked pork or boar meat in the community are of great importance. Trichinellosis usually begins with a sensation of general discomfort and headache, increasing fever, chills and sometimes diarrhoea and/or abdominal pain. Pyrexia, facial and/or eyelid oedema and myalgia represent the principal syndrome of the acute stage. In severe cases, signs of myocarditis, thromboembolic disease and encephalitis can be observed. By laboratory examinations, leucocytosis ($>10{,}000/mm^3$) with marked eosinophilia ($>1{,}000/mm^3$) is observed. Because of the systemic damage of muscles by invading larvae, creatine phosphokinase and lactate dehydrogenase in plasma are increased. A muscle biopsy to detect larvae and the detection of specific antibodies will confirm the diagnosis (Dupouy-Camet and Murell 2007; Gottstein et al. 2009).

The combined treatment of anti-helmintics (mebendazole or albendazole) and glucocorticosteroids is recommended. Mebendazole is usually administered at a daily dose of 5 mg/kg but higher doses (up to 20–25 mg/kg/day) are recommended in some countries. Albendazole is used at 800 mg/day (15 mg/kg/day) administered in two doses. These drugs are usually taken for 10–15 days. The use of mebendazole or albendazole is contraindicated during pregnancy and not recommended in children aged <2 years. The most commonly used steroid is prednisolone at a dose of 30–60 mg/day for 10–15 days, which may alleviate the general symptoms.

5.3 *Filaria*

Filariasis is a vector-borne disease mainly endemic in tropical and subtropical regions and more than 100 million people are affected (WHO data). There are two types of filariasis: one is a lymphatic filariasis caused by *Wuchereria bancrofti* and *Brugia malayi* transmitted by mosquitos; the other type of filariasis is a subcutaneous filariasis caused by *Loa loa* (see Bisoffi et al. 2014), of which the vector is a gadfly (*Chrysops* species), and *Onchocerca volvulus*, of which vector is a blackfly (*Simulium* species). Onchocerciasis and neurological involvement is reviewed in a companion chapter of this book (Njamnshi et al. 2014). In addition, *Dirofilaria immitis*, a dog heart worm, causes VLM in humans.

In lymphatic filariasis, peripheral neuropathy due to elephantiasis is common but CNS involvement is extremely rare (Dumas and Girard 1978). Microfilariae of *W. bancrofti* were found in the cyst fluid of brain tumors (Aron et al. 2002), but its pathogenicity or relation to the clinical symptoms remains unclear. CNS involvement in subcutaneous filariasis is also extremely rare. Encephalopathy (Kamgno et al. 2000; 2008) and meningoencephalitis (Moliner et al. 2011) with the appearance of microfilariae in CSF were observed after ivermectine treatment for loiasis. Similarly, neurological complications with the appearance of microfilariae in the CSF were reported during treatment of onchocerciasis with diethylcarbamazine (Duke et al. 1976). Apart from brain lesions secondary to direct filarial infection, allergic reaction secondary to filarial worm/microfilaria seems to cause recurrent Guillain-Barré syndrome (Bhatia and Misra 1993), meningoencephalitis (Poppert et al. 2009) or acute disseminated encephalomyelitis (ADEM) (Paliwal et al. 2012).

Two dog heart worm species, *Dirofilaria repens* endemic mainly in Europe and *D. immitis* endemic mainly in Asia and the Americas, are both known to infect humans (Simon and Genchi 2000; Simon et al. 2009). While *D. repens* cause cutaneous larva migrans (CLM), *D. immitis* mainly cause VLM in the lungs in humans. Both species are known to cause OLM (Otranto and Eberhard 2011). Infection with dog heartworms of the CNS of humans is extremely rare. A case of intradural dirofilariasis by *D. repens* mimicking a cervical extra-spinal tumor has been reported (Perret-Court et al. 2009). Another case is a German traveler returning from India and Sri Lanka showing a subcutaneous nodular lesion accompanied by signs of meningoencephalitis and aphasia (Poppert et al. 2009). A gravid worm removed from the cutaneous nodule was identified as *D. repens* by PCR sequencing. Although *D. immitis* infection in CNS of humans has never been confirmed, this species is known to migrate into the brain of dogs and cats to cause serious diseases (Suzuki et al. 1970; Donahoe and Holzinger 1974; Hamir 1987).

The definite diagnosis is to detect and identify causative pathogens. This is applicable for subcutaneous filariasis, e.g., loiasis and onchocerciasis or *D. repens* infection in humans. Detection of microfilaria in blood is a definite diagnosis for lymphatic filariasis. However, for CNS filariasis, detection of parasite is extremely difficult and the diagnosis may rely on immunodiagnosis of CSF samples.

6 Conclusion and Future Perspectives

In this chapter, we have gathered and summarized from the published cases of the different types of nematode infection. Apart from filariasis, almost all neuronematodiases, or NLM, are neglected diseases in endemic countries. Therefore, the figures mentioned here are in all likelihood much underestimated. Still, from those literature surveys, we can draw rough figures of the global prevalence of each infection and its geographical distribution. NLM by *Toxocara*, *Ascaris*, *Strongyloides* and *Trichinella* infections are likely to be seen even in the high-income countries. NLM by other nematode species are extremely rare and identification of causative parasite is difficult. Abnormal radioimaging associated with eosinophilia/hyper IgE is strongly suggestive of neuroparasitoses. Immunodiagnosis to detect parasite-specific antibody or parasite antigen in the serum and/or CSF is useful to consider the possibility of parasitic infection. Whenever NLM is diagnosed, association of CLM and VLM should be considered. In strongyloidiasis, neurological manifestations are mostly due to overwhelming co-infections with enterobacteria. Conversely, enterobacterial meningitis is suggestive for disseminated strongyloidiasis.

Until now, availability of immunodiagnostic kits for nematode parasites has been limited and the specificity and/or sensitivity are not well-defined. To perform extensive survey in the developing countries, development of cheap and simple diagnostic system is urgently necessary. Likewise, availability of research laboratories/ institutions where immunodiagnosis for helminth parasites can be performed is limited. Since many case reports have been published in local journals with local languages, a database should be constructed to gather and provide accurate information about NLM. To achieve such tasks, collaborative reference centers should established for neglected parasitic diseases, for which the present International Trichinella Reference Center (http://www.iss.it/site/Trichinella/scripts/serv.asp?lang=2) (Pozio et al. 2001) could be a model.

Acknowledgments Intapan PM. and Maleewong W. are supported by TRF Senior Research Scholar Grant, Thailand Research Fund grant number RTA5580004. Nawa Y. is an invited professor/consultant supported by the Faculty of Medicine, KhonKaen University, Thailand.

References

Akao N, Ohta N (2007) Toxocariasis in Japan. Parasitol Int 56:87–93

Anantaphruti MT, Nuamtanong S, Dekumyoy P (2005) Diagnostic values of IgG4 in human gnathostomiasis. Trop Med Int Health 10:1013–1021

Aquino RTR, Magliari MER, Vital Filho J, Silva MALG, Lima CAC, Rocha AJ, Silva CJ, Rewin JA, Nahas TR, Chieffi PP (2008) Lagochilascariasis leading to severe involvement of ocular globes, ears and meninges. Rev Inst Med Trop Sao Paulo 50:355–358

Aron M, Kapila K, Sarkar C, Verma K (2002) Microfilariae of *Wuchereria bancrofti* in cyst fluid of tumors of the brain: a report of three cases. Diagn Cytopathol 26:158–162

Barbosa CA, Campos DM, de Oliveira JA (1998) Assessment of ivermectin therapeutic efficacy on third-stage larvae of *Lagochilascaris minor* in mice experimentally infected. Rev Inst Med Trop Sao Paulo 40:137–143

Barrera-Pérez M, Manrique-Saide P, Reyes-Novelo E, Escobedo-Ortegón J, Sánchez-Moreno M, Sánchez C (2012) *Lagochilascaris minor* Leiper, 1909 (Nematoda: Ascarididae) in Mexico: three clinical cases from the Peninsula of Yucatan. Rev Inst Med Trop Sao Paulo 54:315–317

Bauer C (2011) Baylisascariosis (*Baylisascaris procyonis*)—a rare parasitic zoonosis in Europe. Berl Munch Tierarztl Wochenschr 124:465–472 (in German)

Bento RF, Mazza Cdo C, Motti EF, Chan YT, Guimarães JR, Miniti A (1993) Human lagochilascariasis treated successfully with ivermectin: a case report. Rev Inst Med Trop Sao Paulo 35:73–75
Bhatia B, Misra S (1993) Recurrent Guillain-Barré syndrome following acute filariasis. J Neurol Neurosurg Psychiatry 56:1133–1134
Bisoffi Z, Buonfrate D, Angheben A (2014) Travel, migration and neglected tropical diseases. In: Bentivoglio M, Cavalheiro EA, Kristensson K, Patel NB (eds) Neglected tropical diseases and conditions of the nervous system. Springer, New York
Boongird P, Phuapradit P, Siridej N, Chirachariyavej T, Chuahirun S, Vejjajiva A (1977) Neurological manifestations of gnathostomiasis. J Neurol Sci 31:279–291
Boschetti A, Kasznica J (1995) Visceral larva migrans induced eosinophilic cardiac pseudotumor: a cause of sudden death in a child. J Forensic Sci 40:1097–1099
Brasil OF, Lewis H, Lowder CY (2006) Migration of *Baylisascaris procyonis* into the vitreous. Br J Ophthalmol 90:1203–1204
Brenes-Madrigal RR, Ruiz A, Frenkel JK (1972) Discovery of *Lagochilascaris* sp. in the larynx of a Costa Rican ocelot (*Felis paradalis mearnsi*). J Parasitol 58:978
Bruschi F, Murrell KD (2002) New aspects of human trichinellosis: the impact of new *Trichinella* species. Postgrad Med J 78:15–22
Bunnag T, Comer DS, Punyagupta S (1970) Eosinophilic myeloencephalitis caused by *Gnatjostoma spinigerum*. Neuropathology of nine cases. J Neurol Sci 10(5):419–434
Campos DM, Freire Filha LG, Vieira MA, Paçô JM, Maia MA (1992) Experimental life cycle of *Lagochilascaris minor* Leiper, 1909. Rev Inst Med Trop Sao Paulo 34:277–287
Chen KM, Liu JY, Lai SC, Hsu LS, Lee HH (2006) Association of plasminogen activators and matrix metalloproteinase-9 proteolytic cascade with blood-CNS barrier damage of angiostrongyliasis. Int J Exp Pathol 87:113–119
Chitanondh H, Rosen L (1967) Fatal eosinophilic encephalomyelitis caused by the nematode *Gnathostoma spinigerum*. Am J Trop Med Hyg 16:638–645
Chotmongkol V, Sawanyawisuth K (2002) Clinical manifestations and outcome of patients with severe eosinophilic meningoencephalitis presumably caused by *Angiostrongylus cantonensis*. Southeast Asian J Trop Med Public Health 33:231–234
Chotmongkol V, Sawanyawisuth K, Thavornpitak Y (2000) Corticosteroid treatment of eosinophilic meningitis. Clin Infect Dis 31:660–662
Chotmongkol V, Wongjitrat C, Sawadpanit K, Sawanyawisuth K (2004) Treatment of eosinophilic meningitis with a combination of albendazole and corticosteroid. Southeast Asian J Trop Med Public Health 35:172–174
Chotmongkol V, Sawadpanitch K, Sawanyawisuth K, Louhawilai S, Limpawattana P (2006) Treatment of eosinophilic meningitis with a combination of prednisolone and mebendazole. Am J Trop Med Hyg 74:1122–1124
Daengsvang S (1981) Gnathostomiasis in Southeast Asia. Southeast Asian J Trop Med Public Health 12:319–332
Dharmkrong-at A, Migasena S, Suntharasamai P, Bunnag D, Priwan R, Sirisinha S (1986) Enzyme-linked immunosorbent assay for detection of antibody to *Gnathostoma* antigen in patients with intermittent cutaneous migratory swelling. J Clin Microbiol 23:847–851
Diao Z, Wang J, Qi H, Li X, Zheng X, Yin C (2011) Human ocular angiostrongyliasis: a literature review. Trop Doct 41:76–78
Diaz JH (2008) Helminthic eosinophilic meningitis: emerging zoonotic diseases in the South. J La State Med Soc 160:333–342
Diaz Camacho S, Ramos MZ, Torrecillas E, Ramirez IO, Velazquez RC, Gaxiola AF, Heredia JB, Willms K, Akahane H, Ogata K, Nawa Y (1998) Clinical manifestation and immunodiagnosis of gnathostomiasis in Culiacan, Mexico. Am J Trop Med Hyg 59:908–915
Donahoe JM, Holzinger EA (1974) *Dirofilaria immitis* in the brains of a dog and a cat. J Am Vet Med Assoc 164:518–519
Dorta-Contreras AJ, Noris-Garcia E, Escobar-Perez X, Duenas-Flores A, Mena-Lopez R (2003) IgG subclasses intrathecal synthesis patterns in eosinophilic meningoencephalitis due to *Angiostrongylus cantonensis*. Rev Neurol 36:506–509

Dorta-Contreras AJ, Noris-Garcia E, Escobar-Perez X, Padilla-Docal B (2005) IgG1, IgG2 and IgE intrathecal synthesis in *Angiostrongylus cantonensis* meningoencephalitis. J Neurol Sci 238:65–70
Draper JW (1963) Infection with *Lagochilascaris minor*. Br Med J 1:931–932
Duke BO, Vincelette J, Moore PJ (1976) Microfilariae in the cerebrospinal fluid, and neurological complications, during treatment of onchocerciasis with diethylcarbamazine. Tropenmed Parasitol 27:123–132
Dumas M, Girard PL (1978) Filariasis of the nervous system. In: Vinken PJ, Bruyn GW, Klawans HL (eds) Handbook of clinical neurology, vol 5, Infections of the nervous system, part III. North-Holland, Amsterdam, pp 161–173
Dupouy-Camet J, Murell KD (2007) FAO/WHO/OIE guidelines for the surveillance, management, prevention and control of trichinellosis. FAO/WHO/OIE, Paris, pp 1–119
Elzi L, Decker M, Battegay M, Rutishauser J, Blum J (2004) Chest pain after travel to the tropics. Lancet 363:1198
Fan CK, Su KE (2004) Cross-reactions with *Ascaris suum* antigens of sera from mice infected with *A. suum*, *Toxocara canis*, and *Angiostrongylus cantonensis*. Parasitol Int 53:263–271
Feydy A, Touze E, Miaux Y, Bolgert F, Martin-Duverneuil N, Laplane D, Chiras J (1996) MRI in a case of neurotrichinosis. Neuroradiology 38(Suppl 1):S80–S82
Finsterer J, Auer H (2012) Parasitoses of the human central nervous system. J Helminthol 10:1–14
Foucan L, Genevier I, Lamaury I, Strobel M (1997) Aseptic purulent meningitis in two patients co-infected by HTLV-1 and *Strongyloides stercoralis*. Med Trop 57:262–264 (in French)
Fox AS, Kazacos KR, Gould NS, Heydemann PT, Thomas C, Boyer KM (1985) Fatal eosinophilc meningoencepalitis and visceral larva migrans caused by the raccoon ascarid *Baylisascaris procyonis*. N Engl J Med 312:1619–1623
Freedman DO (1991) Experimental infection of human subject with *Strongyloides* species. Rev Infect Dis 13:1221–1226
Fukae J, Kawanabe T, Akao N, Kado M, Tokoro M, Yokoyama K, Hattori N (2012) Longitudinal myelitis caused by visceral larva migrans associated with *Toxocara cati* infection: case report. Clin Neurol Neurosurg 114:1091–1094
Gavin PJ, Kazacos KR, Shulman ST (2005) Baylisascariasis. Clin Microbiol Rev 18:703–718
Górgolas M, Santos-O'Connor F, Unzú AL, Fernández-Guerrero ML, Gárate T, Troyas Guarch RM, Grobusch MP (2003) Cutaneous and medullar gnathostomiasis in travelers to Mexico and Thailand. J Travel Med 10:358–361
Gottstein B, Pozio E, Nockler K (2009) Epidemiology, diagnosis, treatment, and control of trichinellosis. Clin Microbiol Rev 22:127–145
Grove DI (1980) Strongyloidiasis in allied ex-prisoners of war in south-east Asia. Br Med J 280:598–601
Grove DI (1996) Human strongyloidiasis. Adv Parasitol 38:251–309
Haider S, Khairnar K, Martin DS, Yang J, Ralevski F, Kazacos KR, Pillai DR (2012) Possible pet-associated baylisascariasis in child, Canada. Emerg Infect Dis 18:347–349
Hajek J, Yau Y, Kertes P, Soman T, Laughlin S, Kanini R, Kazacos K, Dangoudoubiyam S, Opavsky MA (2009) A child with raccoon roundworm meningoencephalitis: a pathogen emerging in your own backyard? Can J Infect Dis Med Microbiol 20:e177–e180
Hale DC, Blumberg L, Frean J (2003) Case report: gnathostomiasis in two travelers to Zambia. Am J Trop Med Hyg 68:707–709
Hamir AN (1987) Heartworm (*Dirofilaria immitis*) in the brain of a dog. Vet Rec 120:207–208
Herman JS, Wall EC, van-Tulleken C, Godfrey-Faussett P, Bailey RL, Chiodini PL (2009) Gnathostomiasis acquired by British tourists in Botswana. Emerg Infect Dis 15:594–597
Hernandez SM, Galbreath B, Riddle DF, Moore AP, Palamar MB, Levy MG, Deperno CS, Correa MT, Yabsley MJ (2013) *Baylisascaris procyonis* in raccoons (*Procyon lotor*) from North Carolina and current status of the parasite in the USA. Parasitol Res 112:693–698
Hoffmeister B, Glaeser S, Flick H, Pornschlegel S, Suttorp N, Bergmann F (2007) Cerebral toxocariasis after consumption of raw duck liver. Am J Trop Med Hyg 76:600–602

Hung T, Neafie RC, Mackenzie IRA (2012) *Baylisascaris procyonis* infection in elderly person, British Columbia, Canada. Emerg Infect Dis 18:341–342

Inatomi Y, Murakami T, Tokunaga M, Ishiwata K, Nawa Y, Uchino M (1999) Encephalopathy caused by visceral larva migrans due to *Ascaris suum*. J Neurol Sci 164:195–199

Intapan PM, Maleewong W, Polsan Y, Sawanyawisuth K, Chotmongkol V (2002) Specific IgG antibody subclasses to *Angiostrongylus cantonensis* in patients with angiostrongyliasis. Asian Pac J Allergy Immunol 20:235–240

Intapan PM, Niwattayakul K, Sawanyawisuth K, Chotmongkol V, Maleewong W (2007) Cerebrospinal fluid eotaxin and eotaxin-2 levels in human eosinophilic meningitis associated with angiostrongyliasis. Cytokine 39:138–141

Intapan PM, Kittimongkolma S, Niwattayakul K, Sawanyawisuth K, Maleewong W (2008) Cerebrospinal fluid cytokine responses in human eosinophilic meningitis associated with angiostrongyliasis. J Neurol Sci 267:17–21

Intapan PM, Khotsri P, Kanpittaya J, Chotmongkol V, Sawanyawisuth K, Maleewong W (2010) Immunoblot diagnostic test for neurognathostomiasis. Am J Trop Med Hyg 83:927–929

Isobe N, Kanamori Y, Yonekawa T, Matsushita T, Shigeto H, Kawamura N, Kira J (2012) First diagnostic criteria for atopic myelitis with special reference to discrimination from myelitis-onset multiple sclerosis. J Neurol Sci 316:30–35

Izumikawa K, Kohno Y, Izumikawa K, Hara K, Hayashi H, Maruyama H, Kohno S (2011) Eosinophilic pneumonia due to visceral larva migrans possibly caused by *Ascaris suum*: a case report and review of recent literatures. Jpn J Infect Dis 64:428–432

Jabbour RA, Kanj SS, Sawaya RA, Awar GN, Hourani MH, Atweh SF (2011) *Toxocara canis* myelitis: clinical features, magnetic resonance imaging (MRI) findings, and treatment outcome in 17 patients. Medicine (Baltimore) 90:337–343

Jitpimolmard S, Sawanyawisuth K, Morakote N, Vejjajiva A, Puntumetakul M, Sanchaisuriya K, Tassaneeyakul W, Tassaneeyakul W, Korwanich N (2007) Albendazole therapy for eosinophilic meningitis caused by *Angiostrongylus cantonensis*. Parasitol Res 100:1293–1296

Kamgno J, Gardon J, Boussinesq M (2000) Analysis of the prevention of post-ivermectin *Loa loa* encephalopathy by administration of initial low dose. Med Trop (Mars) 60:275–277, French

Kamgno J, Boussinesq M, Labrousse F, Nkegoum B, Thylefors BI, Mackenzie CD (2008) Encephalopathy after ivermectin treatment in a patient infected with *Loa loa* and *Plasmodium* spp. Am J Trop Med Hyg 78:546–551

Kanpittaya J, Jitpimolmard S, Tiamkao S, Mairiang E (2000) MR findings of eosinophilic meningoencephalitis attributed to *Angiostrongylus cantonensis*. Am J Neuroradiol 21:1090–1094

Kanpittaya J, Sawanyawisuth K, Intapan PM, Khotsri P, Chotmongkol V, Maleewong W (2012) A comparative study of neuroimaging features between human neuro-gnathostomiasis and angiostrongyliasis. Neurol Sci 33:893–898

Katchanov J, Nawa Y (2010) Helminthic invasion of the central nervous system: many roads lead to Rome. Parasitol Int 59:491–496

Katchanov J, Sawanyawisuth K, Chotmongkol V, Nawa Y (2011) Neurognathostomiasis, a neglected parasitosis of the central nervous system. Emerg Infect Dis 17:1174–1180

Kawamura J, Kohri Y, Oka N (1983) Eosinophilic meningoradiculomyelitis caused by *Gnathostoma spinigerum*. A case report. Arch Neurol 40:583–585

Kira J, Horiuchi I, Suzuki J, Osoegawa M, Tobimatsu S, Murai H, Minohara M, Furue M, Ochi H (2001) Myelitis associated with atopic disorders in Japan: a retrospective clinical study of the past 20 years. Intern Med 40:613–619

Kirsch S, Dekumyoy P, Löscher T, Haberl RL (2008) A case of eosinophilic meningitis in Germany. J Neurol 255:1102–1103

Komiyama A, Hasegawa O, Nakamura S, Ohno S, Kondo K (1995) Optic neuritis in cerebral toxocariasis. J Neurol Neurosurg Psychiatry 59:197–198

Kuberski T, Wallace GD (1979) Clinical manifestations of eosinophilic meningitis due to *Angiostrongylus cantonensis*. Neurology 29:1566–1570

Kwon NH, Oh MJ, Lee SP, Lee BJ, Choi DC (2006) The prevalence and diagnostic value of toxocariasis in unknown eosinophilia. Ann Hematol 85:233–238

Lai SC, Jiang ST, Chen KM, Lee HH (2005) Matrix metalloproteinases activity demonstrated in the infective stage of the nematodes, *Angiostrongylus cantonensis*. Parasitol Res 97:466–471

Laummaunwai P, Sawanyawisuth K, Intapan PM, Chotmongkol V, Wongkham C, Maleewong W (2007) Evaluation of human IgG class and subclass antibodies to a 24 kDa antigenic component of *Gnathostoma spinigerum* for the serodiagnosis of gnathostomiasis. Parasitol Res 101: 703–708

Lee JD, Tsai LY, Chen CH, Wang JJ, Hsiao JK, Yen CM (2006) Blood-brain barrier dysfunction occurring in mice infected with *Angiostrongylus cantonensis*. Acta Trop 97:204–211

Lee AC, Schantz PM, Kazacos KR, Montgomery SP, Bowman DD (2010) Epidemiologic and zoonotic aspects of ascarid infections in dogs and cats. Trends Parasitol 26:155–161

Leiper RT (1909) A new nematode worm from Trinidad: *Lagochilascaris minor* sp.n. Proc Zool Soc (London) 2:742–743

León-Règagnon V, Osorio-Sarabia D, García-Prieto L, Akahane H, Lamothe-Argumedo R, Koga M, Messina-Robles M, Alvarez-Guerrero C (2002) Study of the ethiological agent of gnathostomosis in Nayarit, Mexico. Parasitol Int 51:201–204

Liu IH, Chung YM, Chen SJ, Cho WL (2006) Necrotizing retinitis induced by *Angiostrongylus cantonensis*. Am J Ophthalmol 141:577–579

Maleewong W, Morakote N, Thamasonthi W, Charuchinda K, Tesana S, Khamboonruang C (1988) Serodiagnosis of human gnathostomiasis. Southeast Asian J Trop Med Public Health 19:201–205

Maleewong W, Sombatsawat P, Intapan PM, Wongkham C, Chotmongkol V (2001) Immunoblot evaluation of the specificity of the 29- kDa antigen from young adult female worms *Angiostrongylus cantonensis* for immunodiagnosis of human angiostrongyliasis. Asian Pac J Allergy Immunol 19:267–273

Martínez-Delgado JF, González-Cortiñas M, Tápanes-Cruz TR, Ruiz-Méndez A (2000) Eosinophilic meningoencephalitis in Villa Clara (Cuba). A study of 17 patients. Rev Neurol 31:417–421

Maruyama H, Nawa Y, Noda S, Mimori T, Choi WY (1996) An outbreak of visceral larva migrans due to *Ascaris suum* in Kyushu, Japan. Lancet 347:1766–1777

Mets MB, Noble AG, Basti S, Gavin P, Davis AT, Shulman ST, Kazacos KR (2003) Eye findings of diffuse unilateral subacute neuroretinitis and multiple choroidal infiltrates associated with neural larva migrans due to *Baylisascaris procyonis*. Am J Ophthalmol 135:888–890

Miyazaki I (1960) On the genus *Gnathostoma* and human gnathostomiasis, with special reference to Japan. Exp Parasitol 9:338–370

Moliner JV, Valverde AG, Sorolla JM (2011) [Loa loa meningoencephalitis secondary to treatment with mebendazol]. Med Clin (Barc) 136:228–229, in Spanish

Moncada LI, Alvarez CA, Castellanos C, Caceres E, Nicholls S, Corredor A (1998) *Lagochilascaris minor* in a patient from the Colombian Amazon: a case report. Rev Inst Med Trop Sao Paulo 40:387–389

Moore DA, McCroddan J, Dekumyoy P, Chiodini PL (2003) Gnathostomiasis: an emerging imported disease. Emerg Infect Dis 9:647–650

Moraes MA, Arnaud MV, de Lima PE (1983) [New cases of human infection by *Lagochilascaris minor* Leiper, 1909, found in the State of Pará, Brazil]. Rev Inst Med Trop Sao Paulo 25:139–146, Portuguese

Moraes MA, Arnaud MV, de Macedo RC, Anglada AE (1985) Fatal pulmonary infection caused by *Lagochilascaris* sp., probably *Lagochilascaris minor* Leiper, 1909. Rev Inst Med Trop Sao Paulo 27:46–52

Morimatsu Y, Akao N, Akiyoshi H, Kawazu T, Okabe Y, Aizawa H (2006) A familial case of visceral larva migrans after ingestion of raw chicken livers: appearance of specific antibody in bronchoalveolar lavage fluid of the patients. Am J Trop Med Hyg 75:303–306

Murray WJ, Kazacos KR (2004) Raccoon roundworm encephalitis. Clin Infect Dis 39:1484–1492

Murrell KD, Pozio E (2011) Worldwide occurrence and impact of human trichinellosis, 1986-2009. Emerg Infect Dis 17:2194–2202

Nakamura-Uchiyama F, Tokunaga Y, Suzuki A, Akao N, Hiromatsu K, Hitomi S, Nawa Y (2006) A case of *Ascaris suum* visceral larva migrans diagnosed by using *A. suum* larval excretory-secretory (ES) antigen. Scand J Infect Dis 38:221–224

Nawa Y (1991) Historical review and current status of gnathostomiasis in Asia. Southeast Asian J Trop Med Public Health 22(Suppl):217–219

Neghina R, Neghina AM, Marincu I, Iacobiciu I (2011) Reviews on trichinellosis (II): neurological involvement. Foodborne Pathog Dis 8:579–585

Nishimura K, Hung T-P (1997) Current views on geographic distribution and modes of infection of neurohelminthic diseases. J Neurol Sci 145:5–14

Njamnshi AK, Zoung-Kanyi Bissek A-C, Etya'ale D (2014) Onchocerciasis: Neurological involvement. In: Bentivoglio M, Cavalheiro EA, Kristensson K, Patel N (eds) Neglected tropical diseases and conditions of the nervous system. Springer, New York

Nopparatana C, Setasuban P, Chaicumpa W, Tapchaisri P (1991) Purification of *Gnathostoma spinigerum* specific antigen and immunodiagnosis of human gnathostomiasis. Int J Parasitol 21:677–687

Nuamtanong S (1996) The evaluation of the 29 and 31 kDa antigens in female *A. cantonensis* for serodiagnosis of human angiostrongyliasis. Southeast Asian J Trop Med Public Health 27: 291–296

Nunes CM, Tundisi RN, Garcia JF, Heinemann MB, Ogassawara S, Richtzenhain LJ (1997) Cross-reactions between *Toxocara canis* and *Ascaris suum* in the diagnosis of visceral larva migrans by western blotting technique. Rev Inst Med Trop Sao Paulo 39:253–256

Ollé-Goig JE, Recacoechea M, Feeley T (1996) First case of *Lagochilascaris minor* infection in Bolivia. Trop Med Int Health 1:851–853

Oostburg BF (1971) Thiabendazole therapy of *Lagochilascaris minor* infection in Surinam. Report of a case. Am J Trop Med Hyg 20:580–583

Oostburg BF, Varma AA (1968) *Lagochilascaris minor* infection in Surinam. Report of a case. Am J Trop Med Hyg 17:548–550

Osoegawa M (2004) Diagnosis and treatment of CNS parasite infection with special reference to parasitic myelitis. Rinsho Shinkeigaku 44:961–964 (in Japanese)

Osoegawa M, Matsumoto S, Ochi H, Yamasaki K, Horiuchi I, Kira YO, Ishiwata K, Nakamura-Uchiyama F, Nawa Y (2001) Localised myelitis caused by visceral larva migrans due to *Ascaris suum* masquerading as an isolated spinal cord tumour. J Neurol Neurosurg Psychiatry 70: 265–266

Osoegawa M, Ochi H, Minohara M, Murai H, Umehara F, Furuya H, Yamada T, Kira J (2003) Myelitis with atopic diathesis: a nationwide survey of 79 cases in Japan. J Neurol Sci 209:5–11

Othman AA (2012) Therapeutic battle against larval toxocariasis: are we still far behind? Acta Trop 124:171–178

Otranto D, Eberhard ML (2011) Zoonotic helminths affecting the human eye. Parasit Vectors 4:41

Pacanowski J, Santos MD, Roux A, LE Maignan C, Guillot J, Lavarde V, Cornet M (2005) Subcutaneous ivermectin as a safe salvage therapy in *Strongyloides stercoralis* hyperinfection syndrome: a case report. Am J Trop Med Hyg 73:122–124

Paliwal VK, Goel G, Vema R, Pradhan S, Gupta RK (2012) Acute disseminated encephalomyelitis following filarial infection. Neurol Neurosurg Psychiatry 83:347–349

Perlman JE, Kazacos KR, Imperato GH, Desai RU, Schulman SK, Edwards J et al (2010) *Baylisascaris procyonis* neural larva migrans in an infant in New York City. J Neuroparasitol 1:1–5

Perret-Court A, Coulibaly B, Ranque S, Bouvier C, Lena G et al (2009) Intradural dirofilariasis mimicking a Langerhans cell histiocytosis tumor. Pediatr Blood Cancer 53:485–487

Petithory JC (1996) Can *Ascaris suum* cause visceral larva migrans? Lancet 348:689

Phills JA, Harrold AJ, Whiteman GV, Perelmutter L (1972) Pulmonary infiltrates, asthma and eosinophilia due to *Ascaris suum* infestation in man. N Engl J Med 286:965–970

Pinelli E, Herremans T, Harms MG, Hoek D, Kortbeek LM (2011) *Toxocara* and *Ascaris* seropositivity among patients suspected of visceral and ocular larva migrans in the Netherlands: trends from 1998 to 2009. Eur J Clin Microbiol Infect Dis 30:873–879

Poppert S, Hodapp M, Krueger A, Hegasy G, Niesen WD, Kern WV, Tannich E (2009) *Dirofilaria repens* infection and concomitant meningoencephalitis. Emerg Infect Dis 15(11):1844–1846

Pozio E (2007) World distribution of *Trichinella* spp. infections in animals and humans. Vet Parasitol 149:3–21

Pozio E, Murrell KD (2006) Systematics and epidemiology of Trichinella. Adv Parasitol 63:367–439

Pozio E, La Rosa G, D'Ancona F, Amati M, Mancini Barbieri F, De Giacomo M (2001) Twelve years of activity of the International *Trichinella* Reference Centre. Parasite 8(2 Suppl):S44–S46

Prociv P, Spratt DM, Carlisle MS (2000) Neuro-angiostrongyliasis: unresolved issues. Int J Parasitol 301:295–303

Punyagupta S, Juttijudata P, Bunnag T, Comer DS (1968) Two fatal cases of eosinophilic myelo-encephalitis a newly recognized disease caused by *Gnathostoma spinigerum*. Trans R Soc Trop Med Hyg 62:801–809

Punyagupta S, Juttijudata P, Bunnag T (1975) Eosinophilic meningitis in Thailand: clinical studies of 484 typical cases probably caused by *Angiostrongylus cantonensis*. Am J Trop Med Hyg 24:921–931

Punyagupta S, Bunnag T, Juttijudata P (1990) Eosinophilic meningitis in Thailand. Clinical and epidemiological characteristics of 162 patients with myeloencephalitis probably caused by *Gnathostoma spinigerum*. J Neurol Sci 96:241–256

Quattrocchi G, Nicoletti A, Marin B, Bruno E, Druet-Cabanac M, Preux PM (2012) Toxocariasis and epilepsy: systematic review and meta-analysis. PLoS Negl Trop Dis 6:e1775. doi:10.1371/journal.pntd.0001775

Romero JR, Led JE (1985) [A new case of *Lagochilascaris major* (Leiper 1910) in the Argentine Republic parasitizing the cat (*Felis catus domesticus*)]. Zentralbl Veterinarmed B 32:575–582, Spanish

Rosemberg S, Lopes MB, Masuda Z, Campos R, Vieira Bressan MC (1986) Fatal encephalopathy due to *Lagochilascaris minor* infection. Am J Trop Med Hyg 35:575–578

Rusnak JM, Lucey DR (1993) Clinical gnathostomiasis: case report and review of the English-language literature. Clin Infect Dis 16:33–50

Saffra NA, Perlman JE, Desai RU, Kazacos KR, Coyle CM, Machado FS, Kedhar SR, Engelbert M, Tanowitz HB (2010) *Baylisascaris procyonis* induced diffuse unilateral subacute neuroretinitis in New York City. J Neuroparasitol 1:1, DOI pii:N100401

Sakakibara A, Baba K, Niwa S, Yagi T, Wakayama H, Yoshida K, Kobayashi T, Yokoi T, Hara K, Itoh M, Kimura E (2002) Visceral larva migrans due to *Ascaris suum* which presented with eosinophilic pneumonia and multiple intra-hepatic lesions with severe eosinophil infiltration–outbreak in a Japanese area other than Kyushu. Intern Med 41:574–579

Sakamoto T, Cabrera PA (2002) Subcutaneous infection of *Lagochilascaris minor* in domestic cats from Uruguay. Vet Parasitol 108:145–152

Sanpool O, Intapan PM, Thanchomnang T, Santivarangkana T, Niwattayakul K, Chotmongkol V, Maleewong W (2010) Cerebrospinal fluid u-plasminogen activator and matrix metalloproteinase-9 levels in human eosinophilic meningitis associated with angiostrongyliasis. Cytokine 51:294–297

Sato H, Kamiya H, Furuoka H (2003) Epidemiological aspects of the first outbreak of *Baylisascaris procyonis* larva migrans in rabbits in Japan. J Vet Med Sci 65:453–457

Sato H, Une Y, Kawakami S, Saito E, Kamiya H, Akao N, Furuoka H (2005) Fatal *Baylisascaris* larva migrans in a colony of Japanese macaques kept by a safari-style zoo in Japan. J Parasitol 91:716–719

Sawanyawisuth K, Kitthaweesin K (2008) Optic neuritis caused by intraocular angiostrongyliasis. Southeast Asian J Trop Med Public Health 39:1005–1007

Sawanyawisuth K, Sawanyawisuth K (2008) Treatment of angiostrongyliasis. Trans R Soc Trop Med Hyg 102:990–996

Sawanyawisuth K, Limpawattana P, Busaracome P, Ninpaitoon B, Chotmongkol V, Intapan PM, Tanawirattananit S (2004a) A 1-week course of corticosteroids in the treatment of eosinophilic meningitis. Am J Med 117:802–803

Sawanyawisuth K, Tiamkao S, Kanpittaya J, Dekumyoy P, Jitpimolmard S (2004b) MR imaging findings in cerebrospinal gnathostomiasis. AJNR Am J Neuroradiol 25:446–449

Sawanyawisuth K, Kitthaweesin K, Limpawattana P, Intapan PM, Tiamkao S, Jitpimolmard S, Chotmongkol V (2007) Intraocular angiostrongyliasis: clinical findings, treatments and outcomes. Trans R Soc Trop Med Hyg 101:497–501

Sawanyawisuth K, Takahashi K, Hoshuyama T, Sawanyawisuth K, Senthong V, Limpawattana P, Intapan PM, Wilson D, Tiamkao S, Jitpimolmard S, Chotmongkol V (2009) Clinical factors predictive of encephalitis caused by *Angiostrongylus cantonensis*. Am J Trop Med Hyg 81:698–701

Schmutzhard E (2007) Eosinophilic myelitis, a souvenir from South East Asia. Pract Neurol 7:48–51

Schmutzhard E, Boongird P, Vejjajiva A (1988) Eosinophilic meningitis and radiculomyelitis in Thailand, caused by CNS invasion of *Gnathostoma spinigerum* and *Angiostrongylus cantonensis*. J Neurol Neurosurg Psychiatry 51:80–87

Sexsmith JL, Whiting TL, Green C, Orvis S, Berezanski DJ, Thompson AB (2009) Prevalence and distribution of *Baylisascaris procyonis* in urban raccoons (*Procyon lotor*) in Winnipeg, Manitoba. Can Vet J 50:846–850

Shields CL, Schoenberg E, Kocher K, Shukla SY, Kaliki S, Shields JA (2013) Lesions simulating retinoblastoma (pseudoretinoblastoma) in 604 Cases: results based on age at presentation. Ophthalmology 120:311–316. doi:10.1016/j.ophtha.2012.07.067, pii: S0161-6420(12)00721-X

Simon F, Genchi C (2000) Dirofilariosis and other zoonotic filarioses: an emerging public health problem in developed countries. Res Rev Parasitol 60:1–16

Simon F, Morchon R, Gonzalez-Miguel J, Marcos-Atxutegi C, Mar Siles-Lucas M (2009) What is new about animal and human dirofilariosis? Trend Parasitol 25:404–409

Sinawat S, Sanguansak T, Angkawinijwong T, Ratanapakorn T, Intapan PM, Sinawat S, Yospaiboon Y (2008) Ocular angiostrongyliasis: clinical study of three cases. Eye (Lond) 22:1446–1448

Strady C, Dekumyoy P, Clement-Rigolet M, Danis M, Bricaire F, Caumes E (2009) Long-term follow-up of imported gnathostomiasis shows frequent treatment failure. Am J Trop Med Hyg 80:33–35

Suntharasamai P, Desakorn V, Migasena S, Bunnag D, Harinasuta T (1985) ELISA for immunodiagnosis of human gnathostomiasis. Southeast Asian J Trop Med Public Health 16:274–279

Suzuki Y, Suu S, Satoh H (1970) Cerebrospinal lesions resulting from invasion by canine beartworm larvae. Jpn J Vet Sci 32:11–17

Taira K, Saitoh Y, Kapel CM (2011) *Toxocara cati* larvae persist and retain high infectivity in muscles of experimentally infected chickens. Vet Parasitol 180:287–291

Takashima N, Yazawa S, Ishihara A, Sugimoto S, Shiomi K, Hiromatsu K, Nakazato M (2008) [Fulminant strongyloidiasis successfully treated by subcutaneous ivermectin: an autopsy case]. Rinsho Shinkeigaku 48:30–35, Japanese

Tangchai P, Nye SW, Beaver PC (1967) Eosinophilic meningoencephalitis caused by angiostrongyliasis in Thailand. Autopsy report. Am J Trop Med Hyg 16:454–461

Tapchaisri P, Nopparatana C, Chaicumpa W, Setasuban P (1991) Specific antigen of *Gnathostoma spinigerum* for immunodiagnosis of human gnathostomiasis. Int J Parasitol 21:315–319

Tokiwa T, Harunari T, Tanikawa T, Komatsu N, Koizumi N et al (2012) Phylogenetic relationships of rat lungworm, *Angiostrongylus cantonensis*, isolated from different geographical regions revealed widespread multiple lineages. Parasitol Int 61(3):431–436

Tsai HC, Liu YC, Kunin CM, Lee SS, Chen YS, Lin HH, Tsai TH, Lin WR, Huang CK, Yen MY, Yen CM (2001) Eosinophilic meningitis caused by *Angiostrongylus cantonensis*: report of 17 cases. Am J Med 111:109–114

Tsai HC, Liu YC, Kunin CM, Lai PH, Lee SS, Chen YS, Wann SR et al (2003) Eosinophilic meningitis caused by *Angiostrongylus cantonensis* associated with eating raw snails: correlation of brain magnetic resonance imaging scans with clinical findings. Am J Trop Med Hyg 68:281–285

Tsai HC, Lee SS, Huang CK, Yen CM, Chen ER, Liu YC (2004) Outbreak of eosinophilic meningitis associated with drinking raw vegetable juice in southern Taiwan. Am J Trop Med Hyg 71:222–226

Umehara F, Ookatsu H, Hayashi D, Uchida A, Douchi Y, Kawabata H, Goto R, Hashiguchi A, Matsuura E, Okubo R, Higuchi I, Arimura K, Nawa Y, Osame M (2006) MRI studies of spinal visceral larva migrans syndrome. J Neurol Sci 249:7–12

Vargas-Ocampo F, Alvarado-Aleman FJ (1997) Infestation from *Lagochilascaris minor* in Mexico. Int J Dermatol 36:56–58

Vidal JE, Sztajnbok J, Seguro AC (2003) Eosinophilic meningoencephalitis due to *Toxocara canis*: case report and review of the literature. Am J Trop Med Hyg 69:341–343

Vishwanath S, Baker RA, Mansheim BJ (1982) *Strongyloides* infection and meningitis in an immunocompromised host. Am J Trop Med Hyg 31:857–858

Volcán GS, Medrano CE (1991) Natural infection of *Speothos venaticus* (Carnivora: Canidae) by adult stages of *Lagochilascaris* sp. Rev Inst Med Trop Sao Paulo 33:451–458 (in Spanish)

Volcan GS, Ochoa FR, Medrano CE, de Valera Y (1982) *Lagochilascaris minor* infection in Venezuela. Report of a case. Am J Trop Med Hyg 31:1111–1113

Volcan GS, Medrano CE, Payares G (1992) Experimental heteroxenous cycle of *Lagochilascaris minor* Leiper, 1909 (Nematoda: Ascarididae) in white mice and in cats. Mem Inst Oswaldo Cruz 87:525–532

Walker MD, Zunt JR (2005) Neuroparasitic infections: nematodes. Semin Neurol 25:252–261. doi: 10.1055/s-2005-917662

Wang XT, Huang HJ, Dong QQ, Lin Y, Wang ZM, Li FQ, Nawa Y, Yoshimura K (2002) A clinical study of eosinophilic meningoencephalitis caused by angiostrongyliasis. Chin Med J 115: 1312–1315

Wang LC, Wang IH, Jou JR (2006) Optic neuritis secondary to *Angiostrongylus cantonensis* infection. Ocul Immunol Inflamm 14:189–191

Wang QP, Lai DH, Zhu XQ, Chen XG, Lun ZR (2008) Human angiostrongyliasis. Lancet Infect Dis 8:621–630

Wang QP, Wu ZD, Wei J, Owen RL, Lun ZR (2012) Human *Angiostrongylus cantonensis*: an update. Eur J Clin Microbiol Infect Dis 31:389–395

Weller PF, Liu LX (1993) Eosinophilic meningitis. Semin Neurol 13:161–168

Winkel WE, Treurniet AE (1956) Infestation with *Lagochilascaris minor* (leiper) in man. Doc Med Geogr Trop 8:23–28

Wise ME, Sorvillo FJ, Shafir SC, Ash LR, Berlin OG (2005) Severe and fatal central nervous system disease in humans caused by *Baylisascaris procyonis*, the common roundworm of raccoons: a review of current literature. Microbes Infect 7:317–323

Wongkham C, Maleewong W, Ieamviteevanich K, Intapan PM, Morakote N (2000) Antigenic components of *Gnathostoma spinigerum* recognized by infected human sera by two-dimensional polyacrylamide gel electrophoresis and immunoblotting. Asian Pac J Allergy Immunol 18:47–52

Woodhall D, Starr MC, Montgomery SP, Jones JL, Lum F, Read RW, Moorthy RS (2012) Ocular toxocariasis: epidemiologic, anatomic, and therapeutic variations based on a survey of ophthalmic subspecialists. Ophthalmology 119:1211–1217

Xinou E, Lefkopoulos A, Gelagoti M, Drevelegas A, Diakou A, Milonas I, Dimitriadis AS (2003) CT and MR imaging findings in cerebral toxocaral disease. AJNR Am J Neuroradiol 24: 714–718

Yeitz JL, Gillin CM, Bildfell RJ, Debess EE (2009) Prevalence of *Baylisascaris procyonis* in raccoons (*Procyon lotor*) in Portland, Oregon, USA. J Wildl Dis 45:14–18

Yoshikawa M, Nishiofuku M, Moriya K, Ouji Y, Ishizaka S, Kasahara K, Mikasa K, Hirai T, Mizuno Y, Ogawa S, Nakamura T, Maruyama H, Akao N (2008) A familial case of visceral toxocariasis due to consumption of raw bovine liver. Parasitol Int 57:525–529

Yoshimura K, Aiba H, Oya H, Fukuda Y (1980) *Angiostrongylus cantonensis*: development following pulmonary arterial transfers into permissive and nonpermissive hosts. Exp Parasitol 49:339–352

Amoebiasis: Neurological Involvement and Neurobiology

Ruqaiyyah Siddiqui and Naveed Ahmed Khan

Abstract Granulomatous amoebic encephalitis (GAE) is a rare and severe human disease leading almost always to death. It is caused by two protist pathogens, *Acanthamoeba* and *Balamuthia mandrillaris*. Early diagnosis, followed by aggressive treatment using a combination of drugs is a pre-requisite in successful treatment, but even then the prognosis remains poor (>90 % case fatality rate). Existing drugs have limitations due to a high degree of toxicity and deleterious side effects. In addition, a major concern during the course of treatment is the ability of amoebae to transform into dormant cyst forms, which may provide resistance to drugs and allow amoebae to reactivate following completion of drug therapy, resulting in recurrence of the infection. A complete understanding of the disease, clinical symptoms, available diagnostic methods, possible therapeutic interventions and knowledge of the causative agent would undoubtedly augment our ability to counter this deadly infection.

Keywords Granulomatous amoebic encephalitis • Acanthamoeba • Balamuthia • Central nervous system • Neglected diseases

1 Introduction

Granulomatous amoebic encephalitis (GAE) is a severe human disease leading almost always to death. It is caused by two protist pathogens, *Acanthamoeba* spp. and *Balamuthia mandrillaris*. Both pathogens can produce serious cutaneous and nasopharyngeal infections and GAE which often leads to death (Martinez 1991;

R. Siddiqui • N.A. Khan (✉)
Department of Biological and Biomedical Sciences, Aga Khan University,
Stadium Road, Karachi, Pakistan
e-mail: Ruqaiyyah.Siddiqui@aku.edu; Naveed5438@gmail.com

M. Bentivoglio et al. (eds.), *Neglected Tropical Diseases and Conditions of the Nervous System*,
DOI 10.1007/978-1-4614-8100-3_5,

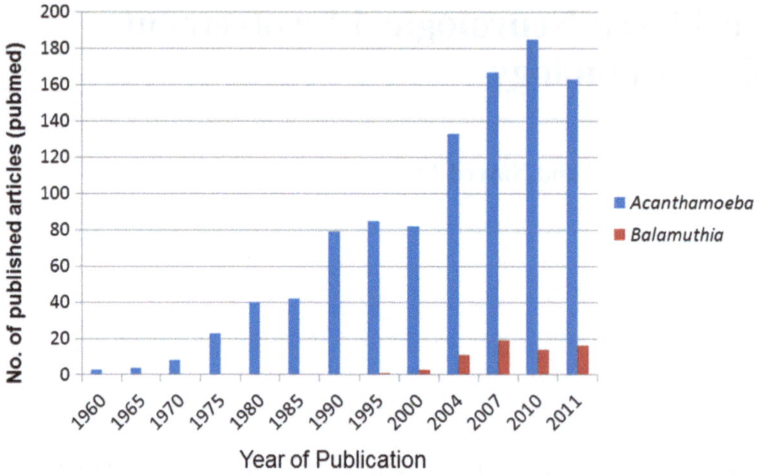

Fig. 1 Increasing scientific interest in *Acanthamoeba* and *Balamuthia mandrillaris*, causative agents of granulomatous amoebic encephalitis as determined by published articles over the last five decades. A PubMed search using the key words "*Acanthamoeba*" and "*Balamuthia*" was carried out

Martinez and Visvesvara 1997; Visvesvara et al. 2007). Over the last few years, GAE has gained increasing attention from the scientific community and health-care professionals (Fig. 1). This is due to increase in the population of susceptible hosts such as human immunodeficiency virus (HIV)-infected patients, or individuals with a weakened immune system due to excessive use of antibiotics/steroids leading to greater susceptibility to pathogenic amoebae (Marciano-Cabral and Cabral 2003; Visvesvara et al. 2007). Moreover, global warming with increased outdoor activities will further augment exposure of susceptible hosts to these pathogens (Schuster and Visvesvara 2004a, b; Khan 2009). Due to the growing HIV pandemic, it is reasonable to predict a boost in the numbers of opportunistic infections such as GAE. Despite our advances in antimicrobial chemotherapy and supportive care, the prognosis for GAE remains extremely poor with more than 90 % case fatality rate. In part, this is due to our incomplete understanding of the pathogenesis and pathophysiology of GAE infections together with difficulties in diagnosis leading to delayed chemotherapy (Khan 2006; Matin et al. 2008). Another major concern during the course of therapy is the ability of amoebae to transform into dormant cyst forms, which may provide resistance against the recommended levels of antimicrobial compounds even during the course of treatment (Ficker et al. 1990; Schuster and Visvesvara 2004a). In addition to drug resistance, cysts may also reactivate following the computation of antimicrobial chemotherapy, leading to recurrence of the disease. The purpose of this review is to present a synopsis of our current understanding of the disease and the infectious agent, diagnostic methods and possible therapeutic interventions.

2 The Infectious Agents

Both *Acanthamoeba* and *B. mandrillaris* are free-living protist pathogens (Fig. 2). *Acanthamoeba* was first observed in yeast (Cryptococcus) cultures by Castellani (1930) and Volkonsky (1931) who created the genus *Acanthamoeba* for these amoebae. The name acanth (Greek: meaning spikes) was added to these amoebae due to the presence of spine-like structures on the surface of these organisms. Later, the pathogenic potential of these organisms was demonstrated for the first time by revealing their ability to produce cytopathic effects on monkey kidney cells *in vitro* and to kill laboratory animals, i.e., monkeys *in vivo* (Culbertson et al. 1958, 1959; Jahnes et al. 1957). The first clearly identified GAE case due to *Acanthamoeba* was observed in 1972 (Jager and Stamm 1972). Soon after, *Acanthamoeba* was identified as a causative agent of blinding keratitis (Nagington et al. 1974) and then was clearly associated with Legionairre's disease by serving as a reservoir for *Legionella pneumophila* (Rowbotham 1980). *B. mandrillaris* has been discovered more recently, in 1986, from the brain of a baboon, who was a victim of meningoencephalitis, and was described as a new genus, i.e., *Balamuthia* (Visvesvara et al. 1990), and then associated with GAE in humans and animals (Anzil et al. 1991; Taratuto et al. 1991; Visvesvara et al. 1993). Since then, more than 150 cases of GAE due to *B. mandrillaris* have been identified (Visvesvara et al. 2007; Schuster et al. 2009). Based on 16S rRNA gene sequences, it is determined that *B. mandrillaris* is closely

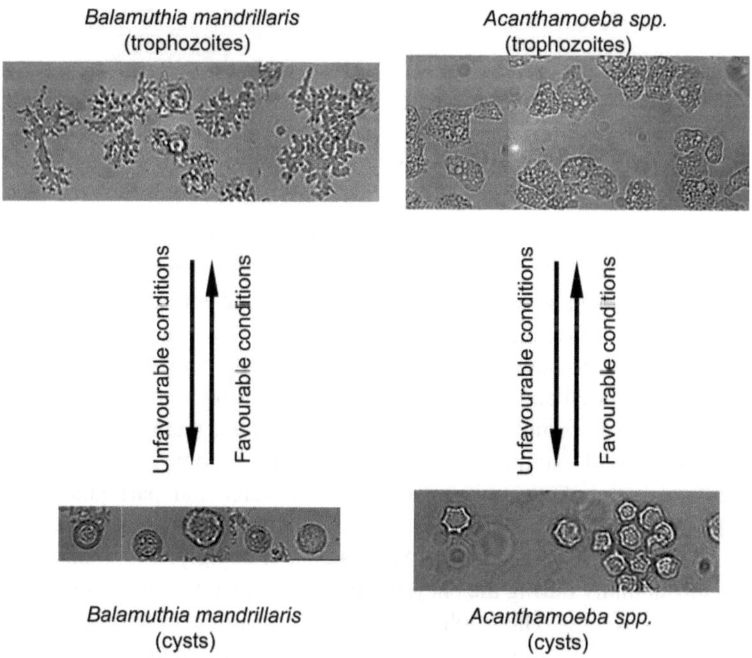

Fig. 2 The life cycle of *Acanthamoeba* and *Balamuthia mandrillaris*. The infective form of the amoebae is known as trophozoites. Under unfavorable conditions, trophozoites differentiate into cysts characterized by double wall as indicated by *arrows*. ×250

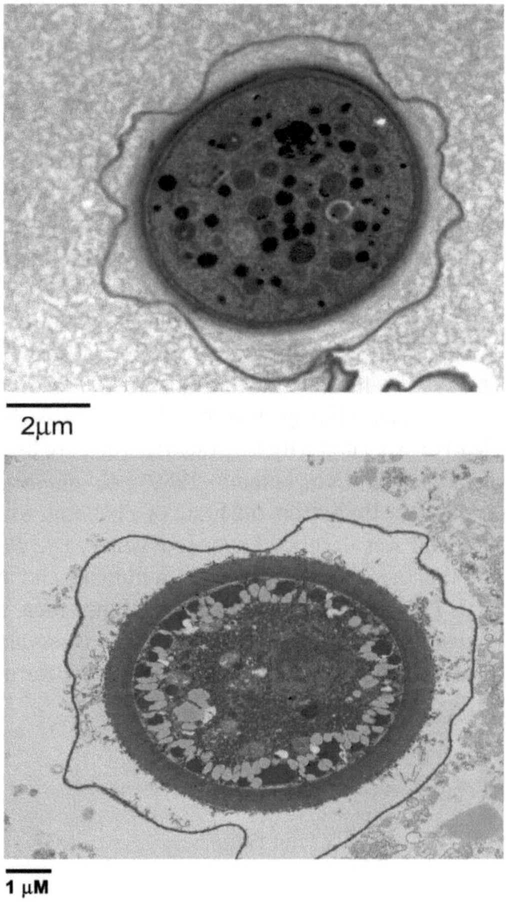

Fig. 3 Representative transmission electron micrographs showing a complete multi-walled cyst of *Acanthamoeba castellanii* belonging to the T4 genotype (**a**) and *Balamuthia mandrillaris* (**b**)

related to *Acanthamoeba* (Booton et al. 2003a, b; Matin et al. 2008; Amaral Zettler et al. 2000). The life cycle of both these pathogens undergoes two stages, a vegetative trophozoite stage and a resistant cyst stage (Figs. 2 and 3) (Matin et al. 2008; Siddiqui et al. 2012; Weisman 1976). The trophozoites are normally in the range of 15–45 μm in diameter, but their size varies significantly between different species. Cell division is asexual and occurs by binary fission. Under harsh conditions (lack of food, extremes in osmolarity, pH and temperatures), the amoebae switch into resistant cyst stage. In simple terms, the trophozoite becomes metabolically inactive (minimal metabolic activity), excess food, water and particulate matter is expelled and the trophozoite encloses itself within a resistant shell to survive harsh periods. The cellular levels of RNA, proteins, triacylglycerides and glycogen declines substantially during the encystment process resulting in decreased cellular volume and dry weight. The trophozoites emerge from the cysts under favourable conditions leaving behind the outer shell and actively reproduce, thus completing the cycle. The ability of these protists to rapidly switch phenotypes into a dormant cyst form is an important property in their defence to resist harsh environmental and physiological conditions as well as drug resistance resulting in disease recurrence.

3 Pathophysiology

The clinical picture of GAE may resemble viral, bacterial or tuberculosis meningitis. For example, the clinical symptoms involve headache, fever, behavioural changes, lethargy, stiff neck, aphasia, ataxia, nausea, cranial nerve palsies, confused state, seizures, coma, finally leading to death (Table 1). Patients exhibit haemorrhagic necrotizing lesions or brain abscess (detected by neuroimaging scans) with severe meningeal irritation and encephalitis (Martinez 1991; Martinez and Visvesvara 1997). Patients with respiratory infections, skin ulcerations or brain abscesses should be suspected for infections due to *Acanthamoeba* and *B. mandrillaris*. Post-mortem examination often shows severe oedema and haemorrhagic necrosis. It is not known whether this necrotic phase is caused by actively feeding trophozoites or inflammatory processes. The characteristic lesions in the parenchyma of GAE are most numerous in the basal ganglia, midbrain, brainstem and cerebral hemispheres. The granuloma is generally composed of macrophages, CD4+ and CD8+ T cells, B-cells and plasma cells. Granulomatous responses may

Table 1 Features of granulomatous amoebic encephalitis due to *Acanthamoeba* and *Balamuthia mandrillaris*

Pathogen	*Acanthamoeba*	*Balamuthia mandrillaris*
Cases reported worldwide	~200	~200
High-risk groups	Generally immunocompromised individuals	Immunocompetent individuals with soil exposure
Pathogenesis	Haematogenous spread from primary site of infection, followed by amoebae traversal of the blood-brain barrier and central nervous system invasion	Haematogenous spread from primary site of infection, followed by amoebae traversal of the blood-brain barrier and central nervous system invasion
Clinical features	Fever, headache, stiff neck, nausea, seizures, disorientation, visual loss, photophobia, coma	Fever, headache, stiff neck, nausea, seizures, disorientation, visual loss, photophobia, coma
Clinical diagnosis using computed tomography or magnetic resonance imaging	The brain image analyses show multiple space occupying lesions indicating brain abscess	The brain image analyses show cerebral edema, hydrocephalus, and multiple space occupying lesions
Laboratory diagnosis	PCR on CSF or biopsy, immunofluorescence assays	PCR on CSF or biopsy, immunofluorescence assays
Treatment	Ketoconazole, fluconazole, itraconazole, pentamidine isethionate, flucytosine, trimethoprim/sulfamethoxazole and rifampin Experimental: miltefosine	Ketoconazole, fluconazole, albendazole, itraconazole, sulfadiazine, pentamidine isethionate, flucytosine, clarithromycin and azithromycin Experimental: phenothiazines-thioridazine, trifluoperazine
Case fatality rate	>90 %	>90 %

be absent or minimal in patients with severe impairment of the cellular immune response (Khan 2009; Martinez and Visvesvara 1997; Visvesvara et al. 2007). The affected organs other than the CNS may include subcutaneous tissue, skin, liver, lungs, kidneys, pancreas, prostate, lymph nodes and bone marrow.

4 Predisposing Factors

The predisposing factors in contracting GAE are not clearly understood. Although the majority of GAE cases due to *Acanthamoeba* are limited to individuals with a weakened immune system, *B. mandrillaris* infections have been shown to occur also in immunocompetent people (Table 1). Thus, GAE can develop in individuals with no history of syphilis, diabetes mellitus, malignancies, fungal and mycobacterial infections and in patients negative for HIV-1 and HIV-2. However, patients suffering from these as well as other diseases including lymphoproliferative disorders, haematologic disorders, pneumonitis, renal failure, liver cirrhosis, rhinitis, pharyngitis, gammaglobulinemia, systemic lupus erythematosus, glucose-6-phosphate deficiency, tuberculosis, chronically alcoholic, radiotherapy, malnourished, chronically ill, or debilitated, are particularly at risk. Patients undergoing organ/tissue transplantation with immunosuppressive therapy, steroids and excessive antibiotics are also at risk of contracting GAE (Fig. 4). Although the reported number of GAE cases is a few hundred, the burden of GAE worldwide will never be known; this is due to lack of awareness of the disease and/or unavailability of diagnostic methods, and poor public health systems especially in the less developed countries.

Unlike *Acanthamoeba*, that is generally limited to immunocompromised patients, GAE due to *B. mandrillaris* can occur in healthy people of any age, though there is a predominance of cases in the young (under 15 years of age) and the elderly (over 60 years of age), probably due to their weaker immune systems (Maciver 2007). The rarity of the disease suggests the presence of predisposing factors. Notably, GAE due to *B. mandrillaris* has been predominantly reported in individuals of Hispanic origin (Siddiqui and Khan 2008; Diaz 2011). In support, it is shown that Hispanics are less able to make antibodies against certain *Acanthamoeba* species (Chappell et al. 2001) and this may be the case also for *B. mandrillaris*. The genetic predisposition of Hispanics to *B. mandrillaris* may play a role in contracting GAE, but it needs further investigation as GAE due to *B. mandrillaris* is reported to occur in a range of mammals including mandrill baboons, monkeys, gibbons, gorillas, sheep, dogs and horses with similar disease presentation to humans (Visvesvara et al. 2007). Other predisposing factors may include working with organic-rich soil (e.g., during agricultural activities), which may explain large number of cases in Hispanics, who are the major workforce in Southern California (Schuster and Visvesvara 2004a, b). Temperature may also be an important factor as the disease seems to be more common in warmer regions, such as Southern California and South America (Deetz et al. 2003; Siddiqui and Khan 2008; Schuster and Visvesvara 2004a, b).

Overall, it is widely accepted that GAE due to free-living protists, *Acanthamoeba* and in particular *B. mandrillaris* can occur in healthy individuals, but

a Host susceptibility factors

- Weak immune system
- HIV-1 and HIV-2
- Organ/tissue transplantation with immunosuppressive therapy
- Steroids and excessive use of antibiotics
- Diabetes mellitus
- Malignancies
- Fungal and mycobacterial infections
- Lymphoproliferative disorders
- Hematologic disorders
- Pneumonitis
- Renal failure
- Liver cirrhosis
- Rhinitis
- Pharyngitis
- Gammalobulinemia
- Systemic lupus erythematosus
- Glucose-6-phosphate deficiency
- Chronically alcoholic

b Environmental factors

- Working in soil/garden
- Swimming in pools (especially unchlorinated)
- Outdoor activities involving water (lakes, ponds)
- Road accidents that result in skin lesions
- Exposure of skin lesions to soil, contaminated water and other objects

Fig. 4 Risk factors associated with granulomatous amoebic encephalitis

immunocompromised or debilitated patients due to HIV infection, diabetes, immunosuppressive therapy, malignancies, malnutrition, and alcoholism are particularly at risk (Schuster and Visvesvara 2004a, b; Visvesvara et al. 2007; Siddiqui and Khan 2008). The risk factors for patients suffering from the above diseases include exposure to contaminated water such as swimming pools, on beaches, or working with soil (garden/compost/agriculture). Future studies will determine the precise host and environmental factors contributing to this fatal infection, which may help design preventive strategies and identify the susceptible populations.

5 Pathogenesis

Although GAE is a rare infection, it almost always proves fatal as mentioned above. The mechanisms associated with pathogenesis of GAE remain unclear; however the pathophysiological complications involving the CNS most likely include the invasion of parasites across the blood-brain barrier (BBB), induction of proinflammatory responses, and neuronal damage leading to brain dysfunction. The routes of entry of the amoeba into the body include the lower respiratory tract leading to

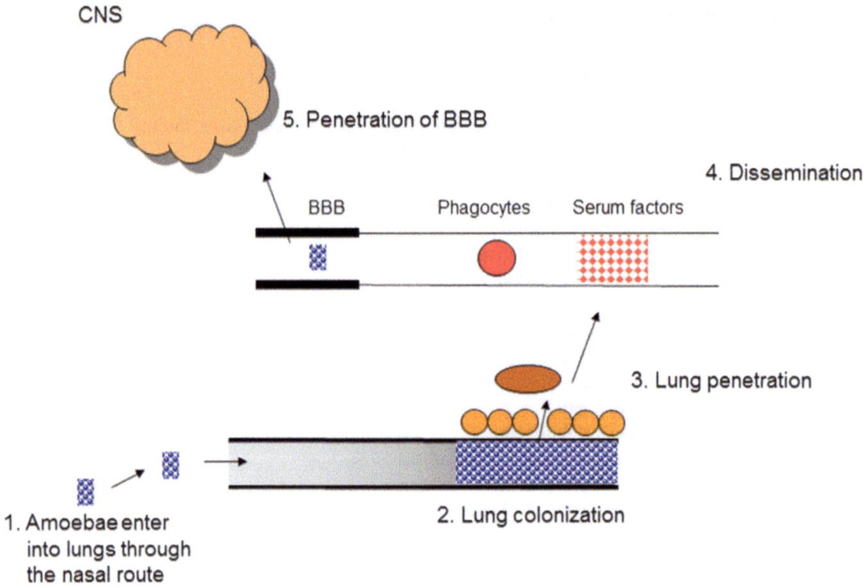

Fig. 5 The model of granulomatous amoebic encephalitis. Amoebae enter lungs via the nasal route. Next, amoebae traverse the lungs into the bloodstream (or directly through skin lesions), followed by haematogenous spread. Finally, amoebae cross the blood-brain barrier and enter into the central nervous system (CNS) to produce disease. The olfactory neuroepithelium may provide an alternative route of entry into the CNS

amoebae invasion of the intravascular space followed by haematogenous spread (Martinez 1991). Skin ulcerations may provide direct amoebae entry into the bloodstream (Fig. 5).

Amoebae entry into the CNS occurs most likely across the BBB (Martinez 1991; Martinez and Visvesvara 1997). The olfactory neuroepithelium route (i.e., invasion of the olfactory part of the nasal epithelium, and migration along nerve fibers, followed by invasion of the olfactory bulb) provides another route of entry into the CNS and has been studied in experimental models (Martinez 1991; Martinez and Visvesvara 1997; Kiderlen and Laube 2004). The cutaneous and/or nasopharyngeal infections can last for months but the involvement of the CNS can result in death within days (Martinez and Visvesvara 1997; Schuster and Visvesvara 2004a, b). Crossing the BBB is a multifactorial process involving pathogen determinants (adhesins, proteases, ecto-ATPases, phospholipases) that allow amoebae to evade the immune system to target the neuronal tissue as well as the host immune responses (interleukin-beta, interleukin-alpha, tumor necrosis factor-alpha, interferon-gamma, host cell apoptosis) that result in the host cell/tissue injury (Table 2). Both *Acanthamoeba* and *B. mandrillaris* are extracellular pathogens and are thus exposed to the host immune system. The immune response plays an active role in protection against these pathogens, as well as contributing to BBB perturbations and disease development. For example, characteristic granulomatous lesions in the CNS are a result of the host immune response and are composed of macrophages, T and B cells

Table 2 Virulence factors and their potential role in the pathogenesis of granulomatous amoebic encephalitis

Pathogen determinant	Role	Reference
Acanthamoeba		
Mannose-binding protein	Cytoadhesion	Garate et al. (2004), Alsam et al. (2003)
Serine proteases	Paracellular transmigration of the blood-brain barrier; Degradation of extracellular matrix membranes; Immune evasion (degradation of antibodies/complement proteins/cytokines)	Alfieri et al. (2000), Mitro et al. (1994), Alsam et al. (2005), Khan and Siddiqui (2009), Benedetto et al. (2003), Marciano-Cabral et al. (2004)
Ecto-ATPases	?Host cell cytotoxicity	Mattana et al. (1997, 2002, 2009), Sissons et al. (2004)
Phospholipases	Unclear	Mortazavi et al. (2011), Mishra et al. (1985)
Balamuthia mandrillaris		
Galactose-binding protein	Cytoadhesion	Matin et al. (2007), Rocha-Azevedo et al. (2007)
Metalloproteases	Paracellular transmigration of the blood-brain barrier; Degradation of extracellular matrix membranes; ?Immune evasion (degradation of antibodies/complement proteins)	Garate et al. (2004), Alsam et al. (2003)
Ecto-ATPases	?Host cell cytotoxicity	Matin and Khan (2008)
Phospholipases	?Endothelial injury	Shadrach et al. (2004), Haider (2007)

as described above, and amoebae. The localization of immune cells in the brain suggests the involvement of pro-inflammatory cytokines in protection as well as in the pathophysiology of complications. The overall outcome is increased permeability and/or apoptosis of the brain microvascular endothelial cells, which promote BBB disruption leading to amoebae invasion of the CNS. Both the parasite and host factors are thought to be important in BBB dysfunction but the precise molecular mechanisms are unclear and should be the subject of future studies. The understanding of cross-talk between amoebae-host interactions as well as between BBB and the CNS in disease will provide insights into the disease neuropathogenesis and may help develop novel therapeutic interventions.

6 Clinical Diagnosis of the GAE

GAE is an infection with a case fatality rate of more than 90 %. Those cases who survived were diagnosed early, followed by aggressive treatment, which led to successful outcome for these patients. However, the symptoms of GAE can be similar to those caused by other CNS pathogens including virus, bacteria and fungi. This makes GAE diagnosis problematic and requires awareness and solid expertise.

Neurological signs are grouped into four categories; (1) impaired consciousness: ranging from confusion to unconsciousness, (2) loss of reflex activity, (3) abnormal speech, e.g. aphasia, and (4) abnormal motor patterns: imbalance or unsteady gait, seizure activities. The advanced stage of the disease is irreversible and includes loss of consciousness, seizures and coma, finally leading to death. Post-mortem examination often shows severe oedema and haemorrhagic necrosis (Martinez 1991; Martinez and Visvesvara 1997; Kiderlen and Laube 2004). Brain image analyses using computed tomography (CT) or magnetic resonance imaging (MRI) may show multifocal areas of altered signal intensities or lesions simulating brain abscesses or tumours. The cerebrospinal fluid findings, although not confirmatory of GAE, are of value in diagnosing CNS involvement. Pleocytosis with lymphocytic predominance is an important feature with elevated polymorphonuclear leukocytes, increased protein concentrations, decreased glucose concentration and slight cloudiness (Martinez and Visvesvara 1997; Schuster and Visvesvara 2004a, b).

The absence of viral and bacterial pathogens should raise the suspicion of GAE. Due to the low density of parasites, the detection of host immune response parameters should be attempted primarily. The demonstration of high levels of *Acanthamoeba*- or *B. mandrillaris*-specific antibodies in the patient's serum may provide a useful and straightforward method to further suspect GAE infection. This is performed using indirect immunofluorescence (IIF) assays. The serial dilutions of the patient's serum are incubated with fixed amoebae-coated slides, followed by incubation with fluorescein isothiocyanate (FITC)-labelled anti-human antibody and visualized under fluorescence microscopy. It is important to remember that the levels of anti-amoebae antibodies in normal populations may be in the range of 1:20–1:60. However, patients with severely impaired immune system may not develop a high titre, thus other clinical findings should be taken into account for accurate diagnosis (Visvesvara et al. 2007). The confirmatory evidence comes from direct microscopic observation of amoebae in the cerebrospinal fluid (after centrifugation at low speed) or in the brain biopsy but requires familiarity of morphological characters. Giemsa-Wright, acridine orange or calcofluor white staining may facilitate morphology-based positive identification of these amoebae. The lack of familiarity with amoebae morphological characteristics may require immunohistochemical studies using antisera made against *Acanthamoeba* and *B. mandrillaris* as aids in the clinical diagnosis of GAE. In addition, it is helpful to inoculate a portion of the cerebrospinal fluid and/or brain biopsy for amoebae culturing. For *Acanthamoeba*, the clinical specimens can be inoculated onto non-nutrient agar plate seeded with Gram-negative bacteria. *Acanthamoeba* feeds on bacteria as food source, and depending on the number of amoebae in the specimen, trophozoites can be observed within a few hours (up to 12 h). However in the absence of amoebae, plates should be monitored for up to 7 days. This method is particularly useful if problems are encountered in differentiating *Acanthamoeba* from monocytes, polymorphonuclear leukocytes and macrophages. *Balamuthia mandrillaris* do not feed on Gram-negative bacteria and should be inoculated onto mammalian cell cultures, but this procedure may require up to several weeks. Alternatively, PCR-based methods have been developed for the clinical diagnosis of amoebae infections and provide rapid diagnosis from clinical specimens (Booton et al. 2003a, b; Jayasekera et al. 2004).

7 Treatment

At present, there is no recommended treatment for GAE and the majority of cases are identified post-mortem. This is possibly due to delayed diagnosis, low sensitivity of amoebae to anti-amoebic agents, and limited efficacy of anti-amoebic agents to cross the BBB and enter the CNS. Current therapeutic agents include a combination of drugs as listed in Table 1, while some have shown promise in experimental studies (Table 1) (Diaz 2011; Kotting et al. 1992; Walochnik 2002). Miltefosine is an alkylphosphocholine drug previously used for treatment against protist diseases, such as leishmaniasis, is an enzyme inhibitor and is well absorbed in humans after oral intake. However, even with treatment, survivors may develop disability such as hearing loss, vision impairment, etc.

To date, there have only been three reported cases of GAE due to *B. mandrillaris* who fully recovered with antimicrobial chemotherapy: a 64-year old male, a 5-year old female and a 72-year old female (Deetz et al. 2003). The 64-year old male was treated with 5-fluorocytosine (flucytosine), fluconazole, sulfadiazine, clarithromycin and trifluoperazine, while the 5-year old female survived after treatment with a similar regimen but without sulfadiazine. In both cases, pentamidine isethionate was also administered initially, but its use was discontinued due to side effects (Deetz et al. 2003). The 72-year old female was treated with sulfadiazine, fluconazole and clarithromycin as well as pentamidine isethionate. Despite the limited success, the prognosis for GAE remains extremely poor. This may be due to variability in the virulence and antimicrobial sensitivity of amoebae isolates causing the infection, the time of drug treatment initiation, infectious load of amoebae, and the limited ability of antimicrobial compounds to cross the BBB (Schuster and Visvesvara 2004a, b). Of interest, the above-mentioned miltefosine (hexadecylphosphocholine) has shown promise against GAE as it can cross the BBB (Schuster and Visvesvara 2004a, b) and exhibits anti-*Acanthamoeba* properties (Kotting et al. 1992; Walochnik et al. 2002).

For GAE due to *Acanthamoeba*, current therapeutic agents include a combination of ketoconazole, fluconazole, sulfadiazine, pentamidine isethionate, amphotericin B, azithromycin, itraconazole, or rifampin that prove effective but with side-effects. For example, ketoconazole and rifampin added to trimethoprim-sulfamethoxazole therapy was used for the treatment of two immunocompetent pediatric patients with GAE due to *Acanthamoeba*, who recovered fully (Singhal et al. 2001). Surgical removal of one localized CNS lesion followed by therapy with fluconazole and sulfadiazine resulted in effective treatment and survival of the GAE patient (Seijo Martinez et al. 2000).

Many of the aforementioned compounds target the ergosterol biosynthesis pathway (Fig. 6). Ergosterol is an important part of fungal membranes, so that its disruption leads to defective membrane function, increased permeability and leakage of ions from the cell. Ergosterol has been shown to be present in the *Acanthamoeba* cell membrane (Smith and Korn 1968). Given that *Balamuthia* is a close relative of *Acanthamoeba*, it is likely that ergosterol is a sterol membrane component of *B. mandrillaris*. This is consistent with previous studies showing that ketoconazole exhibited growth inhibitory effects against *B. mandrillaris* (Schuster and Visvesvara 1996)

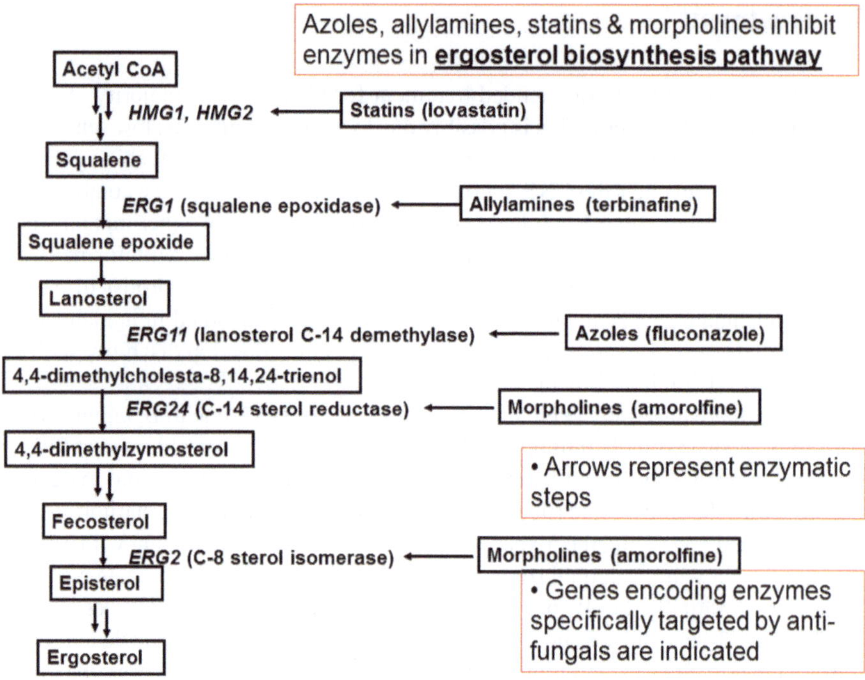

Fig. 6 Ergosterol biosynthesis pathway. *ERG1* codes for squalene epoxidase and allylamines inhibit squalene epoxidase. Allylamines inhibit the enzyme squalene epoxidase, which is another enzyme required for ergosterol synthesis. The azoles inhibit the P450 enzymes responsible for the synthesis of ergosterol in the cell membrane. The azoles act through an unhindered nitrogen, which binds to the iron atom of the heme, preventing the activation of oxygen which is necessary for the demethylation of lanosterol. In addition to the unhindered nitrogen, a second nitrogen in the azoles is thought to interact directly with the apoprotein of lanosterol demethylase. It is thought that the position of this second nitrogen in relation to the apoprotein may determine the specificity of different azole drugs for the enzyme. The resulting depletion of ergosterol alters the fluidity of the membrane and this interferes with the action of membrane-associated enzymes. The overall effect is an inhibition of replication (i.e., the azoles are static drugs). A further repercussion is the inhibition of differentiation of cells (e.g., trophozoites to cysts and vice versa). Since no drug acts with complete specificity, it is not surprising that the azoles also have some effect on the closely related mammalian P450 enzymes

and blocked *B. mandrillaris*-mediated toxicity of human brain microvascular endothelial cells (Siddiqui et al. 2007). Notably, ergosterol biosynthesis is limited to fungal and protists, as human cells contain cholesterol, thus ergosterol biosynthesis can be exploited to design selective compounds to interfere with the biological processes of *Acanthamoeba* and *B. mandrillaris* without harming the host cells. However, the present azole compounds lack complete specificity and also affect the closely related mammalian P450 enzymes resulting in side effects. Another limitation in the use of azole compounds is that their effects are generally considered to be amoebistatic (Schuster and Visvesvara 1996), indicating the need for their prolonged clinical applications, which may lead to the emergence of resistant strains. For example, fungi have been shown to develop azole resistance by increasing activity of efflux

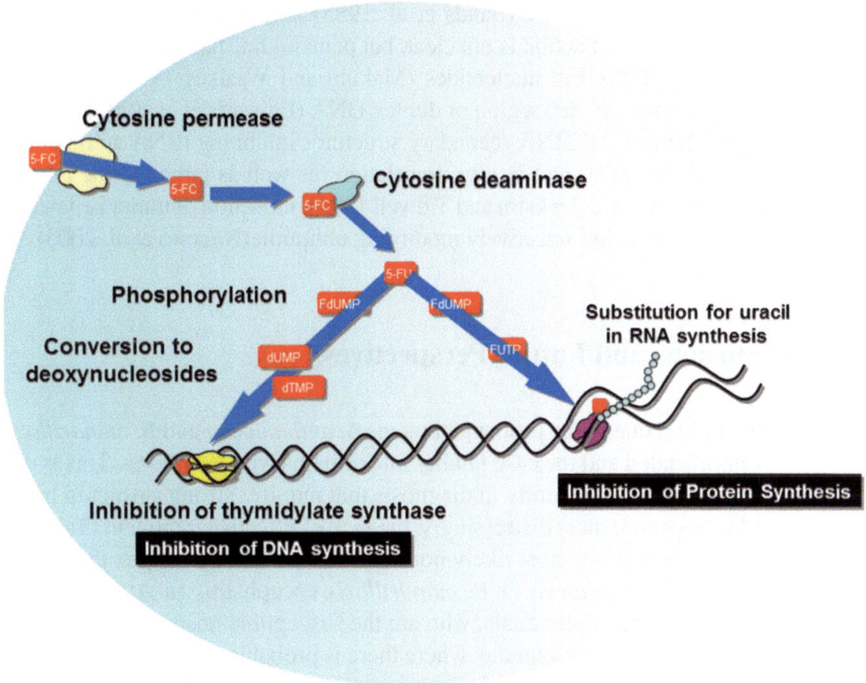

Fig. 7 5-Fluorocytosine (flucytosine), a pyrimidine analogue is taken up by the cell and converted into toxic 5-fluorouracil and 5-fluorodeoxyuracil. Incorporation of 5-fluorouracil into RNA and 5-fluorodeoxyuracil into DNA would lead to errors both at the transcriptional- and translational-level resulting in abnormal proteins. *5-FC* 5-fluorocytosine, *5-FU* 5-fluorouracil, *FdUMP* 5-fluorodeoxyuridine, *FUMP* 5-fluorouridine monophosphate, *FUDP* 5-fluorouridine diphosphate, *FUTP* 5-fluorouridine triphosphate, *dUMP* deoxyuridine monophosphate, *dTMP* deoxythymidine monophosphate

mechanisms to rid intracellular toxins or by inducing changes in the azole targets, i.e., C14 demethylase, or changes in steps of ergosterol biosynthesis. Similar mechanisms may explain variations in antimicrobial sensitivity among various isolates of *Acanthamoeba* and *B. mandrillaris* (Schuster and Visvesvara 2004a). With this in mind, the limited available drugs in the treatment of GAE chemotherapy should be used with caution. Overall, the ergosterol biosynthesis pathway is a potential target in the rational development of targeted therapeutic interventions against fatal GAE due to *Acanthamoeba* and *B. mandrillaris* infections.

Other compounds include 5-fluorocytosine (flucytosine), a pyrimidine analogue. It's a prodrug which is converted into toxic 5-fluorouracil and 5-fluorodeoxyuracil. Incorporation of 5-fluorouracil into RNA and 5-fluorodeoxyuracil into DNA would lead to errors both at the transcriptional- and translational-level resulting in abnormal proteins (Fig. 7), however the therapeutic use of flucytosine is questionable due to its side-effects.

Pentamidine is an aromatic diamidine that is selectively accumulated by the pathogen, rather than the host cell, and used against many protist pathogens. Despite its broad profile of antimicrobial activity, the use of this drug is limited because of its side effects, including nephrotoxicity (Sands et al. 1985) and poor penetration into the CNS. The precise mode of action is not clear, but pentamidine has been found to precipitate with DNA, RNA and nucleotides (Makulu and Waalkes 1975). This drug selectively binds to the AT-rich region of duplex DNA (Edwards et al. 1992), inhibits RNA function, disrupts the tRNA secondary structures inhibiting tRNA aminoacylation (Sun and Zhang 2008) and *in vitro* translation (as well as targeting the *in vitro* activities of topoisomerase (Dykstra and Tidwell 1991), induction of tumor cell necrosis (Pathak et al. 2002), and selectively modifying ubiquitin (Nguewa et al. 2005).

8 Conclusions and Future Perspectives

The burden of GAE due to the protist pathogens *Acanthamoeba* and *B. mandrillaris* is not truly appreciated and the case fatality rate remains well over 90 %. This is due to the lack of awareness, difficulty in diagnosis that requires strong suspicion based on clinical findings and, most distressingly, the lack of effective treatment (Table 3). Numerous GAE cases have most likely not been detected. For example, there is no single report of *Acanthamoeba* or *B. mandrillaris* encephalitis in Africa, despite millions of HIV-infected individuals, who are the susceptible hosts to opportunistic pathogens and live in a warm climate, where there is probable frequent environmental exposure and subordinate sanitation in many places. With the HIV pandemic and other contributing factors, there is a clear increase in the number of susceptible hosts presenting a threat of free-living amoebae to human and animal health. Of particular concern is GAE due to *B. mandrillaris* that has been reported in immunocompromised as well as immunocompetent individuals with few predisposing factors. In addition, warmer climates will undoubtedly add to the increased ubiquity of these pathogens and thus increased exposure to the susceptible population. There is as urgent need for a complete understanding of the pathogenesis and pathophysiology of this disease *in vivo*. In particular, the precise identification of the molecular events involved in pathogen-host cell interactions leading to amoebae translocation of the BBB using *in vitro* and *in vivo* models as well as the selectivity of neuronal tissue should help develop therapeutic interventions against specific targets and/or design new strategies for disease prevention. Understanding of the metabolic pathways in amoebae pathogens will yield information of their ability to encyst and excyst, which is crucial for the development of effective treatment strategies. Identification of anti-amoebic compounds with minimal side effects that can be included in the highly active antiretroviral therapy (HAART), may offer a viable prophylactic strategy against GAE in the highly susceptible population. Additionally, natural products hold promise in the identification of novel anti-amoebal agents against the GAE, however their efficacy to cross the BBB and target pathogens in the CNS must be demonstrated *in vitro* and *in vivo*. Notably, the selectivity of the BBB has led to the ineffectiveness of many of the available drugs to target amoebae in the neuronal tissue, thus contributing to more than 90 % mortality in GAE cases. In addition to intravenous injections and as a last resort, drugs may be administered

Table 3 Challenges in therapy of GAE due to *Acanthamoeba* and *Balamuthia mandrillaris*

- Delayed diagnosis
- Low susceptibility or resistance to amoebae to available drugs
- Selectivity of the blood-brain barrier that impedes drug penetration into the central nervous system to target pathogens
- Lack of specificity/side effects of available drugs
- Treatment costs
- Inadequate knowledge of specific targets to design therapeutic interventions

as (1) trans-cranial drug delivery (intracerebroventricular injection, intracerebral injection or convection-enhanced diffusion), (2) trans-nasal drug delivery to the brain using lipid-soluble small molecules that can be instilled nasally, cross the nasal mucosa and the arachnoid membrane and enter the cerebrospinal fluid with results similar to intracerebroventricular injection), and (3) transient BBB disruption (arterial infusion of hyperosmotic solution or ultrasonic irradiation of the brain). All of the above methods may have complications due to side effects. Other long term approaches include lipidization of small molecules to enhance transport across the BBB, carrier-mediated transport of drugs, or administration of non-viral plasmid DNA encoding antisense RNA against the virulence genes of amoebae but their effectiveness against GAE require further investigations.

There is also a need for educational efforts aimed at neurologists, and other physicians to raise awareness of brain infections due to free-living amoebae to help diagnose such infections at an early stage. The development and availability of rapid, cost-effective, molecular-based assays (antibody, antigen detection and PCR) in diagnostic laboratories with demonstrable efficacy in clinical specimens (tissue biopsies, blood, cerebrospinal fluid) is critical in the early diagnosis of CNS infections due to free-living amoebae. The usefulness of sputum samples as a diagnostic tool for disseminated amoebiasis and GAE needs investigation.

Epidemiology studies are needed to determine the true burden of GAE infections, both in humans and animals and whether disease patterns differ around the globe due to environmental factors (high temperature, low humidity, dust, etc.), demographic factors, socioeconomic factors, and malnutrition. Assessment of the host factors of GAE patients, including hygiene, immunity (type and net state of immune-suppression, or type of transplant, if applicable), nutrition, occupation and underlying disease will identify most likely carriers of free-living amoebae and their physiological state.

Acknowledgements The authors are grateful to Junaid Iqbal, Yousef Aqeel, Department of Biological and Biomedical Sciences, Aga Khan University, Karachi, Pakistan; Selwa Alsam, James Sissons, Ricky Dudley, Samantha Jayasekera and Abdul Matin, School of Biological and Chemical Sciences, Birkbeck, University of London, England, UK; Kwang Sik Kim and Monique Stins, Division of Infectious Diseases, Johns Hopkins University School of Medicine, Baltimore, MD, USA. This work was supported by grants from the Aga Khan University, University of London, University of Nottingham, The Nuffield Foundation, and The Royal Society.

References

Alfieri SC, Correia CE, Motegi SA, Pral EM (2000) Proteinase activities in total extracts and in medium conditioned by *Acanthamoeba* polyphaga trophozoites. J Parasitol 86:220–227
Alsam S, Kim KS, Stins M, Rivas AO, Sissons J, Khan NA (2003) *Acanthamoeba* interactions with human brain microvascular endothelial cells. Microb Pathog 35:235–241
Alsam S, Sissons J, Jayasekera S, Khan NA (2005) Extracellular proteases of *Acanthamoeba castellanii* (encephalitis isolate belonging to T1 genotype) contribute to increased permeability in an in vitro model of the human blood-brain barrier. J Infect 51(2):150–156
Amaral Zettler LA, Nerad TA, O'Kelly CJ, Peglar MT, Gillevet PM, Silberman JD, Sogin ML (2000) A molecular reassessment of the leptomyxid amoebae. Protist 151:275–282
Anzil AP, Rao C, Wrzolek MA, Visvesvara GS, Sher JH, Kozlowski PB (1991) Amebic meningoencephalitis in a patient with AIDS caused by a newly recognized opportunistic pathogen. *Leptomyxid* ameba. Arch Pathol Lab Med 115:21–25
Benedetto N, Rossano F, Gorga F, Folgore A, Rao M, Romano Carratelli C (2003) Defense mechanisms of IFN-gamma and LPS-primed murine microglia against *Acanthamoeba castellanii* infection. Int Immunopharmacol 3:825–834
Booton GC, Carmichael JR, Vivesvara GS, Byers TJ, Fuerst PA (2003a) Identification of *Balamuthia mandrillaris* by PCR assay using the mitochondrial 16S rRNA gene as a target. J Clin Microbiol 41:453–455
Booton GC, Carmichael JR, Vivesvara GS, Byers TJ, Fuerst PA (2003b) Genotyping of *Balamuthia mandrillaris* based on nuclear 18S and mitochondrial 16S rRNA genes. Am J Trop Med Hyg 68:65–69
Castellani A (1930) An amoeba found in culture of yeast: preliminary note. J Trop Med Hyg 33:160
Chappell CL, Wright JA, Coletta M, Newsome AL (2001) Standardized method of measuring *Acanthamoeba* antibodies in sera from healthy human subjects. Clin Diagn Lab Immunol 8:724–730
Culbertson CG, Smith JW, Miner JR (1958) *Acanthamoeba*: observation on animal pathogenicity. Science 127:1506
Culbertson CG, Smith JW, Cohen HK, Miner JR (1959) Experimental infection of mice and monkeys by *Acanthamoeba*. Am J Pathol 35:185–197
Deetz TR, Sawyer MH, Billman G, Schuster FL, Visvesvara GS (2003) Successful treatment of *Balamuthia* amoebic encephalitis: presentation of 2 cases. Clin Infect Dis 37:1304–1312
Diaz JH (2011) The public health threat from *Balamuthia mandrillaris* in the southern United States. J La State Med Soc 163:197–204
Dykstra CC, Tidwell RR (1991) Inhibition of topoisomerases from *Pneumocystis carinii* by aromatic dicationic molecules. J Protozool 38:78S–81S
Edwards KJ, Jenkins TC, Neidle S (1992) Crystal structure of a pentamidine-oligonucleotide complex: implications for DNA-binding properties. Biochemistry 31:7104–7109
Ficker L, Seal D, Warhurst D, Wright P (1990) *Acanthamoeba* keratitis: resistance to medical therapy. Eye 4:835–838
Garate M, Cao Z, Bateman E, Panjwani N (2004) Cloning and characterization of a novel mannosebinding protein of *Acanthamoeba*. J Biol Chem 279:29849–29856
Haider R (2007) Colourimertic determination of phospholipase activities in *Balamuthia mandrillaris*. Am J Biochem Biotechnol 3:171–179
Jager BV, Stamm WP (1972) Brain abscesses caused by free-living amoeba probably of the genus *Hartmannella* in a patient with Hodgkin's disease. Lancet 2:1343–1345
Jahnes WG, Fullmer HM, Li CP (1957) Free-living amoebae as contaminants in monkey kidney tissue culture. Proc Soc Exp Biol Med 96:484–488
Jayasekera S, Sissons J, Tucker J, Rogers C, Nolder D, Warhurst D, Alsam S, White JML, Higgins EM, Khan NA (2004) Post-mortem culture of *Balamuthia mandrillaris* from the brain and cerebrospinal fluid of a case of granulomatous amoebic encephalitis, using human brain microvascular endothelial cells. J Med Microbiol 53:1007–1012

Khan NA (2006) *Acanthamoeba*: biology and increasing importance in human health. FEMS Microbiol Rev 30:564–595
Khan NA (2009) *Acanthamoeba*: biology and pathogenesis. Caister Academic, Norfolk, UK
Khan NA, Siddiqui R (2009) *Acanthamoeba* affects the integrity of human brain microvascular endothelial cells and degrades the tight junction proteins. Int J Parasitol 39(14):1611–1616
Kiderlen AF, Laube U (2004) *Balamuthia mandrillaris*, an opportunistic agent of granulomatous amebic encephalitis, infects the brain via the olfactory nerve pathway. Parasitol Res 94:49–52
Kotting J, Berger MR, Unger C, Eibl H (1992) Alkylphosphocholines: influence of structural variations on biodistribution of antineoplastically active concentrations. Cancer Chemother Pharmacol 30:105–112
Maciver SK (2007) The threat from *Balamuthia mandrillaris*. J Med Microbiol 56:1–3
Makulu DR, Waalkes TP (1975) Interaction between aromatic diamidines and nucleic acids: possible implications for chemotherapy. J Natl Cancer Inst 54:305–309
Marciano-Cabral F, Cabral G (2003) *Acanthamoeba* spp. as agents of disease in humans. Clin Microbiol Rev 16:273–307
Marciano-Cabral F, Ludwick C, Puffenbarger RA, Cabral GA (2004) Differential stimulation of microglial pro-inflammatory cytokines by *Acanthamoeba culbertsoni* versus *Acanthamoeba castellanii*. J Eukaryot Microbiol 51:472–479
Martinez AJ (1991) Infections of the central nervous system due to *Acanthamoeba*. Rev Infect Dis 13:S399–S402
Martinez AJ, Visvesvara GS (1997) Free-living, amphizoic and opportunistic amebas. Brain Pathol 7:583–598
Matin A, Khan NA (2008) Demonstration and partial characterization of ecto-ATPase in *Balamuthia mandrillaris* and its possible role in the host-cell interactions. Lett Appl Microbiol 47:348–354
Matin A, Siddiqui R, Jayasekera S, Khan NA (2008) Increasing importance of Balamuthia mandrillaris. Clin Microbiol Rev 21:435–448
Matin A, Siddiqui R, Jung SY, Kim KS, Stins M, Khan NA (2007) Balamuthia mandrillaris interactions with human brain microvascular endothelial cells in vitro. J Med Microbiol 56(Pt 8):1110–1115
Mattana A, Alberti L, Delogu G, Fiori PL, Cappuccinelli P (2009) In vitro activityof *Acanthamoeba castellanii* on human platelets and erythrocytes. Infect Immun 77:733–738
Mattana A, Bennardini F, Usai S, Fiori PL, Franconi F, Cappuccinelli P (1997) *Acanthamoeba castellanii* metabolites increase the intracellular calcium level and cause cytotoxicity in wish cells. Microb Pathog 23(2):85–93
Mattana A, Cappai V, Alberti L, Serra C, Fiori PL, Cappuccinelli P (2002) ADP and other metabolites released from *Acanthamoeba castellanii* lead to human monocytic cell death through apoptosis and stimulate the secretion of proinflammatory cytokines. Infect Immun 70:4424–4432
Mishra SK, Jaiswal AK, Garg NK (1985) Purification and properties of phospholipase A2 of *Acanthamoeba culbertsoni*. Indian J Biochem Biophys 22:344–349
Mitro K, Bhagavathiammai A, Zhou OM, Bobbett G, McKerrow JH, Chokshi R, Chokshi B, James ER (1994) Partial characterization of the proteolytic secretions of *Acanthamoeba* polyphaga. Exp Parasitol 78:377–385
Mortazavi PN, Keisary E, Loh LN, Jung SY, Khan NA (2011) Possible roles of phospholipase A(2) in the biological activities of *Acanthamoeba castellanii* (T4 Genotype). Protist 161:168–176
Nagington J, Watson PG, Playfair TJ, McGill J, Jones BR, Steele AD (1974) Amoebic infections of the eye. Lancet 2:1537–1540
Nguewa PA, Fuertes MA, Cepeda V, Iborra S, Carrion J, Valladares B, Alonso C, Perez JM (2005) Pentamidine is an antiparasitic and apoptotic drug that selectively modifies ubiquitin. Chem Biodivers 2:1387–1400
Pathak MK, Dhawan D, Lindner DJ, Borden EC, Farver C, Yi T (2002) Pentamidine is an inhibitor of PRL phosphatases with anticancer activity. Mol Cancer Ther 1:1255–1264
Rocha-Azevedo B, Jamerson M, Cabral GA, Silva-Filho FC, Marciano-Cabral F (2007) The interaction between the amoeba *Balamuthia mandrillaris* and extracellular matrix glycoproteins in vitro. Parasitology 134(Pt 1):51–58

Rowbotham TJ (1980) Preliminary report on the pathogenicity of *Legionella pneumophila* for freshwater and soil amoebae. J Clin Pathol 33:1179–1183

Sands M, Kron MA, Brown RB (1985) Pentamidine: a review. Rev Infect Dis 7:625–634

Schuster FL, Visvesvara GS (1996) Axenic growth and drug sensitivity studies of *Balamuthia mandrillaris*, an agent of amebic meningoencephalitis in humans and other animals. J Clin Microbiol 34:385–388

Schuster FL, Visvesvara GS (2004a) Opportunistic amoebae: challenges in prophylaxis and treatment. Drug Resist Updat 7:41–51

Schuster FL, Visvesvara GS (2004b) Free-living amoebae as opportunistic and non-opportunistic pathogens of humans and animals. Int J Parasitol 34:1001–1027

Schuster FL, Yagi S, Gavali S, Michelson D, Raghavan R, Blomquist I, Glastonbury C, Bollen AW, Scharnhorst D, Reed SL, Kuriyama S, Visvesvara GS, Glaser CA (2009) Under the radar: *Balamuthia* amebic encephalitis. Clin Infect Dis 48:879–887

Seijo Martinez M, Gonzalez-Mediero G, Santiago P, Rodriguez De Lope A, Diz J, Conde C, Visvesvara GS (2000) Granulomatous amebic encephalitis in a patient with AIDS: isolation of *Acanthamoeba* sp. Group II from brain tissue and successful treatment with sulfadiazine and fluconazole. J Clin Microbiol 38:3892–3895

Shadrach WS, Rydzewski K, Laube U, Holland G, Ozel M, Kiderlen AF, Flieger A (2005) *Balamuthia mandrillaris*, free-living ameba and opportunistic agent of encephalitis, is a potential host for *Legionella pneumophila* bacteria. Appl Environ Microbiol 71(5):2244–2249

Siddiqui R, Khan NA (2008) *Balamuthia* amoebic encephalitis: an emerging disease with fatal consequences. Microb Pathog 44:89–97

Siddiqui R, Matin A, Warhurst D, Stins M, Khan NA (2007) Effect of antimicrobial compounds on *Balamuthia mandrillaris* encystment and human brain microvascular endothelial cell cytotoxicity. Antimicrob Agents Chemother 51:4471–4473

Siddiqui R, Dudley R, Khan NA (2012) *Acanthamoeba* differentiation: a two-faced drama of Dr Jekyll and Mr Hyde. Parasitology 139:826–834

Singhal T, Bajpai A, Kalra V, Kabra SK, Samantaray JC, Satpathy G, Gupta AK (2001) Successful treatment of *Acanthamoeba* meningitis with combination oral antimicrobials. Pediatr Infect Dis J 20:623–627

Sissons J, Alsam S, Jayasekera S, Khan NA (2004) Ecto-ATPases of clinical and non-clinical isolates of *Acanthamoeba*. Microb Pathog 37:231–239

Smith FR, Korn ED (1968) 7-Dehydrostigmasterol and ergosterol: the major sterol of an amoebae. J Lipid Res 9:405–408

Sun T, Zhang Y (2008) Pentamidine binds to tRNA through non-specific hydrophobic interactions and inhibits aminoacylation and translation. Nucleic Acids Res 36:1654–1664

Taratuto AL, Monges J, Acefe JC, Meli F, Parades A, Martinez AJ (1991) *Leptomyxid amoeba* encephalitis: report of the first case in Argentina. Trans R Soc Trop Med Hyg 85:77

Visvesvara GS, Martinez AJ, Schuster FL, Leitch GJ, Wallace SV, Sawyer TK, Anderson M (1990) *Leptomyxid* ameba, a new agent of amebic meningoencephalitis in humans and animals. J Clin Microbiol 28:2750–2756

Visvesvara GS, Schuster FL, Martinez AJ (1993) *Balamuthia mandrillaris* agent of amebic meningoencephalitis in humans and other animals. J Eukaryot Microbiol 40:504–514

Visvesvara GS, Moura H, Schuster FL (2007) Pathogenic and opportunistic free-living amoebae: *Acanthamoeba* spp., *Balamuthia mandrillaris*, *Naegleria fowleri*, and *Sappinia diploidea*. FEMS Immunol Med Microbiol 50:1–26

Volkonsky M (1931) *Hartmanella castellanii* Douglas, et classification des hartmannelles. Arch Zool Exp Gen 72:317–339

Walochnik J, Duchêne M, Seifert K, Obwaller A, Hottkowitz T, Wiedermann G, Eibl H, Aspöck H (2002) Cytotoxic activities of alkylphosphocholines against clinical isolates of *Acanthamoeba* spp. Antimicrob Agents Chemother 46:695–701

Weisman RA (1976) Differentiation in *Acanthamoeba castellanii*. Annu Rev Microbiol 30: 189–219

Neuroschistosomiasis

Teresa Cristina de Abreu Ferrari

Abstract Neuroschistosomiasis, the involvement of the central nervous system by schistosome, is a neglected and under-recognized disease. This chapter provides an overview of the clinical features, pathogenetic mechanisms of this infection, as well as perspectives. The presentation of neuroschistosomiasis ranges from asymptomatic to different clinical pictures of varying severity. Acute schistosomal encephalopathy presents as an acute diffuse encephalopathy. Pseudotumoral encephalic schistosomiasis manifests as a slow-growth tumor-like lesion, and spinal cord schistosomiasis as an acute/subacute lower spinal cord syndrome. The latter two forms result mostly from egg deposition in the central nervous system, while the pathogenesis of acute schistosomal encephalopathy is unclear. *Schistosoma mansoni* or *S. haematobium* infection involves more frequently the spinal cord, while *S. japonicum* infection targets preferentially the brain. The diagnosis of neuroschistosomiasis is largely based on clinical features. Imaging methods demonstrate nonspecific findings. Serum anti-schistosome antibody tests are of limited diagnostic value, but results in the cerebrospinal fluid are promising. Neuroschistosomiasis is treated with praziquantel and a steroid. Surgery is performed in specific cases. The outcome of neuroschistosomiasis is largely dependent on early treatment and is worse in spinal cord schistosomiasis than in the other forms. Several aspects of neuroschistosomiasis remain to be clarified.

Keywords Neglected tropical diseases • Spinal cord schistosomiasis • Cerebral schistosomiasis • *Schistosoma* • Central nervous system • Parasites

T.C.d.A. Ferrari (✉)
Departamento de Clínica Médica, Faculdade de Medicina, Universidade Federal de Minas Gerais, Av. Professor Alfredo Balena 190, 30130-100 Belo Horizonte, MG, Brazil
e-mail: tferrari@medicina.ufmg.br

M. Bentivoglio et al. (eds.), *Neglected Tropical Diseases and Conditions of the Nervous System*, DOI 10.1007/978-1-4614-8100-3_6,

1 Introduction

Schistosomiasis is a tropical parasitic disease caused by digenetic trematode platyhelminths of the genus *Schistosoma*. These blood-dwelling flukes use man and other mammals as definitive hosts, and aquatic and amphibian snails as intermediate hosts. Five major *Schistosoma* species—*S. mansoni*, *S. haematobium*, *S. japonicum*, *S. intercalatum*, and *S. mekongi*—infect humans. The first three species are the most widely distributed, whereas the last two occur in a few geographical areas and are only of local importance (Ross et al. 2002; Gryseels et al. 2006; Gryseels 2012).

Although schistosomiasis has been successfully controlled in many countries over the last 30 years, it remains one of the most prevalent parasitic diseases in the world, and has significant economic and public health consequences in many low-income countries especially in sub-Saharan Africa. It is estimated that approximately 207 million people are infected by schistosome in 74 countries; 120 million of them have symptoms; and 20 million have severe disease (Chitsulo et al. 2000; Gryseels 2012). Disease caused by *S. mansoni* infection is endemic in large regions in Africa, part of Brazil, Suriname, Venezuela, Caribbean islands, and the Arabian Peninsula. *S. haematobium* infection occurs in most of the African and Middle Eastern countries. *S. japonicum* is reported in China, Indonesia, the Philippines, and Thailand. *S. intercalatum* infection is reported in a few countries in Africa, and *S. mekongi* only in Laos and Cambodia (Chitsulo et al. 2000).

Schistosome infection is more common in rural areas, but is also reported in some countries at the periphery of urban centers associated with poor sanitation. Despite the advances in control in many areas, the transmission of schistosomiasis remains high, mostly in low income countries. The construction of water schemes to meet agricultural requirements, and population growth and migration have contributed to increased transmission and introduction of schistosomiasis into new areas (Steinmann et al. 2006). Furthermore, the growth of eco and adventure tourism has resulted in increasing number of infections in travelers and tourists (Corachan 2002; Meltzer and Schwartz 2013).

The schistosome species differ from each other in several features and these differences are major determinants of the clinical presentations of the infection. *S. mansoni*, *S. japonicum*, *S. mekongi* and *S. intercalatum* are primarily localized in the portal and mesenteric veins, and cause liver and intestinal disease; *S. haematobium* inhabits the pelvic plexuses and damages the urinary tract (Gryseels et al. 2006; Gryseels 2012).

Neuroschistosomiasis (NS), i.e. the involvement of the central nervous system (CNS) by schistosome, is a neglected and under-recognized condition. Although considered rare, NS has been increasingly diagnosed, but its true prevalence is unknown. Except for two cases of brain involvement by *S. mekongi* (Houston et al. 2004; Carmody et al. 2008), all the other published cases of NS are due to one of the three most widespread species. The neural lesions depend basically on the presence of the parasite eggs in the nervous tissue and on the host's immune response (Pittella 1997). Virtually all districts of the CNS may be affected. *S. mansoni* and *S. haematobium* infection involves more frequently the spinal cord, whereas *S. japonicum* usually causes encephalopathy. The neurological manifestations are diverse, ranging from a completely asymptomatic course to various symptoms and syndromes—depending on the location of the CNS lesions—of varying severity (Ferrari 1999; Carod-Artal 2008, 2010, 2012; Ferrari and Moreira 2011). During the last decades, several studies on NS have provided a better understanding of its pathogenesis, natural history, clinical features, diagnosis and treatment. However, several aspects of this disease are still unknown and deserve investigation.

2 The Essentials of the Natural History of Schistosome Infection

In order to facilitate the description and comprehension of some features of NS, the general aspects of the natural history of schistosome infections are here briefly described.

The *Schistosoma* swim as larvae (cercariae) in freshwater, infect humans percutaneously and transform into schistosomula. They migrate to the lungs, and then to the liver where they grow into adult worms, mate and descend into the mesenteric venules of the colon (*S. mansoni*), small intestine (*S. japonicum*), or pelvic plexus, especially the vesical plexus (*S. haematobium*), where eggs are laid. About 25–30 % of the eggs are excreted in the faeces or urine. Eggs that are not excreted remain trapped in the intestinal or bladder wall or are carried by the bloodstream to the liver and other sites, including the CNS (Pittella 1997; Ferrari and Moreira 2011). Adult worms do little damage to the host and usually do not cause symptoms. Most of pathological changes result from the host's immune response to retained eggs (Ross et al. 2002; Gryseels et al. 2006; Gryseels 2012; Hams et al. 2013).

After egg deposition into the tissues, the embryos (miracidium) reach maturity and secrete antigens that evoke a cell-mediated periovular granulomatous reaction, which evolves through three successive stages. The necrotic-exudative granulomas are characteristic of the early phase of schistosomiasis (up to 110 days post-infection), when the immune response to the antigens released by the eggs reaches maximum intensity. As the infection progresses into the chronic phase, the granulomatous response is down-modulated originating the smaller productive-stage granulomas, and then the granulomas in the healing-by-fibrosis stage (Pittella 1997; Hams et al. 2013).

Schistosome infections are classified into acute and chronic phases and different clinical forms (Ferrari and Moreira 2011), as briefly described below.

2.1 *Acute Phase*

The acute phase of schistosomiasis is usually unapparent in people who live in endemic areas. However, clinical manifestations of varying severity occur in about 65 % of non-immune people exposed to transmission (Jauréguiberry et al. 2010).

A temporary maculopapular eruption may arise at the site of penetration of the cercariae (swimmers' itch). It usually develops within hours and might persist for days. Nonspecific constitutional symptoms may occasionally be observed (symptomatic pre-egg-laying phase) (Lambertucci 2010; Ferrari and Moreira 2011).

The symptoms of acute schistosomiasis (acute toxaemic form—*S. mansoni*, Katayama syndrome—*S. japonicum*) usually begin with the deposition of the eggs into host tissues, which occurs 28–110 days after infection depending on the schistosome species (symptomatic post-egg-laying phase) (Ross et al. 2007; Ferrari and Moreira 2011). It is a systemic hypersensitivity reaction against migrating schistosomula and early oviposition, and has been attributed to circulating immune complexes (Wynn et al. 2004). Symptoms usually arise suddenly and subside spontaneously over a period of a few weeks. The severity of the clinical manifestations varies according to the cercarial burden and immune response to the released antigens (Jauréguiberry et al. 2010). Fever, headache, fatigue, urticaria, diarrhoea, abdominal pain, hepatomegaly, and non-productive cough with pulmonary infiltrates on chest X-ray are common manifestations. Pathological examination of involved organs discloses granulomas uniformly at the necrotic-exudative stage. Eosinophilia, usually marked, is a very frequent finding.

2.2 *Chronic Phase*

The chronic phase of the infection is usually asymptomatic, but individuals who live in endemic areas may present clinical manifestations that begin insidiously and progress without specific treatment. The intensity and duration of the infection determine the amount of antigen released and the severity of chronic fibro-obstructive disease (Ross et al. 2002; Ferrari and Moreira 2011).

S. mansoni or *S. japonicum* eggs deposited into the gut wall provoke mucosal granulomatous inflammation that may cause chronic or intermittent diarrhea with or without blood, loss of appetite, and abdominal pain and discomfort (intestinal form). Hepatomegaly, due to granulomatous inflammation around the eggs embolized to the presinusoidal periportal spaces of the liver may occur early in the course of the chronic phase (hepatointestinal form). In patients with sustained heavy infection, periportal collagen deposits lead to fibrosis and progressive obstruction of blood flow with portal hypertension (hepatosplenic form) (Gryseels et al. 2006; Gryseels 2012).

Granulomatous inflammation around *S. haematobium* eggs in the bladder and ureteral walls may cause hematuria, dysuria, and urinary frequency (mild urinary form). With progressive involvement, fibrosis and calcification may occur resulting in hydroureter and hydronephrosis (obstructive uropathy) (Ross et al. 2002; Gryseels et al. 2006; Gryseels 2012).

The intestinal, hepatointestinal and non-obstructive urinary forms are considered mild chronic forms, whereas hepatosplenic schistosomiasis and obstructive uropathy represent severe chronic forms (Ferrari and Moreira 2011).

3 Pathogenesis of Neuroschistosomiasis

The pathogenesis of NS remains largely unknown; however, available evidence strongly suggests that the CNS lesions depend mostly on the presence of the parasite eggs in the nervous tissue and on host's immune response (Pittella 1997; Ferrari et al. 2008b). It has been accepted that schistosome eggs reach the CNS either by embolization or by *in situ* oviposition following anomalous migration of adult worms to sites close to the CNS, via different routes. One route is provided by the arterial system, especially if pulmonary arteriovenous fistulas exist or portopulmonary anastomosis had been formed as a consequence of portal hypertension. Another route is represented by retrograde venous flow into the vertebral Batson's plexus, which connects the portal venous system and inferior venae cavae to spinal cord and encephalic veins (Pittella 1997).

The CNS may be involved at any time during the infection, but there are marked differences of neurological involvement depending on the phase and clinical form of schistosomiasis. Acute schistosomal encephalopathy (ASE) is the most common form of NS that occurs during or immediately after acute symptomatic schistosomiasis (ASS). In association with the mild chronic forms, both pseudotumoral encephalic schistosomiasis (PES) and spinal cord schistosomiasis (SCS) may be observed (Ferrari and Moreira 2011). Finally, CNS involvement during the severe chronic forms is usually asymptomatic (Pittella 1997; Ferrari et al. 2008b).

ASE occurs more frequently in people without any previous exposure to schistosome. Although *S. japonicum* is the most frequent agent of this form of NS, cases attributed to *S. mansoni* and *S. haematobium* have been increasingly reported (Clerinx et al. 2006; Jauréguiberry et al. 2007; Christo et al. 2010). As there are no pathological studies of ASE, its pathogenic mechanisms have been inferred from clinical and imaging findings. The pathogenic mechanisms suggested for ASE are described below.

Some cases of ASE have been ascribed to an ADEM-like disorder (Clerinx et al. 2006; Devine et al. 2008; Houdon et al. 2010). According to these case reports, the patients developed, a few weeks after ASS, a monophasic disorder with multifocal neurological symptoms and encephalopathy associated with multiple contrast-enhanced white matter lesions on magnetic resonance imaging (MRI) resembling ADEM. However, other authors strongly disagree on the similarities between these ASE cases and cases of ADEM (Caumes and Vidailhet 2010).

Based on brain and cardiac findings on imaging methods, Sarazin et al. (2004) suggested that cardiogenic emboli due to hypereosinophilic-related endomyocardial fibrosis may be responsible for the development of ASE. However, since eosinophilic leucocyte-mediated cardiac damage requires sustained eosinophilia for its development, this mechanism is unlikely (Ferrari and Moreira 2011).

Eosinophilic leukocyte-mediated toxicity, leading to vasculitis and small-vessel thrombosis, has been proposed as the most likely pathogenic mechanism of ASE (Jauréguiberry et al. 2007; Jauréguiberry and Caumes 2008). According to this interpretation, brain MRI suggestive of border zone infarcts observed in some cases and rapid response to steroid therapy described in some patients provide strong evidence of the role of vasculitis in the pathogenesis of ASE. Peptides and peroxide released from eosinophil granules exert direct toxicity on endothelial cells and nerve tissue (Jauréguiberry et al. 2007). As discussed in the review by Ferrari and Moreira (2011), although vasculitis is not a pathological feature on biopsy specimens from patients with ASS, skin lesions ascribed to immune complex-mediated vasculitis, based on pathological findings, have already been described in a patient with *S. mansoni* ASS. Furthermore, vasculitis plays a role in the pathogenesis of other forms of NS (see below). These data, together with the exacerbated immune response leading to disseminated immune complex formation and deposition, which characterizes ASS, support the possibility that eosinophilic leukocyte- and/or immune complexes-mediated vasculitis plays a major role in the pathogenesis of ASE.

As shown by histological examination, *S. mansoni* ASS is associated with intense miliary dissemination of eggs surrounded by necrotic-exudative granulomas to virtually any organ. Although the organs located in the abdominal and thoracic cavities are the most frequently involved, other organs including the brain may also be affected. Thus, eggs and granulomas could play a major role in the pathogenesis of ASE (Ferrari and Moreira 2011). The fact that ASE usually develops within a few days or weeks after ASS (during the post-egg-laying phase), and the existence of a few histologically confirmed cases of *S. mansoni* SCS that developed during ASS are in agreement with this hypothesis. However, considering that the post-egg-laying phase is associated with a powerful immune response and extensive immune complex formation, the immune complex- and/or eosinophilic leukocyte-mediated vasculitis, or even other immune-mediated mechanism related to oviposition, and not the eggs themselves, may represent the chief pathogenic mechanism of ASE (Ferrari and Moreira 2011).

In summary, the pathogenesis of ASE remains unclear. The intense immune response which characterizes ASS points towards a major role of immune-related mechanisms. However, the participation of these and/or other mechanisms remains to be clarified. It is also still unknown when the eggs and granulomas become the major mechanism of CNS injury in NS. Additionally, the possibility of development of ASE during the pre-egg-laying phase can not be completely ruled out, as systemic symptoms might start during this phase and early CNS involvement has already been reported (Ferrari and Moreira 2011).

According to available evidence, PES and SCS are due to the same pathogenetic mechanisms. A large number of eggs surrounded by granulomas lodged in circumscribed areas of the cerebral hemispheres, cerebellum, brain stem or spinal cord can damage the nervous tissue by the mass effect and by the inflammatory reaction itself (Fig. 1) (Pittella 1997). *S. mansoni* and *S. haematobium* affect more frequently the spinal cord, and *S. japonicum* the brain. The small, round eggs of *S. japonicum* travel all the way though the Batson's plexus reaching the brain. On the other hand, *S. mansoni* and *S. haematobium* eggs, which are larger and bear protruding spines, are usually retained in the lower portions of the spinal cord.

Based on histopathological findings in SCS cases, it has been suggested that ischemic lesions due to immune complex-mediated vasculitis also play a role in the

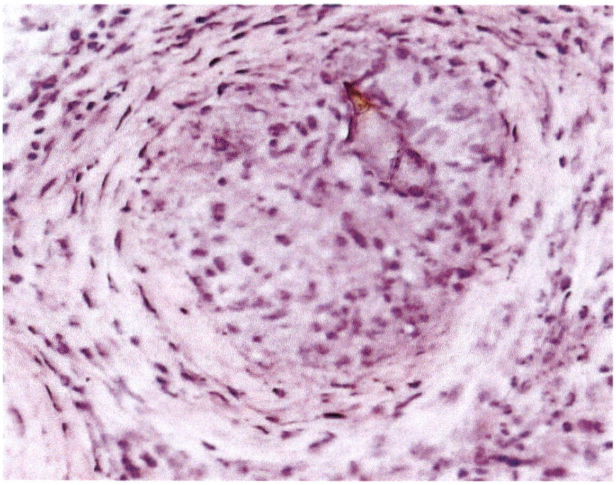

Fig. 1 *Schistosoma mansoni* egg surrounded by productive stage granuloma in a spinal cord section. Haematoxylin and eosin stain (courtesy of Prof. E A Bambirra, Universidade Federal de Minas Gerais, Faculdade de Medicina, Departamento de Anatomia Patológica e Medicina Legal, Belo Horizonte, Brazil)

genesis of symptoms in NS (Pittella 1997). Immune complexes containing soluble egg antigen of *S. mansoni* (SEA) have been demonstrated in the cerebrospinal fluid (CSF) of SCS patients (Ferrari et al. 2011). In this context, investigations of cytokine profiles in CSF and serum of SCS patients suggest the occurrence of an inflammatory and skewed type-2 immune response in the CNS and also systemically (Ferrari et al. 2006; Sousa-Pereira et al. 2006).

The high parasite burden, intense oviposition, and continuous embolization of eggs through collateral vessels of the portal-systemic circulation may explain the higher frequency of egg deposition in ectopic sites, including the CNS, that occurs in association with the most severe chronic forms of schistosomiasis as demonstrated in post-mortem studies (Gonçalves et al. 1995). Although frequent, egg deposition in the CNS of individuals with these severe forms is usually asymptomatic. The lack of symptoms is attributed to the sparse distribution of the eggs and less intense granulomatous reaction that occurs in long-standing schistosomiasis (Pittella 1997).

3.1 Experimental Investigations on Neuroschistosomiasis

Considering the important problems posed by NS, it is surprising that only a few attempts have been made to experimentally study it. Aloe et al. (1996) observed decreased expression of the neurotrophin nerve growth factor in mice infected percutaneously with *S. mansoni* cercareae, which presented schistosome egg, verified histologically, in the brain. In contrast, Silva et al. (2002) using a similar model, did not succeed in finding *S. mansoni* eggs in the brain in spite of widespread distribution of eggs in several organs; thus, these authors concluded that the murine model did not appear to be suitable for experimental studies on NS. Recently, Wang et al. (2011) have been able to observe the three stages of the granulomatous reaction and different neurological symptoms, including seizures and motor weakness, after direct injection of *S. japonicum* egg suspension into rabbit brain. Similarly, neurological symptoms and schistosome egg granuloma in the brain were observed in 90 % of the rabbits which received injection of *S. japonicum* eggs into their brain (Xu et al. 2013). However, these last two models are different from the natural course of the infection.

Researches aimed at developing a suitable animal model of NS need to be implemented and would be especially useful for the investigation of pathogenetic mechanisms of this disease.

4 Clinical Features, Laboratory Investigations and Diagnosis of Neuroschistosomiasis

All clinical forms of NS may affect men or women at any age, but they are more common in male young adults, teenagers and children.

4.1 Acute Schistosomal Encephalopathy

ASE is more frequent in people without any previous contact with schistosome. Typically, the disease begins about 3 weeks after ASS. The onset of neurological manifestations is usually acute. Headache and some degree of impaired mental status have been observed in virtually all patients. Other common symptoms include focal and generalized seizures, sensory disturbances, weakness of the extremities and cerebellar syndrome. Visual and speech disturbances may also occur. Some of these symptoms may improve and disappear spontaneously in a few days or weeks (Jauréguiberry and Caumes 2008; Ferrari and Moreira 2011).

Computed tomography (CT) and MRI demonstrate multiple focal, small, contrast-enhanced lesions surrounded by oedema. CSF may be normal or may show nonspecific alterations. It is often difficult to detect schistosome eggs in stools or urine, since in recent infection the parasite burden is low. Peripheral marked eosinophilia is a frequent finding. As ASE usually develops in people from non-endemic areas, anti-schistosome antibody detection in serum is useful for the diagnosis. However, these tests may be negative initially. The reliability in detecting early infection by searching for somatic schistosome antigens using monoclonal antibodies remains to be determined (Ross et al. 2007). Polymerase chain reaction (PCR) techniques are being developed for the diagnosis of schistosomiasis with promising results.

In summary, the diagnosis of ASE is largely based on epidemiological and clinical data—especially the history of exposure to contaminated water and presence of ASS a few days or weeks before the neurological symptoms—and positive schistosomal serology.

4.2 Pseudotumoral Encephalic Schistosomiasis

PES typically occurs in people who live in endemic areas, and the patients usually do not have any other manifestation of schistosomiasis. The neurological symptoms develop slowly and are due to tumor-like lesions that cause intracranial hypertension and focal neurological signs, which vary according to the site of the lesion. The pseudotumor may be located in any brain region, but the most common site is the cerebellum, supporting the hypothesis that eggs and/or worms reach the brain through the Batson's venous plexus (Ferrari et al. 2008b). Extracerebral lesion attached to the dura mater has also been reported (Rommel et al. 2005). PES often presents as a single lesion, but cases of two distinct lesions have been described (Pittella et al. 1996; Roberts et al. 2006).

Headache, seizures, altered mental status, visual abnormalities, speech disturbances, sensory impairment, motor deficits, nystagmus, vertigo, ataxia, nausea, vomiting and papilledema are commonly reported. Focal motor seizures and a cerebral lesion on CT or MRI may be the only evidence of the disease. Symptoms are present from a few weeks to more than 1 year before the diagnosis (Pittella et al. 1996; Ferrari 2004; Betting et al. 2005; Li et al. 2011).

CT and MRI show a nodular, space-occupying lesion with surrounding oedema, and heterogeneous contrast enhancement. T1-weighted MRI may show a central linear enhancement surrounded by multiple enhancing punctate nodules, with an "arborized" appearance (Sanelli et al. 2001). Schistosome eggs may or may not be present in stool or urine. Reports on CSF examination are rare. Peripheral eosinophilia is commonly mild or absent. A positive serum schistosomal antibody test is not sufficient to confirm active disease, as it is only evidence of previous exposure to schistosome antigens. Thus, serology is often devoid of diagnostic value in individuals born and raised in schistosome endemic areas (Ross et al. 2002). Techniques to detect parasite antigens have been developed (Deelder et al. 1994), but they are not yet commercially available.

The diagnose of PES may be difficult because the clinical findings are indistinguishable from those caused by any other tumor-like lesions of slow growth, and laboratory tests are of little help. Therefore, diagnosis of PES relies on the demonstration of eggs and granulomas in nervous tissue biopsy (Ferrari and Moreira 2011).

4.3 Spinal Cord Schistosomiasis

SCS is also more frequent in people who live in endemic areas. The patients usually do not have any other symptom of the infection except for hepatomegaly, which has been observed in about one-fourth of the patients with *S. mansoni* SCS (Ferrari 2004). There are only a few reported cases during (or soon after) ASS (Pittella 1991) or in association with hepatosplenic schistosomiasis (Ferrari et al. 2001; Araújo et al. 2006).

Typically, SCS presents as a low cord syndrome with acute or subacute progression, usually associated with involvement of the cauda equina roots. The most frequent initial complaint is low back pain, which usually irradiates to the lower limbs, followed by weakness and sensory impairment in the lower limbs, and autonomic dysfunction, particularly bladder dysfunction. Pain intensity varies from mild to very severe, and lower limb weakness can prevent walking in about 63 % of the cases (Ferrari 2004; Ferrari et al. 2008a). Other common symptoms of SCS are lower limbs paresthesia, hypoesthesia or anesthesia, deep tendon reflexes abnormalities, constipation and sexual impotence. The level of the lesion determined by the clinical examination is usually equal or below T6, especially at T11-L1. The low localization in the spinal cord, the acute or subacute progression of symptoms, and the involvement of both the spinal cord and spinal roots are the most characteristic clinical features of SCS (Ferrari 1999).

Atypical presentations of SCS have occasionally been reported and include a slowly progressive form, a waxing and waning myeloradiculopathy that takes months to reach its complete development (Ferrari 1999; Chen et al. 2006; Lighter et al. 2008), and a high thoracic or even cervical localization (Ferrari 1999; Junker et al. 2001).

MRI demonstrates findings of inflammatory myelopathy. The most frequent abnormalities are enlargement of the spinal cord, especially the conus medullaris, and thickening of the cauda equina roots with a heterogeneous pattern of contrast enhancement (Fig. 2). Stool and urine examination fails in revealing eggs in more than half of patients. Rectal biopsy with ovogram allows identification of *S. mansoni* eggs in 95–100 % of the cases (Ferrari 2004; Silva et al. 2004; Araújo et al. 2006). CSF examination usually shows a slight-to-moderate increase in both total protein concentration and cell count. Mononuclear cells usually predominate and eosinophils may or may not be present. As stated above, a positive serology in serum is virtually devoid of diagnostic value in people from endemic areas. In contrast, because of the segregation of the CNS from the rest of the body due to physiological barriers, the finding of antibodies against schistosome antigens in the CSF has been considered a potentially useful tool in the diagnostic approach to SCS (Pammenter et al. 1991; Ferrari et al. 1995).

In this context, the measurement of IgG against SEA in the CSF by ELISA for the diagnosis of SCS has demonstrated a high specificity, though with limited sensitivity (Pammenter et al. 1991; Ferrari et al. 1995; Ferrari 2010). Reduction of CSF anti-SEA antibodies after treatment was also observed in SCS patients (Haribhai et al. 1991; Magalhães-Santos et al. 2003). Intrathecal IgG synthesis was demonstrated in SCS patients by both estimation of the IgG index (Ferrari et al. 1999) and identification of oligoclonal bands in the CSF that were absent in paired serum samples (Pammenter et al. 1991). Higher concentrations of IgG1 in the CSF of SCS patients when compared to the paired serum levels was also observed (Magalhães-Santos et al. 2003). An index based on the ratio between CSF and serum levels of antibodies against schistosome antigens was developed with the aim of distinguishing SCS cases from other CNS disorders (de Jongste et al. 2010), but this index has been evaluated in only one patient affected by SCS.

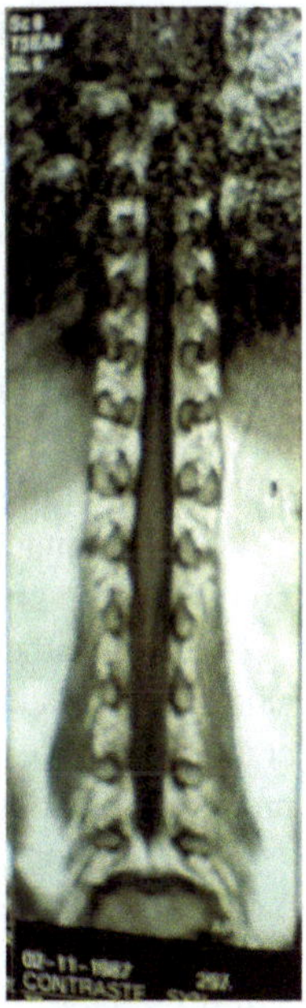

Fig. 2 Coronal gadolinium-enhanced T1-weighted MRI of the spinal cord showing enlargement of the conus medullaris in a patient with spinal cord schistosomiasis due to *Schistosoma mansoni*

Taken together, these findings provide strong evidence that methods involving the search for anti-schistosome antibodies in the CSF are likely to be useful for the diagnosis of SCS, and even the encephalic forms. However, they need further investigation. It is necessary to standardize and validate the immunoassays in other populations of NS patients to assess their actual diagnostic value, as well the most effective test or combination of tests. Techniques to detect parasite antigens in the CSF should also be considered. As NS is an uncommon condition, multicentric studies could facilitate the achievement of the adequate sample size.

Currently, a definite diagnosis of SCS can only be established by the demonstration of the eggs and granulomas in nervous tissue. As spinal cord biopsy should be avoided because of the risks (Case 1996; Ferrari 1999), the diagnosis of this disorder is essentially presumptive and based on evidence of spinal cord (low thoracic, lumbar and/or sacral) and/or cauda equina lesions, confirmation of active schistosome infection by a direct method, and exclusion of other causes of myeloradiculopathy (Center for Disease Control and Prevention 1984; Ferrari et al. 1993). This reinforces the need of a non-invasive test (or tests) to enable a more accurate diagnosis of this disease.

5 Treatment and Outcome

Treatment of NS is based on the administration of an antischistosomal drug—usually praziquantel (PZQ)—associated with a corticosteroid. PZQ kills adult worms preventing *de novo* egg deposition and subsequent granuloma formation. The steroid counters the inflammatory response, reducing compression and destruction of the nervous tissue. No consensus has been reached to date on the optimal therapeutic regimen to treat the different forms of NS. Controlled trials to access the efficacy of different therapeutic regimens have not been hitherto performed. However, some general recommendations can be made.

ASE should be treated initially with steroid (e.g., prednisone 1 mg/kg/day) to suppress the hypersensitivity reaction, followed by PZQ. The optimal timing for this drug is still unclear. Based on rare reports of clinical worsening of patients with ASE after taking PZQ (Jauréguiberry et al. 2007), it has been suggested to prescribe this drug soon after the stabilization of the neurological picture, when the patient is still receiving high steroid doses to avoid clinical worsening (Ferrari and Moreira 2011). As immature worms are not susceptible to PZQ, a second run of administration should be given after 6–12 weeks (Gryseels et al. 2006).

Although either surgical resection (followed by the use of PZQ) or medical therapy alone has been successfully used to treat PES, the clinical treatment is preferred (Fowler et al. 1999). PZQ and a steroid (equivalent to 1 mg/kg/day of prednisone) should be promptly started, and the steroid should be continued upon clinical monitoring.

The common SCS treatment schedule is Brazil foresees PZQ (60 mg/kg/day for 3 days—maximum daily and total dose of 5 g and 15 g, respectively) administered in two daily doses at a 4-h interval, and intravenous methylprednisolone (15 mg/kg/day for 5 days—maximum dose 1 g/day) divided into two daily doses followed by prednisone (1.5–2 mg/kg/day) given in three daily doses for about 3–4 weeks, and then replaced with a single daily dose, which is gradually tapered until its complete discontinuation within 3–4 months (Ferrari 2004; Ferrari et al. 2008a). The steroid should be started immediately in highly suspected cases and PZQ soon after the confirmation of schistosome infection.

Surgical treatment of SCS should be limited to highly selected cases particularly those with evidence of spinal compression and rapidly worsening of motor function in lower limbs despite medical treatment (Ferrari and Moreira 2011; Carod-Artal 2012).

Anticonvulsants are frequently necessary in ASE and PES. Mannitol may help in reducing the intracranial pressure in selected cases (Li et al. 2011). Symptomatic treatment of pain, spasticity, and autonomic dysfunctions may improve patients' quality of life. Physical therapy is important for rehabilitation (Araújo et al. 2010).

The need to establish an optimal treatment for the different forms of NS is obvious. However, controlled clinical trials could require long period of time to achieve adequate number of patients. Multicentric investigations certainly will facilitate this endeavor. Until such studies are performed, experts should consider developing guidelines in an attempt to reduce the great heterogeneity of doses and treatment time employed.

The prognosis of NS depends largely on early treatment, but features of the disease itself are also related to the outcome. ASE and PES are associated with a better outcome; most treated patients recover without any significant deficit. About 65 % of SCS patients treated early recover completely or are left with some negligible deficit, while the remaining patients show sequelae that vary in severity (Silva et al. 2004; Ferrari et al. 2008a). In general, outcome is less favorable in cases with predominance of spinal cord involvement, and more satisfactory in patients with predominant involvement of the cauda equina roots (Ferrari 1999; Ferrari 2004).

6 Conclusions and Future Perspectives

NS is a neglected and under-diagnosed condition, but is increasingly recognized. Its presentation varies from a completely asymptomatic course to different symptoms and syndromes of varying severity. Virtually any part of the CNS may be involved. ASE is the usual form of presentation of CNS involvement during or soon after ASS, whereas PES and SCS usually occur simultaneously with the mild chronic clinical forms of the infection.

Many questions regarding the pathogenesis of NS remain unanswered. Although PES and SCS depend mostly on the presence of eggs and granulomas in the nervous tissue, the pathogenesis of ASE remains unclear. Experimental studies in animals should help to investigate pathogenetic mechanisms.

Outcome of symptomatic NS is largely dependent on early diagnosis and treatment. As its diagnosis is largely based on circumstantial evidence, the importance of developing non-invasive tools to enable a more accurate diagnosis is a major challenge. In this context, CSF tests based on PCR methodology and/or immunological techniques are an important area of investigation. Immunological studies could also provide deeper understanding of NS immunopathogenesis.

NS is treated with PZQ and a steroid, but there is no consensus on the optimal therapeutic regimen. Multicentric controlled clinical trials could allow to achieve this goal. Until these studies are developed, experts should consider developing guidelines in an attempt to reduce the great heterogeneity of doses and duration of treatment.

References

Aloe L, Moroni R, Fiore M, Angelucci F (1996) Chronic parasite infection in mice induces brain granulomas and differentially alters brain nerve growth factor levels and thermal responses in paws. Acta Neuropathol 92:300–305

Araújo KCGM, Silva CR, Barbosa CS, Ferrari TCA (2006) Clinical-epidemiological profile of children with schistosomal myeloradiculopathy attended at the Instituto Materno-Infantil de Pernambuco (IMIP). Mem Inst Oswaldo Cruz 101:149–156

Araújo KCGM, Silva CR, Santos AGA, Barbosa CS, Ferrari TCA (2010) Clinical-epidemiologic profile of the schistosomal myeloradiculopathy in Pernambuco, Brazil. Mem Inst Oswaldo Cruz 105:454–459

Betting LD, Pirani-Jr C, Queiroz LS, Damasceno BP, Cendes F (2005) Seizures and cerebral schistosomiasis. Arch Neurol 62:1008–1010

Carmody D, Nolan C, Allcutt D (2008) Intracranial *Schistosoma mekongi* infection. Ir Med J 101:315

Carod-Artal FJ (2008) Neurological complications of schistosoma infection. Trans R Soc Trop Med Hyg 8:1307–1318

Carod-Artal FJ (2010) Neuroschistosomiasis. Expert Rev Anti Infect Ther 8:1307–1318

Carod-Artal FJ (2012) Cerebral and spinal schistosomiasis. Curr Neurol Neurosci Rep 12:666–674

Case records of the Massachusetts General Hospital. Weekly clinicopathological exercises. Case 4–1996 (1996) A 40-year-old woman with the rapid onset of flaccid paraplegia. N Engl J Med 334:382–389

Caumes E, Vidailhet M (2010) Acute neuroschistosomiasis: a cerebral vasculitis to treat with corticosteroids not praziquantel. J Travel Med 17:359

Center for Disease Control and Prevention (1984) Acute schistosomiasis with transverse myelitis in American students returning from Kenya. MMWR Morb Mortal Wkly Rep 33:446–447

Chen AWY, Alam MH, Williamson JML, Brawn LA (2006) An unusually late presentation of neuroschistosomiasis. J Infect 53:155–158

Chitsulo L, Engels D, Montresor A, Savioli L (2000) The global status of schistosomiasis and its control. Acta Trop 77:41–51

Christo PP, Zambelli VA, Amâncio FF, Oliveira M, Carvalho SH, Said RC (2010) Acute encephalitis caused by *Schistosoma mansoni*. Arq Neuropsiquiatr 68:968–970

Clerinx J, van Gompel A, Lynen L, Ceulemans B (2006) Early neuroschistosomiasis complicating Katayama syndrome. Emerg Infect Dis 12:1465–1466

Corachan M (2002) Schistosomiasis and international travel. Clin Infect Dis 35:446–450

de Jongste AH, Tilanus AM, Bax H, Willems MH, van der Feltz M, van Hellemond JJ (2010) New insights in diagnosing *Schistosoma* myelopathy. J Infect 60:244–247

Deelder AM, Qian ZL, Kremsner PG, Acosta L, Rabello AL, Enyong P, Simarro PP, van Etten EC, Krijger FW, Rotmans JP et al (1994) Quantitative diagnosis of *Schistosoma* infections by measurement of circulating antigens in serum and urine. Trop Geogr Med 46:233–238

Devine MJ, Wilkinson PA, Doherty JF, Jarman PR (2008) Neuroschistosomiasis presenting as brainstem encephalitis. Neurology 70:2262–2264

Ferrari TCA (1999) Spinal cord schistosomiasis: a report of 2 cases and review emphasizing clinical aspects. Medicine 78:176–190

Ferrari TCA (2004) Involvement of the central nervous system in the schistosomiasis. Mem Inst Oswaldo Cruz 99:59–62

Ferrari TCA (2010) A laboratory test for the diagnosis of neuroschistosomiasis. Neurol Res 32:252–262

Ferrari TCA, Moreira PRR (2011) Neuroschistosomiasis: clinical symptoms and pathogenesis. Lancet Neurol 10:853–864

Ferrari TCA, Moreira PRR, Ferrari MLA, Viana IR, Correa-Oliveira R, Gazzinelli G, Cunha AS (1993) Clinical and immunological study of schistosomal myeloradiculopathy. Ann Trop Med Parasitol 87:295–297

Ferrari TCA, Moreira PRR, Oliveira RC, Ferrari MLA, Gazzinelli G, Cunha AS (1995) The value of an enzyme-linked immunosorbent assay (ELISA) for the diagnosis of schistosomal mansoni myeloradiculopathy. Trans R Soc Trop Med Hyg 89:496–500

Ferrari TCA, Correa-Oliveira R, Xavier MP, Gazzinelli G, Cunha AS (1999) Estimation of the local synthesis of immunoglobulin G (IgG) in the central nervous system of patients with spinal cord schistosomiasis by the IgG index. Trans R Soc Trop Med Hyg 93:558–559

Ferrari TCA, Moreira PRR, Cunha AS (2001) Spinal cord involvement in the hepatosplenic form of *Schistosoma mansoni* infection. Ann Trop Med Parasitol 95:633–635

Ferrari TCA, Moreira PRR, Cunha AS (2004) Spinal cord schistosomiasis: a prospective study of 63 cases emphasizing clinical and therapeutic aspects. J Clin Neurosci 11:246–253

Ferrari TCA, Moreira PRR, Sampaio MJ, da Cunha AS, de Oliveira JT, Gazzinelli G, Correa-Oliveira R (2006) Intrathecal cytokines in spinal cord schistosomiasis. J Neuroimmunol 77:136–141

Ferrari TCA, Moreira PRR, Cunha AS (2008a) Clinical characterization of neuroschistosomiasis due to *Schistosoma mansoni* and its treatment. Acta Trop 108:89–97

Ferrari TCA, Correa-Oliveira R, Gazzinelli G (2008b) Immune response and pathogenesis of neuroschistosomiais *mansoni*. Acta Trop 108:83–88

Ferrari TCA, Faria LC, Vilaça TS, Correa CR, Góes AM (2011) Identification and characterization of immune complexes in the cerebrospinal fluid of patients with spinal cord schistosomiasis. J Neuroimmunol 230:188–190

Fowler R, Lee C, Keystone JS (1999) The role of corticosteroids in the treatment of cerebral schistosomiasis caused by *Schistosoma mansoni*: case report and discussion. Am J Trop Med Hyg 6:47–50

Gonçalves EC, Fonseca AP, Pittella JE (1995) Frequency of schistosomiasis *mansoni*, of its clinicopathological forms and of the ectopic locations of the parasite in autopsies in Belo Horizonte, Brazil. J Trop Med Hyg 98:289–295

Gryseels B (2012) Schistosomiasis. Infect Dis Clin North Am 26:383–397

Gryseels B, Polman K, Clerinx J, Kestens L (2006) Human schistosomiasis. Lancet 368:1106–1118

Hams E, Aviello G, Fallon PG (2013) The schistosoma granuloma: friend or foe? Front Immunol 4:89

Haribhai HC, Bhigjee AI, Bill PL, Pammenter MD, Modi G, Hoffmann M, Kelbe C, Becker P (1991) Spinal cord schistosomiasis: a clinical, laboratory and radiologic study, with a note on therapeutic aspects. Brain 114:709–726

Houdon L, Flodrops H, Rocaboy M, Bintner M, Fériot JP, Tournebize P, Renouil M, Ezzedine K, Malvy D (2010) Two patients with imported acute neuroschistosomiasis due to *Schistosoma mansoni*. J Travel Med 17:274–277

Houston S, Kowalewska-Grochowska K, Naik S, McKean J, Johnson ES, Warren K (2004) First report of *Schistosoma mekongi* infection with brain involvement. Clin Infect Dis 38:e1–e6

Jauréguiberry S, Caumes E (2008) Neurological involvement during Katayama syndrome. Lancet Infect Dis 8:9–10

Jauréguiberry S, Ansart S, Perez L, Danis M, Bricaire F, Caumes E (2007) Acute neuroschistosomiasis: two cases associated with cerebral vasculitis. Am J Trop Med Hyg 76:964–966

Jauréguiberry S, Paris L, Caumes E (2010) Acute schistosomiasis, a diagnostic and therapeutic challenge. Clin Microbiol Infect 16:225–231

Junker J, Eckardt L, Husstedt I (2001) Cervical intramedullar schistosomiasis as a rare cause of acute tetraparesia. Clin Neurol Neurosurg 103:39–42

Lambertucci JR (2010) Acute schistosomiasis *mansoni*: revisited and reconsidered. Mem Inst Oswaldo Cruz 105:422–435

Li Y, Ross AG, Hou X, Lou Z, McManus DP (2011) Oriental schistosomiasis with neurological complications: case report. Ann Clin Microbiol Antimicrob 10:5–10

Lighter J, Kim M, Krasinski K (2008) Intramedullary schistosomiasis presenting in an adolescent with prolonged intermittent back pain. Pediatr Neurol 39:44–47

Magalhães-Santos IF, Lemaire DC, Andrade-Filho AS, Queiroz AC, Carvalho OM, Carmo TM, Siqueira IC, Andrade DM, Rego MF, Guedes AP, Reis MG (2003) Antibodies to *Schistosoma mansoni* in human cerebrospinal fluid. Am J Trop Med Hyg 68:294–298

Meltzer E, Schwartz E (2013) Schistosomiasis: current epidemiology and management in travelers. Curr Infect Dis Rep 15:211–215

Pammenter MD, Haribhai HC, Epstein SR, Rossouw EJ, Bhigjee AI, Bill PLA (1991) The value of immunological approaches to the diagnosis of schistosomal myelopathy. Am J Trop Med Hyg 44:329–335

Pittella JEH (1991) The relation between involvement of the central nervous system in schistosomiasis *mansoni* and the clinical forms of the parasitosis: a review. J Trop Med Hyg 94:15–21

Pittella JEH (1997) Neuroschistosomiasis. Brain Pathol 7:649–662

Pittella JEH, Gusmão SNS, Carvalho GTC, Silveira RL, Campos GF (1996) Tumoral form of cerebral schistosomiasis *mansoni*: a report of four cases and a review of the literature. Clin Neurol Neurosurg 98:15–20

Roberts M, Cross J, Pohl U, Lucas S, Dean A (2006) Cerebral schistosomiasis. Lancet Infect Dis 6:820

Rommel D, Ragé M, Duprez T, Parent M, Sindic CJ (2005) Paucisymptomatic brainstem lesions revealing CNS schistosomiasis. Acta Neurol Belg 105:89–93

Ross AGP, Bartley PB, Sleigh AC, Olds GR, Li Y, Williams GM, McManus DP (2002) Schistosomiasis. N Engl J Med 346:1212–1220

Ross AGP, Vickers D, Olds GR, Shah SM, McManus DP (2007) Katayama syndrome. Lancet Infect Dis 7:218–224

Sanelli PC, Lev MH, Gonzalez RG, Schaefer PW (2001) Unique linear and nodular MR enhancement pattern in schistosomiasis of the spinal cord: report of three patients. AJR Am J Roentgenol 177:1471–1474

Sarazin M, Caumes E, Cohen A, Amarenco P (2004) Multiple microembolic borderzone brain infarctions and endomyocardial fibrosis in idiopathic hypereosinophilic syndrome and in *Schistosoma mansoni* infestation. J Neurol Neurosurg Psychiatry 75:305–307

Silva LM, Oliveira CN, Andrade ZA (2002) Experimental neuroschistosomiasis: inadequacy of the murine model. Mem Inst Oswaldo Cruz 97:599–600

Silva LCS, Maciel PE, Ribas JG, Souza-Pereira SR, Antunes CM, Lambertucci JR (2004) Treatment of schistosomal myeloradiculopathy with praziquantel and corticosteroids and evaluation by magnetic resonance imaging: a longitudinal study. Clin Infect Dis 39:1619–1624

Sousa-Pereira SR, Teixeira AL, Silva LCS (2006) Serum and cerebrospinal fluid levels of chemokines and Th2 cytokines in *Schistosoma mansoni* myeloradiculopathy. Parasite Immunol 28:473–478

Steinmann P, Keiser J, Bos R, Tanner M, Utzinger J (2006) Schistosomiasis and water resources development: systematic review, meta-analysis, and estimates of people at risk. Lancet Infect Dis 6:411–425

Wang P, Wang D, Chen SJ (2011) Establishment of a cerebral schistosomiasis experimental model in rabbits. Neurosci Bull 27:91–98

Wynn TA, Thompson RW, Cheever AW, Mentink-Kane MM (2004) Immunopathogenesis of schistosomiasis. Immunol Rev 201:156–167

Xu J, Lu XJ, Wang D, Wu MC, Chen SJ, Li JC, Wang P (2013) Preliminary study on establishing an animal model of neuroschistosomiasis by direct injection of *Schistosoma japonicum* eggs through skull [article in Chinese]. Zhongguo Xue Xi Chong Bing Fang Zhi Za Zhi 25:28–30

Neurocysticercosis: Neurology and Neurobiology

Arturo Carpio and Agnès Fleury

Abstract Neurocysticercosis, the most common parasitic brain disease worldwide, is due to the larvae infestation of *Taenia solium*. It is an endemic, neglected disease in poor countries with deprived sanitation, and is increasingly being reported in wealthy countries due to migration. Humans are the only definitive host of *T. solium*, while pigs are the intermediate hosts. Humans may become intermediate host by ingesting food or water contaminated by *T. solium* eggs.

Infection is associated with local and systemic immune-inflammatory responses modulated by the developmental stage of the parasite in the host (vesicular, colloidal. granular-nodular, and calcified stages) and by the central nervous system compartment where the parasites are located. Genetic diversity of cysticerci has been studied and the genome of *T. solium* is currently being sequenced.

The clinical manifestations are heterogeneous and depend mainly on the localization of cysts and immune response to the host. Seizures, headache, focal deficits and cognitive abnormalities are the most frequent manifestations. The prognosis is good; nevertheless, it may lead to long-term neurological sequels such as epilepsy and hydrocephalus.

Diagnosis is made mainly by neuroimaging, which is useful in the detection of evolutionary stage, number and localization of cysts. Immunological testing can be helpful; nonetheless, a negative test does not rule out the diagnosis.

A. Carpio (✉)
School of Medicine, University of Cuenca, Cuenca, Ecuador

G.H. Sergievsky Center, Faculty of Medicine, Columbia University, New York, NY, USA

Consultorios Sta Inés, F. Proaño y D. Cordova (esquina) 4to piso, Cuenca, Ecuador
e-mail: arturocarpio@etapanet.net

A. Fleury
Instituto de Investigaciones Biomédicas, Universidad Nacional Autónoma de México, México, D.F., México

Instituto Nacional de Neurologia y Neurocirugía, SS, México, D.F., México
e-mail: agnesf_2@hotmail.com; afleury@biomedicas.unam.mx

M. Bentivoglio et al. (eds.), *Neglected Tropical Diseases and Conditions of the Nervous System*, DOI 10.1007/978-1-4614-8100-3_7,

Treatment is complex and should be individualized, based on location and viability of the parasites. In most cases treatment is only symptomatic. Antihelminthic drugs are effective in approximately one-third of patients with parenchymal viable cysts. The most effective approach to Taeniasis/cysticercosis is prevention. This should be a primary public health focus for poor countries.

Keywords Seizures • Epilepsy • Imaging • Albendazole • Praziquantel • Parasitic infection • Taeniasis/cysticercosis • *Taenia solium*

1 Introduction

Neurocysticercosis (NC), caused by the presence in the central nervous system (CNS) of the larval phase of *Taenia solium* is a neglected disease still endemic in most developing countries, mainly in the low-income ones, of Latin America, sub-Saharan Africa, the Indian subcontinent, and Southern Asia (Winkler et al. 2008; Blocher et al. 2011; Fleury et al. 2011). NC is currently an emergent disease in some high-income countries due to the increase of migration and tourism. It still causes high morbidity and mortality, the burden of which is difficult to state due to underdiagnosis (Bhattarai et al. 2012). The economic losses it entails are also important for the population and the society in general (Croker et al. 2012).

2 Life Cycle: Transmission

Cysticercosis is caused by ingestion of the eggs of *T. solium* shed in the feces of a human tapeworm carrier. Pigs, that are coprophages, become infected when they are reared under a free-range system, a condition still frequent in the rural communities of endemic countries, where disposal of human feces is deficient.

After ingestion, *T. solium* eggs hatch in the intestine, liberating motile oncospheres that invade the intestinal wall and migrate through the bloodstream to different tissues, where they develop into cysticerci. The parasite life cycle is completed when humans ingest undercooked pork containing cysticerci, resulting in human tapeworm infection. Cysts evaginate and attach to the small intestine by their scolex. Adult tapeworms develop, and can reside in the human small intestine for years.

Incidentally, infection of humans by cysticerci is possible mainly through the ingestion of food or water contaminated with *T. solium* eggs. Self-infection can also occur, mainly by orofaecal contamination and possibly by reverse peristalsis. In humans, at least in Latin America, the preferential localization of cysticerci is the central nervous system (CNS).

Infection may be prevented by proper disposal of human feces around pigs, use of latrines, and keeping pigs indoors (Goodman et al. 1999). In the case of infected pork meat, this must be cut into slices of 5 cm or less in thickness, and fried for at least 1 h or boiled for 2 or 3 h (Aluja et al. 1987). Freezing of the meat before consumption at –5 °C for 4 days or at –15 °C for 3 days is also efficient (Sotelo and Marin 1987).

3 Causative Agent

Adult *T. solium*, like other species of the gender Taenia is a plane worm. Its head (scolex) consists of four suckers and a rostellum with two rows of hooks, which are essential to fix the parasite in the human small intestine. The head thins to form a neck from which the proglottids are produced. These form a chain, named strobila (1.5–5 m in length). The proglottids nearest to the neck are immature, while those located at the end of the strobila are differentiated and contain about 50,000 eggs. After 2 or 3 months of infection, 4 or 5 gravid proglottids are released every day in the host's faeces. Eggs contained in each proglottid are at different grades of maturation; almost 50 % of them contain an infective completely developed oncosphere. The immature ones can mature out of the host and can persist for weeks, viable and infective, in water, soil, and vegetation. Eggs are round and measure around 20–40 μm. They are morphologically indistinguishable from the eggs of other *Taeniae* and are protected from the environment by a rigid structure called the embryophore (Laclette et al. 1982). In the larval phase, there are two morphologically distinct types of cysticerci (Rabiela et al. 1989). The most common one is the "cellulosae", a small vesicle (0.5–2 cm in diameter) containing a small invaginated scolex similar to the scolex of the adult taenia. This form of cysticerci is found in muscle, subcutaneous tissue, and cerebral parenchyma. The "racemosus" form constitutes a hydropic change that leads to a large parasite structure composed of a conglomeration of vesicles of different sizes. It has no evident scolex and is mainly located in the cerebral ventricles and in the basal subarachnoid space of the CNS. As cestodes lack a digestive track, they obtain their nutrients and excrete their waste through their tegument surface by absorption and diffusion.

Four developmental phases have been described in the life cycle of cysticerci (Escobar 1983). In the first one (vesicular phase), the parasite is alive, and the cyst contains the invaginated larva of 4–5 mm that lies in a transparent liquid (Figs. 1 and 4a). In the following one, (colloidal phase, Figs. 2 and 4b), the cyst is adherent and surrounded by a connective capsule. The content of the vesicle loses fluidity and becomes milky. In the third stage (nodular granular phase), the size of the cyst has diminished, and its membrane cannot be easily identified as it is completely attached to the collagen capsule. Finally, in the last stage (calcified phase, Figs. 3 and 5b) a hard, completely calcified nodule is identified, reduced to less than half of its original size. In adjacent tissue, astrogliosis and low-grade inflammatory reaction are present. It is interesting to note that this temporal sequence varies according to the localization of the parasites. In particular, when parasites are in the cisterns of the subarachnoid space or in the ventricular system, the vesicular phase is much longer than in other locations, facilitating parasite growth. It is also interesting to note that calcification of parasites in these locations is extremely rare.

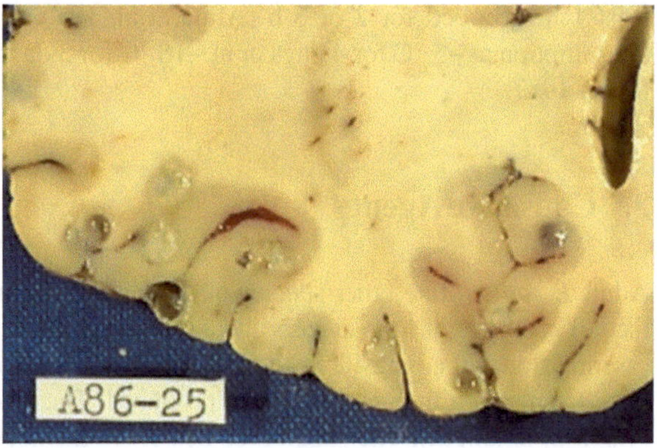

Fig. 1 Coronal section of the brain showing several vesicular cysticerci

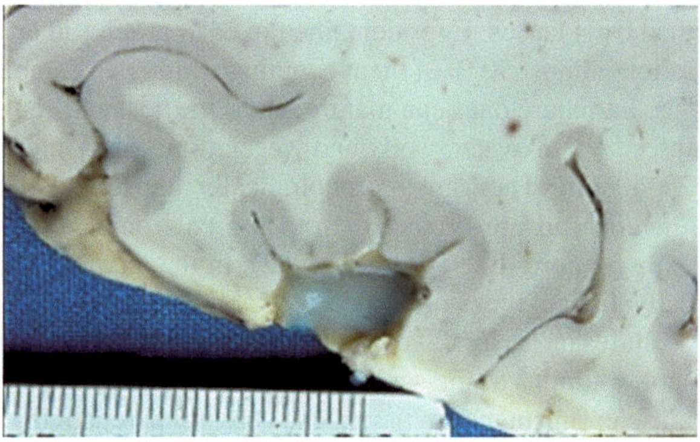

Fig. 2 Coronal section of the brain showing colloidal cysticerci

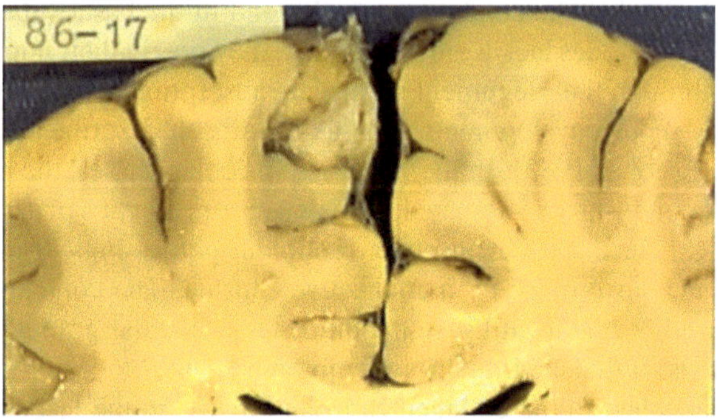

Fig. 3 Coronal section of the brain showing a calcified cysticercus

Genetic diversity of cysticerci collected in pork has been investigated. The comparison of parasites from different continents shows significant genetic differentiation, suggesting diverse evolutionary paths (Vega et al. 2003; Maravilla et al. 2003). On the other hand, when cysticerci of the nearest geographical regions or from different tissues of the same pig were studied, DNA diversity was within the range expected from a recombination process (Vega et al. 2003; Bobes et al. 2010).

The genome of *T. solium* is currently being sequenced in Mexico and only partial results have been published at the time of this review. The information generated will provide a powerful resource for the study of taeniasis/cysticercosis helping to address fundamental questions such as the molecular basis of host-parasite specificity and mechanisms of parasite pathogenesis, among others (Aguilar-Díaz et al. 2006). Genome size has been estimated to be between 251 and 270 Mb and only about 24 % of the sequences are homologous to mammalian genes, in particular humans. The sequences lacking homologues in humans are strong candidates for investigation of treatment, diagnosis and vaccination (Aguilar-Díaz et al. 2006).

4 Immunological Response

In their larval phase, cysticerci express a complex and diverse set of antigens. A few of them are specific of *T. solium*, but many others cross-react with those of other helminths. Some of them are present in all the phases of the parasites, while others are phase-specific. Most patients have detectable antibodies, generally of the IgG class, against parasites in serum and cerebrospinal fluid (CSF). They can also be detected in the saliva (Bueno et al. 2000), The four subclasses of IgG are increased in symptomatic NC patients compared to the asymptomatic patients. The relevance of antibodies to parasite damage has not been extensively studied. Although some observations indicate that antibodies can have a protective action, the demonstration of a net beneficial effect is still lacking.

Characteristics of the local inflammatory reaction associated with the presence of the parasite depend on the CNS compartment where the parasites are located. When located in the parenchyma or in the subarachnoid space, inflammation is confined around the parasite. In these cases, the CSF is in general normal and immunohistochemical studies of the inflammatory infiltrate around the cysticerci reveals the presence of specific IgM and plasma cells, natural killer (NK) lymphocytes, macrophages, granulocytes, and T cells. Intensity of the inflammatory reaction varies, based upon parasite evolutionary phase, being almost absent in the vesicular phase and intense in the colloidal one (Alvarez et al. 2002). When the parasites are located in the basal subarachnoid space or in the ventricular system, inflammation is mainly seen in the CSF with the presence of lymphocytic pleocytosis, mild elevation of proteins, and hypoglycorrhachia, with increased levels of all IgG subclasses as well as interleukin (IL) -6, -5, and -10 (Chavarría et al. 2005). This reaction can affect the leptomeninges with the formation of a dense exudate containing collagen fibers, lymphocytes, and multinucleated giant cells resulting in meningeal thickening as well as occlusion of Luschka and Magendie foramina that can lead to chronic hydrocephalus (Escobar 1983). In these cases, inflammation can cause damage in vascular and nervous structures remote from the parasites.

Systemic immunological changes are also observed. In the blood of patients with inflammatory parasites located in the subarachnoid space or in the ventricular system, a decreased proliferation of specific cells without production of cytokines, and an increase of the proportion of T regulatory cells, significantly correlated with the increase of these cells in CSF, have been observed (Chavarria et al. 2006; Adalid-Peralta et al. 2012). On the other hand, patients with asymptomatic calcified NC present a specific proliferative reaction, with production of cytokines predominantly of the Th_2 type (IL-4, IL-5, IL-13, IL-12) and a higher level of specific IgG4, compared to persons exposed to parasites but non-infected (Chavarría et al. 2003).

Parasites have developed different protective mechanisms to evade the immune surveillance of the host. Particularly, antigen B, the most frequent antigen recognized by the serum of patients, can bind factor C1q from the complement system, a property that can prevent the potential toxicity of antibody-mediated parasite damage (Laclette et al. 1992). The presence of a great amount of immunoglobulin on the surface of the cysticercus can mask its presence in the immunological system (Flisser et al. 1986).

5 Clinical Manifestations

One of the most intriguing aspects of NC is that presumably a high percentage of the individuals harboring NC remain asymptomatic. Among the symptomatic group, clinical manifestations of NC are determined mainly by the evolutive phase and location of the parasite within the CNS, as well as by the intensity of the immunological response of the patient (Loureiro das Chagas et al. 2003; Patel et al. 2006). The combination of these factors, among others, explains the great heterogeneity and the absence of a specific clinical picture. A systematic review was conducted to estimate the clinical manifestation frequencies of symptomatic NC (Carabin et al. 2011). Among NC patients seen in neurology clinics, about 79 % had seizures/epilepsy, 38 % severe headaches, 16 % focal deficits, and 12 % signs of increased intracranial pressure. Several other symptoms were also reported in less than 10 % of the patients.

5.1 *Parenchymal Neurocysticercosis*

The most common clinical manifestation of parenchymal NC is epileptic seizure, which occurs in 60–90 % of cases. The symptomatology of altered mental state and psychiatric manifestations remains poorly described in the literature (Carabin et al. 2011). In two studies (Forlenza et al. 1997; Carpio et al. 2008) which provided definitions of clinical manifestations, depression was reported in about 53 % and 15 % of the patients, respectively. A recent study reports a spectrum of cognitive abnormalities, including dementia (Rodrigues et al. 2012).

5.1.1 Neurocysticercosis and Epilepsy

Seizures associated with NC may be categorized as either acute symptomatic (Carpio et al. 1998; Murthy and Yangala 1999) or as unprovoked, remote symptomatic (epilepsy, if recurrent). This differentiation is very important, due to its implications concerning treatment and prognosis. The classification of seizure types in patients with NC varied among studies (Cukiert et al. 1994). It seems that either generalized seizures or partial seizures with secondary generalization are most commonly reported, while complex partial seizures are less frequent (Carpio et al. 2013). Seizures may occur at any evolutionary phase of the parasite. In a recent prospective cohort study (Kelvin et al. 2011), transitional cysts have been found to be associated with a significantly higher probability of seizure in the univariate analysis. However, this association was lost after adjusting for patient age and gender as well as for number and location of the cysts. Patients with cysts in the parietal and frontal lobes were also more likely to present seizures.

5.1.2 Epileptogenesis and Neurocysticercosis

So far, the mechanism by which the calcified neurocysticercal lesions (CNL) cause seizures or epilepsy is not known (Antoniuk et al. 2001; Carpio et al. 2013; Rathore et al. 2012). This has been attributed to residual perilesional gliosis that results in chronic epileptogenic foci (Leite et al. 2000). CNL are frequently encountered in computed tomography (CT) scans of asymptomatic individuals, and studies from Latin American countries report that the majority are incidental lesions (Fleury et al. 2010; Leite et al. 2000; Kowacs et al. 2006). These observations would question the epileptogenicity of CNL.

Edema associated with CNL has also been suggested to be implicated in epileptogenicity. In fact, episodic appearance of edema surrounding the CNL after seizures has been described (Nash et al. 2008). In this study, the authors argue that episodic release of cysticercal antigens from the calcified lesions can lead to inflammation, perilesional edema, and seizures. However, it is not clear whether this edema is the cause or the consequence of seizure (Leite et al. 2000).

5.1.3 Neurocysticercosis as Etiology of Epilepsy

Although it is clear that epilepsy is the main symptom of NC, the relevance of NC as a generator of epilepsy is debated. The percentage of patients with NC in epileptic patients of endemic countries varies among studies from 14 to 54 % (Ndimubanzi et al. 2010). It is interesting, however, to note that some observations may question the causal relationship between NC and epilepsy (Sakamoto et al. 1999). Parasite location may be remote from the apparent epileptogenic region, so that there is no correlation between the NC burden of lesions and the severity of epilepsy; patients with severe refractory seizures may have only one calcified lesion, while, on the other hand, there are patients with multiple cysts or calcifications but no seizures (Carpio et al. 1998; Ferreira et al. 2002).

It is important to note that both NC and epilepsy are common diseases in most developing countries, suggesting both causal and fortuitous relationships between these pathological conditions (Sakamoto et al. 1999; Terra-Bustamante et al. 2005). In particular, in cross-sectional studies investigating the etiology of intractable epilepsy in Brazil, NC turned out to mostly represent a coexistent pathology (Pal et al. 2000; Velasco et al. 2006).

5.1.4 Risk of Seizure Recurrence in Patients with Neurocysticercosis

Some studies have reported that NC patients with acute symptomatic seizures have a good prognosis in terms of remission of seizures (Carpio and Hauser 2009; Singhi et al. 2000; Manreza 2000; Ferreira et al. 2001; Goel et al. 2010). Other studies have reported that most patients have a high risk of seizure recurrence, and suggest that prognosis improves after antihelminthic treatment (Garcia et al. 2004). Prospective cohort studies have determined a risk between 17 and 56 % of seizure recurrence after a first seizure due to NC, depending on the viability of the parasite. The risk is greater in the transitional forms and diminishes in the viable or calcified forms (Carpio and Hauser 2002; De Souza et al. 2009; Sharma et al. 2011; Thussu et al. 2008).

5.2 Extraparenchymal Neurocysticercosis (Basal Subarachnoid Space and Ventricular System)

At these locations (around 15–30 % of cases) headache and signs of elevated intracranial pressure due to the obstruction of CSF circulation are the most frequent symptoms. These occur in 88 % of the cases, in comparison with 10 % of cases with parenchymal location (Carpio et al. 1994). Inflammation of meninges can also generate cranial nerve dysfunction, chiasmatic syndrome, and cerebral infarcts secondary to vasculitis (Agapejev et al. 2007; Cárdenas et al. 2010a, b). When hydrocephalus is present (Fig. 5a), the mortality rate is high (50 %), and most patients die within 2 years after CSF shunting (Cárdenas et al. 2010a, b; Sotelo et al. 1988). This is why ventricular and basal cisternal locations are considered to be malignant forms of NC (Estañol et al. 1986).

Spinal cord cysticercosis is rare. Patients experience nonspecific clinical manifestations, such as nerve root pain or spinal cord compression syndromes, according to the level of the lesion (Alsina et al. 2002). Severe forms of NC may exceptionally occur, including cysticercotic encephalitis, and result in permanent neurological sequels, such as amaurosis.

6 Host and Parasite Factors Modulating Clinical Presentation

6.1 Age

In addition to being less common in children, NC clinical manifestations clearly vary in different age groups (Kelvin et al. 2009a, b; Rosenfeld et al. 1996). Most cases of childhood NC present mild to moderate symptomatology and single lesions (Ruiz-Garcia et al. 1997; Kelvin et al. 2011). A study specifically carried out to compare the clinical manifestations between pediatric and adult NC patients (Sáenz et al. 2006) has reported statistically significant differences: seizures were more frequent in children (80.4 % vs. 56.1 %) and intracranial hypertension and headaches were more frequent in adults (27.2 % vs. 15.2 % and 35.1 % vs. 21.7 %, respectively). Although these age-related differences seem clear, a single effect of age is difficult to demonstrate, since various confounding factors are probably involved (Kelvin et al. 2009a, b).

Most paediatric cases show a single enhancing lesion, also named solitary cysticercus granuloma (Singh et al. 2010). This lesion is a common finding in patients with newly identified seizures in developing countries. These patients have some benign and transitory clinical manifestations, predominantly partial or partial secondary generalized seizures. Single enhancing lesions tend to resolve spontaneously, without anticysticercal treatment or surgery, since the parasite is already in the degenerative phase and will eventually disappear or become calcified.

6.2 Gender

Inflammation surrounding parenchymal cysticerci is more intense in women (Kelvin et al. 2009a, b), and multiple degenerating parasites localized in the CNS parenchyma are also more frequently reported in young women. Regardless of the localization of the parasite, the inflammatory response, as expressed by CSF cellularity, is more intense in women (Fleury et al. 2010). There are significant gender and age differences in the local immune response (Kelvin et al. 2009a, b). It has been suggested that both age and gender influence the strength of the host's immune response. The odds of having transitional cysts are higher for female patients than for males (Kelvin et al. 2011).

6.3 Genes

Clinical heterogeneity across geographical areas is well documented. Most cases from the Indian subcontinent present single degenerative lesions, whereas those from Latin America present few viable cysts (Singh et al. 2010). These differences are probably due to complex interactions between the host, parasite, and environmental factors (Singh 1997; Fleury et al. 2010). Genetic differences in *T. solium* cysticerci have been reported from different countries (Maravilla et al. 2008; Vega et al. 2003) and may contribute to clinical variations among countries. Genetic susceptibility to NC has been suggested on the basis of positive association of HLA-DRBII 13 with single, contrast-enhancing CT lesions (Jain et al. 1999). However, neither familial aggregation of seizures in first degree relatives of NC patients with seizures (Kelvin et al. 2009a, b) nor significant aggregation of NC cases in families have been found (Fleury et al. 2006), arguing in favor of the involvement of multiple genes.

7 Diagnosis

Diagnosis of NC cannot rely only on clinical grounds, since there are no specific clinical manifestations of NC.

7.1 Neuroimaging

Diagnosis of NC is mainly made by neuroimaging. MRI is more sensitive than CT for the diagnosis of viable and degenerating cysticerci, since it improves recognition of the perilesional edema and degenerative changes of the parasite, as well as of cysts located inside the ventricles or the subarachnoid space. However, CT is more sensitive than MRI for the detection of calcifications.

CT or MRI can identify the four developmental phases of cysticerci when located in the brain parenchyma. In the vesicular phase, the CT scan depicts circumscribed, round, hypodense areas, varying in size and number, without enhancement by contrast media (Zee et al. 2000). In the MRI, the vesicular larva appears with a CSF-like intensity signal on all sequences, with no surrounding high signal on T2-weighted images (Lucato et al. 2007; Mont'Alverne Filho et al. 2011). Both MRI and CT may show a high intensity or hyperdense, 2–3 mm mural nodule depicting the scolex, within some vesicular cysts (Fig. 4a). MRI sequences such as axial fluid attenuated inversion recovery (FLAIR) detect a significantly higher number of scolices than other sequences, which is helpful for improving the diagnosis of NC (Lucato et al. 2007). As the cyst degenerates, the contrast-enhanced CT scan shows an annular (colloidal phase) or nodular (nodular phase) enhancement surrounded by irregular perilesional edema (Fig. 4b). In this phase, the fluid content gives a slightly higher signal than CSF and is sometimes isodense with the parenchyma on MRI-T1 and/or proton density-weighted, and high signal on T2 images. The capsule shows a higher signal than the adjacent brain tissue, with thick ring enhancement on T1 images, while on T2 images there is a low ring signal surrounded by high signal lesion, due mostly to edema (Dumas et al. 1997; Zee et al. 2000). When the cyst dies it may disappear or become an inactive calcified nodule with homogeneous high density on CT or low intensity on proton-weighted MRI (Fig. 5b).

When the parasites are located in the subarachnoid space or within the ventricular system, recognition of parasites with MRI is more difficult, as parasites emit an intensity signal similar to that of the CSF, generally do not enhance after intravenous administration of contrast, and commonly lack a scolex. Thus, often only indirect signs of the

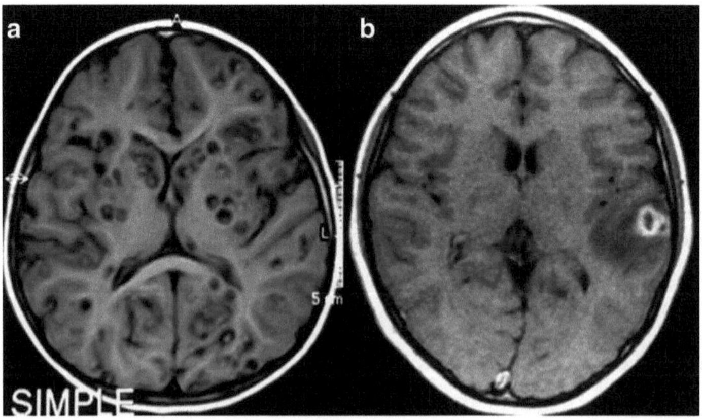

Fig. 4 MRI of parenchymal neurocysticercosis. (**a**) Viable cysts showing the scolex. (**b**) Colloidal cyst appearing as a ring-enhancing lesion with perilesional edema

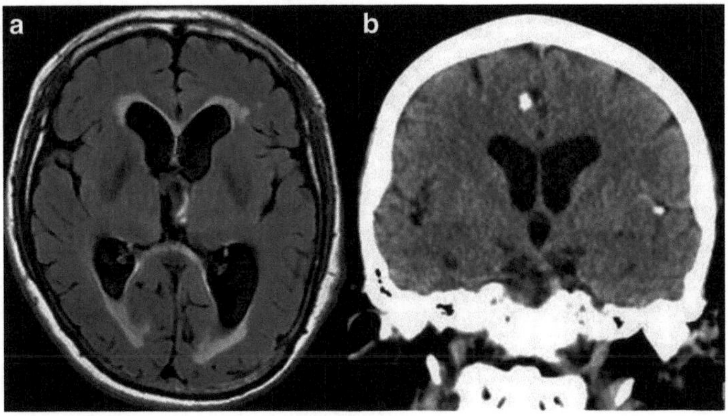

Fig. 5 Imaging findings in patients with neurocysticercosis. (**a**) MRI showing hydrocephalus with transependymal edema. (**b**) CT showing calcified cysticerci associated with ventriculomegalia

presence of the parasite are available, such as the unilateral enlargement of the basal cistern (Lucato et al. 2007) (Fig. 6). Specific MRI sequences including diffusion-weighted MRI magnetization transfer ratio (MTR), 3D constructive interference in steady state (3DCISS), fast imaging employing steady state acquisition sequences (FIESTA; Fig. 6a), and FLAIR sequences have proven to be more sensitive tools to visualize the cyst wall (Braga et al. 2004; Fleury et al. 2011). In case of meningeal inflammatory process, gadolinium enhancement of MRI or contrast-enhanced CT may depict leptomeningeal thickening (Kioumehr et al. 1995) and hydrocephalus (Fig. 5a).

7.2 *Immunodiagnosis*

No optimal immunological test for NC diagnosis is yet available. The difficulties of developing a sensitive and specific immunological test for NC diagnosis are mainly the result of the proper characteristics of the disease. An immunological test useful for medical practice must be specific in terms of CNS localization and should differentiate between viable and non-viable forms of the parasite.

Different immunological tests have been developed. The most widely used tests aim at the detection of specific antibodies. In these cases, different types of antigens have been used: crude antigens or partially purified antigenic extracts of *T. solium* or of the related parasites *T. crassiceps* or *T. saginata*, and recombinant or synthetic proteins. Techniques have also evolved, from complement fixation test, indirect hemagglutination, to the enzyme-linked immunosorbent assay (ELISA) and enzyme-linked immunoelectrotransfer blot (EITB) assay (Tsang et al. 1989; Brandt et al. 1992), which are the two main techniques currently in use. Also, detection of antigens by monoclonal or polyclonal antibodies using the ELISA technique has been developed (Brandt et al. 1992).

Comparisons of these tests have given divergent results, in part due to differences in methodology between studies. Despite these sources of variations, it seems that EITB in sera has higher sensibility and specificity than ELISA to detect antibodies, at least when

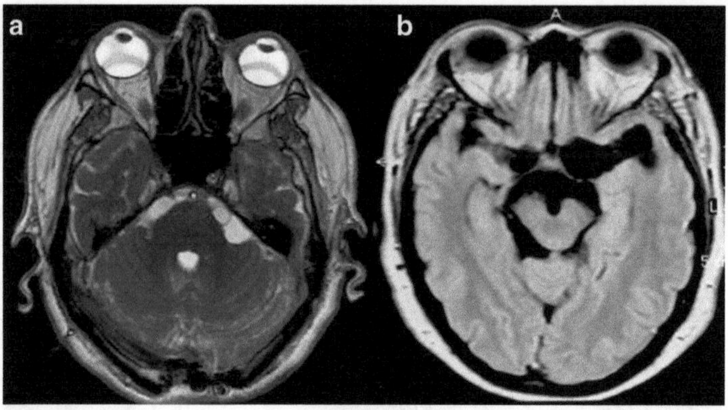

Fig. 6 MRI of extraparenchymal neurocysticercosis. (**a**) Cysticerci located in the cerebellopontine cistern (FIESTA sequence). (**b**) Cysticerci located in the basal subarachnoideal cisterns

carried out in non-endemic areas (Ramos-Kuri et al. 1992). It is also clear that sensitivity of both tests falls in cases with a single parasite, when parasites are located in the brain parenchyma or are calcified (Wilson et al. 1991; Singh et al. 1999). In CSF, performance of Ab-ELISA and EITB seems to be quite similar (Proaño-Narvaez et al. 2002; Michelet et al. 2011). Antigen detection in sera or CSF is highly specific for detecting viable extra-parenchymal parasites (Fleury et al. 2007). Another important point is the feasibility of carrying out ELISA and EITB techniques in endemic countries. In fact, while ELISA can be accomplished in 20 h and costs around $2.00 per sample, EITB assay requires almost 2 weeks, sophisticated equipment, highly skilled personnel, and its cost is up to ten times greater than ELISA (Proaño-Narvaez et al. 2002).

Despite the current immunological and imaging advances, the diagnosis of NC is still a challenge in many patients. Del Brutto et al. (2001) proposed diagnostic criteria for NC based on clinical, imaging, immunological and epidemiological features. This proposal, though not validated so far, may be useful to identify patients with parenchymal, but not extraparenchymal, forms of NC (Machado 2010). Such diagnostic criteria should be revised to incorporate current scientific knowledge, in order to achieve a new consensus on the diagnosis of NC.

8 Treatment

The treatment of NC should be individualized, based on the pathogenesis and natural history of the disease in each patient. Therapy in most cases is limited to symptomatic treatment with anti-seizure medication (ASM) for patients with seizures. Regarding duration of the ASM following an acute NC episode, some clinicians routinely continue ASM for 1 year, but shorter or longer intervals have also been recommended (Carpio et al. 2013; Takayanagui et al. 2011). It is assumed that the risk of seizures is substantial as long as there is an active ongoing process as characterized by persistence of edema around the degenerating lesion. Because of this, CT scan is a useful tool for these treatment decisions. It is appropriate to monitor cyst activity with CT scanning or MRI and to continue ASM until resolution of the acute lesion (Carpio et al. 2013). Seizures occurring in individuals after resolution of edema and reabsorption or calcification of the degenerating cyst should be considered unprovoked; in this situation, long-term ASM is warranted (Fig. 7).

Mannitol or oral glycerol is used if high intracranial pressure is a feature; analgesics should be given for headache. Corticosteroids are often administered for NC, on the premise that they reduce inflammation and edema around dying parenchymal cysts, and are also recommended for treatment of large subarachnoid cysts and arachnoiditis. However, the doses, duration, form and, importantly, timing of administration of corticosteroids are not clear. Alternative non-hormonal anti-inflammatory agents are dextrochlorpheniramine, ketoprofen, and immunosuppressants such as azathioprine and methotrexate. The efficacy of these drugs in NC has not, however, been widely confirmed.

Surgery is currently restricted mainly to placement of ventricular shunts for hydrocephalus and, in occasional cases of accessible racemose subarachnoid cysts and intraventricular cysts, mainly by endoscopic approach (Goel et al. 2008; Proano et al. 2009; Torres-Corzo et al. 2010). Transitional or degenerative cysts, regardless of their size or location should not be biopsied or removed if differential diagnosis has been discarded, since the parasite is dead and will disappear or become calcified spontaneously (Singh et al. 2010).

Treatment for NC with antihelminthic drugs (AHD) has been available for at least 25 years, but their use has always been controversial. Praziquantel (PZQ) was used for the first time in México (Robles and Chavarrıa 1979) and albendazole (ALB) was used for

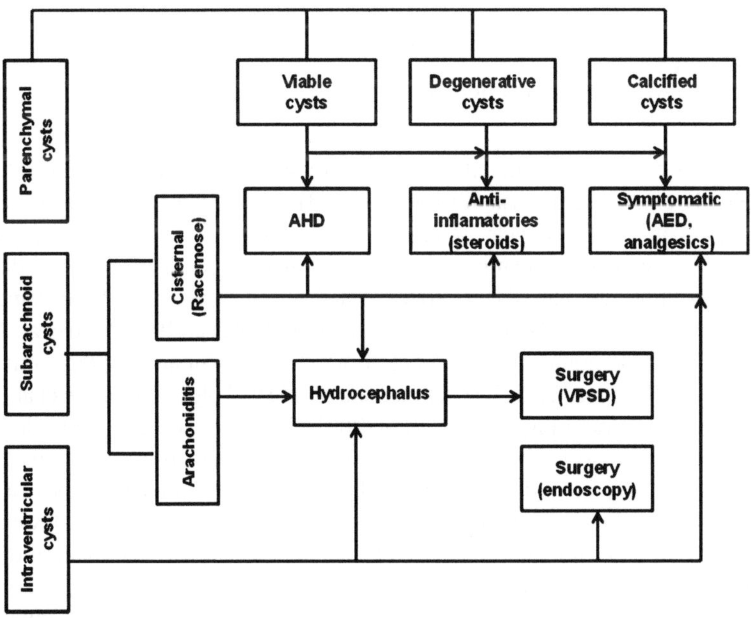

Fig. 7 Scheme for treatment of neurocysticercosis. *AHD* antihelminthic drugs, *AED* antiepileptic drugs, *VPS* ventriculoperitoneal shunt

the first time in China (Xiao et al. 1986) to cure NC. PZQ is an acylated isoquinole-pyrazine with broad antihelminthic activity. Its mechanism of action is not fully understood, but it is assumed that PZQ changes metabolism and intracellular calcium, with the main side effect of inhibition of muscle movements (Garcia-Dominguez et al. 1991). ALB is a benzimidazole with a broad antihelminthic spectrum, whose anticysticercal effect is by inhibition of glucose uptake by parasitic membranes, causing energy depletion (Lacey 1990). To date there are no controlled clinical trials to establish definitive doses and duration of treatment. The most frequent treatment scheme for PZQ is 50 mg/kg/day for 15 days and for ALB, 15 mg/kg/day for 8 days (Carpio et al. 2013).

A meta-analysis of treatment of NC reported disappearance of viable parenchymal cysts in 44 % of patients who were treated with ALB vs. 19 % of the placebo group. In contrast, disappearance of degenerative cysts in 72 % of patients of the ALB group and in 63 % of the placebo group was not statistically significant (Del Brutto et al. 2006). An Editorial comment of this meta-analysis (Del Brutto et al. 2006) stated that selected studies were small and heterogeneous, and only 5 of 11 were of good quality. The Editorial comment concluded that studies provided limited evidence of a modest efficacy of NC treatment. A recent Cochrane review (Abba et al. 2010) concluded that in adults with viable cysts the use of ALB is associated with a decrease in the number of cysts, but with no difference in seizure recurrence between ALB and no treatment.

Regarding extraparenchymal cysts, the management is even more unclear. Although AHD have demonstrated efficacy in some cases, it is also clear that not all the cases respond to current treatment (Das et al. 2007; Carpio et al. 2008; Cárdenas et al. 2010a, b). This is why it is a priority to search for new treatment alternatives (Jung-Cook 2012; Diazgranados-Sánchez et al. 2008).

9 Eradication of the Disease

Cysticercosis has been considered a neglected “tools-ready disease” (WHO 2007) and a potentially eradicable disease (Task Force 1993). Simple approaches, such as interrupting the parasite´s life cycle by placing fences to avoid the contact of pigs with human feces eventually contaminated with *T. solium* eggs could eradicate the disease. Some other strategies have been proposed and tested: massive AHD treatment of humans to reduce the number of tapeworm carriers (Allan et al. 1997; Sarti et al. 2000), health education programs, improvement of pig-management and sanitary conditions (Ngowi et al. 2008; Sarti et al. 2000), treatment of infected pigs (Sikasunge et al. 2008) and vaccination of rural pigs (Huerta et al. 2001; Sciutto et al. 2007; Gonzalez et al. 2001). Although almost all these strategies have shown relative efficacy, they have been evaluated only in studies on small cohorts and during limited periods of time. It is relevant to mention that a pilot control program in the poorest states of Mexico is ongoing, based on health and sanitary education associated with vaccination of pigs, and the preliminary results are encouraging (De Aluja et al. 2012).

The main and urgent strategy to be carried out for NC is the implementation of specific national and international health policies. Also, since pathogen transmission does not respect borders, the implementation of multinational and regional networks is indispensable to fight against this parasitic disease. Currently, networks of specialists have been organized in Africa, Asia, Europe and Latin America. These efforts must be encouraged, but as eradication depends mainly on social, economic and political factors, their real impact will depend on convincing national governments to make of this objective a public health priority.

10 Conclusions and Future Perspectives

NC is still endemic in countries with poor sanitation, and is being increasingly reported in high income countries due to migration and tourism. The morbidity and mortality caused by this preventable disease are unacceptable in the twenty-first century. As shown in this chapter, many questions are now solved, although problems related to immunology, diagnosis and treatment are still open. One of the reasons of the slow progress is that this disease occurs mainly in low income countries, where diagnosis tools are not available to all the population and where only few researchers are interested in the problem. Also, the epidemiologic transition, present in most of these countries, boosts the authorities to disregard it and to focus efforts to diseases of the "first world". Multidisciplinary research is needed, mainly to identify sensitive and specific diagnostic tools of low cost, to investigate the factors involved in the lack of response to treatment of some patients and to propose new therapeutic compounds. Knowledge of the *T. solium* genome will certainly represent an important step forward for many of these issues, inspiring optimism. In addition, close communication with local authorities is indispensable regarding prevention, because control programs will be successful only if they are involved in their application.

Acknowledgments We thanks the Neuropathological department of the Instituto Nacional de Neurologia y Neurocirugia (Mexico, DF) for the macroscopic images of cysticerci (Figs. 1, 2, and 3).

References

(1993) Recommendations of the International Task Force for Disease Eradication. MMWR Recomm Rep 42:1–38

Abba K, Ramaratnam S, Ranganathan LN (2010) Antihelmintics for people with neurocysticercosis. Cochrane Database Syst Rev 3, CD000215

Adalid-Peralta L, Fleury A, García-Ibarra TM, Hernández M, Parkhouse M, Crispín JC, Voltaire-Proaño J, Cárdenas G, Fragoso G, Sciutto E (2012) Human neurocysticercosis: in vivo expansion of peripheral regulatory T cells and their recruitment in the central nervous system. J Parasitol 98:142–148

Agapejev S, Pouza AF, Bazan R, Faleiros AT (2007) Clinical and evolutive aspects of hydrocephalus in neurocysticercosis. Arq Neuropsiquiatr 65:674–680

Aguilar-Díaz H, Bobes RJ, Carrero JC et al (2006) The genome project of *Taenia solium*. Parasitol Int 55(Suppl):S127–S130

Allan JC, Velasquez-Tohom M, Fletes C, Torres-Alvarez R, Lopez-Virula G, Yurrita P, Soto de Alfaro H, Rivera A, Garcia-Noval J (1997) Mass chemotherapy for intestinal *Taenia solium* infection: effect on prevalence in humans and pigs. Trans R Soc Trop Med Hyg 91:595–598

Alsina GA, Johnson JP, McBride DQ, Rhoten PRL, Mehringer CM, Stokes JK (2002) Spinal neurocysticercosis. Neurosurg Focus 12:1–7

Aluja A, Escobar A, Escobedo F, Flisser A et al (1987) Cisticercosis: una recopilación actualizada de los conocimientos básicos para el manejo y control de la cisticercosis causada por *Taenia solium*. Fondo de Cultura Económica, Biblioteca de la salud, México

Alvarez JI, Colegial CH, Castaño CA et al (2002) The human nervous tissue in proximity to granulomatous lesions induced by *Taenia solium* metacestodes displays an active response. J Neuroimmunol 127:139–144

Antoniuk SA, Bruck I, Dos Santos LH et al (2001) Seizures associated with calcifications and edema in neurocysticercosis. Pediatr Neurol 25:309–311

Bhattarai R, Budke CM, Carabin H et al (2012) Estimating the non-monetary burden of neurocysticercosis in Mexico. PLoS Negl Trop Dis 6:e1521

Blocher J, Schmutzhard E, Wilkins PP, Gupton PN, Schaffert M, Auer H, Gotwald T, Matuja W, Winkler AS (2011) A cross-sectional study of people with epilepsy and neurocysticercosis in Tanzania: clinical characteristics and diagnostic approaches. PLoS Negl Trop Dis 5:e1185

Bobes RJ, Fragoso G, Reyes-Montes M del R, Duarte-Escalante E, Vega R, de Aluja AS, Zúñiga G, Morales J, Larralde C, Sciutto E (2010) Genetic diversity of *Taenia solium* cysticerci from naturally infected pigs of central Mexico. Vet Parasitol 168:130–135

Braga F, Rocha AJ, Gomes HR, Filho GH, Silva CJ, Fonseca RB (2004) Noninvasive MR cisternography with fluid-attenuated inversion recovery and 100% Supplemental O2 in the evaluation of neurocysticercosis. AJNR Am J Neuroradiol 25:295–297

Brandt JR, Geerts S, De Deken R et al (1992) A monoclonal antibody-based ELISA for the detection of circulating excretory-secretory antigens in *Taenia saginata* cysticercosis. Int J Parasitol 22:471–477

Bueno EC, Vaz AJ, Machado LD, Livramento JA (2000) Neurocysticercosis: detection of IgG, IgA and IgE antibodies in cerebrospinal fluid, serum and saliva samples by ELISA with *Taenia solium* and *Taenia crassiceps* antigens. Arq Neuropsiquiatr 58:18–24

Carabin H, Ndimubanzi PC, Budke CM, Nguyen H, Qian Y, Cowan LD, Stoner JA, Rainwater E, Dickey M (2011) Clinical manifestations associated with neurocysticercosis: a systematic review. PLoS Negl Trop Dis 5:e1152

Cárdenas G, Carrillo-Meza R, Jung H et al (2010a) Subarachnoidal neurocysticercosis non-responsive to cysticidal drugs: a case series. BMC Neurol 10:16

Cárdenas G, Jung H, Ríos C, Fleury A, Soto-Hernández JL (2010b) Severe cysticercal meningitis: clinical and imaging characteristics. Am J Trop Med Hyg 82:121–125

Carpio A, Fleury A. Hauser WA (2013) Neurocysticercosis: five new things. Neurol Clin Pract 3:118–125

Carpio A, Hauser W (2002) Prognosis for seizure recurrence in patients with newly diagnosed neurocysticercosis. Neurology 59:1730–1734

Carpio A, Hauser WA (2009) Epilepsy in the developing world. Curr Neurol Neurosci Rep 9:319–326

Carpio A, Placencia M, Santillan F, Escobar A (1994) A proposal for a classification of neurocysticercosis. Can J Neurosci 21:43–47

Carpio A, Escobar A, Hauser WA (1998) Epilepsy and cysticercosis: a critical review. Epilepsia 39:1025–1040

Carpio A, Kelvin E, Bagiella E, Leslie D, Leon P, Andrews H, Hauser WA, Ecuadorian Neurocysticercosis Group (2008) Effects of albendazole treatment on neurocysticercosis: a randomized controlled trial. J Neurol Neurosurg Psychiatry 79:1050–1055

Chavarría A, Roger B, Fragoso G, Tapia G, Fleury A, Dumas M, Dessein A, Larralde C, Sciutto E (2003) TH2 profile in asymptomatic *Taenia solium* human neurocysticercosis. Microbes Infect 5:1109–1115

Chavarría A, Fleury A, García E, Márquez C, Fragoso G, Sciutto E (2005) Relationship between the clinical heterogeneity of neurocysticercosis and the immune-inflammatory profiles. Clin Immunol 116:271–278

Chavarría A, Fleury A, Bobes RJ, Morales J, Fragoso G, Sciutto E (2006) A depressed peripheral cellular immune response is related to symptomatic neurocysticercosis. Microbes Infect 8:1082–1089

Croker C, Redelings M, Reporter R, Sorvillo F, Mascola L, Wilkins P (2012) The impact of neurocysticercosis in California: a review of hospitalized cases. PLoS Negl Trop Dis 6:e1480

Cukiert A, Puglia P, Scapolan HB, Vilela MM, Marino Júnior R (1994) Congruence of the topography of intracranial calcifications and epileptic foci. Arq Neuropsiquiatr 52:289–294

Das K, Mondal GP, Banerjee M, Mukherjee BB, Singh OP (2007) Role of antiparasitic therapy for seizures and resolution of lesions in neurocysticercosis patients: an 8 year randomised study. J Clin Neurosci 14:1172–1177

De Aluja A, Morales Soto J, Sciutto E (2012) A programme to control taeniosis-cysticercosis (*Taenia solium*) in Mexico. In: Rodriguez-Morales A (ed) Current topics in tropical medicine. InTech, New York, available from: http://www.intechopen.com/books/current-topics-in-tropical-medicine/a-program-to-control-taeniosis-cysticercosis-taenia-solium-in-mexico

De Souza A, Thennarasu K, Yeshraj G, Kovoor JM, Nalini A (2009) Randomized controlled trial of albendazole in new onset epilepsy and MRI confirmed solitary cerebral cysticercal lesion: effect on long-term seizure outcome. J Neurol Sci 276:108–114

Del Brutto OH, Rajshekhar V, White AC Jr et al (2001) Proposed diagnostic criteria for neurocysticercosis. Neurology 57:177–183

Del Brutto OH, Roos KL, Coffey CS, Garcıa HH (2006) Meta-analysis: cysticidal drugs for neurocysticercosis: albendazole and praziquantel. Ann Intern Med 45:43–51

Diazgranados-Sánchez JA, Barrios-Arrázola G, Costa JL, Burbano-Pabon J, Pinzón-Bedoya J (2008) Ivermectin as a therapeutic alternative in neurocysticercosis that is resistant to conventional pharmacological treatment. Rev Neurol 46:671–674

Dumas JL, Visy JM, Belin C, Gaston A, Goldlust D, Dumas M (1997) Parenchymal neurocysticercosis: follow-up and staging by MRI. Neuroradiology 39:12–18

Escobar A (1983) The pathology of neurocysticercosis. In: Palacios E, Rodriguez-Carbajal J, Taveras JM (eds) Cysticercosis of central nervous system. Charles C Thomas, Springfield

Estañol B, Corona T, Abad P (1986) A prognostic classification of cerebral cysticercosis: therapeutic implications. J Neurol Neurosurg Psychiatry 49:1131–1134

Ferreira LS, Zanardi VA, Scotoni AE, Li LM, Guerreiro MM (2001) Childhood epilepsy due to neurocysticercosis: a comparative study. Epilepsia 42:1438–1444

Ferreira LS, Li LM, Zanardi VA, Guerreiro MM (2002) Number and viability of parasite influence seizure frequency in children with neurocysticercosis. Arq Neuropsiquiatr 60:909–911

Fleury A, Morales J, Bobes RJ, Dumas M, Yánez O, Piña J, Carrillo-Mezo R, Martínez JJ, Fragoso G, Dessein A, Larralde C, Sciutto E (2006) An epidemiological study of familial neurocysticercosis in an endemic Mexican community. Trans R Soc Trop Med Hyg 100:551–558

Fleury A, Hernández M, Avila M, Cárdenas G, Bobes RJ, Huerta M, Fragoso G, Uribe-Campero L, Harrison LJ, Parkhouse RM, Sciutto E (2007) Detection of HP10 antigen in serum for diagnosis and follow-up of subarachnoidal and intraventricular human neurocysticercosis. J Neurol Neurosurg Psychiatry 78:970–974

Fleury A, Escobar A, Fragoso G, Sciutto E, Larralde C (2010) Clinical heterogeneity of human neurocysticercosis results from complex interactions among parasite, host and environmental factors. Trans R Soc Trop Med Hyg 104:243–250

Fleury A, Carrillo-Mezo R, Flisser A, Sciutto E, Corona T (2011) Subarachnoid basal neurocysticercosis: a focus on the most severe form of the disease. Expert Rev Anti Infect Ther 9: 123–133

Flisser A, Espinoza B, Tovar A, Plancarte A, Correa D (1986) Host–parasite relationship in cysticercosis: immunologic study in different compartments of the host. Vet Parasitol 20:95–102

Forlenza OV, Filho AH, Nobrega JP, dos Ramos Machado L, de Barros NG, de Camargo CH, da Silva MF (1997) Psychiatric manifestations of neurocysticercosis: a study of 38 patients from a neurology clinic in Brazil. J Neurol Neurosurg Psychiatry 62: 612–616

Garcia HH, Pretell EJ, Gilman RH, Martinez SM, Moulton LH, Del Brutto OH, Herrera G, Evans CA, Gonzalez AE; Cysticercosis Working Group in Peru (2004) A trial of antiparasitic treatment to reduce the rate of seizures due to cerebral cysticercosis. N Engl J Med 350:249–258

Garcia-Dominguez C, Correa D, Rabiela MT, Flisser A (1991) Praziquantel treatment of muscle *Taenia solium* cysticercosis. 4. Reversible in vitro effect. Parasitol Res 77:691–696

Goel RK, Ahmad FU, Vellimana AK, Suri A, Chandra PS, Kumar R, Sharma BS, Mahapatra AK (2008) Endoscopic management of intraventricular neurocysticercosis. J Clin Neurosci 15:1096–1101

Goel D, Mittal M, Bansal KK, Singhal A (2010) Natural history of solitary cerebral cysticercosis cases after albendazole therapy: a longitudinal follow-up study from India. Acta Neurol Scand 121:204–208

Gonzalez AE, Gavidia C, Falcon N, Bernal T, Verastegui M, Garcia HH, Gilman RH, Tsang VC; Cysticercosis Working Group in Peru (2001) Protection of pigs with cysticercosis from further infections after treatment with oxfendazole. Am J Trop Med Hyg 65:15–18

Goodman K, Ballagh SA, Carpio A (1999) Case control study of seropositivity for cysticercosis in Cuenca, Ecuador. Am J Trop Med Hyg 60:70–74

Huerta M, de Aluja AS, Fragoso G, Toledo A, Villalobos N, Hernández M, Gevorkian G, Acero G, Díaz A, Alvarez I, Avila R, Beltrán C, Garcia G, Martinez JJ, Larralde C, Sciutto E (2001) Synthetic peptide vaccine against *Taenia solium* pig cysticercosis: successful vaccination in a controlled field trial in rural Mexico. Vaccine 20:262–266

Jain S, Padma MV, Kanga U, Mehra NK, Puri A, Maheshwari MC (1999) Family studies and human leukocyte antigen class II typing in Indian probands with seizures in association with single small enhancing computed tomography lesions. Epilepsia 40:232–238

Jung-Cook H (2012) Pharmacokinetic variability of antihelmintics: implications for the treatment of neurocysticercosis. Expert Rev Clin Pharmacol 5:21–30

Kelvin EA, Carpio A, Bagiella E, Leslie D, Leon P, Andrews H, Hauser WA; Ecuadorian Neurocysticercosis Group (2009a) The association of host age and gender with inflammation around neurocysticercosis cysts. Ann Trop Med Parasitol 103:487–499

Kelvin EA, Carpio A, Hesdorffer DC, Bagiella E, Leslie D, Leon P, Andrews H, Hauser WA; Ecuadorian Neurocysticercosis Group (2009b) Investigation of familial aggregation of seizures in neurocysticercosis patients. Epilepsy Res 84:67–71

Kelvin EA, Carpio A, Bagiella E, Leslie D, Leon P, Andrews H, Hauser WA; Ecuadorian Neurocysticercosis Group (2011) Seizure in people with newly diagnosed active or transitional neurocysticercosis. Seizure 20:119–125

Kioumehr F, Dadsetan MR, Feldman N, Mathison G, Moosavi H, Rooholamini SA, Verma RC (1995) Postcontrast MRI of cranial meninges: leptomeningitis versus pachymeningitis. J Comput Assist Tomogr 19:713–720

Kowacs PA, Rogacheski E, Muzzio J, Werneck LC (2006) The role of the irritative zone and of the number and distribution of calcifications in the severity of epilepsy associated with intracranial calcifications. Arq Neuropsiquiatr 64:905–911

Lacey E (1990) Mode of action of benzimidazoles. Parasitol Today 6:112–115

Laclette JP, Ornelas Y, Merchant MT, Willms K (1982) Ultrastructure of the surrounding envelopes of *Taenia solium* eggs. In: Flisser A, Willms K, Laclette JP et al (eds) Cysticercosis: present state of knowledge and perspectives. Academic, New York

Laclette JP, Shoemaker CB, Richter D, Arcos L, Pante N, Cohen C, Bing D, Nicholson-Weller A (1992) Paramyosin inhibits complement C1. J Immunol 148:124–128

Leite JP, Terra-Bustamante VC, Fernandes RM, Santos AC, Chimelli L, Sakamoto AC, Assirati JA, Takayanagui OM (2000) Calcified neurocysticercotic lesions and postsurgery seizure control in temporal lobe epilepsy. Neurology 55:1485–1491

Loureiro das Chagas MG, D'Oliveira A Jr, Tavares-Neto J (2003) Clinical aspects of neurocysticercosis at semi-desert region of the Brazilian northeast. Arq Neuropsiquiatr 61:398–402

Lucato LT, Guedes MS, Sato JR, Bacheschi LA, Machado LR, Leite CC (2007) The role of conventional MR imaging sequences in the evaluation of neurocysticercosis: impact on characterization of the scolex and lesion burden. AJNR Am J Neuroradiol 28:1501–1504

Machado LR (2010) The diagnosis of neurocysticercosis: a closed question? Arq Neuropsiquiatr 68:1–2

Manreza ML (2000) Epilepsia e neurocisticerose. In: Guerreiro CAM, Guerreiro MM, Cendes F, Lopes-Cendes I (eds) Epilepsia. Lemos Editorial, Sao Paulo

Maravilla P, Souza V, Valera A et al (2003) Detection of genetic variation in *Taenia solium*. J Parasitol 89:1250–1254

Maravilla P, Gonzalez-Guzman R, Zuñiga G, Peniche A, Dominguez-Alpizar JL, Reyes-Montes R, Flisser A (2008) Genetic polymorphism in *Taenia solium* cysticerci recovered from experimental infections in pigs. Infect Genet Evol 8:213–216

Michelet L, Fleury A, Sciutto E, Kendjo E, Fragoso G, Paris L, Bouteille B (2011) Human neurocysticercosis: comparison of different diagnostic tests in cerebrospinal fluid. J Clin Microbiol 49:195–200

Mont'Alverne Filho FE, Machado Ldos R, Lucato LT, Leite CC (2011) The role of 3D volumetric MR sequences in diagnosing intraventricular neurocysticercosis: preliminary results. Arq Neuropsiquiatr 69:74–78

Murthy JM, Yangala R (1999) Acute symptomatic seizures – incidence and etiological spectrum: a hospital-based study from South India. Seizure 8:162–165

Nash TE, Pretell EJ, Lescano AG, Bustos JA, Gilman RH, Gonzalez AE, Garcia HH; Cysticercosis Working Group in Peru (2008) Perilesional brain oedema and seizure activity in patients with calcified neurocysticercosis: a prospective cohort and nested case-control study. Lancet Neurol 7:1099–1105

Ndimubanzi PC, Carabin H, Budke CM, Nguyen H, Qian YJ, Rainwater E, Dickey M, Reynolds S, Stoner JA (2010) A systematic review of the frequency of neurocyticercosis with a focus on people with epilepsy. PLoS Negl Trop Dis 4:e870

Ngowi HA, Carabin H, Kassuku AA, Mlozi MR, Mlangwa JE, Willingham AL 3rd (2008) A health-education intervention trial to reduce porcine cysticercosis in Mbulu District, Tanzania. Prev Vet Med 85:52–67

Pal DK, Carpio A, Sander JWAS (2000) Neurocysticercosis and epilepsy. J Neurol Neurosurg Psychiatry 68:137–143

Patel R, Jha S, Yadav RK (2006) Pleomorphism of the clinical manifestations of neurocysticercosis. Trans R Soc Trop Med Hyg 100:134–141

Proaño JV, Torres-Corzo J, Rodríguez-Della Vecchia R, Guizar-Sahagun G, Rangel-Castilla L (2009) Intraventricular and subarachnoid basal cisterns neurocysticercosis: a comparative study between traditional treatment versus neuroendoscopic surgery. Childs Nerv Syst 25:1467–1475

Proaño-Narvaez JV, Meza-Lucas A, Mata-Ruiz O, García-Jerónimo RC, Correa D (2002) Laboratory diagnosis of human neurocysticercosis: double-blind comparison of enzyme-linked immunosorbent assay and electroimmunotransfer blot assay. J Clin Microbiol 40:2115–2118

Rabiela MT, Rivas A, Flisser A (1989) Morphological types of *Taenia solium* cysticerci. Parasitol Today 5:357–359

Ramos-Kuri M, Montoya RM, Padilla A, Govezensky T, Díaz ML, Sciutto E, Sotelo J, Larralde C (2002) Immunodiagnosis of neurocysticercosis. Disappointing performance of serology (enzyme-linked immunosorbent assay) in an unbiased sample of neurological patients. Arch Neurol 49:633–636

Rathore C, Thomas B, Kesavadas C, Radhakrishnan K (2012) Calcified neurocysticercosis lesions and hippocampal sclerosis: potential dual pathology? Epilepsia 53:e60-62. doi: 10.1111/j.1528-1167.2011.03386.x

Robles C, Chavarrıa M (1979) Presentacion de un caso clınico de cisticercosis cerebral tratado medicamente con un nuevo farmaco: Praziquantel. Salud Pub Mex XXI:603–618

Rodrigues CL, de Andrade DC, Livramento JA, Machado LR, Abraham R, Massaroppe L, Lucato LT, Caramelli P (2012) Spectrum of cognitive impairment in neurocysticercosis: differences according to disease phase. Neurology 78:861–866

Rosenfeld EA, Byrd SE, Shulman ST (1996) Neurocysticercosis among children in Chicago. Clin Infect Dis 23:262–268

Ruiz-Garcia M, Gonzalez-Astiazarán A, Rueda-Franco F (1997) Neurocysticercosis in children. Clinical experience in 122 patients. Childs Nerv Syst 13:608–612

Sáenz B, Ruíz-Garcia M, Jiménez E, Hernández-Aguilar J, Suastegui R, Larralde C, Sciutto E, Fleury A (2006) Neurocysticercosis: clinical. radiologic, and inflammatory differences between children and adults. Pediatr Infect Dis J 25:801–803

Sakamoto AC et al (1999) Cysticercosis and epilepsy. In: Kotagal P, Luders HO (eds) The epilepsies: etiologies and prevention. Academic, San Diego

Sarti E, Schantz PM, Avila G, Ambrosio J, Medina-Santillán R, Flisser A (2000) Mass treatment against human taeniasis for the control of cysticercosis: a population-based intervention study. Trans R Soc Trop Med Hyg 94:85–89

Sciutto E, Rosas G, Hernández M, Morales J, Cruz-Revilla C, Toledo A, Manoutcharian K, Gevorkian G, Blancas A, Acero G, Hernández B, Cervantes J, Bobes RJ, Goldbaum FA, Huerta M, Diaz-Orea A, Fleury A, de Aluja AS, Cabrera-Ponce JL, Herrera-Estrella L, Fragoso G, Larralde C (2007) Improvement of the synthetic tri-peptide vaccine (S3Pvac) against porcine *Taenia solium* cysticercosis in search of a more effective, inexpensive and manageable vaccine. Vaccine 25:1368–1378

Sharma P, Garg RK, Verma R, Singh MK, Shukla R (2011) Risk of seizure recurrence in patients of new-onset partial seizure having a solitary cysticercus granuloma of brain or normal neuro-imaging. J Neurol Sci 301:21–26

Sikasunge CS, Johansen MV, Willingham AL 3rd, Leifsson PS, Phiri IK (2008) *Taenia solium* porcine cysticercosis: viability of cysticerci and persistency of antibodies and cysticercal antigens after treatment with oxfendazole. Vet Parasitol 158:57–66

Singh G (1997) Neurocysticercosis in South-Central America and the Indian subcontinent. A comparative evaluation. Arq Neuropsiquiatr 55:349–356

Singh G, Kaushal V, Ram S, Kaushal RK, Dhanuka AK, Khurana S (1999) Cysticercus immunoblot assay in patients with single, small enhancing lesions and multilesional neurocysticercosis. J Assoc Physicians India 47:476–479

Singh G, Rajshekhar V, Murthy JM, Prabhakar S, Modi M, Khandelwal N, Garcia HH (2010) A diagnostic and therapeutic scheme for a solitary cysticercus granuloma. Neurology 75:2236–2245

Singhi P, Ray M, Singhi S, Khandelwal N (2000) Clinical spectrum of 500 children with neurocysticercosis and response to albendazole therapy. J Child Neurol 15:207–213

Sotelo J, Marin C (1987) Hydrocephalus secondary to cysticercotic arachnoiditis. A long-term follow-up review of 92 cases. J Neurosurg 66:686–689

Sotelo J, Guerrero J, Rubio F (1988) A new classification based on active and inactive forms. Arch Intern Med 148:442–445

Takayanagui OM, Odashima NS, Bonato PS, Lima JE, Lanchote VL (2011) Medical management of neurocysticercosis. Expert Opin Pharmacother 12:2845–2856

Terra-Bustamante VC, Coimbra ER, Rezek KO, Escorsi-Rosset SR, Guarnieri R, Dalmagro CL, Inuzuka LM, Bianchin MM, Wichert-Ana L, Alexandre V, Takayanagui OM, Araújo D, dos Santos AC, Carlotti CG, Walz R, Markowitsch HJ, Sakamoto AC (2005) Cognitive performance of patients with mesial temporal lobe epilepsy and incidental calcified neurocysticercosis. J Neurol Neurosurg Psychiatry 76:1080–1083

Thussu A, Chattopadhyay A, Sawhney IM, Khandelwal N (2008) Albendazole therapy for single small enhancing CT lesions (SSECTL) in the brain in epilepsy. J Neurol Neurosurg Psychiatry 79:272–275

Torres-Corzo JG, Tapia-Pérez JH, Vecchia RR, Chalita-Williams JC, Sánchez-Aguilar M, Sánchez-Rodríguez JJ (2010) Endoscopic management of hydrocephalus due to neurocysticercosis. Clin Neurol Neurosurg 112:11–16

Tsang VC, Brand JA, Boyer AE (1989) An enzyme-linked immunoelectrotransfer blot assay and glycoproteins antigens for diagnosing human cysticercosis (*Taenia solium*). J Infect Dis 174: 1007–1009

Vega R, Piñero D, Ramanankandrasana B, Dumas M, Bouteille B, Fleury A, Sciutto E, Larralde C, Fragoso G (2003) Population genetic structure of *Taenia solium* from Madagascar and Mexico: implications for clinical profile diversity and immunological technology. Int J Parasitol 33:1479–1485

Velasco TR, Zanello PA, Dalmagro CL, Araújo D Jr, Santos AC, Bianchin MM, Alexandre V Jr, Walz R, Assirati JA, Carlotti CG Jr, Takayanagui OM, Sakamoto AC, Leite JP (2006) Calcified cysticercotic lesions and intractable epilepsy: a cross sectional study of 512 patients. J Neurol Neurosurg Psychiatry 77:485–488

WHO (2007) Global plan to combat neglected tropical diseases 2008-2015. World Health Organization, Geneva

Wilson M, Gryan R, Fried J, Ware D, Schantz P, Pilcher JB, Tsang VCW (1991) Clinical evaluation of the cysticercosis enzyme-linked immunoelectrotransfer blot in patients with neurocysticercosis. J Infect Dis 164:1007–1009

Winkler AS, Blocher J, Auer H, Gotwald T, Matuja W, Schmutzhard E (2008) Anticysticercal and antitoxocaral antibodies in people with epilepsy in rural Tanzania. Trans R Soc Trop Med Hyg 102:1032–1038

Xiao ZX, Zhao CY, Liu LP (1986) Albendazole treatment in cerebral cysticercosis. Chin J Int Med 25:100–102

Zee CS, Go JL, Kim PE, DiGiorgio CM (2000) Imaging of neurocysticercosis. Neuroimaging Clin N Am 10:391–407

Onchocerciasis: Neurological Involvement

Alfred K. Njamnshi, Anne-Cécile Zoung-Kanyi Bissek, and Daniel Etya'ale

Abstract Onchocerciasis or "river blindness" is an endemic parasitic disease caused by the filarial nematode *Onchocerca volvulus*, transmitted by a blackfly of the genius *Simulium*. It has a serious public health impact in 39 countries worldwide, particularly in sub-Saharan Africa where the burden is highest. The neurological involvement in onchocerciasis includes both the peripheral and central nervous system. The clinical manifestations constitute a spectrum of disorders spanning from the very stigmatizing onchocercal itch to optic nerve disease, epilepsy and post-treatment encephalopathy. There has been significant progress in recent years in the understanding of the neuroscience of itching in general, with the discovery of itch-specific neurons and their pathways. However, the exact mechanisms of the onchocercal itch, as well as the pathogenesis of optic nerve disease, epilepsy or encephalopathy are yet to be fully understood. Nevertheless, the relatively new concepts such as "pruritoception" on one hand, and "river epilepsy" on the other hand, may become future active areas of neuroscience research. This may

A.K. Njamnshi (✉)
Faculty of Medicine and Biomedical Sciences, Ministry of Public Health,
The University of Yaoundé I, Yaoundé, Cameroon

Department of Neurology, Central Hospital Yaoundé, PO Box 25265, Yaoundé, Cameroon
e-mail: aknjamnshi@yahoo.co.uk

A.-C.Z.-K. Bissek
Faculty of Medicine and Biomedical Sciences, Ministry of Public Health,
The University of Yaoundé I, Yaoundé, Cameroon

D. Etya'ale
University of Cape Town, Cape Town, South Africa and Academia Ophthalmologica
Internationalis, San Francisco, California, USA

M. Bentivoglio et al. (eds.), *Neglected Tropical Diseases and Conditions of the Nervous System*,
DOI 10.1007/978-1-4614-8100-3_8, © Springer Science+Business Media New York 2014

lead eventually not only to novel basic and translational knowledge but also to drug development and improved control strategies for onchocerciasis and associated nervous system complications.

Keywords River blindness • Filarial • Nematodes • Itch • Optic nerve disease • Epilepsy • Encephalopathy • *Wolbachia*

1 Introduction: Disease Distribution, Clinical Forms and Control Programs

Onchocerciasis or "river blindness" is an endemic parasitic disease caused by the filarial nematode *Onchocerca volvulus* transmitted by a blackfly of the genus *Simulium* which breeds near fast-flowing rivers. Onchocerciasis occurs in 39 countries of the world (Crump et al. 2012), 31 of which are in sub-Saharan Africa, 6 in Latin America, and 2 in the Arabian Peninsula (Enk 2006; Boatin and Richards 2006; Siddiqui and Al-Khawajah 1991) (Fig. 1). Some 42 million people were infected

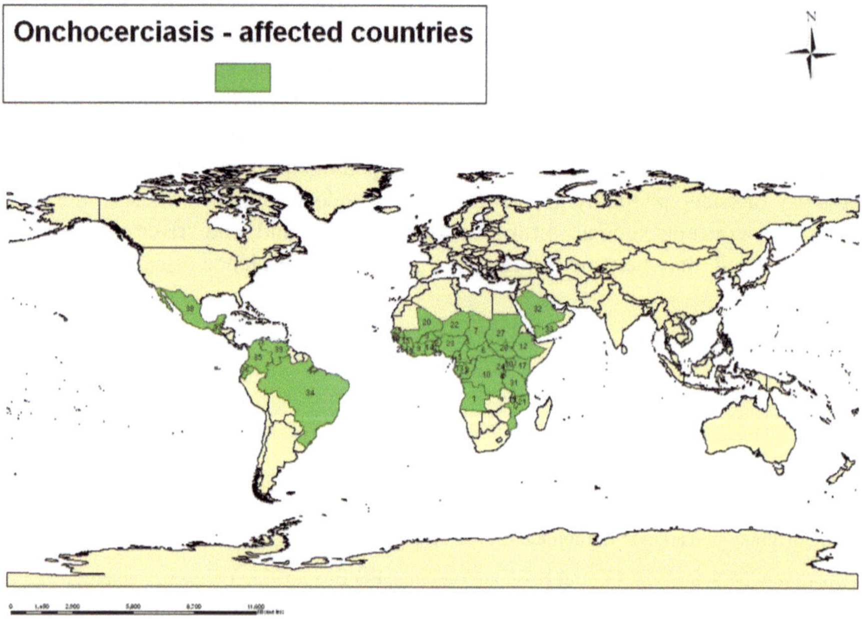

Fig. 1 Onchocerciasis-affected countries of the world (drawn by Njamnshi EN, BSc Geography). *Countries in Africa*: 1. Angola, 2. Benin, 3. Burkina Faso, 4. Burundi, 5. Cameroon, 6. Central African Republic, 7. Chad, 8. Congo, 9. Côte d'Ivoire, 10. Democratic Republic of Congo, 11. Equatorial Guinea, 12. Ethiopia, 13. Gabon, 14. Ghana, 15. Guinea, 16. Guinea-Bissau, 17. Kenya, 18. Liberia, 19. Malawi, 20. Mali, 21. Mozambique, 22. Niger, 23. Nigeria, 24. Rwanda, 25. Senegal, 26. Sierra Leone, 27. Sudan, 28. South Sudan, 29. Togo, 30. Uganda, 31. United Republic of Tanzania. *Countries in Asia*: 32. Saudi Arabia, 33. Yemen. *Countries in Latin America*: 34. Venezuela, 35. Mexico, 36. Guatemala, 37. Ecuador, 38. Colombia, 39. Brazil

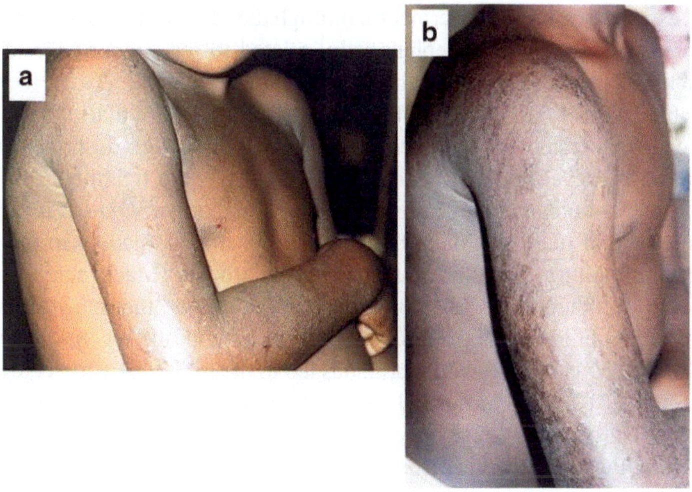

Fig. 2 Onchodermatitis in the acute (a) and chronic (b) forms

before the mass treatment programme in 1995 but this number reduced to nearly 26 million in 2008 (WHO 2010). The disease burden is highest in Africa in terms of disability-adjusted life years (DALYs). Sub-Saharan Africa is home to more than 99 % of infected persons and a further 87 million people are at risk of infection (Amazigo et al. 2006).

Classically, descriptions of the clinical features of onchocerciasis have been dominated by two groups of symptoms, namely skin and eye symptoms. The skin condition which manifests initially as a severe, stigmatizing and troublesome itch in the acute phase usually progresses to a chronic skin disease (Fig. 2). A clinical classification and grading system is now being used as follows: acute papular onchodermatitis (APOD), chronic papular onchodermatitis (CPOD), lichenified onchodermatitis (LOD), atrophy and depigmentation.

Onchocercal skin disease is more frequent in the forest regions, whereas onchocercal ocular disease and blindness are more prevalent in the savannah regions. These clinical differences are thought to be due to a variety of factors including differences in *Onchocerca volvulus* strains, vector type, host immunity and co-infection with other parasites (Enk 2006; Amazigo et al. 2006). Onchocercal ocular disease involves mainly keratitis, chorioretinitis and optic nerve disease (OND). In a study by Coffeng et al. (2012) in Cameroon, it was found that dermatological and ocular onchocercal diseases significantly occur concurrently. Also the association between onchocercal visual impairment and skin depigmentation could be partly explained, among other factors, by duration of exposure to infection and host characteristics, which may play a role in the pathogenesis of these co-morbidities (Coffeng et al. 2012).

In the last 15 years, there has been growing research interest in the association between onchocerciasis and epilepsy (Pion and Boussinesq 2012; Kaiser et al. 2010, 2011; Konig et al. 2010; Pion et al. 2009). Recently, with the introduction of mass control programmes for onchocerciasis by the World Health Organisation and its partners, especially the use of ivermectin through the Mectizan Donation Programme (Thylefors 2008), cases of encephalopathy related to the concurrence of *Loa loa* infection have been described in Central Africa (Gonzalez et al. 2012; Hoerauf et al. 2011; Babalola 2011; see also Bisoffi et al. 2014). *Loa loa* is a filarial nematode that causes recurrent, transitory skin edema and eosinophilia. As described further in this chapter, the development of encephalopathy following treatment of onchocerciasis has been associated with certain levels of concomitant *Loa loa* infection.

The efficacy of mass treatment has been demonstrated, leading to the hope for possible elimination of onchocerciasis from the tropics (Mackenzie et al. 2012; Thylefors 2008; Sauerbrey 2008; Boatin 2008). However, the many logistic challenges observed in these control programs, coupled with the recent observation of cases of filarial resistance to ivermectin, has led to the trial of new molecules (Nana-Djeunga et al. 2012; Mackenzie et al. 2012; Takesue et al. 2012; Osei-Atweneboana et al. 2011). These issues suggest that elimination may not be achievable in the near future. Thus, there is still need for research on clinical manifestations for better disease classification or staging for management purposes, and on pathogenetic mechanisms that will contribute to new drug development. In this chapter, we have attempted to summarize the major clinical nervous system manifestations of onchocerciasis, highlighting where possible, the current state of knowledge of the disease mechanisms and knowledge gaps.

2 Involvement of the Peripheral Nervous System: Pruritus (Onchocercal Itch)

2.1 *Definition and Clinical Assessment of Itch*

Itch or pruritus has been described for many years as an unpleasant sensation that evokes the urgent desire to scratch (Ikoma et al. 2011). The clinical measurement of pruritus intensity can be done using self-reporting scales similar to those used in pain medicine. These include the visual analogue scale (VAS), which is the most commonly used tool, the numerical rating scale (NRS) and the verbal rating scale (VRS) (Reich et al. 2011; Phan et al. 2011; Furue et al. 2013).

2.2 *Clinical Importance of Onchocercal Itch*

Although itch is the most disturbing symptom associated with onchocercal skin disease, it has not received a high priority. Nevertheless, several studies in Africa have shown the importance of this symptom. In a review by Murdoch (2010), troublesome itching was found to affect 32 % of the population aged 5 years and above. In another study in Nigeria, onchocercal itching (40 %) and onchocercal skin manifestations (34.3 %) were identified as the most troublesome signs and symptoms

(Mbanefo et al. 2010). In a multicenter study comprising seven sites in five different African countries to evaluate the impact of the community-directed treatment with ivermectin, it was found that severe onchocercal itching varied between 2 and 38 % before ivermectin mass treatment and dropped to 0.2–12 % after 5–6 years of treatment, indicating the importance of this treatment on symptom alleviation (Ozoh et al. 2011). This observation has interesting implications for the patient's quality of life given that the most worrisome consequence of onchocerciasis may be social exclusion or stigmatization (see Tabah et al. 2014). In the Nigerian study, about 35 % of the participants complained of stigmatization and psychological impact of the disease affecting almost all facets of their lives (Mbanefo et al. 2010).

2.3 *The Neuroscience of Itching:* Scratching the Brain for an Understanding of Itching

Although some evidence from the study of the neurobiology, neurophysiology and functional neuroimaging of itch has contributed progressively to a better understanding of the peripheral and central mechanisms of itching (Ikoma et al. 2011), the origin, transmission and perception of this socially embarrassing symptom is still a subject of discussion. Other open issues concern nerves and receptors involved in itch induction, relevant neural pathways and brain encoding of itch impulses. The answers to these questions are important not only for the understanding of the mechanisms but also for the development of novel, effective therapeutic agents (Tey and Yosipovitch 2011).

Concerning itch induction, the "intensity theory" hypothesizes that depending on the signal intensity, signal transduction on the same nerves leads to either pain (high intensity) or itch (low intensity) (Ikoma et al. 2011). On the other hand, the "labeled-line coding theory" hypothesizes the complete separation of pain and itch pathways (Ikoma et al. 2011).

The pathogenesis of acute and chronic (>6 weeks duration) pruritus is complex and involves the skin network of resident cells such as sensory neurons and transient inflammatory cells (lymphocytes) (Grundmann and Ständer 2011). Several classes of histamine-sensitive or histamine insensitive C fibers are involved in itch transmission in the skin. Ringkamp et al. (2011) have recently investigated, in humans and in an experimental model, the role of small nociceptive, myelinated fibers in non-histaminergic itch sensation. They tested the effect of a differential nerve block on itch produced by intradermal insertion of spicules from the pods of a cowhage plant (*Mucuna pruriens*) in psychophysical studies and also carried out electrophysiological experiments in the anesthetized monkey to investigate the responsiveness of cutaneous, nociceptive, myelinated afferents to different chemical stimuli (cowhage spicules, histamine, capsaicin). Their findings provided some evidence that activity in nociceptive, myelinated afferents contributes to cowhage-induced sensations, and that non-histaminergic itch is mediated through activity in both unmyelinated and myelinated afferents.

Itch-dedicated nociceptor neurons have been recently discovered (Greaves and Khalifa 2004). Specific receptors have been discovered on cutaneous and spinal neurons to be exclusively involved in the processing of pruritic signals. Some intracutaneous itch mediators have also been identified such as endovanilloids, proteases, cannabinoids, opioids, neurotrophins and cytokines. Relevant receptors are vanilloid receptor channels and proteinase-activated, cannabinoid, opioid, cytokine, and new histamine receptors. New data indicate that specific pruritic ligands carrying both itch and pain information are selectively recognized by different G protein-coupled receptors (GPCRs), and this information may be transduced through different intracellular circuits in the same neuron (Han and Simon 2011). These findings raise questions about the intracellular mechanisms that process and perhaps encode GPCR-mediated signals. In rodents, second-order neurons expressing gastrin-releasing peptide receptor (GRPR) and spinothalamic tract neurons are also involved (Jeffry et al. 2011). Some data suggest that the newly identified itch-specific neuronal pathways in the spinothalamic tract are distinct from pain pathways and relay to CNS regions that process peripheral pruritogenic stimuli (Paus et al. 2006).

Recent studies have suggested that GRPR-positive neurons constitute a long-sought labeled line for itch sensation in the spinal cord (Sun et al. 2009). It has been demonstrated that different dorsal horn neurons respond to histamine and allergic itch stimuli (Nakano et al. 2008). Recent studies also demonstrated that various neuronal receptors in the spinal cord are involved in pruritus (Cevikbas et al. 2011). Glutamate is the principal excitatory transmitter between C fibers and gastrin-releasing peptide (GRP)-positive dorsal horn neurons (Koga et al. 2011). Spinal bombesin-recognized neurons are critical to both the histamine-dependent and histamine-independent pathways for itch, and they mediate more non-histaminergic than histaminergic sensation of itch in mice (Han et al. 2012).

It has even been demonstrated that genetically susceptible persons would develop itch more intensely than others when exposed to visual cues, suggesting that interpersonal social cues can dramatically alter the subjective sensory experience of itch (Papoiu et al. 2011). The perception of the sensation of itch depends on various processes by a network of different brain regions contributing to the encoding of sensory, emotional, attention-dependent, cognitive-evaluative and motivational patterns (Pfab et al. 2012). These networks are yet to be fully understood.

3 Optic Neuropathy: *The Blind Spot of River Blindness?*

3.1 Epidemiology and Burden

The ocular disease associated with onchocercal infection has been well described, with sclerosing keratitis (Fig. 3), iridocyclitis, optic neuritis or atrophy, and chorioretinitis or chorioretinal atrophy accounting for visual morbidity (McKechnie et al. 1997). In a well illustrated study of the fundus oculi of 244 patients in Cameroon,

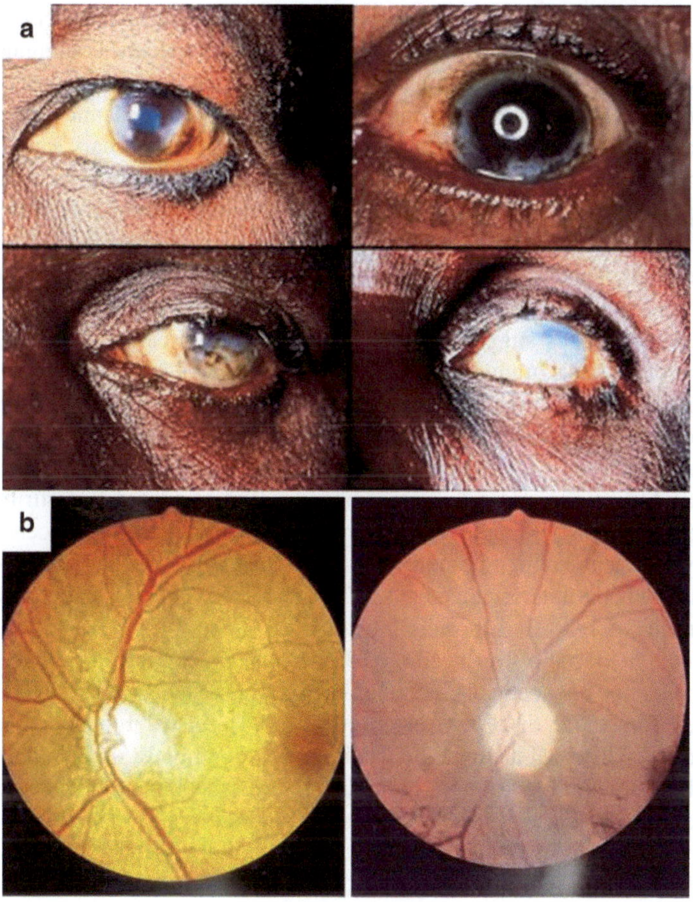

Fig. 3 (**a**) Ocular onchocerciasis: sclerosing keratitis (reproduced with permission from Etya'ale 2001). (**b**) Optic atrophy (APOC) and optic atrophy plus chorioretinitis (APOC) in onchocerciasis

Bird et al. (1976) working in Cameroon found that optic nerve disease, alone or in the presence of choroido-retinal changes, was responsible for a large proportion (about 88 %) of the blindness due to posterior segment lesions in onchocerciasis. In a forest-saving mosaic zone of south-eastern Nigeria, endemic for onchocerciasis, onchocerciasis-related eye disease was present in about 14 % of the investigated patients' cohort (Umeh et al. 1996) and constituted 21 % of the total number of eye disorders. In this study, a total of 78 of 235 subjects with visual impairment had onchocerciasis-related eye lesions, 35 of whom were blind in both eyes, and onchocerciasis-induced eye disease was the cause in 28. The prevalence of bilateral blindness from all causes in the study area was 4.1 %, while that from onchocerciasis-related causes was 3.3 %. The commonest onchocerciasis-induced lesions responsible for visual impairment and blindness were choroidoretinitis and OND.

A study in Nigeria (Yang et al. 2001) reported that the mean intraocular pressure was 1.58 mmHg lower in the individuals from the meso-endemic communities compared with those from the non-endemic communities despite the prevalence of peripheral anterior synechiae being higher in the meso-endemic communities. Glaucomatous optic nerve damage may therefore not be the primary cause of visual loss in ocular onchocerciasis as this occurs late and is probably preceded by other blinding onchocercal pathology. In the meso-endemic savannah communities of Kaduna State in Nigeria with a relatively high prevalence (9 %) of OND, it has been suggested that onchocercal infection was the single most important cause of OND, accounting for 50 % of all cases (Cousens et al. 1997). About 13 % of cases were associated with signs suggestive of glaucoma. Di-ethyl carbamazine use for the treatment of filariasis might be responsible for up to 30 % of all OND.

A survey of onchocercal eye disease was performed in the hyperendemic area of a rain forest focus of onchocerciasis in Esmeraldas Province in Ecuador (Cooper et al. 1995). A total of 785 skin snip positive individuals from black and Chachi American Indian communities were examined. The blindness rate attributable to onchocerciasis was 0.4 and 8.2 % were visually impaired. Onchocercal ocular lesions were seen in a high proportion of the study cases: punctate keratitis (34 %), microfilariae in the anterior chamber (29 %) and cornea (34 %), iridocyclitis (1.5 %), optic atrophy (5.1 %), and chorioretinopathy (28 %).

During a field trial of ivermectin, 6,831 persons age 5 years and above, living in 34 mesoendemic onchocercal communities in Kaduna State, northern Nigeria, were examined for ocular disease (Abiose et al. 1994). Visual function assessments included tests of visual acuity and visual fields. A total of 185 individuals (2.7 %) were bilaterally blind by acuity criteria with a further 28 blind by field constriction. The overall prevalence of blindness was 3.1 %. Additional 118 individuals were visually impaired by WHO criteria. The cause of blindness in 43 % of eyes in bilaterally blind patients was onchocerciasis. A further 11 % were blind from optic atrophy probably mostly onchocercal in origin. Glaucoma was the next most common cause of blindness in the bilaterally blind (11 %).

3.2 *Effect of Mass Treatment of Onchocerciasis on Optic Nerve Disease*

Reduction in incidence of OND with annual ivermectin to control onchocerciasis was reported by Abiose et al. (1993). There was evidence that ivermectin reduced the incidence of OND in subjects with microfilarial loads above 10 mf/mg but had little effect in those with lower loads. An updated Cochrane review (Ejere et al. 2012) showed evidence for a protective effect of mass treatment with ivermectin on visual field loss and OND in communities meso-endemic for the savannah strain of *O. volvulus*. However, whether these findings can be applied to communities with different endemicity and affected by the forest strain is unclear.

3.3 Etiopathogenesis of Optic Nerve Disease

If the anterior segment eye pathology has been associated with the intensity of onchocercal infection and host responses to the inflammatory process from degrading microfilariae, the posterior segment pathology is less well understood (McKechnie et al. 1997) although an immunogenetic mechanism has been proposed based on experimental models. The pathogenesis of OND in onchocerciasis is not well known, though it could involve host inflammatory responses to the degrading parasite antigens, as postulated in anterior ocular disease. A review by Hall and Pearlman (1999) pointed out that onchocerca-mediated keratitis results from an inflammatory response in the anterior portion of the eye. Furthermore, based on experimental models, the pathogenesis could be due to the host inflammatory response to degenerating parasites in the eye, with sensitized T helper cells and cytokines playing an important role (Hall and Pearlman 1999). Glaucoma may not be a major contributor to OND in onchocerciasis as it is not very common and occurs late (Yang et al. 2001). It may be suspected that, as in skin disease (Murdoch et al. 1997), the spectrum of onchocercal ocular disease, including OND, also has an immunogenetic basis possibly with specific HLA-DQ molecules associated with the level of immune response to parasite antigens present locally or systemically.

Thus, to date the basic mechanisms of onchocercal OND are unknown. A few experimental studies suggest an immunogenetic process but this needs to be confirmed. Due to this state of knowledge, we consider this the "blind spot of river blindness".

4 Epilepsy

4.1 Epidemiology and the Concept of "River Epilepsy"

Following the Lancet report of an improvement of epileptic seizures after ivermectin treatment (Kipp et al. 1992), there has been a growing amount of literature, based primarily on epidemiological data, on the association of onchocerciasis and epilepsy, ranging from classical seizures to more exotic forms of paroxystic manifestations like head nodding syndrome (Williams 2012; Pion and Boussinesq 2012; Kaiser et al. 1996, 1998, 2000, 2007, 2008, 2009, 2010, 2011; Konig et al. 2010; Pion et al. 2009; Winkler et al. 2008; Prischich et al. 2008; Katabarwa et al. 2008; Marin et al. 2006; Dozie et al. 2006; Twum-Danso 2004; Druet-Cabanac et al. 1999, 2004; Kamgno et al. 2003; Boussinesq et al. 2002; Farnarier et al. 2000; Arborio and Dozon 2000; Maegga 1998; Duke 1998; Burnham 1998; Newell et al. 1997; Kabore et al. 1996; Kipp et al. 1994; Kilian 1994; Ovuga et al. 1992). Although this association has been an issue of debate in the past two decades with some of the controversy arising probably from methodological issues in some of the studies, a recent systematic review and meta-analysis and other studies have

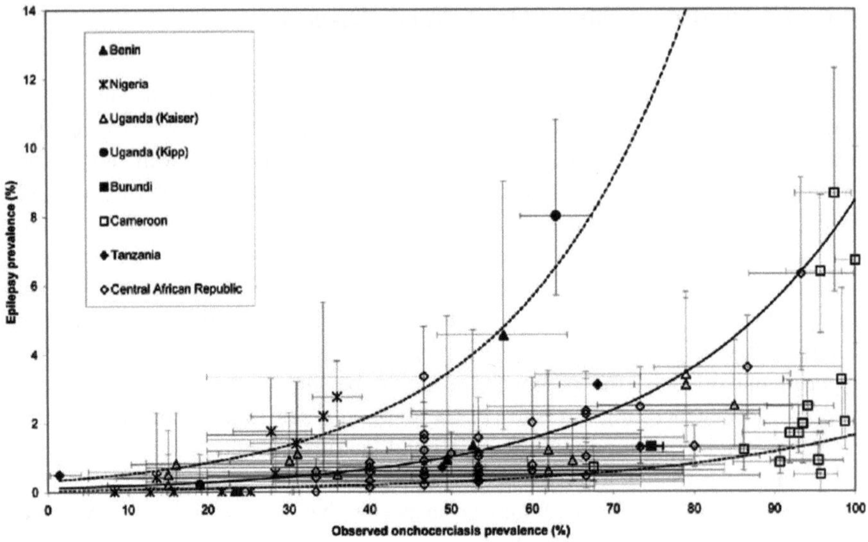

Fig. 4 Epilepsy prevalence versus onchocerciasis prevalence (reproduced with permission from Pion et al. 2009). *Error bars* represent 95 % exact confidence intervals. *Solid line*: predicted relationship estimated by random-effect logistic regression; *dashed lines*: 95 % confidence interval of the model predictions

provided further evidence of the existence of an association between onchocerciasis and epilepsy (Pion and Boussinesq 2012; Kaiser et al. 2010, 2011; Konig et al. 2010; Pion et al. 2009).

A review of eight studies (Pion et al. 2009) included a total of 79,270 individuals screened for epilepsy from west Africa (Benin and Nigeria), central Africa (Cameroon and Central African Republic) and east Africa (Uganda, Tanzania and Burundi). The prevalence of epilepsy ranged from 0 to 8.7 % whereas that of onchocerciasis ranged from 5.2 to 100 %. According to this review, the variation in epilepsy prevalence was consistent with a logistic function of onchocerciasis prevalence, with epilepsy prevalence being increased, on average, by 0.4 % for each 10 % increase in onchocerciasis prevalence (Fig. 4). Data from these studies on the association of the two disease conditions led to suggest the concept of "river epilepsy" as a parallel of "river blindness" (Pion et al. 2009).

A trend associating the presence and the mean number of subcutaneous onchocerca nodules has been demonstrated in onchocerciasis-associated epilepsy (Pion and Boussinesq 2012; Kaiser et al. 2011, 2012), suggesting some contribution of the severity of the infection to the pathogenesis of associated epilepsy. It has been suggested that other factors, including parasitic infections such as intestinal worms, may play a role in this association (Njamnshi et al. 2007).

4.2 Seizures and Evolution Following Onchocerciasis Treatment

One clinical and electroencephalographic study of epilepsy in an onchocercal endemic area in west Uganda revealed that 78 % of the seizures were partial while 22 % were generalized (Kaiser et al. 2000). The types of epileptic seizures and syndromes associated with onchocerciasis do not appear to show specific characteristics but it is worth cautioning that excellent clinical descriptions are rare.

There have been some reports of reduction of seizures following treatment of onchocerciasis with ivermectin (Kipp et al. 1992, 1994). Nevertheless, large cohort studies are needed to clarify these observations.

4.3 Open Questions on Onchocerciasis-Associated Epilepsy

The types of epileptic seizures and epileptic syndromes associated with onchocerciasis are yet to be well characterized in large studies. Furthermore, only few studies have involved the correlation of clinical and electrophysiological characteristics of "river epilepsy". What we now need to know is whether there is any causal relationship in the onchocerca-epilepsy association. If this suspected causality is confirmed, the involved pathogenetic mechanisms, the role of onchocerciasis severity in inducing epilepsy and relevant determining factors remain to be clarified. We also do not know the role of parasitic infections other than onchocerciasis and co-morbidities in inducing epilepsy and if these modify the course of epilepsy.

5 Post-treatment Encephalopathy

5.1 Epidemiology and Clinical Manifestations

The first reports of serious reactions, including encephalopathy, following mass treatment of onchocerciasis with ivermectin have been from Cameroon in areas with *Loa loa* co-endemicity (Gardon et al. 1997; Boussinesq et al. 1998; Kamgno et al. 2000; Bockarie and Deb 2010). These cases of encephalopathy are rare but can be fatal (Kamgno et al. 2008) and have been reported to be similar to those described following treatment with diethyl carbamazine (Gardon et al. 1997).

The clinical picture is characterized by disorders of consciousness progressing to coma, agitation, involuntary movements, urinary incontinence, motor and sensory deficits, hypertonia, sometimes absent myotatic reflexes, paraesthesia, transient grasping reflex, and brisk tendon reflexes (Gardon et al. 1997; Kamgno et al. 2008).

The electroencephalogram shows diffuse abnormalities which disappear after 3–6 months (Boussinesq et al. 2003). A pretreatment *Loa loa* microfilarial count of more than 50,000 mf/mL (*Loa loa* microfilariae were found in the cerebrospinal fluid), has been shown to predispose to the development of encephalopathy, with an incidence of 7 % (Gardon et al. 1997), decreasing to 0.7 % with a microfilarial count of 30,000 mf/mL (Kamgno et al. 2008). Lower doses of ivermectin have failed to prevent *Loa loa* encephalopathy in onchocerciasis (Kamgno et al. 2000). Genetic predisposition is thought to play a role in the development of encephalopathy as haplotypes associated with altered drug bioavailability especially in brain tissue were present as homozygotes in two of the patients with encephalopathy (50 %), but absent in controls (Bourguinat et al. 2010).

5.2 *Neurobiology of* Loa loa *Encephalopathy*

The exact physiopathological mechanisms of *Loa loa* encephalopathy associated with mass treatment of onchocerciasis are not well known. However, observations from a few autoptic cases have revealed vascular changes with a thickened basement membrane, perivascular inflammatory reaction and some neuronal degeneration (Kamgno et al. 2008). Suggested possible mechanisms include microfilarial obstruction of cerebral blood vessels leading to ischemia and degeneration, as retinal hemorrhages have been described in these patients (Moussala et al. 2004; Boussinesq et al. 2003), or microfilarial invasion of brain parenchyma as microfilariae escape from blood vessels following treatment.

As an alternative, an inflammatory process triggered by the release of degradation products of bacteria that live in a symbiotic relationship in and with microfilaria (*Wolbachia* endosymbionts) has also been suggested (Keiser et al. 2002; Tamarozzi et al. 2011). In fact, this novel understanding of the role of *Wolbachia* bacteria in the biology of microfilariae has led in recent years to new control strategies employing anti-*Wolbachia* agents, such as tetracyclines, with very promising results (Tamarozzi et al. 2011).

Since most onchocerciasis patients affected by *Loa loa* co-infection do not develop post-treatment encephalopathy, several co-factors in the development of encephalopathy have been proposed including trypanosomiasis, syphilis, *Plasmodium* infection and common cold (Kamgno et al. 2008). One could also question the possible role of HIV and tuberculosis co-infection, which have a significant effect on the immune system.

Besides secondary encephalopathy, primary encephalopathy could also occur in onchocerciasis, as in other conditions such as African trypanosomiasis (see Masocha et al. 2014) and HIV infection, potentially causing sleep and cognitive alterations. To the best of our knowledge, there are no studies on these aspects of neurological involvement in onchocerciasis.

6 Conclusions and Perspectives

The involvement of the nervous system in onchocerciasis leads to a wide spectrum of peripheral and central neurological signs and symptoms, some with very significant impact on the patient's quality of life and some even lethal. The involved basic neuronal and molecular mechanisms are largely unknown. The relatively new concepts such as "pruritoception" and "river epilepsy" may lead to further neuroscience research and eventual drug development and control strategies.

There is need for neurophysiological, neuroimaging, neuropathological and neurocognitive studies to better characterize the spectrum of ocular onchocerciasis, especially OND. The correlation of neuroscience data with immuno-genetic mechanisms could provide a better understanding of OND pathogenesis and guide future management and prevention of what we have referred to as the blind spot of river blindness.

Although considerable progress has been made in the description of the association of onchocerciasis and epilepsy, more clinical and neurophysiological studies are needed to characterize the different epileptic syndromes in this association. Further neuroscience research using neuroimaging, neuropathology, neuroimmunology and neurogenetics approaches will in the future help us to better understand the nature of the relationship between these conditions and possibly determine causality. Neuroepidemiological studies to determine the outcome of epilepsy in areas where onchocerciasis would have been eradicated will pave the way for more effective control strategies.

It is clear that a better understanding of the etiopathogenetic mechanisms of *Loa loa*-related post-onchocerciasis treatment encephalopathy and the exact role of *Wolbachia* or other factors in the next few years will open up the way for neuroscience research leading to more specific and effective management and preventive strategies.

Lastly, with the recent development of simple and user-friendly technology such as actigraphy, it has been possible to study sleep disorders in human African trypanosomiasis even in the difficult field conditions (Njamnshi et al. 2012) (see also Buguet et al. 2014) and this technology could be applied to onchocerca patients. Similarly, with the adaptation of western neuropsychological tests and development of normative data in tropical countries (Sacktor et al. 2005; Njamnshi et al. 2008; Kanmogne et al. 2010) it would be of interest to apply these tests in areas where sophisticated neuroimaging technology is not available for the study of possible cognitive disorders in onchocerciasis.

References

Abiose A, Jones BR, Cousens SN, Murdoch I, Cassels-Brown A, Babalola OE, Alexander ND, Nuhu I, Evans J, Ibrahim UF et al (1993) Reduction in incidence of optic nerve disease with annual ivermectin to control onchocerciasis. Lancet 341(8838):130–134

Abiose A, Murdoch I, Babalola O, Cousens S, Liman I, Onyema J, Evans J, Gregory W, Jones B (1994) Distribution and aetiology of blindness and visual impairment in mesoendemic onchocercal communities, Kaduna State, Nigeria. Kaduna Collaboration for Research on Onchocerciasis. Br J Ophthalmol 78(1):8–13

Amazigo U, Noma M, Bump J, Benton B, Liese B, Yaméogo L, Zouré H, Seketeli A (2006) Onchocerciasis. In: Jamison DT, Makgoba FR, Boser MW, Baingana FK, Hofman KJ, Rogo KO (eds) Disease and mortality in sub-Saharan Africa, 2nd edn. World Bank, Washington, DC

Arborio S, Dozon JP (2000) Sociocultural dimension of epilepsy (kirikirimasien) in a Bambara rural area (Mali). Bull Soc Pathol Exot 93:241–246

Babalola OE (2011) Ocular onchocerciasis: current management and future prospects. Clin Ophthalmol 5:1479–1491

Bird AC, Anderson J, Fuglsang H (1976) Morphology of posterior segment lesions of the eye in patients with onchocerciasis. Br J Ophthalmol 60:2–20

Bisoffi Z, Buonfrate D, Angheben A (2014) Travel, migration and neglected tropical diseases. In: Bentivoglio M, Cavalheiro EA, Kristensson K, Patel N (eds) Neglected tropical diseases and conditions of the nervous system. Springer, New York

Boatin B (2008) The Onchocerciasis Control Programme in West Africa (OCP). Ann Trop Med Parasitol 102(Suppl 1):13–17

Boatin BA, Richards FO Jr (2006) Control of onchocerciasis. Adv Parasitol 61:349–394

Bockarie MJ, Deb RM (2010) Elimination of lymphatic filariasis: do we have the drugs to complete the job? Curr Opin Infect Dis 23:617–620

Bourguinat C, Kamgno J, Boussinesq M, Mackenzie CD, Prichard RK, Geary TG (2010) Analysis of the mdr-1 gene in patients co-infected with *Onchocerca volvulus* and *Loa loa* who experienced a post-ivermectin serious adverse event. Am J Trop Med Hyg 83:28–32

Boussinesq M, Gardon J, Gardon-Wendel N, Kamgno J, Ngoumou P, Chippaux JP (1998) Three probable cases of *Loa loa* encephalopathy following ivermectin treatment for onchocerciasis. Am J Trop Med Hyg 58:461–469

Boussinesq M, Pion SD, Demanga N, Kamgno J (2002) Relationship between onchocerciasis and epilepsy: a matched case-control study in the Mbam Valley, Republic of Cameroon. Trans R Soc Trop Med Hyg 96:537–541

Boussinesq M, Gardon J, Gardon-Wendel N, Chippaux JP (2003) Clinical picture, epidemiology and outcome of Loa-associated serious adverse events related to mass ivermectin treatment of onchocerciasis in Cameroon. Filaria J 2(Suppl 1):S4

Buguet A, Mpanzou G, Bentivoglio M (2014) Human African trypanosomiasis: a highly neglected neurological disease. In: Bentivoglio M, Cavalheiro EA, Kristensson K, Patel NB (eds) Neglected tropical diseases and conditions of the nervous system. Springer, New York

Burnham G (1998) Onchocerciasis. Lancet 351:1341–1346

Cevikbas F, Steinhoff M, Ikoma A (2011) Role of spinal neurotransmitter receptors in itch: new insights into therapies and drug development. CNS Neurosci Ther 17(6):742–749

Coffeng LE, Fobi G, Ozoh G, Bissek AC, Nlatté BO, Enyong P, Olinga JM, Zouré HG, Habbema JD, Stolk WA, de Vlas SJ, Boussinesq M, Noma M (2012) Concurrence of dermatological and ophthalmological morbidity in onchocerciasis. Trans R Soc Trop Med Hyg 106(4):243–251

Cooper PJ, Proano R, Beltran C, Anselmi M, Guderian RH (1995) Onchocerciasis in Ecuador: ocular findings in *Onchocerca volvulus* infected individuals. Br J Ophthalmol 79:157–162

Cousens SN, Yahaya H, Murdoch I, Samaila E, Evans J, Babalola OE, Zakari M, Abiose A, Jones BR (1997) Risk factors for optic nerve disease in communities mesoendemic for savannah onchocerciasis, Kaduna State, Nigeria. Trop Med Int Health 2:89–98

Crump A, Morel CM, Omura S (2012) The onchocerciasis chronicle: from the beginning to the end? Trends Parasitol 28:280–288

Dozie IN, Onwuliri CO, Nwoke BE, Chukwuocha UM, Chikwendu CI, Okoro I, Njemanze PC (2006) Onchocerciasis and epilepsy in parts of the Imo river basin, Nigeria: a preliminary report. Public Health 120:448–450

Druet-Cabanac M, Preux PM, Bouteille B, Bernet-Bernady P, Dunand J, Hopkins A, Yaya G, Tabo A, Sartoris C, Macharia W, Dumas M (1999) Onchocerciasis and epilepsy: a matched case-control study in the Central African Republic. Am J Epidemiol 149:565–570

Druet-Cabanac M, Boussinesq M, Dongmo L, Farnarier G, Bouteille B, Preux PM (2004) Review of epidemiological studies searching for a relationship between onchocerciasis and epilepsy. Neuroepidemiology 23:144–149
Duke BO (1998) Onchocerciasis, epilepsy and hyposexual dwarfism. Trans R Soc Trop Med Hyg 92:236
Ejere HO, Schwartz E, Wormald R, Evans JR (2012) Ivermectin for onchocercal eye disease (river blindness). Cochrane Database Syst Rev 8:CD002219
Enk CD (2006) Onchocerciasis-river blindness. Clin Dermatol 24:176–180
Etya'ale D (2001) Vision 2020: update on onchocerciasis. Community Eye Health 14:19–21
Farnarier G, Diop S, Coulibaly B, Arborio S, Dabo A, Diakite M, Traore S, Banou A, Nimaga K, Vaz T, Doumbo O (2000) Onchocerciasis and epilepsy. Epidemiological survey in Mali. Med Trop (Mars) 60:151–155
Furue M, Ebata T, Ikoma A, Takeuchi S, Kataoka Y, Takamori K, Satoh T, Saeki H, Augustin M, Reich A, Szepietowski J, Fleischer A, Blome C, Phan NQ, Weisshaar E, Yosipovitch G, Stander S (2013) Verbalizing extremes of the visual analogue scale for pruritus: a consensus statement. Acta Derm Venereol 93:214–215
Gardon J, Gardon-Wendel N, Demanga N, Kamgno J, Chippaux JP, Boussinesq M (1997) Serious reactions after mass treatment of onchocerciasis with ivermectin in an area endemic for *Loa loa* infection. Lancet 350:18–22
Gonzalez P, Gonzalez FA, Ueno K (2012) Ivermectin in human medicine, an overview of the current status of its clinical applications. Curr Pharm Biotechnol 13:1103–1109
Greaves MW, Khalifa N (2004) Itch: more than skin deep. Int Arch Allergy Immunol 135:166–172
Grundmann S, Ständer S (2011) Chronic pruritus: clinics and treatment. Ann Dermatol 23:1–11
Hall LR, Pearlman E (1999) Pathogenesis of onchocercal keratitis (River blindness). Clin Microbiol Rev 12:445–453.
Han SK, Simon MI (2011) Intracellular signaling and the origins of the sensations of itch and pain. Sci Signal 4:pe38
Han N, Zu JY, Chai J (2012) Spinal bombesin-recognized neurones mediate more nonhistaminergic than histaminergic sensation of itch in mice. Clin Exp Dermatol 37:290–295
Hoerauf A, Pfarr K, Mand S, Debrah AY, Specht S (2011) Filariasis in Africa – treatment challenges and prospects. Clin Microbiol Infect 17:977–985
Ikoma A, Cevikbas F, Kempkes C, Steinhoff M (2011) Anatomy and neurophysiology of pruritus. Semin Cutan Med Surg 30:64–70
Jeffry J, Kim S, Chen ZF (2011) Itch signaling in the nervous system. Physiology (Bethesda) 26:286–292
Kabore JK, Cabore JW, Melaku Z, Druet-Cabanac M, Preux PM (1996) Epilepsy in a focus of onchocerciasis in Burkina Faso. Lancet 347:836
Kaiser C, Kipp W, Asaba G, Mugisa C, Kabagambe G, Rating D, Leichsenring M (1996) The prevalence of epilepsy follows the distribution of onchocerciasis in a west Ugandan focus. Bull World Health Organ 74:361–367
Kaiser C, Asaba G, Leichsenring M, Kabagambe G (1998) High incidence of epilepsy related to onchocerciasis in West Uganda. Epilepsy Res 30:247–251
Kaiser C, Benninger C, Asaba G, Mugisa C, Kabagambe G, Kipp W, Rating D (2000) Clinical and electro-clinical classification of epileptic seizure in west Uganda. Bull Soc Pathol Exot 93:255–259
Kaiser C, Asaba G, Kasoro S, Rubaale T, Kabagambe G, Mbabazi M (2007) Mortality from epilepsy in an onchocerciasis-endemic area in West Uganda. Trans R Soc Trop Med Hyg 101:48–55
Kaiser C, Pion S, Preux PM, Kipp W, Dozie I, Boussinesq M (2008) Onchocerciasis, cysticercosis, and epilepsy. Am J Trop Med Hyg 79:643–644, author reply 644–645
Kaiser C, Pion S, Boussinesq M (2009) Head nodding syndrome and river blindness: a parasitologic perspective. Epilepsia 50:2325–2326

Kaiser C, Pion S, Boussinesq M (2010) Do helminths cause epilepsy? The case of *Onchocerca volvulus*. Parasite Immunol 32:79–80

Kaiser C, Rubaale T, Tukesiga E, Kipp W, Kabagambe G, Ojony JO, Asaba G (2011) Association between onchocerciasis and epilepsy in the Itwara hyperendemic focus, West Uganda: controlling for time and intensity of exposure. Am J Trop Med Hyg 85:225–228

Kaiser C, Rubaale T, Ojony JO, Asaba G, Tukesiga E, Kipp W, Kabagambe G (2012) In response: a call for more case-controlled studies on onchocerciasis and epilepsy. Am J Trop Med Hyg 86:558

Kamgno J, Gardon J, Boussinesq M (2000) Analysis of the prevention of post-ivermectin *Loa loa* encephalopathy by administration of initial low dose. Med Trop (Mars) 60:275–277

Kamgno J, Pion SD, Boussinesq M (2003) Demographic impact of epilepsy in Africa: results of a 10-year cohort study in a rural area of Cameroon. Epilepsia 44:956–963

Kamgno J, Boussinesq M, Labrousse F, Nkegoum B, Thylefors BI, Mackenzie CD (2008) Encephalopathy after ivermectin treatment in a patient infected with *Loa loa* and *Plasmodium* spp. Am J Trop Med Hyg 78:546–551

Kanmogne GD, Kuate CT, Cysique LA, Fonsah JY, Eta S, Doh R, Njamnshi DM, Nchindap E, Franklin DR Jr, Ellis RJ, McCutchan JA, Binam F, Mbanya D, Heaton RK, Njamnshi AK (2010) HIV-associated neurocognitive disorders in sub-Saharan Africa: a pilot study in Cameroon. BMC Neurol 10:60. doi:10.1186/1471-2377-10-60

Katabarwa M, Lakwo T, Habumogisha P, Richards F, Eberhard M (2008) Could neurocysticercosis be the cause of "onchocerciasis-associated" epileptic seizures? Am J Trop Med Hyg 78:400–401

Keiser PB, Reynolds SM, Awadzi K, Ottesen EA, Taylor MJ, Nutman TB (2002) Bacterial endosymbionts of *Onchocerca volvulus* in the pathogenesis of posttreatment reactions. J Infect Dis 15:805–811

Kilian AH (1994) Onchocerciasis and epilepsy. Lancet 343:983

Kipp W, Burnham G, Kamugisha J (1992) Improvement in seizures after ivermectin. Lancet 340:789–790

Kipp W, Kasoro S, Burnham G (1994) Onchocerciasis and epilepsy in Uganda. Lancet 343:183–184

Koga K, Chen T, Li XY, Descalzi G, Ling J, Gu J, Zhuo M (2011) Glutamate acts as a neurotransmitter for gastrin releasing peptide-sensitive and insensitive itch-related synaptic transmission in mammalian spinal cord. Mol Pain 7:47. doi: 10.1186/1744-8069-7-47

Konig R, Nassri A, Meindl M, Matuja W, Kidunda AR, Siegmund V, Bretzel G, Loscher T, Jilek-Aall L, Schmutzhard E, Winkler AS (2010) The role of *Onchocerca volvulus* in the development of epilepsy in a rural area of Tanzania. Parasitology 137:1559–1568

Mackenzie CD, Homeida MM, Hopkins AD, Lawrence JC (2012) Elimination of onchocerciasis from Africa: possible? Trends Parasitol 28:16–22

Maegga B (1998) Onchocerciasis in Tanzania. Santé 8:49–50

Marin B, Boussinesq M, Druet-Cabanac M, Kamgno J, Bouteille B, Preux PM (2006) Onchocerciasis-related epilepsy? Prospects at a time of uncertainty. Trends Parasitol 22:17–20

Mbanefo EC, Eneanya CI, Nwaorgu OC, Oguoma VM, Otiji MO, Ogolo BA (2010) Onchocerciasis in Anambra State, Southeast Nigeria: clinical and psychological aspects and sustainability of community directed treatment with ivermectin (CDTI). Postgrad Med J 86:573–577

Mckechnie NM, Gurr W, Braun G (1997) Immunization with the cross-reactive antigens Ov39 from *Onchocerca volvulus* and hr44 from human retinal tissue induces ocular pathology and activates retinal microglia. J Infect Dis 176:1334–1343

Moussala M, Fobi G, Ongolo-Zogo P, Bella Hiag LA, Bengono G, McMoli TE (2004) Retinal hemorrhages after ivermectin treatment for onchocerciasis in a patient with *Loa loa* microfilaremia. J Fr Ophtalmol 27:63–66

Murdoch ME (2010) Onchodermatitis. Curr Opin Infect Dis 23:124–131

Murdoch ME, Payton A, Abiose A, Thomson W, Panicker VK, Dyer PA, Jones BR, Maizels RM, Ollier WE (1997) HLA-DQ alleles associate with cutaneous features of onchocerciasis. The Kaduna-London-Manchester Collaboration for Research on Onchocerciasis. Hum Immunol 55:46–52

Nakano T, Andoh T, Lee JB, Kuraishi Y (2008) Different dorsal horn neurons responding to histamine and allergic itch stimuli. Neuroreport 19:723–726

Nana-Djeunga HBC, Pion SD, Kamgno J, Gardon J, Njiokou F, Boussinesq M, Prichard RK (2012) Single nucleotide polymorphisms in β-tubulin selected in *Onchocerca volvulus* following repeated ivermectin treatment: possible indication of resistance selection. Mol Biochem Parasitol 185:10–18

Newell ED, Vyungimana F, Bradley JE (1997) Epilepsy, retarded growth and onchocerciasis, in two areas of different endemicity of onchocerciasis in Burundi. Trans R Soc Trop Med Hyg 91:525–527

Njamnshi AK, Sini V, Djientcheu V de P, Ongolo-Zogo P, Mapoure N, Yepnjio FN, Echouffo TJB, Zebaze R, Meli J, Atchou G, Dongmo L, Muna W (2007) Risk factors associated with epilepsy in a rural area in Cameroon: a preliminary study. Afr J Neurol Sci 26(2):18–26

Njamnshi AK, Djientcheu Vde P, Fonsah JY, Yepnjio FN, Njamnshi DM, Muna WE (2008) The International HIV Dementia Scale is a useful screening tool for HIV-associated dementia/cognitive impairment in HIV-infected adults in Yaoundé-Cameroon. J Acquir Immune Defic Syndr 49(4):393–397

Njamnshi AK, Perrig S, Seke Etet PF, Acho A, Fonsah JY, Mumba DN, Muyembe J-J Kristensson K, Bentivoglio M (2012) Actigraphy in human African trypanosomiasis: a new tool to evaluate disease severity. PLoS Negl Trop Dis 6:e1525

Osei-Atweneboana MY, Awadzi K, Attah SK, Boakye DA, Gyapong JO, Prichard RK (2011) Phenotypic evidence of emerging ivermectin resistance in *Onchocerca volvulus*. PLoS Negl Trop Dis 5:e998. doi:10.1371/journal.pntd.0000998

Ovuga E, Kipp W, Mungherera M, Kasoro S (1992) Epilepsy and retarded growth in a hyperendemic focus of onchocerciasis in rural western Uganda. East Afr Med J 69:554–556

Ozoh GA, Murdoch ME, Bissek AC, Hagan M, Ogbuagu K, Shamad M, Braide EI, Boussinesq M, Noma MM, Murdoch IE, Sékétéli A, Amazigo UV (2011) The African Programme for Onchocerciasis Control: impact on onchocercal skin disease. Trop Med Int Health 16:875–883

Papoiu A, Wang H, Coghill R, Chan YH, Yosipovitch G (2011) Contagious itch in humans. A study of visual "transmission" of itch in atopic dermatitis and healthy subjects. Br J Dermatol 164:1299–1303

Paus R, Schmelz M, Bíró T, Steinhoff M (2006) Frontiers in pruritus research: scratching the brain for more effective itch therapy. J Clin Invest 116:1174–1186

Pfab F, Valet M, Napadow V, Tölle TR, Behrendt H, Ring J, Darsow U (2012) Itch and the brain. Chem Immunol Allergy 98:253–265

Phan NQ, Blome C, Fritz F, Gerss J, Reich A, Ebata T, Augustin M, Szepietowski JC, Stander S (2011) Assessment of pruritus intensity: prospective study on validity and reliability of the visual analogue scale, numerical rating scale and verbal rating scale in 471 patients with chronic pruritus. Acta Derm Venereol 92:502–507

Pion SD, Boussinesq M (2012) Significant association between epilepsy and presence of onchocercal nodules: case-control study in Cameroon. Am J Trop Med Hyg 86:557, author reply 558

Pion SD, Kaiser C, Boutros-Toni F, Cournil A, Taylor MM, Meredith SE, Stufe A, Bertocchi I, Kipp W, Preux PM, Boussinesq M (2009) Epilepsy in onchocerciasis endemic areas: systematic review and meta-analysis of population-based surveys. PLoS Negl Trop Dis 3:e461

Prischich F, De Rinaldis M, Bruno F, Egeo G, Santori C, Zappaterreno A, Di Fattouch J, Bonaventura C, Bada J, Russo G, Pizzuti A, Cardona F, Sa'a Vullo V, Giallonardo AT, D'erasmo E, Pelliccia A, Vanacore N (2008) High prevalence of epilepsy in a village in the Littoral Province of Cameroon. Epilepsy Res 82:200–210

Reich A, Heisig M, Phan NQ, Taneda K, Takamori K, Takeuchi S, Furue M, Blome C, Augustin M, Stander S, Szepietowski JC (2011) Visual analogue scale: evaluation of the instrument for the assessment of pruritus. Acta Derm Venereol 92:497–501

Ringkamp M, Schepers RJ, Shimada SG, Johanek LM, Hartke TV, Borzan J, Shim B, LaMotte RH, Meyer RA (2011) A role for nociceptive, myelinated nerve fibers in itch sensation. J Neurosci 31(42):14841–14849.

Sacktor NC, Wong M, Nakasujja N, Skolasky RL, Selnes OA, Musisi S, Robertson K, McArthur JC, Ronald A, Katabira E (2005) The International HIV Dementia Scale: a new rapid screening test for HIV dementia. AIDS 19:1367–1374

Sauerbrey M (2008) The Onchocerciasis Elimination Program for the Americas (OEPA). Ann Trop Med Parasitol 102(Suppl 1):25–29

Siddiqui MA, Al-Khawajah MM (1991) The black disease of Arabia. Sowda-onchocerciasis. New findings. Int J Dermatol 30:130–133

Sun YG, Zhao ZQ, Meng XL, Yin J, Liu XY, Chen ZF (2009) Cellular basis of itch sensation. Science 325:1531–1534

Tabah EN, Yepnjio FN, Njamnshi AK (2014) Stigma in neurological diseases in the tropics. In: Bentivoglio M, Cavalheiro EA, Kristensson K, Patel N (eds) Neglected tropical diseases and conditions of the nervous system. Springer, New York

Takesue AHY, Ono K, Murakami A (2012) Towards the elimination of onchocerciasis at the global scale: history, challenges and future prospects. Nihon Ganka Gakkai Zasshi 116:547–553

Tamarozzi F, Halliday A, Gentil K, Hoerauf A, Pearlman E, Taylor MJ (2011) Onchocerciasis: the role of *Wolbachia* bacterial endosymbionts in parasite biology, disease pathogenesis, and treatment. Clin Microbiol Rev 24:459–468

Tey HL, Yosipovitch G (2011) Targeted treatment of pruritus: a look into the future. Br J Dermatol 165:5–17

Thylefors B (2008) The Mectizan Donation Program (MDP). Ann Trop Med Parasitol 102:39–44

Twum-Danso NA (2004) Mass treatment of onchocerciasis with ivermectin: should people with epilepsy and/or growth-retardation syndromes be excluded? Ann Trop Med Parasitol 98:99–114

Umeh RE, Chijioke CP, Okonkwo PO (1996) Eye disease in an onchocerciasis-endemic area of the forest-savanna mosaic region of Nigeria. Bull World Health Organ 74:95–100

World Health Organisation (2010) First report on neglected tropical diseases: working to overcome the global impact of neglected tropical diseases. WHO Press, WHO, 20 Av Appia, Geneva.

Williams SC (2012) Nodding syndrome leaves baffled scientists shaking their heads. Nat Med 18:334

Winkler AS, Friedrich K, Konig R, Meindl M, Helbok R, Unterberger I, Gotwald T, Dharsee J, Velicheti S, Kidunda A, Jilek-Aall L, Matuja W, Schmutzhard E (2008) The head nodding syndrome-clinical classification and possible causes. Epilepsia 49:2008–2015

Yang YF, Cousens S, Murdoch IE, Babalola OE, Abiose A, Jones B (2001) Intraocular pressure and gonioscopic findings in rural communities mesoendemic and nonendemic for onchoceriasis, Kaduna State, Nigeria. Eye (Lond) 15:756–759

Human African Trypanosomiasis: A Highly Neglected Neurological Disease

Alain Buguet, Ghislain Mpanzou, and Marina Bentivoglio

Abstract Human African trypanosomiasis (HAT) or sleeping sickness is a severe vector-borne disease, with marked involvement of the peripheral and central nervous system. Still endemic in sub-Saharan Africa, HAT is caused by transmission of subspecies of the protozoan parasites *Trypanosoma brucei* (*T. b.*) through bites of tsetse flies (genus *Glossina*). Foci of HAT are reported mostly in remote, resource-poor settings, and areas of political instability. The disease has a chronic form caused by *T. b. gambiense* in Western and Central Africa, and an acute form caused by *T. b. rhodesiense* in Eastern and Southern Africa. Both forms, almost invariably fatal without treatment, evolve from a first, hemolymphatic stage to a second, meningoencephalitic stage due to *T. b.* brain invasion. Clinical features involve a constellation of sensory, motor and neuropsychiatric signs and symptoms, with a characteristic sleep disorder leading to sleep-wake cycle disorganization and sleep structure alterations. Therapy currently available to cure the second stage of both HAT forms is toxic. Stage biomarkers and safer therapy are urgently needed. Clinical objective evaluation is essential for diagnostic purposes, treatment assessment and patients' follow-up. The recent decline in the number of reported new HAT cases should not foster further neglect of this highly neglected nervous system infection.

Keywords Sleeping sickness • *Trypanosoma brucei* • Parasitic brain infection • Sensory disorders • Neuroinflammation • Sleep • Pain • Behavioral disorders

---

A. Buguet
Polyclinique Marie-Louise Poto-Djembo, Pointe-Noire, République du Congo

G. Mpanzou
Clinique Municipale Albert Leyono, Brazzaville, République du Congo

M. Bentivoglio (✉)
Department of Neurological and Movement Sciences, School of Medicine,
University of Verona, Strada Le Grazie 8, 37134 Verona, Italy
e-mail: marina.bentivoglio@univr.it

M. Bentivoglio et al. (eds.), *Neglected Tropical Diseases and Conditions of the Nervous System*,
DOI 10.1007/978-1-4614-8100-3_9,

1 Introduction

Human African trypanosomiasis (HAT), also known as sleeping sickness, is one of the most neglected tropical diseases, with marked involvement of the peripheral nervous system and central nervous system (CNS). This chapter presents an overview of the disease prevalence and clinical features (see for recent reviews Brun et al. 2010; Malvy and Chappuis 2011; Bouteille and Buguet 2012; Welburn and Maudlin 2012; Kennedy 2013; Lejon et al. 2013), with emphasis on neurological signs and symptoms.

The disease is caused by infection with the extracellular protozoan parasites *Trypanosoma brucei* (*T. b.*). These unicellular flagellates are transmitted by bites of infected tsetse flies (genus *Glossina*) in sub-Saharan Africa (Fig. 1). After a complex cycle in the vector's midgut, the parasites reach the salivary glands and are transmitted through the saliva of tsetse flies.

Two subspecies of *T. b.* are responsible for HAT forms with different geographical distribution (Fig. 1) and clinical course. Infection with *T. b. gambiense* causes the chronic form of HAT in Western and Central Africa; infected humans are the primary reservoir, though animals (including pigs and sheep) may also be infected. *T. b. gambiense* is transmitted by three species of tsetse flies, and mainly by *Glossina palpalis*. Gambian HAT has an average course of 3 years; parasitaemia is in general low and occurs in cycles, rendering difficult the detection of parasites in blood smears. Infection with *T. b. rhodesiense*, transmitted mainly by the *Glossina morsitans* group of tsetse flies, causes the acute, zoonotic form of HAT in Eastern and Southern Africa. A number of wildlife animal species (including those in game parks) and domestic animal species (especially cattle) act as reservoirs for *T. b. rhodesiense*. This infection progresses rapidly, with an average course of 6–8 weeks and high parasitaemia. The vast majority of cases of HAT (about 97 %) are due to *T. b. gambiense* infection, while *T. b. rhodesiense* infection is responsible for about 3 % of the cases.

Both forms of HAT evolve in two stages: an early, hemolymphatic stage and a subsequent, meningoencephalitic stage, due to *T. b.* invasion of the CNS. The parasite passage across the blood-brain barrier to invade the CNS parenchyma is an active multistep process, dealt with in the companion chapter of this volume (see Masocha et al. 2014).

Due to rapid switching of the expression of genes encoding variant surface glycoproteins that compose the parasite coat, *T. b.* can evade the immune system of the host. For this high degree of antigenic variation, no vaccine to prevent HAT is available at present. As mentioned further (see Sect. 4), the therapy currently available to cure CNS disease in both forms of HAT has severe adverse effects. Although disease staging is, therefore, essential for therapeutic decisions, precise stage biomarkers have not been identified as yet.

As stated by Gadelha et al. (2011), "the African trypanosome is among the deadliest of human pathogens". If left untreated, both forms of HAT are almost invariably fatal. However, a limited number of cases of human trypanotolerance, possibly associated with genetic variability of the parasite and/or the host, has been reported (Jamonneau et al. 2012; Kennedy 2013).

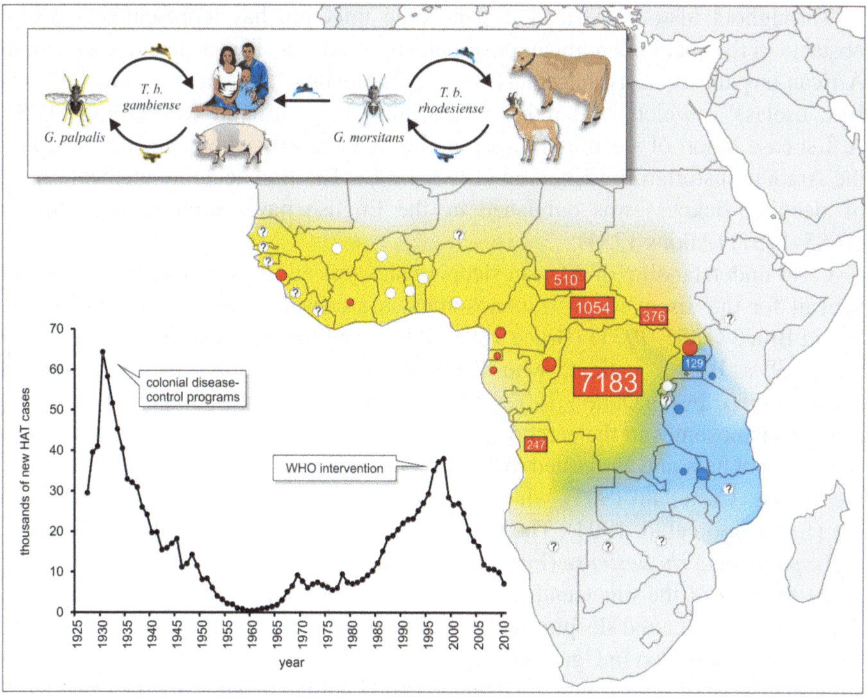

Fig. 1 Transmission and epidemiology of human African trypanosomiasis (HAT). *Top*: the two subspecies of *Trypanosoma brucei* (*T. b.*), *T. b. gambiense* and *T. b. rhodesiense*, causing the two forms of HAT, respectively, are transmitted by the bite of different species of tsetse flies (genus *Glossina*, *G.*) in endemic regions of sub-Saharan Africa. Humans are the primary reservoir for *T. b. gambiense*, which may also infect animals. Cattle and wild animals are the primary reservoirs for *T. b. rhodesiense*. *Bottom left*: the chart indicates the number of reported new HAT cases in the last nine decades. Note the epidemic peaks of the disease in the 1930s (when control and surveillance by mobile teams were instituted during the colonial period), and in the late 1990s. The World Health Organization (WHO) and partner organizations, including national control programs, launched then initiatives for HAT control and surveillance, resulting in a steady decrease of new reported cases. *Bottom right*: geographical distribution of *T. b. gambiense* infection (*yellow shading*), which causes the vast majority of HAT cases, and of *T. b. rhodesiense* infection (*light blue shading*). The *superimposed markers* indicate new HAT cases reported in 2009 for each of the countries under WHO observation, according to the "HAT atlas initiative" (Simarro et al. 2010). *Blank circles* represent countries with no reported new cases; *question marks* indicate countries with incomplete data. *Colored circles* represent foci with less than 100 new reported cases; circle size is roughly proportional to the number of reported cases. The number of new reported HAT cases is indicated. Modified from Kristensson et al. (2010), courtesy of Giuseppe Bertini

2 Notes on the History of Human African Trypanosomiasis

The appearance of African trypanosomes, dating back to prehistory, far preceded that of mankind. Trypanosome resistance probably played an important role in the selection of early hominids (Steverding 2008).

Throughout history, African trypanosome infection has represented a severe obstacle to the socio-economical development of Africa. This is also due to animal African trypanosomiasis, nagana (from the Zulu word "N'gana" meaning "powerless, useless"), which rendered stock farming very difficult (Steverding 2008). A first case report of sleeping illness, fatal to the Emperor of Mali, is described by the Arabian historian Ibn Khaldun (1332–1406). The first accurate medical report of sleeping sickness was published by the English naval surgeon John Atkins (1685–1757) (Atkins 1734).

Clear understanding of African sleeping sickness etiology and transmission had to wait for the discovery of trypanosomes as a causal agent of cattle nagana by David Bruce (1855–1931) (Bruce 1895, 1897), and that of trypanosomes from the blood of a British patient, who lived in Gambia, affected by the so-called *Trypanosoma* fever (Dutton 1902). One year later, Aldo Castellani (1878–1971) found trypanosomes in the cerebrospinal fluid (CSF) of patients with symptoms of sleeping sickness and suggested that these parasites could be the causative agent of the disease (Castellani 1903), which initiated a debate on the priority of the discovery (Bentivoglio et al. 1994). The track for causal agents ended with the discovery of *Trypanosoma rhodesiense* (Fantham and Thomson 1910).

At the end of the nineteenth century, two entities were considered separately: *Trypanosoma* fever and sleeping sickness. In 1903, the report of the British Sleeping Sickness Commission in Uganda (Anonymous 1903) reached the following conclusions (1) trypanosomes found in Ugandan and Gambian patients are the same parasite; (2) *Trypanosoma* fever represents the first stage of sleeping sickness; (3) monkeys contract a sleeping sickness-like disease when they are infected with either blood from *Trypanosoma* fever patients or CSF from sleeping sickness patients; (4) trypanosomes are transmitted by *Glossina palpalis* tsetse flies; and (5) sleeping sickness is a human disease due to transmission of trypanosomes from patients to healthy individuals by tsetse fly bite. Therefore, *Trypanosoma* fever and sleeping sickness were then considered as two consecutive stages of the same disease caused by African trypanosomes (Greig and Gray 1904).

Severe epidemics of sleeping sickness ravaged sub-Saharan Africa, affecting mainly Uganda and the Congo basin at the turn of the twentieth century. Led by Charles Louis Alphonse Laveran (1845–1922), the French Commission created a central laboratory at Brazzaville (Congo) and launched mobile teams along the Congo River and affluent rivers. In 4 years, the work accomplished was impressive and was reported in an extensive publication (Martin et al. 1909). The clinical aspects of the disease were presented in detail and its course in two "periods" confirmed. The first period is the "time lapse since the parasites appear in the blood or lymph glands and their discovery in the CSF" (Martin and Leboeuf 1908); "the second period of the disease starts with the appearance of the flagellates in the sub-arachnoid spaces". The French Commission stressed the importance of clinical signs in stage determination, and distinguished also a third terminal period with cachexia, permanent torpor, convulsions, and final coma. The conclusions of the French Commission were corroborated by Belgian, Portuguese and German reports. At the same time, Broden and Rodhain (1908) used the absence or presence of white blood cells (WBC) in the CSF to distinguish between the two stages of sleeping sickness.

After World War I, extensive control operations established by the colonial authorities "fearing an unpopulated continent and a shortage of human labour to exploit natural resources" (Simarro et al. 2011a), promoted the search and treatment of HAT patients by mobile teams. By the 1960s, sleeping sickness was almost eliminated (Fig. 1). The control of the disease thus lost priority, also due to political instability of many African countries after independence. In the 1980s HAT re-emerged, and the number of reported cases kept increasing. Towards the end of the twentieth century HAT almost reached the levels of epidemics seen at the beginning of the century (Fig. 1), and the World Health Organization (WHO) estimated 300,000 new infected cases every year. At the turn of the twenty-first century, the WHO, with partner organizations including national control programs in endemic countries, reinforced HAT control and surveillance, and the number of reported cases started to decline (Fig. 1). Between 50,000 and 70,000 infected patients were reported by WHO in 2006.

3 Prevalence and Geographical Distribution

Transmission of HAT occurs in discrete areas of endemicity or foci (Fig. 1). Approximately 70 million people (10 % of the total population in endemic countries) are estimated to be at various levels of risk of HAT infection over an area of 1.55 million km^2 (7.4 % of the total area of endemic countries) in sub-Saharan Africa (Simarro et al. 2012). The vast "tsetse belt" extends between the Sahara and Kalahari deserts, 14° North and 20° South to the Equator (Fig. 1). Twenty-one million people are currently at moderate to high risk of infection, the largest proportion (82.2 %) being at risk of *T. b. gambiense* infection (Simarro et al. 2012). While the geographical distribution of the two forms of HAT is roughly delimited by the Rift Valley, there are areas at risk of merges (e.g. lake Tanganyika) between the two forms (Simarro et al. 2010).

Tsetse flies are attracted to dark colors (blue and black) and moving vehicles. Their ecology requires shadow, coolness and warm blood diet. Tsetse flies occur predominantly along lakes and riversides, forest galleries (where they can move beyond 2 km), plantations and village surroundings. However, houses can also attract tsetse flies in their host-seeking behavior (early morning and late afternoon), and during hot hours (late morning and early afternoon) to find a cool refuge (Vale et al. 2013). The natural habitat for the *T. b. rhodesiense* vector is the savannah. Passive dispersion through movements of domestic animals and humans can cause reinvasion of cleared zones. In particular, movements of cattle along trade routes and through livestock markets represent a big challenge for the control of Rhodesian HAT (Simarro et al. 2010). Population displacements may also spread the disease by new cycles of transmission in previously uninfected vectors (Tong et al. 2011).

Transmission of HAT occurs primarily in rural areas, during activities such as farming, fishing and hunting. Reports of HAT in peri-urban Kinshasa (Democratic Republic of Congo, DRC) indicate that transmission can also occur in peri-urban belts (Fèvre et al. 2008), where tsetse flies have adapted to high population density.

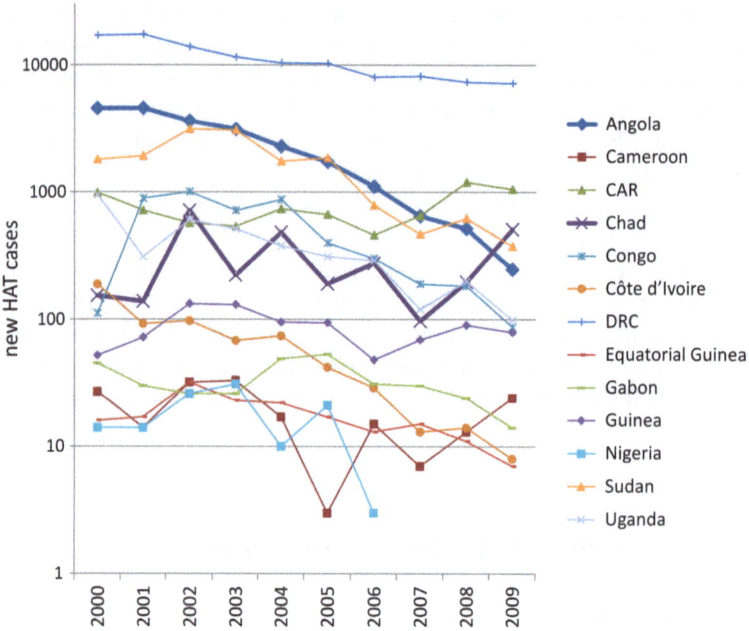

Fig. 2 Graphic representation of the number of reported new HAT cases caused by *T. b. gambiense* infection in 13 endemic countries in 10 years (2000–2009), based on the "Atlas of human African trypanosomiasis" (Simarro et al. 2010). The *thicker lines* indicate two paradigmatic situations: a country (Angola) in which new HAT cases dropped, and the worrying situation of another country (*Chad*), which reported a recent increase of new cases, with foci at the border with the Central African Republic (CAR) in which an increase of new cases is also reported. Courtesy of Giuseppe Bertini

The "HAT atlas initiative", based on geospatial analysis allowed by modern technologies, was launched in 2008 by WHO with joint implementation by partner organization (Simarro et al. 2010). Geographical-epidemiological maps of the HAT atlas can be found in the WHO website (http://www.who.int). Interestingly, many of the foci of HAT cross national borders, indicating that partnerships in the governmental initiatives of the involved countries are needed for surveillance and control of the disease.

The number of currently reported HAT cases due to *T. b. gambiense* infection shows a steady decrease in most endemic countries, but still an increase in others, and is highest in DRC (Fig. 2), indicating that sustained surveillance and targeted actions are needed. In 2009, the number of reported new HAT cases has dropped "below the symbolic number of 10,000" (Simarro et al. 2011a). These figures should, however, be regarded with extreme caution due to under-detection and under-reporting. The rate of under-reporting (not formally quantified for Gambian HAT) is as high as 40 % in some foci of Rhodesian HAT (Fèvre et al. 2008), also due to the acuteness and rapid progression of *T. b. rhodesiense* infection (Simarro et al. 2012).

Foci of HAT are currently mostly confined to resource-poor settings: remote rural areas, areas of political instability or armed conflict, where it may be difficult for patients to access treatment (Tong et al. 2011). Qualitative and quantitative analyses of 30 year data have confirmed that clusters of HAT incidence have predominantly coincided with periods of conflict and civil wars (Berrang-Ford et al. 2011). Importantly, these analyses have also indicated a lag period of approximately 10 years between the beginning of conflict events and a peak of HAT incidence (Berrang-Ford et al. 2011). This further implies that the effort for surveillance and control of HAT should be maintained for decades for a chance of success in the elimination of this disease from the African continent.

Active screening (through mobile teams travelling to remote villages to test the population) and passive case-finding (testing at fixed health facilities), followed by treatment and vector control are essential for interventions against HAT (Simarro et al. 2010; Tong et al. 2011). Passive screening is less efficient than active screening, as access to health facilities is very difficult for most HAT patients (Tong et al. 2011). Under-reporting and under-detection of HAT cases can occur both in areas covered by active or passive surveillance, and in areas not covered by surveillance due to remoteness or instability (Simarro et al. 2012).

Disease control can be very difficult in health resource-poor regions, as shown by a recent survey in ten health institutions in Zambia (a country with low scale HAT endemicity). In this survey, based on questionnaires on general knowledge on HAT targeted to health workers, most respondents scored very low, and none of them knew how to differentiate HAT stages (Mwanakasale et al. 2013). Also the follow-up (24 months with control visits every 6 months, recommended by WHO) can be difficult to achieve in resource-limited settings.

Cases of HAT have also been reported in non-endemic areas in Africa, as well as in travelers returning to Europe or North America (Brun et al. 2010; Blum et al. 2011; Migchelsen et al. 2011; Simarro et al. 2011b; Malvy and Chappuis 2011; Kennedy 2013; see also Bisoffi et al. 2014). Gambian HAT has been reported in immigrants, refugees, expatriates resident for long periods in rural areas, and is very rare in short-term travelers. Rhodesian HAT has instead been reported in short-term tourists travelling to game reserves in East Africa. The diagnosis of these cases poses serious challenges (see Sect. 4.2). Awareness of the fact that *T. b.* infection still represents a risk for travelers and migrants is mandatory. Because of its long incubation time, Gambian HAT should be considered for diagnosis even after long intervals from the stay in an endemic country.

4 Clinical Features

Pathogenetic mechanisms and experimental data on animal models are beyond the scope of the present chapter, as they are discussed in the companion chapter of this volume on the neurobiology of African trypanosomiasis (see Masocha et al. 2014).

Detailed accounts on the current status of the diagnostics and therapy of HAT is also beyond the limits of the present treatise, and are presented in recent publications (Brun et al. 2010; Blum et al. 2011; Bouteille and Buguet 2012; Kennedy 2013; Lejon et al. 2013). Concerning diagnostics, we only wish to recall here that the card agglutination test for trypanosomiasis (CATT), developed in 1978 (Magnus et al. 1978), is still the primary screening tool used in areas endemic for Gambian HAT. This test has 95 % specificity but may result in false positivity, and cannot be used for follow-up or relapses. No similar test is available for Rhodesian HAT, during which, however, there are higher chances to detect the parasites in blood smears due to the high parasitaemia.

Treatment of stage 1 HAT utilizes pentamidine (developed in 1937) and suramin (developed in 1916), which are relatively well tolerated though they may have adverse effects. These drugs, however, do not cross the blood-brain barrier and cannot cure CNS disease. The arsenical compound melarsoprol (developed in 1949) is still used to cure stage 2 of both forms of HAT. Melarsoprol injections are very painful (the patients even refuse the therapy) and adverse effects of the treatment are very severe. In 10 % of the patients, melarsoprol treatment causes post-treatment reactive arsenical encephalopathy, which is fatal in half of the cases. Eflornithine, and more recently the nifurtimox-eflornithine combination therapy have been introduced for CNS disease caused by *T. b. gambiense*, but require cycles of intravenous administration and have side effects. The search for new drugs for treatment of CNS disease in HAT is ongoing in different laboratories (see Kennedy 2013), but has not been successful up to now.

The issue of HAT staging, which is obviously crucial for therapeutic decisions, is currently highly debated, and also research for effective biomarkers is actively ongoing. Waiting for new, substantial discoveries, the criteria still in use for HAT staging are those indicated by WHO (1998, 2007) guidelines related to CSF parameters (which therefore require a lumbar puncture): the presence of trypanosomes in the CSF and/or ≤5 WBC per μl for stage 1, and >5 WBC per μl for stage 2. Due to different cut-off criteria in different HAT foci, the definition of an intermediate stage (>5 up to ≤20 WBC per μl in the CSF) is adopted in many endemic countries before treatment for HAT stage 2 is initiated (see, for example, Chappuis et al. 2005; Blum et al. 2006; Wastling and Welburn 2011).

An objective clinical evaluation, very much needed for the assessment of HAT severity, is made difficult by the fact that signs and symptoms of HAT are heterogeneous and variable, and would therefore require standardization. Variability of the clinical features of HAT, including the predominance of different neurological signs and symptoms, is also related to different foci (Blum et al. 2006; Kennedy 2013). Clinical features, often described as early and late symptoms, actually evolve as a continuum, with an insidious progression.

A skin lesion (trypanosomal chancre), with erythema and edema, appears from a few days to 2 weeks at the site of the tsetse bite, especially in *T. b. rhodesiense* infection, while it is rarely seen in *T. b. gambiense* infection.

Early lesions can include facial edema, fugacious and mobile, mainly localized at the external corner of the lower eyelids which gives the patient a puffy face. It is more common in *T. b. gambiense* infection than in *T. b. rhodesiense* infection, during which, however, edema of the extremities can occur.

Skin rash (trypanides) may be observed as polymorphic papulo-erythematous eruptions, often edematous and itchy, distributed in large plaques or in one to several cm-wide striated formations over the trunk or at the thigh, or even all over the body.

Adenopathies, splenomegaly and/or hepatomegaly may develop. Hematological alterations, such as anemia, and heart involvement (myocarditis, pericarditis and/or other severe alterations) may occur. Ophthalmological disorders (iritis, keratitis, conjunctivitis) may also develop.

Posterior cervical lymphadenopathy appears early, is a typical sign of Gambian HAT, and its detection is part of the diagnostic procedure in the field. It is called Winterbottom sign after the English physician Thomas Winterbottom (1766–1859) who described swollen lymph nodes in the back of the neck at an early stage of the disease (Winterbottom 1803). Microscopic examination of lymph node aspirate for parasite detection is in use for HAT diagnosis in the field since colonial times.

Chilliness is commonly reported, alternating with feverish feelings. Fever, resistant to quinine and other anti-malaria drugs, is irregular and accompanied by other general malaise symptoms such as headache, prostration and insomnia (Forde 1902). Body temperature oscillates and may present hypothermic events reaching 34–35 °C. In the rat model of HAT infected with *T. b. brucei*, the circadian oscillation of abdominal core temperature is preserved for 10 days after infection. Then, a sudden drop in core temperature occurs (as low as 33 °C), accompanied by disturbances in the temperature circadian rhythmicity, as well as anorexia and body weight loss (Darsaud et al. 2004). Hypothermic events have been related to nitric oxide synthase inhibition (Scammell et al. 1996), which was observed in Gambian HAT patients (Buguet et al. 2002, 2009) and in *T. b. brucei*-infected experimental rats (Amrouni et al. 2010, 2011).

Different types of endocrine dysfunction (amenorrhea, dysmenorrhea; impotence, loss of libido; sterility, abortion, stillbirths) can occur early after infection and are reversed by treatment. As an anecdote, patients we observed (Buguet et al. 2009) believed that pentamidine is aphrodisiac as libido recovered after treatment.

Headache is a very frequent symptom (Blum et al. 2006), which becomes increasingly persistent and severe with disease progression.

Pruritus and pain are important sign of peripheral nervous system involvement. Pruritus remains intermittent and discrete at an early stage, but becomes overwhelming as the disease worsens. Skin lesions are then provoked by scratching.

Back and lumbar pain are frequently complained by the patients. Deep muscle and bone pain has no preferential localization, and can be spontaneous or provoked, revealed by the Kérandel's sign, represented by deep delayed hyperesthesia provoked by pressure of soft tissues. The sign was described by the French physician Jean F. Kérandel (1873–1934), affected by the disease, who suffered from a sudden and intense palm pain when turning the key to open his door (Kérandel 1910).

Sleep abnormalities, which gave the disease its alternative name of sleeping sickness, are dealt with below (see Sect. 4.1). Neurological signs and symptoms include motor disturbances: gait alterations, cerebellar ataxia, speech disturbances, choreiform movements, spasticity due to involvement of the pyramidal system. Lower limb paralysis may be caused by myelopathy or myelitis, or peripheral motor neuropathies. Abnormal primitive reflexes (pout, palmo-mental, sucking reflexes) can be present.

Mood disorders are dominated by anxiety and modify affective contacts. Alternation of depression, apathy and excitation is often reported by the family. The patients become instable, irritable, sad, indifferent to themselves and their surroundings, and have difficulties to work. The physician General Léon Lapeyssonnie (1915–2001), in his lectures at the Pharo Institute of Tropical Medicine in Marseille (France) used to say that any neurologic or psychiatric or behavioral symptom in an area endemic for HAT must evoke the diagnosis of this disease.

The overall health status of the patient worsens as HAT progresses. The initial malaise develops into deep asthenia, fatigue, anorexia and body weight loss that may reach cachexia and doldrums.

In a case series (159 patients infected by *T. b. gambiense*) diagnosed during a 5-year investigation (2005–2009) in Congo (Brazzaville) (Buguet et al. 2009; Bouteille and Buguet 2012), applying a clinical protocol designed by the Limoges Institute of Tropical Neurology, signs and symptoms have been classified according to the disease stages (stage 1, intermediate stage, stage 2) defined on the basis of diagnostic procedures including WBC counts in the CSF (see above). The most frequently reported symptom, at all disease stages, was represented by headache. Feverish feeling, fatigue, back and lumbar pain, sexual disorders, pruritus were also evenly reported at all stages. However, certain symptoms affected mostly stage 2 patients: fever, fatigue, anorexia and pruritus. Stage 1 patients complained more often from back and lumbar pain than stage 2 patients, and this may be related to the fact that patients in stage 1 are still very active in their farmland.

Adenopathies were observed very frequently. Enlarged lymph nodes were most often encountered in intermediate stage patients, although they were present in almost half of stage 1 and stage 2 patients. Splenomegaly and hepatomegaly were rarely observed, and they were more frequently observed at stage 1 than at more advanced stages. Stage 2 patients presented more frequently with facial edema and body weight loss, and with trypanides and other cutaneous signs. Interestingly, endocrine dysfunction such as amenorrhea, dysmenorrhea, decreased libido and impotence occurred at all stages of HAT.

Neurological signs were mostly encountered at stage 2, although they were also present in stage 1 and in the intermediate stage. This was the case for abnormal tendon reflexes, pyramidal signs (Babinski and Hoffman signs), abnormal primitive reflexes, tremor, myoclonus and hyperesthesia. However, a number of signs were almost exclusively observed in stage 2 patients: gait disorder, cerebellar ataxia with Romberg sign, tics, choreiform movements, and convulsions.

Mental and behavioral manifestations were predominant in stage 2, but did also occur in the intermediate stage and, to a lesser extent, at stage 1. Lack of interest was the most frequently observed sign, followed by indifference and agitation (the nineteenth century practitioners considered these traits as characteristic of sleeping sickness patients). However, other symptoms were encountered almost exclusively in stage 2 patients: ideomotor slowing, confusion and stupor, fantasizing, excessive familiarity, irritability, aggressive behavior, impudicity and bulimia, dementia and delirium.

Sleep complaints predominated in stage 2 patients, but were not absent from the other two stages. Interestingly, in patients at stage 1, sleep disorders reports came third after headaches and back and bone pains.

This case series further strengthens the view that neurological symptoms and signs do not necessarily indicate that HAT patients are in stage 2. In the largest published study on clinical aspects of HAT (2,541 patients in stage 2 from seven countries), the most frequently observed symptoms were consistently represented by sleep disorders, followed by adenopathies and splenomegaly. It should also be considered that the individual competence of the examiner may lead to emphasize certain signs and neglect others. For instance, primitive reflexes are frequently found by neurologists (Giordano et al. 1977), while they are rarely described by general practitioners (Ginoux et al. 1982). Furthermore, symptoms such as pruritus may also be related to other tropical parasitic diseases. There is therefore a need to harmonize and standardize the clinical examination of HAT patients, establishing also objective, clinical scales of severity.

4.1 Focus on Sleep-Wake Disorder

The characteristic sleep disorder eponymous for the disease occurs in the majority of HAT cases (74 % of *T. b. gambiense*-infected patients in the study by Blum et al. 2006). As sleep disturbances may be overlooked by the patients and the family, specific questions concerning sleep pattern should be asked to the patients at presentation.

Sleep-wake disorders, with nocturnal insomnia and excessive daytime somnolence, can be observed in stage 1 HAT patients and exacerbate, or can appear, with worsening of the illness.

The first detailed description of sleep alterations in HAT at an advanced stage was reported by Stephen Mackenzie at the end of the nineteenth century, in a 22-year old man from the Congo Free State: "he had many little sleeps in the daytime, but rarely slept for long together either by night or day, so that the total amount of sleep did not exceed the amount usually taken by healthy people" (Mackenzie 1890). Such report of a major disturbance in the 24-h sleep and wake alternation ruled out hypersomnia since this first description. The clinical description of sleep disturbances of HAT was enriched by the memoir of the French neuropsychiatrist Jean-Jacques Lhermitte (1877–1959) on the "narcolepsies of sleeping sickness" (Lhermitte 1910).

Clinical observations on sleep disorder in HAT found an objective confirmation with clinical neurophysiological studies of the sleep-wake cycle (Buguet et al. 1989) based on the use of polysomnography (PSG) (Fig. 3). This technique involves simultaneous recording of the electroencephalogram, gravity muscle tone with electromyography, and eye movements with electrooculography.

The two components of the so-called "PSG syndrome" in HAT (Fig. 3) were thus defined in *T. b. gambiense*-infected patients (Buguet et al. 2001) (1) a circadian disruption of sleep-wake alternation and/or a marked fragmentation of sleep; and (2) alterations of the sleep structure due to the occurrence of "sleep onset rapid eye movement (REM) sleep periods" (SOREMP). Buguet et al. (2005) hypothesized that the full blown PSG could be used to assess HAT severity and even discriminate

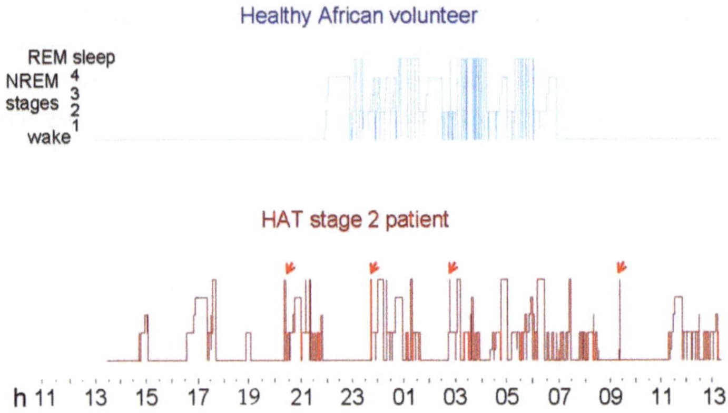

Fig. 3 Polysomnographic syndrome of the sleep-wake disruption in HAT illustrated by hypnograms recorded from a healthy African volunteer, and from a stage 2 HAT patient infected by *T. b. gambiense*, respectively. Note in the healthy subject the normal, nycthemeral distribution of the sleep-wake cycle, with one major sleep period during the night, as well as the normal sleep structure, with rapid eye movement (REM) sleep occurring after non-REM (NREM) sleep. Note in the HAT patient the major circadian disruption of sleep-wake distribution throughout the 24-h recording, and the abnormal sleep structure with sleep-onset REM periods not preceded by NREM sleep (*arrows*)

between stages 1 and 2. The validity of PSG for the determination of disease severity in patients suffering from *T. b. gambiense* infection, especially during the post-therapeutic follow-up, has been recently confirmed (Buguet et al. 2012).

Fewer PSG data have been collected in *T. b. rhodesiense*-infected patients. Polysomnography was performed in two French soldiers infected by *T. b. rhodesiense* in Rwanda (Montmayeur et al. 1994). Sent back to France, they were examined 4, 6 and 11 months later. On the first examination, the patients presented a complete PSG syndrome with sleep-wake cycle disturbances and SOREMP. The abnormalities disappeared progressively after therapy during the last two examinations.

In a recent pilot study in *T. b. gambiense*-infected patients, the technique of actigraphy proved to be useful for an objective assessment of day/night disturbances during the disease (Njamnshi et al. 2012). This technique is based on the use of wrist-worn watch-like devices (actigraphs) which record the subject's rest and activity, from which sleep and wake are derived with widely standardized algorithms. Although PSG is the gold standard for sleep studies, actigraphy is a very user-friendly tool. In this investigation (Njamnshi et al. 2012), the use of actigraphy in HAT was validated with simultaneous nocturnal PSG recording. Although SOREMP were not revealed by actigraphy, the technique proved to be very promising for an objective clinical evaluation of HAT severity (including HAT-affected children), suited also for long-term assessment and follow-up investigations. Interestingly, as with PSG, sleep-wake disturbances were revealed by actigraphy also in stage 1 HAT patients. In addition, clinical disturbances revealed by actigraphic data did not consistently match WBC counts in the CSF, further demonstrating the need for objective clinical evaluation to be correlated with disease biomarkers (Njamnshi et al. 2012).

The alterations of sleep structure represented by SOREMP also occur in other neurological diseases and disorders and are especially characteristic of narcolepsy with cataplexy (see, for example, Dauvilliers et al. 2007). Loss of muscle tone, and especially sudden drops in neck muscle tone, during HAT are evocative of cataplexy and may be related to SOREMP (Bouteille and Buguet 2012). In narcolepsy with cataplexy, the CSF levels of the neuropeptide orexin/hypocretin, which plays a key role in the maintenance of wakefulness (Sakurai et al. 2010), are decreased. Orexin/hypocretin levels have been measured in the CSF of *T. b. gambiense*-infected patients, and they have been found to be higher than in narcoleptic patients, though with high interindividual variability (Dauvilliers et al. 2008).

4.2 Clinical Features of HAT in Travelers and Migrants

In travelers and immigrants, the clinical features of HAT can vary and the presentation can be atypical (Blum et al. 2011; Urech et al. 2011; Kennedy 2013).

In travelers the disease presents as an acute febrile illness, which becomes irregular if left untreated, with a predominance of nonspecific gastrointestinal symptoms (nausea, vomiting, diarrhea) and hepatomegaly in both forms of HAT. Jaundice has been reported in about one-third of *T. b. rhodesiense* infections. In addition, in HAT cases of travelers and expatriates from non-endemic countries, lymphadenopathy and sleep disturbances have been reported only occasionally, and the progression of the disease is rapid. In particular, daytime somnolence and nighttime insomnia are reported in a minority of travelers (5–7 %) with *T. b. gambiense* infections and in about 21 % of cases of *T. b. rhodesiense* infection (Urech et al. 2011). This could be due to the facts that most travelers present in stage 1 of HAT, with a short duration of disease, and that nighttime insomnia might have been under-reported, having been considered a minor problem in the context of an otherwise severe disease (Urech et al. 2011). Low-grade fever and neurological and psychiatric signs and symptoms predominate instead in HAT-affected African immigrants, who have even been admitted in some cases to psychiatric clinics.

The incubation period of *T. b. rhodesiense* infection in travelers is short, in all cases less than 1 month, and in the majority of cases of a few days (Blum et al. 2011; Urech et al. 2011). Also in *T. b. gambiense* infection of travelers the incubation period is usually shorter than 1 month, but can be very long in immigrants (Blum et al. 2011). Interestingly, since the incubation time for *T. b. gambiense* infection is difficult to assess in endemic areas and is therefore unknown, cases in immigrants indicate that the incubation period may last even 7 years (Urech et al. 2011).

5 Concluding Remarks and Perspectives

Basic and clinical research should go hand in hand in the fight against HAT, motivating all stakeholders for vector control, as well as effective control and surveillance of cases.

Clinical evaluation of HAT patients should regain a lost priority in combating this dreadful disease. Medical doctors are not included nowadays in most modern mobile team campaigns, so that clinical examination is mostly not applied by mobile teams for diagnosis and staging of HAT. This is due to several reasons. First, a thorough investigation of symptoms and signs takes time, and mobile teams, by definition, must move rapidly from one campsite to the next village, due to the cost of prospection. Second, medical doctors are rarely present in field activities, and, when present, they are mainly involved in the coordination of activities. It would, however, be important to introduce in mobile teams a protocol (standardized together with neurologists) for a thorough clinical examination by a medical doctor. The same protocol should be adopted by all health facilities in endemic areas.

The importance of clinical (and in particular neurological) evaluation of HAT patients should not be underscored: objective tests and scales of HAT clinical features, totally lacking at present, are needed for clinical assessment and follow-up, evaluation of relapses, assessment of the efficacy of novel treatments.

References

Amrouni D, Gautier-Sauvigné S, Meiller A, Vincendeau P, Bouteille B, Buguet A, Cespuglio R (2010) Cerebral and peripheral changes occurring in nitric oxide (NO) synthesis in a rat model of sleeping sickness: identification of brain iNOS expressing cells. PLoS One 5:e9211

Amrouni D, Meiller A, Gautier-Sauvigné S, Piraud M, Bouteille B, Vincendeau P, Buguet A, Cespuglio R (2011) Cerebral changes occurring in arginase and dimethylarginine dimethylaminohydrolase (DDAH) in a rat model of sleeping sickness. PLoS One 6:216891

Anonymous (1903) Reports of the sleeping sickness commission, No 1. Harrison and Sons, London

Atkins J (1734) The Navy surgeon or a practical system of surgery. Caesar Ward and Richard Chandler, London

Bentivoglio M, Grassi-Zucconi G, Kristensson K (1994) From trypanosomes to the nervous system, from molecules to behavior: a survey, on the occasion of the 90th anniversary of Castellani's discovery of the parasites in sleeping sickness. Ital J Neurol Sci 15:77–89

Berrang-Ford L, Lundine J, Breau S (2011) Conflict and human African trypanosomiasis. Soc Sci Med 72:398–407

Bisoffi Z, Buonfrate D, Angheben A (2014) Travel, migration and neglected tropical diseases. In: Bentivoglio M, Cavalheiro EA, Kristensson K, Patel N (eds) Neglected tropical diseases and conditions of the nervous system. Springer, New York

Blum J, Schmid C, Burri C (2006) Clinical aspects of 2541 patients with second stage human African trypanosomiasis. Acta Trop 97:55–64

Blum JA, Neumayr AL, Hatz CE (2011) Human African trypanosomiasis in endemic populations and travelers. Eur J Clin Microbiol Infect Dis. doi:10.1007/s10096-011-1403-y

Bouteille B, Buguet A (2012) The detection and treatment of human African trypanosomiasis. Res Rep Trop Med 3:35–45

Broden A, Rodhain J (1908) Rapport sur les travaux faits au Laboratoire de la Société belge d'études coloniales à l'hôpital des Noirs et au Lazaret pour trypanosomiés à Léopoldville (Congo belge) 1907-1908. Société belge d'études coloniales (ed) Hayez, Imprimeur des Académies royales de Belgique, Bruxelles

Bruce D (1895) Preliminary report on the tsetse fly disease or nagana in Zululand. Ubombo, December 1895. Bennett and Davis, Durban, South Africa

Bruce D (1897) Further report on the tsetse fly disease or nagana in Zululand. Ubombo, 29 May 1896. London, UK

Brun R, Blum J, Chappuis F, Burri C (2010) Human African trypanosomiasis. Lancet 375:148–159

Buguet A, Gati R, Sèvre JP, Develoux M, Bogui P, Lonsdorfer J (1989) 24-hour polysomnographic evaluation in a patient with sleeping sickness. Electroencephalogr Clin Neurophysiol 72:471–478

Buguet A, Bourdon L, Bouteille B, Chapotot F, Radomski MW, Dumas M (2001) The duality of sleeping sickness: focusing on sleep. Sleep Med Rev 5:139–153

Buguet A, Banzet S, Bouteille B, Vincendeau P, Tapie P, Doua F, Bogui P, Cespuglio R (2002) NO a cornerstone in sleeping sickness: voltammetric assessment in mouse and man. In: Moravec J, Takeda N, Singal PK (eds) Adaptation biology and medicine, vol 3, New frontiers. Narosa, New Dehli

Buguet A, Bisser S, Josenando T, Chapotot F, Cespuglio R (2005) Sleep structure: a new diagnostic tool for stage determination in sleeping sickness. Acta Trop 93:107–117

Buguet A, Bouteille B, Mpandzou G, Chapotot F, Ngampo S, Obanda F, Vincendeau P, Cespuglio R (2009) La recherche sur la maladie du sommeil (trypanosomose humaine africaine) en République du Congo de 2004 à 2009. Editions Les Manguiers, Brazzaville

Buguet A, Chapotot F, Ngampo S, Bouteille B, Cespuglio R (2012) Management of African trypanosomiasis of the CNS: polysomnography as non-invasive staging tool. Future Neurol 7:453–472

Castellani A (1903) On the discovery of a species of trypanosome in the cerebrospinal fluid of cases of sleeping sickness. Lancet 161:1735–1736

Chappuis F, Loutan L, Simarro P, Lejon V, Büscher P (2005) Options for field diagnosis of human African trypanosomiasis. Clin Microbiol Rev 18:133–146

Dauvilliers Y, Arnulf I, Mignot E (2007) Narcolepsy with cataplexy. Lancet 369:499–511

Dauvilliers Y, Bisser S, Chapotot F, Vatunga G, Cespuglio R, Josenando T, Buguet A (2008) Hypocretin and human African trypanosomiasis. Sleep 31:348–354

Darsaud A, Bourdon L, Mercier S, Chapotot F, Bouteille B, Cespuglio R, Buguet A (2004) Twentyfour hour disruption of the sleep-wake cycle and sleep-onset REM-like episodes in a rat model of African trypanosomiasis. Sleep 27:42–46

Dutton JE (1902) Note on *Trypanosoma* occurring in the blood of man. Br Med J 123:881–884

Fantham HB, Thomson JG (1910) Enumerative studies on *Trypanosoma gambiense* and *T. rhodesiense* in rats, guinea-pigs, and rabbits: periodic variations disclosed (preliminary note). Proc R Soc Lond B 83:206–211

Fèvre EM, Wissmann BV, Welburn SC, Lutumba P (2008) The burden of human African trypanosomiasis. PLoS Negl Trop Dis 2(12):e333

Forde RM (1902) Some clinical notes on a European patient in whose blood a *Trypanosoma* was observed. J Trop Med 5:261–263

Gadelha C, Holden JM, Allison HC, Field MC (2011) Specializations in a successful parasite: what makes the bloodstream-form African trypanosome so deadly? Mol Biochem Parasitol 179:51–58

Ginoux PY, Frézil JL, Alary JC (1982) La trypanosomiase humaine au moment du dépistage en République populaire du Congo. Distribution des signes cliniques. Med Trop 42:281–287

Giordano C, Clerc M, Doutriaux C, Doucet J, Nozais JP, Bureau JP, Piquemal M (1977) Le diagnostic neurologique au cours des différentes phases de la trypanosomiase humaine africaine. Ann Soc Belg Med Trop 57:213–225

Greig EDW, Gray ACH (1904) Notes on lymphatic glands in sleeping sickness. Lancet 163:1570

Jamonneau V, Ilboudo H, Kaboré J, Kaba D, Koffi M, Solano P, Garcia A, Courtin D, Laveissière C, Lingue K, Büscher P, Bucheton B (2012) Untreated human infections by *Trypanosoma brucei gambiense* are not 100% fatal. PLoS Negl Trop Dis 6(6):e1691

Kennedy PGE (2013) Clinical features, diagnosis, and treatment of human African trypanosomiasis (sleeping sickness). Lancet Neurol 12:186–194

Kérandel J (1910) Un cas de trypanosomiase chez un médecin (auto-observation). Bull Soc Pathol Exot Filiales 3:642–662

Kristensson K, Nygård M, Bertini G, Bentivoglio M (2010) African trypanosome infections of the nervous system: parasite entry and effects on neuronal functions. Prog Neurobiol 91:152–171

Lejon V, Bentivoglio M, Franco JR (2013) Human African trypanosomiasis. Handb Clin Neurol 114:169–181

Lhermitte J (1910) La maladie du sommeil et les narcolepsies. Etablissements d'imprimerie L. Severeyns, Bruxelles

Mackenzie S (1890) Negro lethargy. Lancet 136:1100–1101

Magnus E, Vervoot T, Van Meirvenne N (1978) A card agglutination test with stained trypanosomes (CATT) for the serological diagnosis of *T.b. gambiense* trypanosomiasis. Ann Soc Belg Med Trop 58:169–176

Malvy D, Chappuis F (2011) Sleeping sickness. Clin Microbiol Infect 17:986–995

Martin G, Leboeuf P (1908) Etude clinique sur la trypanosomiase humaine (maladie du sommeil). Ann Hyg Med Colon 11:381–393

Martin G, Leboeuf A, Roubaud E (1909) Rapport de la mission d'études de la maladie du sommeil au Congo français 1906-1908. Masson et Cie Editeurs, Paris

Masocha W, Kristensson K, Rottenberg ME (2014) Neurobiology of African trypanosomiasis. In: Bentivoglio M, Cavalheiro EA, Kristensson K, Patel N (eds) Neglected tropical diseases and conditions of the nervous system. Springer, New York

Migchelsen SJ, Büscher P, Hoepelman AI, Schallig HD, Adams ER (2011) Human African trypanosomiasis: a review of non-endemic cases in the past 20 years. Int J Infect Dis 15:517–524

Montmayeur A, Brosset C, Imbert P, Buguet A (1994) Cycle veille-sommeil au décours d'une trypanosomose humaine africaine à *Trypanosoma brucei rhodesiense* chez deux parachutistes français. Bull Soc Pathol Exot 87:368–371

Mwanakasale V, Sogolo P, Daka V (2013) Challenges in the control of human African trypanosomiasis in the Mpika district of Zambia. BMC Res Notes 6:180. doi:10.1186/1756-0500-6-180

Njamnshi AK, Seke Etet PF, Perrig S, Acho A, Fonsah JY, Mumba D, Muyembe JJ, Kristensson K, Bentivoglio M (2012) Actigraphy in human African trypanosomiasis as a tool for objective clinical evaluation and monitoring: a pilot study. PLoS Negl Trop Dis 6:e1525

Sakurai T, Mieda M, Tsujino N (2010) The orexin system roles in sleep/wake regulation. Ann N Y Acad Sci 1200:149–161

Scammell TE, Elmquist JK, Saper CB (1996) Inhibition of nitric oxide synthase produces hypothermia and depresses lipopolysaccharide fever. Am J Physiol 271:R333–R338

Simarro PP, Cecchi G, Paone M, Franco JR, Diarra A, Ruiz JA, Fèvre EM, Courtin F, Mattioli RC, Jannin JG (2010) The Atlas of human African trypanosomiasis: a contribution to global mapping of neglected tropical diseases. Int J Health Geogr 9:57

Simarro PP, Diarra A, Ruiz Postigo JA, Franco JR, Jannin JJ (2011a) The human African trypanosomiasis control and surveillance programme of the World Health Organization 2000-2009: the way forward. PLoS Negl Trop Dis 5(2):e1007

Simarro PP, Franco JR, Cecchi G, Paone M, Diarra A, Ruiz Postigo JA, Jannin JG (2011b) Human African trypanosomiasis in non-endemic countries (2000-2010). J Travel Med 19:44–53

Simarro PP, Cecchi G, Franco JR, Paone M, Diarra A, Ruiz-Postigo JA, Fèvre EM, Mattioli RC, Jannin JG (2012) Estimating and mapping the population at risk of sleeping sickness. PLoS Negl Trop Dis 6(10):e1859

Steverding D (2008) The history of African trypanosomiasis. Parasit Vectors 1:3. doi:10.1186/1756-3305-1-3

Tong J, Valverde O, Mahoudeau C, Yun O, Chappuis F (2011) Challenges of controlling sleeping sickness in areas of violent conflict: experience in the Democratic Republic of Congo. Confl Health 5:7. doi:10.1186/1752-1501-5e7

Urech K, Neumayr A, Blum J (2011) Sleeping sickness in travellers – do they really sleep? PLoS Negl Trop Dis 5(11):e1358

Vale GA, Chamisa A, Mangwiro C, Torr SJ (2013) A neglected aspect of the epidemiology of sleeping sickness: the propensity of the tsetse fly vector to enter houses. PLoS Negl Trop Dis 7(2):e2086

Wastling SL, Welburn SC (2011) Diagnosis of human sleeping sickness: sense and sensitivity. Trends Parasitol 27:394–402

Welburn SC, Maudlin I (2012) Priorities for the elimination of sleeping sickness. Adv Parasitol 79:299–337

WHO (1998) Control and surveillance of African trypanosomiasis. Report of a WHO Expert Committee. World Health Organ Tech Rep Ser 881:1–114

WHO (2007) Recommendations of the informal consultation on issues for clinical product development for human African trypanosomiasis. WHO/CDS/NTD/IDM/2007.1, Geneva, Switzerland

Winterbottom T (1803) An account of the native Africans in the neighbourhood of Sierra Leone to which is added an Account of the present state of medicine among them. C. Whittingham and John Hatchard, London, second edition (1969): Frank Cass & Co. Ltd., London

Neurobiology of African Trypanosomiasis

Willias Masocha, Krister Kristensson, and Martin E. Rottenberg

Abstract Invasion of the central nervous system (CNS) by the extracellular parasite *Trypanosoma brucei* is a major pathogenic event, causing a number of neurological dysfunctions, which makes African trypanosomiasis (sleeping sickness) almost always fatal if untreated. We here present our current understanding on the complex cellular and molecular host-parasite interactions, which underlie the neurobiology of African trypanosomiasis. Such interactions involve parasite- and host-derived molecules that play a role in the parasite invasion of the CNS. This invasion occurs in two phases: first through fenestrated vessels in the choroid plexus, circumventricular organs and peripheral nerve root ganglia, and then as a multistep process across post-capillary venules in the brain parenchyma. Potential mechanisms mediating the control of parasite accumulation within the brain parenchyma and the nervous system dysfunctions that include pain and sleep pattern disruptions are discussed. The need for better understanding of the neurobiology of the pathogenesis of these interactions in order to improve the diagnosis and management of sleeping sickness is also highlighted.

Keywords *Trypanosoma brucei* • Sleeping sickness • Immune responses • Blood-brain barrier • Sleep • Pain • Inflammation • Nervous system • Parasites

W. Masocha (✉)
Department of Pharmacology and Therapeutics, Faculty of Pharmacy, Kuwait University, P.O. Box 24923, Safat 13110, Kuwait
e-mail: masocha@hsc.edu.kw

K. Kristensson
Department of Neuroscience, Karolinska Institutet, Stockholm, Sweden
e-mail: krister.kristensson@ki.se

M.E. Rottenberg
Microbiology and Tumor Biology Center, Karolinska Institutet, Stockholm, Sweden

M. Bentivoglio et al. (eds.), *Neglected Tropical Diseases and Conditions of the Nervous System*,
DOI 10.1007/978-1-4614-8100-3_10,

1 Introduction

One of the most important pathogenic event of *Trypanosoma brucei (T. b.)* infections is the trypanosomes' ability to invade the central nervous system (CNS) after going through waves of parasitemia. These waves of parasitemia in the infected host are the result of an intricate interplay between the parasites and the host immune system, in which the parasites manage to "dribble" past the immune system in the periphery and invade the brain. The invasion is not passive, but rather a complex multistep active process. Paradoxically, trypanosomes utilize the immune response molecules aimed at controlling or eliminating the parasite to facilitate their entry into the brain. As described in Buguet et al. (2014), and also reviewed by Kennedy (2008), invasion of the nervous system results in peculiar sensory and pain syndromes, disturbances in sleep patterns, hence the common term, sleeping sickness, and extrapyramidal Parkinsonism-like signs of disease. The involvement of the CNS makes the disease intractable and almost always fatal if not treated. Understanding the neurobiology of African trypanosomiasis is important in order to devise better therapeutic options to manage the disease. Several drugs capable of eliminating the parasites in the periphery fail to do so when the parasites are in the CNS, and drugs effective for CNS infection are either toxic or difficult to administer. Thus, it is important to determine when to use CNS-penetrating drugs or not (described in Buguet et al. 2014). In this chapter, we will give an overview of the neurobiology of African trypanosomiasis, the interaction of the parasites with the host cellular and humoral components of the immune system, the mechanisms by which the parasites may invade and grow in the nervous system and how the parasites can cause dysfunctions in various regions of the nervous system to cause the clinical signs of the disease. The vast majority of the data on neuropathological changes associated with trypanosome infection and drug treatment is derived from animal models of the disease. Since the first description (Mott 1905) only limited numbers of post-mortem reports from human cases of African trypanosomiasis are available (reviewed by Kristensson and Bentivoglio 1999). Many animal species have been used in investigations, but most of the information comes from rodent and primate models (Kennedy 2007).

2 The Trypanosome Parasites

As described in the clinical chapter (Buguet et al. 2014), the lifecycle of the extracellular *T. b.* is complex and involves changes to adapt to the different environments of the intermediate host and vector, the *Glossina* tsetse fly species and the mammalian definitive hosts: humans for *T. b. gambiense* and wild as well as domestic animals for *T. b. rhodesiense* and *T. b. brucei*. The tsetse fly inoculates non-proliferative metacyclic forms of the parasite into the mammalian host, which transform into bloodstream slender forms that proliferate in the blood. When parasites reach a certain density, their bloodstream slender form differentiates into a stumpy form (Figarella et al. 2005; Vassella et al. 1997). The transitions between the slender and stumpy forms of the parasites can also be seen *in vitro*. The level of a parasite-secreted soluble factor (stumpy induction factor, whose chemical nature is unknown) in plasma regulates the differentiation into the stumpy form (Duszenko et al. 2006). The stumpy form is non-proliferative, so that transformation from bloodstream slender to stumpy form limits parasite numbers, which enhances the probability of the host to survive longer. In a chronic infection, the stumpy forms probably constitute the majority of circulating parasites (MacGregor et al. 2012), while most parasites that have invaded the brain parenchyma have a slender appearance (Mulenga et al. 2001). Importantly, the stumpy form is the tsetse fly infective form of the parasite, and thus the differentiation into stumpy forms is required for transmission to tsetse flies during a blood-meal. In the tsetse fly, the parasites transform through various stages and are eventually transferred to the fly's salivary glands in a metacyclic form that has the capacity to infect a mammalian host.

3 Interactions Between Trypanosomes and the Host Immune System

Like several other pathogenic microbes, African trypanosomes have evolved means to evade the immune responses in order to survive in the host (MacGregor et al. 2011). Establishing an infection is facilitated by activation of trypanosome-derived adenylate cyclases, which reduce the innate immune responses (Salmon et al. 2012). Long-term infections are promoted by antigenic variations of the trypanosome surface proteins that evade the humoral immune responses. The major proteins that constitute the *T. b.* surface coat, the variant surface glycoproteins (VSGs), are very antigenic, but are periodically changed. Thus, during an infection different VSG types sequentially emerge and each parasite variant is controlled by antibodies with distinct specificities. Through a pre-programmed switching between different VSG variants the trypanosomes are able to escape the effector mechanism of destruction mediated by the anti-VSG antibodies. *VSG* genes occupy 10 % of the trypanosome genome, but only one *VSG* gene is expressed at a time (Rudenko 2011). The switching between slender and stumpy forms, and the antibody control of the various VSG variants create the characteristic waves of parasitemia seen during the infection. Trypanosomal infections through these mechanisms can persist in humans for months (Rhodesian form of human African trypanosomiasis; HAT) or even years (Gambian form of HAT). Moreover, this efficient mechanism of immune evasion is reflected by the absence of protective immune memory, since animals cured from a *T. b. brucei* infection have the same susceptibility to re-infection as naïve mice (Radwanska et al. 2008).

Although tightly controlled, the innate immune system is activated during trypanosome infection. Macrophages release cytokines such as tumor necrosis factor (TNF)-α, interferon (IFN)-α/β, interleukin (IL)-12 amongst others, as well as nitric oxide (NO), after stimulation by trypanosomal CpG containing unmethylated DNA (ligand for the toll-like receptor (TLR) 9) released from phagocytosed trypanosomes (Harris et al. 2006; Pan et al. 2006).

The adaptive immune responses show a predominant and robust Th1 pattern with elevated levels of Th1 cytokines such as IFN-γ. This is paradoxical, because Th1 responses are commonly elicited to combat intracellular not extracellular pathogens. IFN-γ mediates the control of parasitemia and resistance to the infection (Hertz et al. 1998). This probably induces macrophage-mediated digestion and clearance of parasites in the liver, whereby parasite adenylate cyclases from lysed trypanosomes might hamper further TNF-α secretion (Salmon et al. 2012). The Th1 immune responses also contribute to the immunopathology of African trypanosomiasis. Over time, the infection eventually leads to an extensive immunosuppression leading to increased susceptibility of the patients to secondary infectious pathogens such as bacteria (Baral 2010; Vincendeau and Bouteille 2006).

4 Trypanosome Invasion of the Central Nervous System

In the CNS under physiological conditions, different barriers prevent the non-selective passage of macromolecules from the blood into the brain parenchyma in order to maintain a constant internal environment appropriate for neuronal function. These barriers are represented by the blood-brain barrier (BBB) and the blood–cerebrospinal fluid (CSF) barriers in the choroid plexus and the leptomeningeal vessels, and they also prevent passage of most microbes into the brain. However, as explained in other chapters of this book, some pathogens can cross or circumvent them and this is also true for African trypanosomes.

In rodents infected with *T. b. brucei*, the parasites appear to reach the nervous system in two phases. During an infection the parasites appear relatively early in the leptomeninges as well as in areas that lack a BBB, i.e. in the choroid plexus, the circumventricular organs (CVOs) and peripheral nerve root ganglia. In a later phase of the infection (weeks or months), the parasites penetrate the BBB and enter the brain parenchyma. An early invasion of the brain parenchyma (hours), as recently visualized by confocal microscopy of intravenously injected transgenic fluorescent trypanosomes in mice after the trauma of craniotomy or vibratome sectioning of un-fixed brain tissue (Frevert et al. 2012; MacLean et al. 2013), has not been observed in several other studies (Kristensson et al. 2010; Wolburg et al. 2012).

4.1 *Early Invasion of Trypanosomes into the Leptomeninges, Choroid Plexus, CVOs and Peripheral Nerve Ganglia*

The endothelial cells in the blood vessels of the leptomeninges (arachnoid membrane and pia mater) are linked by tight junctions, but they are more permeable to macromolecules than brain parenchymal vessels (Broadwell and Sofroniew 1993).

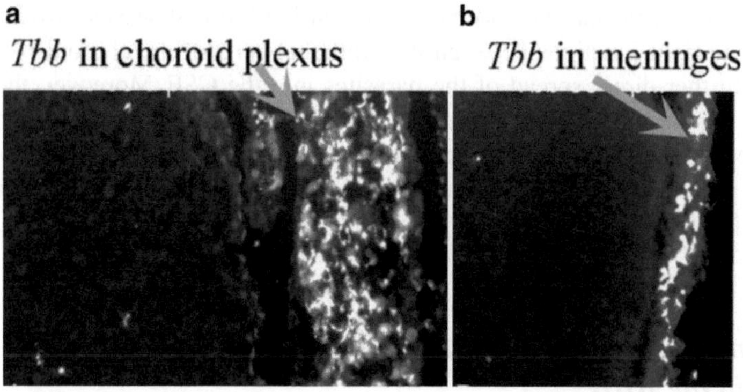

Fig. 1 Invasion of (**a**) the choroid plexus of a lateral ventricle and (**b**) the leptomeninges by *T. b. brucei* parasites in rats infected with the parasite can occur early during the infection before invasion the brain parenchyma. Note that there is no gradient of parasites from these two areas into the brain parenchyma. Reproduced from Mulenga et al. (2001), with permission

These vessels lack an astrocyte-derived parenchymal basement membrane (see Sect. 4.2) and this may facilitate the passage of a number of pathogens into the meninges and subarachnoidal space. In rodent models, *T. b.* and white blood cells (WBCs) localize in the meninges early after the infection, before they are observed in the brain parenchyma. In an *in vivo* imaging study parasites could be killed by a stage 1 drug when observed in the meninges as early as 5 days after infection, while parasites that had entered the brain parenchyma 21 days after infection survived (Myburgh et al. 2013). At this later stage when the parasite has entered the brain parenchyma, there is no gradient of trypanosome density from the superficial to the deep layers of the cortex (Fig. 1b), suggesting that the parasites do not migrate from the meninges into the cortex.

The blood-CSF barrier is also formed by the epithelial cells of the choroid plexus, the structure responsible for CSF formation (Engelhardt and Sorokin 2009). The endothelial cells of capillaries in the choroid plexus are fenestrated and allow passage of macromolecules from the bloodstream into the choroid plexus stroma. However, the tight junctions that link the epithelial cells of the choroid plexus prevent further non-selective passage of macromolecules into the CSF (Czosnyka et al. 2004).

Other nervous system structures whose blood vessels are fenestrated and therefore are permeable to macromolecules are the peripheral nerve root ganglia and the CVOs around the third and fourth ventricles, including area postrema, median eminence, organum vasculosum of the lamina terminalis, pineal gland, posterior pituitary, and subcommissural and subfornical organs. The CVOs are enclosed by special cells called tanycytes that may restrict movement of macromolecules from the CVOs into the CSF and/or brain parenchyma.

As with leptomeninges, *T. b. brucei* and WBCs localize in the choroid plexus (Fig. 1a), CVOs and peripheral nerve root ganglia early after infection, before invasion of the brain parenchyma (Schultzberg et al. 1988). A particularly long slender form of the parasites, which appear mostly after three to four waves of parasitemia, has been associated with choroid plexus invasion, and this may explain why the parasites do not invade the choroid plexus during the first waves (Wolburg et al. 2012). The tight junctions between the epithelial cells of the choroid plexus may prevent further direct spread of the parasites into the CSF. Moreover, the CSF is toxic to trypanosomes (Wolburg et al. 2012) and no parasite gradient around the ventricles in the brain is observed indicating that trypanosomes do not spread through the ventricular CSF into the brain parenchyma (Masocha et al. 2004; Mulenga et al. 2001).

4.2 Invasion of the Brain Parenchyma by Trypanosomes via the BBB

The BBB, formed by cerebral endothelial cells, pericytes, basement membranes and astrocytic end-feet, is the largest and most restrictive barrier in the CNS, which prevents non-selective passage of molecules from the blood into the parenchyma (Abbott et al. 2006; Sixt et al. 2001). Tight junctions linking endothelial cells form a structural barrier, which prevents passage of macromolecules between the blood vessel and brain via the paracellular route, while the low endocytotic activity of endothelial cells, restricts passage of molecules via the transcellular route (transcytosis); this forms a functional barrier selective for specific molecules (Abbott et al. 2010; Wilhelm et al. 2011). The exchange of metabolites across the BBB mainly occurs at the level of capillaries, whereas the infiltration of WBCs during inflammation occurs at the level of post-capillary venules (Owens et al. 2008). Post-capillary venule endothelial cells are separated from the perivascular astrocytic end-feet by two basement membranes: the endothelial and the parenchymal basement membrane also called the astrocytic basement membrane (Owens et al. 2008; Sixt et al. 2001; Sorokin 2010). The composition of laminin molecules makes these two basement membranes differ with respect to WBCs penetration. The endothelial basement membranes composed of laminin $\alpha 4$ permit penetration of WBCs while those composed of $\alpha 5$ chains restrict. The parenchymal basement membranes, which have laminin $\alpha 1$ and $\alpha 2$ chains, must be "opened" before WBCs can pass (Sixt et al. 2001). Infiltrating WBCs, as occur in experimental allergic encephalomyelitis, are trapped between the endothelial and the parenchymal basement membranes before matrix metalloproteases (MMPs) are activated (Owens et al. 2008; Sorokin 2010). The focal activation of MMPs 2 and 9 in macrophages to cleave the dystroglycan receptors that anchor the astrocytic end-feet to the parenchymal basement membrane is required for further infiltration of WBCs (Agrawal et al. 2006; Sorokin 2010). Pericytes enwrapped by the endothelial basement membrane have recently been shown to be important for regulation of transcytosis across the BBB (Allt and Lawrenson 2001; Armulik et al. 2010), but their role in inflammatory processes are not yet clear.

Similar to WBCs, trypanosomes cross the BBB in a multistep process (Mulenga et al. 2001). First they cross the endothelial cell layer (Masocha et al. 2004) where capillaries are surrounded by a permissive laminin $\alpha 4$ containing basement membrane, followed by passage across the parenchymal basement membrane, a process regulated by different immune molecules as described further (see Sect. 4.3.2). It should be noted that brain invasion of trypanosomes and WBCs appears predominantly in the white matter, septal nuclei and diencephalic nuclei around the ventricles in experimental rodent models. This is consistent with neuropathological findings in brains from HAT victims, in which the inflammatory cell infiltration is mainly seen in the white matter. HAT is therefore considered a "leukoencephalitis".

4.3 Molecules That Facilitate Parasite Invasion of the Brain Parenchyma

Several *in vitro* and *in vivo* studies suggest that both parasite- and host-derived factors have a role in the trypanosome invasion of the CNS (Table 1).

4.3.1 Parasite-Derived Molecules That Facilitate Trypanosome BBB Crossing

Similar to other parasite species, such as *Schistosoma* and *T. cruzi*, that secrete proteases to enable them to enter the host or the host cells, the *T. b.* subspecies express phosphatases and proteases on their external surfaces. Such molecules have been suggested to facilitate *T. b.* passage across the BBB. *In vitro* models of the BBB have shown that cysteine proteases play a role in parasite's passage across cerebral endothelial cells either via the paracellular or the transcellular route (Abdulla et al. 2008; Grab et al. 2004, 2009; Nikolskaia et al. 2006a, b). The parasite-derived cysteine protease-mediated penetration of BBB depends on the activation of endothelial cell G protein-coupled receptors (i.e. protease-activated receptor-2, PAR2) and calcium signaling (Grab et al. 2009, 2011). The inhibition of parasite proteases by RNA interference resulted in prolonged survival of infected mice in one study (Abdulla et al. 2008), but whether these manipulated parasites can invade the brain was not explored. It is important to note that cerebral endothelial cells used *in vitro* show marked differences in the expression of molecules involved in the BBB as compared to the cells *in vivo*. Moreover, the structure of post-capillary venules *in vitro* still has to be established.

4.3.2 Host-Derived Molecules That Facilitate Trypanosome BBB Crossing

The same isolate of *T. b. brucei* invades the brain parenchyma of C57BL/6 mice earlier and more extensively than in SV-129/Ev mice even though the latter strain of mice higher levels of parasites in the blood (Masocha et al. 2008). Brains of *T. b. brucei-infected* C57BL/6 mice also express higher levels of inflammatory

Table 1 Molecules/factors involved in *Trypanosoma brucei* spp. crossing of the BBB

Molecule/factor	Proposed roles	References
Parasite-derived molecules		
Cysteine proteases (i.e. brucipain)	Increase in BBB permeability which promote parasite crossing of the BBB	Abdulla et al. (2008), Grab et al. (2009, 2011), Nikolskaia et al. (2006a, b)
Host derived-molecules		
CXCL10	Attraction or retention of T cells and trypanosomes in the brain parenchyma	Amin et al. (2009)
IFNα/β	Induces a limited release of CXCL10 from astrocytes and endothelial cells, which is enough for penetration of sensitized T cells and some trypanosomes into the brain parenchyma (initiation of T cell and parasite crossing the BBB into the brain parenchyma)	Amin et al. (2012)
IFN-γ	Facilitate T cell and parasite crossing of the BBB by inducing the expression of CXCL10 and other molecules whose nature is not yet determined	Masocha et al. (2004)
PAR2	Brucipain interacts with PAR2 on endothelial cells to increase BBB permeability which promote parasite crossing of the BBB	Grab et al. (2009, 2011)
TLR9-MyD88 signalling pathway	Induces release of the innate immune response molecules TNF-α and IFN-α/β that facilitate T cell and parasite crossing of the BBB	Amin et al. (2012)
TNF-α	Facilitate T cell and parasite crossing of the BBB by increasing expression of adhesion molecules i.e. ICAM-1	Amin et al. (2012)

BBB blood-brain barrier, *CXCL10* C-X-C motif chemokine 10, *ICAM-1* intercellular adhesion molecule 1, *IFN* interferon, *MyD88* myeloid differentiation primary response gene (88), *PAR2* protease-activated receptor-2, *TLR9* toll-like receptor 9

molecules, including pro-inflammatory cytokines and adhesion molecules than SV-129/Ev mice (Masocha et al. 2008), which suggests that genetic differences in the immune response regulation play an important role in parasite invasion of the CNS.

The role of selected immune response molecules in *T. b. brucei* passage across the BBB has been studied in genetically engineered mice and can be summarized as follows. The innate immune system, which recognizes pathogen-associated molecular patterns of trypanosomes via TLRs, is activated by the intracellular adaptor molecule MyD88 and plays a role in the control of parasitemia (Amin et al. 2012; Drennan et al. 2005). The induction of the cytokines TNF-α and IFN-α/β during trypanosome infection is dependent on TLR2/9-MyD88-mediated signaling (Amin et al. 2012). After *T. b. brucei* infection, the number of both parasites and T cells in the brain parenchyma is reduced in mice deficient in either TNF-α or IFN-α/β

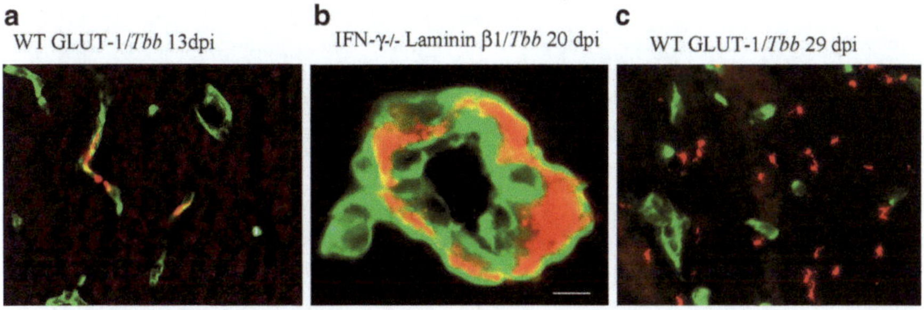

Fig. 2 Invasion of the brain parenchyma by *T. b. brucei* parasites (in *red*) in mice infected with the parasite. (**a**) At an early stage of the infection the parasites are confined in the lumen of cerebral blood vessels (cerebral endothelial cells in *green*), (**b**) but at a later stage they cross the endothelial cell layer and the endothelial basement membrane into the perivascular space (noticeable in the absence of IFN-γ) (basement membrane in *green*), (**c**) followed by crossing the parenchymal basement membrane to appear in the brain parenchyma (cerebral endothelial cells in *green*). Reproduced from Masocha et al. (2004), with permission

signaling compared to wild-type mice (Amin et al. 2012). However, T cell number is lower but parasite accumulation is increased in the brain parenchyma of TLR2 and 9 or MyD88 deficient mice compared to wild type mice (Amin et al. 2012). Thus, neuroinvasion and growth control (see Sect. 5) of the trypanosomes in the brain are regulated by different sets of molecules.

The importance of adaptive immune response molecules, and in particular T cell-derived IFN-γ, for trypanosome crossing the parenchymal basement membrane, was discovered in a series of experiments. *T. b. brucei* and T cell penetration of the brain parenchyma is reduced in mice deficient in IFN-γ signaling despite increased parasite levels in the blood (Masocha et al. 2004). In the absence of IFN-γ, parasites could cross the endothelium and the endothelial, but not the parenchymal, basement membrane and accumulate as cuffs in the perivascular space (Fig. 2). In accordance, IFN-γ levels are very low after *T. b. brucei* infection and parasite invasion of the CNS is reduced in RAG-1 deficient mice (lacking mature T and B cells) (Masocha et al. 2004). IFN-γ is secreted during the infection by antigen-specific T cells which tentatively recognize parasite epitopes on microglia or perivascular macrophages; macrophages can phagocytize parasites (Wolburg et al. 2012). IFN-γ and also IFN--α/β, albeit the latter less potently, regulate the expression of various chemokines including the T cell attractant CXCL10. Cerebral endothelial cells and astrocytes produce CXCL10 in the CNS of trypanosome infected mice (Amin et al. 2009). Mice deficient in CXCL10 or its receptor CXCR3 on T cells had less parasites in the brain parenchyma compared to wild type mice, albeit similar parasitemia levels. However the cuffing of trypanosomes and T cells around cerebral vessels observed in IFN-γ and in RAG-deficient mice was not seen in these mice (Amin et al. 2009). This suggests that different IFN-γ-regulated genes are involved in the attraction or retention of T cells and trypanosomes in the brain on one hand and in the opening of the parenchymal basement membrane for their passage on the other hand.

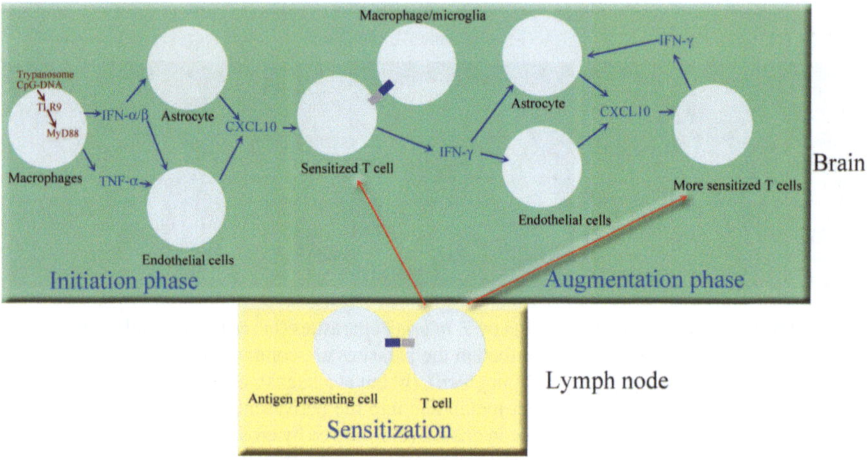

Fig. 3 Schematic diagram of the cells and molecules thought to play a role in parasite and T cell invasion of the brain parenchyma during trypanosome infection. Macrophages, after phagocytosing and lysing trypanosomes, release trypanosome CpG-DNA that activates TLR9-MyD88 signalling pathways systemically. This event induces secretion of the innate immune response molecules TNF-α and IFN-α/β. TNF-α activates endothelial cells possibly leading to expression of adhesion molecules, and IFN-α/β induces a limited expression of CXCL10 from endothelial cells and astrocytes. This could enable a few sensitized T cells and some trypanosomes to enter the brain (initiation phase). Trypanosome-derived antigens taken up and expressed by perivascular macrophages or microglia are then recognized by sensitized T cells to cause IFN-γ release. This will augment the process, since IFN-γ is a very potent inducer of CXCL10 and other molecules that lead to enhanced attraction of both T cells and trypanosomes into the brain parenchyma (augmentation phase)

A scheme of cells and molecules thought to play a role in parasite and T cell neuroinvasion during trypanosome infection (Fig. 3) and a hypothetical sequence of events that may lead to invasion of predominantly the white matter (Fig. 4) can be derived from these data.

5 Control of Trypanosome Growth in the CNS

Patients infected by *T. b. gambiense* often have a long course of disease (months or several years) with signs of nervous system involvement. A recent study found that occasionally patients infected with *T. b. gambiense* develop trypanotolerance i.e., the infection may spontaneously cure without treatment, so that the disease is not 100 % fatal as previously thought (Jamonneau et al. 2012). Although, as mentioned above, the CSF is hostile for trypanosome survival (Wolburg et al. 2012), experimental models have suggested that trypanosomes may remain in the brain following treatment that eliminate the parasites from the rest of the body (Jennings et al. 1979). Relapses may then occur several months later, possibly from the pool of parasites in the brain. Field isolates of *T. b. gambiense* that cause chronic infections

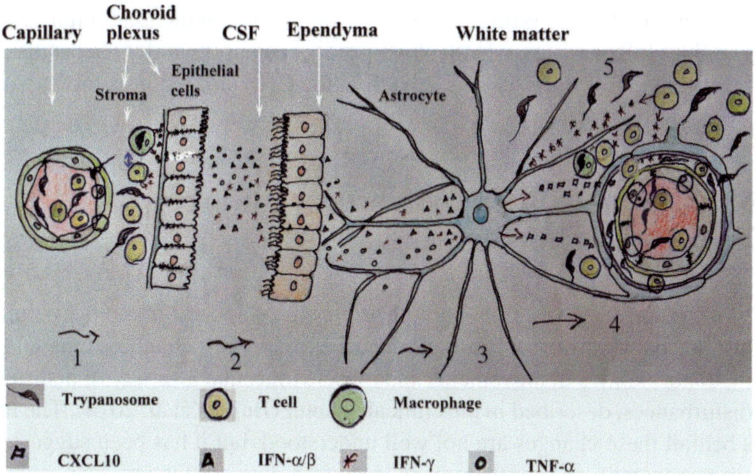

Fig. 4 A hypothetical model of a series of events leading to preferential infiltration of T cells and *T. b. brucei* into the white matter of the brain. (1) White blood cells and parasites cross the fenestrated endothelial cell layer of capillary vessels into the choroid plexus stroma early during the infection. (2) TNF-α released by the subsequent inflammation may increase the permeability of the choroid plexus epithelial cell layer, enhancing the diffusion of cytokines into the CSF; cytokines have been found elevated in the CSF of HAT patients (Amin et al. 2009). From the ventricles, such cytokines could readily diffuse past the ependymal cells, which are not linked by tight junction, into the white matter that, in contrast to the cerebral cortex, have wide extracellular spaces allowing a rapid diffusion of macromolecules throughout. (3) IFN-α/β (and later IFN-γ) induces a limited release of CXCL10 by astrocyte and endothelial cells, (4) which permit attraction and/or retention of a few sensitized T cells and some trypanosomes into the brain parenchyma to start the initiation phase which is augmented when more IFN-γ is released as described in Fig. 3. This series of events is similar to that described for the invasion of T cells into the white matter of brain of mice infected with lymphocytic choriomeningitis virus, which is a virus that first infects cells in the choroid plexus and then microglia in the white matter (Thomsen 2009)

for 6–8 months in mice show preferential localization in the olfactory bulbs and cerebellum (Giroud et al. 2009). This suggests that growth of the trypanosomes, which in their slender form replicate rapidly, might be controlled in the brain.

It is often difficult to distinguish between the immune responses that regulate the passage of trypanosomes across the BBB from factors that control parasite replication within the brain parenchyma. As described above, we have observed that recognition of parasites via TLR2 and TLR9, but not the release of TNF-α and IFN-α/β, is required for the control of trypanosome growth in the brain (Amin et al. 2012), although the nature of the effector molecules for this are still unknown. In this context it is of interest to note that *in vitro* studies have shown that endogenous neuropeptides, such as vasoactive intestinal peptide (VIP), adrenomedullin, alpha-melanocyte-stimulating hormone, urocortin, corticotropin-releasing hormone and ghrelin, have potent trypanolytic activity *in vitro*. Administration of VIP intraperitoneally to *T. b. brucei*-infected mice reduces parasitemia and prolongs animal survival and interestingly, infection with *T. b. brucei* increases serum levels of VIP

(Delgado et al. 2009). While these *in vitro* studies show that high levels of neuropeptides kill parasites, whether host-produced neuropeptides reach sufficiently high levels to control trypanosome growth in the CNS still requires investigation.

6 Mechanisms of Brain Dysfunctions During Trypanosome Infections

HAT is associated with symptoms both from the peripheral nervous system, e.g., pruritus and pain, and the CNS, e.g., disturbances in circadian rhythms and sleep pattern, abnormalities in movements and muscle tone, and a variety of neuropsychiatric disturbances, described in the clinical chapter (Buguet et al. 2014). The mechanisms behind these changes are not well understood, but it has been suggested that both trypanosome- and host-immune response-derived factors play roles. For instance, prostaglandin (PG) D_2, which can be synthesized by both the trypanosomes and the host in response to infection, has been implicated in the pathogenesis of the pain and sleep disturbances. The immune response molecules TNF-α and IFN-γ that, as mentioned above, play roles in facilitating trypanosome crossing of the BBB, can affect synaptic functions and are involved pain and sleep pattern disruptions, similarly to some other inflammation-released molecules. Such possible mechanisms are summarized below (review, see Kristensson et al. 2010).

PGs can sensitize nociceptive neurons to painful stimuli at the peripheral and spinal cord level. Both PGE_2 and PGD_2 can induce hyperalgesia (Ohkubo et al. 1983), and induce or modulate allodynia (a painful response to a normally innocuous stimulus) (Eguchi et al. 1999; Telleria-Diaz et al. 2008). The cytokines TNF-α and IFN-γ can induce neuropathic pain in rodents acting possibly both peripherally and centrally (Robertson et al. 1997; Vikman et al. 2003, 2007; Zimmermann 2001). Similarly to HAT patients, hyperalgesia occurs in rats early after infection with *T. b. brucei* (Wiesenfeld-Hallin et al. 1991) when trypanosomes have invaded the dorsal root ganglia (Schultzberg et al. 1988), but not the brain parenchyma, and when the rats are still sensitive to drugs used for treatment of the early stage of the disease. It is therefore very important to distinguish between symptoms and signs from the peripheral and the central nervous system in clinical evaluation protocols of HAT patients, which, unfortunately, is currently not the case.

PGD_2 is a potent somnogenic molecule (Huang et al. 2007; Urade and Hayaishi 2011). PGD_2 receptors are located mainly in the leptomeninges covering the basal forebrain (Mizoguchi et al. 2001; Urade and Hayaishi 2011), that trypanosomes infiltrate early in infection. PGD_2 is also involved in the regulation of the non-rapid eye movement (NREM) sleep (Ueno et al. 1982; Urade and Hayaishi 2011). The somnogenic effects of PGD_2 are mediated through binding to its receptor causing release of adenosine that binds to adenosine A_{2A} receptors in ventrolateral preoptic nuclei neurons (Scammell et al. 1998, 2001) and to adenosine A_1 receptors in tuberomammillary nuclei neurons (Oishi et al. 2008). This promotes NREM sleep, and PGD_2 synthesized by trypanosomes has been suggested to cause the sleep disturbances in HAT (Figarella et al. 2005; Kubata et al. 2000). However, HAT is characterized by disruption of the sleep rhythm, with appearance of sleep-onset-REM sleep episodes during daytime (rapid transitions from wakefulness into sleep that occurs in narcolepsy) and wakefulness episodes during night, rather than by an increase in time spent in NREM sleep, which suggests that other factors are also involved in causing the sleep disturbances in HAT.

Since trypanosomes target the CVOs before entering the brain parenchyma, inflammatory mediators in the CVOs could, hypothetically, sensitize axon terminals of hypothalamic neurons reaching these organs (outside the BBB) to transmit information by retrograde signals to their cell bodies (inside the BBB). For instance, the arcuate nucleus-median eminence complex (AMC) is a site of heavy accumulation of parasites and lymphocytes in both rats and mice. The AMC is connected to several areas involved in sleep/wakefulness and circadian rhythm regulation, including the suprachiasmatic nucleus (SCN) through which the AMC provides a feedback system for the control of sleep rhythms as well as corticosterone and melatonin secretion rhythms (Buijs et al. 2006). The secretion of these molecules is disturbed in HAT. Both TNF-α and IFN-γ affect circadian functions in the SCN, and the function of sleep-wakefulness regulating hypothalamic and brainstem neurons are also influenced by other inflammatory molecules (review, see Kristensson et al. 2010). Thus, by affecting CVOs, and in particular the AMC, the trypanosome infections may cause changes in circadian rhythms and sleep-wake regulatory networks, changes which may initiate even before the parasite penetrates into the BBB (see Buguet et al. 2014).

7 Perspectives and Role of Neuroscience for Better Understanding of Sleeping Sickness

Neuroscience research can immensely contribute to the understanding of sleeping sickness since the involvement of the CNS defines the disease and makes it difficult to manage. On the other hand, the study of the disease can lead to knowledge progress in neuroscience.

Trypanosomes and the BBB have a long history in common. The BBB was discovered 100 years ago by Goldmann (1913), who was inspired by Paul Ehrlich, the founder of chemotherapy. Discovery of the BBB was a follow-up of Ehrlich's experiments with trypanosomes. After Ehrlich had discovered that living trypanosomes, in contrast to living cells of the host, were stained with trypan dyes, he tried unsuccessfully in 1904–1905 to treat trypanosome-infected rodents with trypan dyes. It was noted that the brain and spinal cord were not stained by the dyes at variance with other body tissues. Goldmann (1913) then observed that dyes injected into the cerebral ventricles stained the surrounding brain tissue. The finding showed that the failure of staining of the brain after systemic injection was due to a permeability problem: the BBB was thus discovered.

Experimental models will be crucial to correlate objective measurements of neurological signs of disease and new biomarkers with passage of trypanosomes across the BBB and the outcome of various treatments. Currently, the number of WBCs and presence of trypanosomes in the CSF is used to distinguish between the first and second stage of the disease, but this is not an accurate criteria. Since there is no gold standard to distinguish these stages, it is difficult to validate new biomarkers in clinical studies, and animal studies are therefore needed to guide these endeavors. Interestingly, in recent multi-center studies neopterin, which is a stable product of IFN-γ-stimulated macrophages, and the IFN-γ-inducible CXCL10 has been suggested as cerebrospinal fluid markers for late stage human African trypanosomiasis (Tiberti et al. 2012, 2013a, b); these molecules are biologically meaningful since they play a crucial role in trypanosome invasion of the brain parenchyma, Sect. 4.3.2.

Neither in HAT patients nor in experimental models are seizures and neurodegenerative changes prominent in spite of the presence of neuroinflammation. However, treatment of trypanosome-infected rats with a non-steroidal anti-inflammatory drug induced severe neurodegeneration (Quan et al. 2000). The role of inflammatory molecules in the pathogenesis of seizures and of neurodegenerative diseases is extensively investigated in neuroscience. Trypanosome infections can provide interesting models of chronic neuroinflammation for studies of factors that may disturb the balance between neuroprotection and brain dysfunctions in such conditions. For instance, although nitric oxide (NO) in general is regarded as a pro-inflammatory molecule, in preliminary experiments we have found that it may paradoxically impede brain invasion of both T cells and trypanosomes, but that trypanosome growth in the brain is under the control of other, still unknown, factors that cause no or only minimal neurodegeneration. Thus, studies on identifying factors that regulate pathogen-host interactions could have a broader utility of providing new knowledge on not only control of a microbe but also on protection from neurodegeneration and seizures.

Acknowledgements The studies have been supported by grants from the US NIH/Fogarty (1R21NS064888-01A1) and the Swedish Research Council (04480).

References

Abbott NJ, Ronnback L, Hansson E (2006) Astrocyte-endothelial interactions at the blood-brain barrier. Nat Rev Neurosci 7:41–53

Abbott NJ, Patabendige AA, Dolman DE, Yusof SR, Begley DJ (2010) Structure and function of the blood-brain barrier. Neurobiol Dis 37:13–25

Abdulla MH, O'Brien T, Mackey ZB, Sajid M, Grab DJ, McKerrow JH (2008) RNA interference of Trypanosoma brucei cathepsin B and L affects disease progression in a mouse model. PLoS Negl Trop Dis 2:e298

Agrawal S, Anderson P, Durbeej M, van Rooijen N, Ivars F, Opdenakker G, Sorokin LM (2006) Dystroglycan is selectively cleaved at the parenchymal basement membrane at sites of leukocyte extravasation in experimental autoimmune encephalomyelitis. J Exp Med 203:1007–1019

Allt G, Lawrenson JG (2001) Pericytes: cell biology and pathology. Cells Tissues Organs 169:1–11
Amin DN, Rottenberg ME, Thomsen AR, Mumba D, Fenger C, Kristensson K, Buscher P, Finsen B, Masocha W (2009) Expression and role of CXCL10 during the encephalitic stage of experimental and clinical African trypanosomiasis. J Infect Dis 200:1556–1565
Amin DN, Vodnala SK, Masocha W, Sun B, Kristensson K, Rottenberg ME (2012) Distinct Toll-like receptor signals regulate cerebral parasite load and interferon alpha/beta and tumor necrosis factor alpha-dependent T-cell infiltration in the brains of Trypanosoma brucei-infected mice. J Infect Dis 205:320–332
Armulik A, Genove G, Mae M, Nisancioglu MH, Wallgard E, Niaudet C, He L, Norlin J, Lindblom P, Strittmatter K, Johansson BR, Betsholtz C (2010) Pericytes regulate the blood-brain barrier. Nature 468:557–561
Baral TN (2010) Immunobiology of African trypanosomes: need of alternative interventions. J Biomed Biotechnol 2010:389153
Broadwell RD, Sofroniew MV (1993) Serum proteins bypass the blood-brain fluid barriers for extracellular entry to the central nervous system. Exp Neurol 120:245–263
Buguet A, Mpanzou G, Bentivoglio M (2014) Human African trypanosomiasis: a highly neglected neurological disease. In: Bentivoglio M, Cavalheiro EA, Kristensson K, Patel N (eds) Neglected tropical diseases and conditions of the nervous system. Springer, New York
Buijs RM, Scheer FA, Kreier F, Yi C, Bos N, Goncharuk VD, Kalsbeek A (2006) Organization of circadian functions: interaction with the body. Prog Brain Res 153:341–360
Czosnyka M, Czosnyka Z, Momjian S, Pickard JD (2004) Cerebrospinal fluid dynamics. Physiol Meas 25:R51–R76
Delgado M, Anderson P, Garcia-Salcedo JA, Caro M, Gonzalez-Rey E (2009) Neuropeptides kill African trypanosomes by targeting intracellular compartments and inducing autophagic-like cell death. Cell Death Differ 16:406–416
Drennan MB, Stijlemans B, Van den Abbeele J, Quesniaux VJ, Barkhuizen M, Brombacher F, De Baetselier P, Ryffel B, Magez S (2005) The induction of a type 1 immune response following a Trypanosoma brucei infection is MyD88 dependent. J Immunol 175:2501–2509
Duszenko M, Figarella K, Macleod ET, Welburn SC (2006) Death of a trypanosome: a selfish altruism. Trends Parasitol 22:536–542
Eguchi N, Minami T, Shirafuji N, Kanaoka Y, Tanaka T, Nagata A, Yoshida N, Urade Y, Ito S, Hayaishi O (1999) Lack of tactile pain (allodynia) in lipocalin-type prostaglandin D synthase-deficient mice. Proc Natl Acad Sci U S A 96:726–730
Engelhardt B, Sorokin L (2009) The blood-brain and the blood-cerebrospinal fluid barriers: function and dysfunction. Semin Immunopathol 31:497–511
Figarella K, Rawer M, Uzcategui NL, Kubata BK, Lauber K, Madeo F, Wesselborg S, Duszenko M (2005) Prostaglandin D2 induces programmed cell death in Trypanosoma brucei bloodstream form. Cell Death Differ 12:335–346
Frevert U, Movila A, Nikolskaia OV, Raper J, Mackey ZB, Abdulla M, McKerrow J, Grab DJ (2012) Early invasion of brain parenchyma by African trypanosomes. PLoS One 7:e43913
Giroud C, Ottones F, Coustou V, Dacheux D, Biteau N, Miezan B, Van Reet N, Carrington M, Doua F, Baltz T (2009) Murine models for Trypanosoma brucei gambiense disease progression—from silent to chronic infections and early brain tropism. PLoS Negl Trop Dis 3:e509
Goldmann EE (1913) Vitalfärbung am Zentralnervensystem. Beitrag zur Physio-Pathologie des Plexus Chorioideus und der Hirnhäute. Verlag der königlichen Akademie der Wissenschaften, Berlin
Grab DJ, Nikolskaia O, Kim YV, Lonsdale-Eccles JD, Ito S, Hara T, Fukuma T, Nyarko E, Kim KJ, Stins MF, Delannoy MJ, Rodgers J, Kim KS (2004) African trypanosome interactions with an in vitro model of the human blood-brain barrier. J Parasitol 90:970–979
Grab DJ, Garcia-Garcia JC, Nikolskaia OV, Kim YV, Brown A, Pardo CA, Zhang Y, Becker KG, Wilson BA, de A Lima AP, Scharfstein J, Dumler JS (2009) Protease activated receptor signal-

ing is required for African trypanosome traversal of human brain microvascular endothelial cells. PLoS Negl Trop Dis 3:e479

Grab DJ, Chakravorty SJ, van der Heyde H, Stins MF (2011) How can microbial interactions with the blood-brain barrier modulate astroglial and neuronal function? Cell Microbiol 13:1470–1478

Harris TH, Cooney NM, Mansfield JM, Paulnock DM (2006) Signal transduction, gene transcription, and cytokine production triggered in macrophages by exposure to trypanosome DNA. Infect Immun 74:4530–4537

Hertz CJ, Filutowicz H, Mansfield JM (1998) Resistance to the African trypanosomes is IFN-gamma dependent. J Immunol 161:6775–6783

Huang ZL, Urade Y, Hayaishi O (2007) Prostaglandins and adenosine in the regulation of sleep and wakefulness. Curr Opin Pharmacol 7:33–38

Jamonneau V, Ilboudo H, Kabore J, Kaba D, Koffi M, Solano P, Garcia A, Courtin D, Laveissiere C, Lingue K, Buscher P, Bucheton B (2012) Untreated human infections by Trypanosoma brucei gambiense are not 100% fatal. PLoS Negl Trop Dis 6:e1691

Jennings FW, Whitelaw DD, Holmes PH, Chizyuka HG, Urquhart GM (1979) The brain as a source of relapsing Trypanosoma brucei infection in mice after chemotherapy. Int J Parasitol 9:381–384

Kennedy PG (2007) Animal models of human African trypanosomiasis—very useful or too far removed? Trans R Soc Trop Med Hyg 101:1061–1062

Kennedy PG (2008) The continuing problem of human African trypanosomiasis (sleeping sickness). Ann Neurol 64:116–126

Kristensson K, Bentivoglio M (1999) Pathology of African trypanosomiasis. In: Dumas M, Bouteille B, Buguet A (eds) Progress in human African trypanosomiasis, sleeping sickness. Springer, Paris

Kristensson K, Nygard M, Bertini G, Bentivoglio M (2010) African trypanosome infections of the nervous system: parasite entry and effects on sleep and synaptic functions. Prog Neurobiol 91:152–171

Kubata BK, Duszenko M, Kabututu Z, Rawer M, Szallies A, Fujimori K, Inui T, Nozaki T, Yamashita K, Horii T, Urade Y, Hayaishi O (2000) Identification of a novel prostaglandin f(2alpha) synthase in Trypanosoma brucei. J Exp Med 192:1327–1338

MacGregor P, Savill NJ, Hall D, Matthews KR (2011) Transmission stages dominate trypanosome within-host dynamics during chronic infections. Cell Host Microbe 9:310–318

MacGregor P, Szoor B, Savill NJ, Matthews KR (2012) Trypanosomal immune evasion, chronicity and transmission: an elegant balancing act. Nat Rev Microbiol 10:431–438

MacLean L, Myburgh E, Rodgers J, Price HP (2013) Imaging African trypanosomes. Parasite Immunol 35:283–294

Masocha W, Robertson B, Rottenberg ME, Mhlanga J, Sorokin L, Kristensson K (2004) Cerebral vessel laminins and IFN-gamma define Trypanosoma brucei brucei penetration of the blood-brain barrier. J Clin Invest 114:689–694

Masocha W, Amin DN, Kristensson K, Rottenberg ME (2008) Differential invasion of Trypanosoma brucei brucei and lymphocytes into the brain of C57BL/6 and 129Sv/Ev mice. Scand J Immunol 68:484–491

Mizoguchi A, Eguchi N, Kimura K, Kiyohara Y, Qu WM, Huang ZL, Mochizuki T, Lazarus M, Kobayashi T, Kaneko T, Narumiya S, Urade Y, Hayaishi O (2001) Dominant localization of prostaglandin D receptors on arachnoid trabecular cells in mouse basal forebrain and their involvement in the regulation of non-rapid eye movement sleep. Proc Natl Acad Sci U S A 98:11674–11679

Mott FW (1905) Observations on the brains of men and animals infected with various forms of trypanosomes. Preliminary note. Proc R Soc Lond B 76:235–242

Mulenga C, Mhlanga JD, Kristensson K, Robertson B (2001) Trypanosoma brucei brucei crosses the blood-brain barrier while tight junction proteins are preserved in a rat chronic disease model. Neuropathol Appl Neurobiol 27:77–85

Myburgh E, Coles JA, Ritchie R, Kennedy PG, McLatchie AP, Rodgers J, Taylor MC, Barrett MP, Brewer JM, Mottram JC (2013) In vivo imaging of trypanosome-brain interactions and development of a rapid screening test for drugs against CNS stage trypanosomiasis. PLoS Negl Trop Dis 7:e2384

Nikolskaia OV, de A Lima AP, Kim YV, Lonsdale-Eccles JD, Fukuma T, Scharfstein J, Grab DJ (2006a) Blood-brain barrier traversal by African trypanosomes requires calcium signaling induced by parasite cysteine protease. J Clin Invest 116:2739–2747

Nikolskaia OV, Kim YV, Kovbasnjuk O, Kim KJ, Grab DJ (2006b) Entry of Trypanosoma brucei gambiense into microvascular endothelial cells of the human blood-brain barrier. Int J Parasitol 36:513–519

Ohkubo T, Shibata M, Takahashi H, Inoki R (1983) Effect of prostaglandin D2 on pain and inflammation. Jpn J Pharmacol 33:264–266

Oishi Y, Huang ZL, Fredholm BB, Urade Y, Hayaishi O (2008) Adenosine in the tuberomammillary nucleus inhibits the histaminergic system via A1 receptors and promotes non-rapid eye movement sleep. Proc Natl Acad Sci U S A 105:19992–19997

Owens T, Bechmann I, Engelhardt B (2008) Perivascular spaces and the two steps to neuroinflammation. J Neuropathol Exp Neurol 67:1113–1121

Pan W, Ogunremi O, Wei G, Shi M, Tabel H (2006) CR3 (CD11b/CD18) is the major macrophage receptor for IgM antibody-mediated phagocytosis of African trypanosomes: diverse effect on subsequent synthesis of tumor necrosis factor alpha and nitric oxide. Microbes Infect 8:1209–1218

Quan N, Mhlanga JD, Whiteside MB, Kristensson K, Herkenham M (2000) Chronic sodium salicylate treatment exacerbates brain neurodegeneration in rats infected with Trypanosoma brucei. Neuroscience 96:181–194

Radwanska M, Guirnalda P, De Trez C, Ryffel B, Black S, Magez S (2008) Trypanosomiasis-induced B cell apoptosis results in loss of protective anti-parasite antibody responses and abolishment of vaccine-induced memory responses. PLoS Pathog 4:e1000078

Robertson B, Xu XJ, Hao JX, Wiesenfeld-Hallin Z, Mhlanga J, Grant G, Kristensson K (1997) Interferon-gamma receptors in nociceptive pathways: role in neuropathic pain-related behaviour. Neuroreport 8:1311–1316

Rudenko G (2011) African trypanosomes: the genome and adaptations for immune evasion. Essays Biochem 51:47–62

Salmon D, Vanwalleghem G, Morias Y, Denoeud J, Krumbholz C, Lhomme F, Bachmaier S, Kador M, Gossmann J, Dias FB, De Muylder G, Uzureau P, Magez S, Moser M, De Baetselier P, Van Den Abbeele J, Beschin A, Boshart M, Pays E (2012) Adenylate cyclases of Trypanosoma brucei inhibit the innate immune response of the host. Science 337:463–466

Scammell T, Gerashchenko D, Urade Y, Onoe H, Saper C, Hayaishi O (1998) Activation of ventrolateral preoptic neurons by the somnogen prostaglandin D2. Proc Natl Acad Sci U S A 95:7754–7759

Scammell TE, Gerashchenko DY, Mochizuki T, McCarthy MT, Estabrooke IV, Sears CA, Saper CB, Urade Y, Hayaishi O (2001) An adenosine A2a agonist increases sleep and induces Fos in ventrolateral preoptic neurons. Neuroscience 107:653–663

Schultzberg M, Ambatsis M, Samuelsson EB, Kristensson K, van Meirvenne N (1988) Spread of Trypanosoma brucei to the nervous system: early attack on circumventricular organs and sensory ganglia. J Neurosci Res 21:56–61

Sixt M, Engelhardt B, Pausch F, Hallmann R, Wendler O, Sorokin LM (2001) Endothelial cell laminin isoforms, laminins 8 and 10, play decisive roles in T cell recruitment across the blood-brain barrier in experimental autoimmune encephalomyelitis. J Cell Biol 153:933–946

Sorokin L (2010) The impact of the extracellular matrix on inflammation. Nat Rev Immunol 10:712–723

Telleria-Diaz A, Ebersberger A, Vasquez E, Schache F, Kahlenbach J, Schaible HG (2008) Different effects of spinally applied prostaglandin D2 on responses of dorsal horn neurons with knee input in normal rats and in rats with acute knee inflammation. Neuroscience 156:184–192

Thomsen AR (2009) Lymphocytic choriomeningitis virus-induced central nervous system disease: a model for studying the role of chemokines in regulating the acute antiviral CD8+ T-cell response in an immune-privileged organ. J Virol 83:20–28

Tiberti N, Hainard A, Lejon V, Courtioux B, Matovu E, Enyaru JC, Robin X, Turck N, Kristensson K, Mumba Ngoyi D, Vatunga GML, Krishna S, Büscher P, Bisser S, Ndung'u JM, Sanchez J-C

(2012) Cerebrospinal fluid neopterin as marker of the meningo-encephalitic stage of *Trypanosoma brucei gambiense* sleeping sickness. PLoS One 7:e40909. doi:10.1371/journal.pone.0040909

Tiberti N, Matovu E, Hainard A, Enyaru JC, Lejon V, Robin X, Turck N, Mumba Ngoyi D, Krishna S, Bisser S, Courtioux B, Büscher P, Kristensson K, Ndung'u JM, Sanchez J-C (2013a) New biomarkers for stage determination in *Trypanosoma brucei rhodesiense* sleeping sickness patients. Clin Transl Med 2:1. doi:10.1186/2001-1326-2-1

Tiberti N, Lejon V, Hainard A, Courtioux B, Robin X, Turck N, Kristensson K, Matovu E, Enyaru JC, Mumba Ngoyi D, Krishna S, Bisser S, Ndung'u JM, Büscher P, Sanchez J-C (2013b) Neopterin is a new cerebrospinal fluid marker for treatment outcome evaluation in patients affected by Trypanosoma brucei gambiense sleeping sickness. PLoS Negl Trop Dis 7:e2088. doi:10.1371/journal.pntd.0002088

Ueno R, Ishikawa Y, Nakayama T, Hayaishi O (1982) Prostaglandin D2 induces sleep when microinjected into the preoptic area of conscious rats. Biochem Biophys Res Commun 109:576–582

Urade Y, Hayaishi O (2011) Prostaglandin D2 and sleep/wake regulation. Sleep Med Rev 15:411–418

Vassella E, Reuner B, Yutzy B, Boshart M (1997) Differentiation of African trypanosomes is controlled by a density sensing mechanism which signals cell cycle arrest via the cAMP pathway. J Cell Sci 110(Pt 21):2661–2671

Vikman KS, Hill RH, Backstrom E, Robertson B, Kristensson K (2003) Interferon-gamma induces characteristics of central sensitization in spinal dorsal horn neurons in vitro. Pain 106:241–251

Vikman KS, Duggan AW, Siddall PJ (2007) Interferon-gamma induced disruption of GABAergic inhibition in the spinal dorsal horn in vivo. Pain 133:18–28

Vincendeau P, Bouteille B (2006) Immunology and immunopathology of African trypanosomiasis. An Acad Bras Cienc 78:645–665

Wiesenfeld-Hallin Z, Kristensson K, Samuelsson EB, Schultzberg M (1991) Studies of hyperalgesia induced by Trypanosoma brucei brucei infection in rats. Acta Trop 48:215–222

Wilhelm I, Fazakas C, Krizbai IA (2011) In vitro models of the blood-brain barrier. Acta Neurobiol Exp (Wars) 71:113–128

Wolburg H, Mogk S, Acker S, Frey C, Meinert M, Schonfeld C, Lazarus M, Urade Y, Kubata BK, Duszenko M (2012) Late stage infection in sleeping sickness. PLoS One 7:e34304

Zimmermann M (2001) Pathobiology of neuropathic pain. Eur J Pharmacol 429:23–37

Chagas Disease: Neurology and Neurobiology

José Rodrigues Coura

Abstract Chagas disease or American trypanosomiasis, caused by the intracellular protozoan *Trypanosoma cruzi*, is characterized by (1) a systemic acute phase with, in the most severe cases, myocarditis and/or meningoencephalitis that can be fatal, especially in infants; (2) a chronic phase, with an indeterminate (asymptomatic) form, a cardiac form, a digestive form (with megaesophagus and/or megacolon), and a mixed form combining the cardiac and digestive forms. Concerning nervous system damage, in the chronic phase the peripheral nervous system and especially the autonomic nervous system are involved, with denervation of the heart and gastrointestinal tract. *T. cruzi* brain infections, which attack glial cells, represent an increasing problem during immunosuppressive therapies and diseases, especially after the emergence of HIV-AIDS. Disease pathogenesis, and in particular mechanisms by which parasite-derived molecules and the host immune response interact with neurons that innervate the heart and the gastrointestinal tract remain to be clarified and deserve an engagement of the neuroscience community.

Keywords American trypanosomiasis • *Trypanosoma cruzi* • Meningoencephalitis • Parasites • Megaviscera • Denervation • Heart block and arrhythmias • Autonomic nervous system • Peripheral nervous system

J.R. Coura (✉)
Laboratory of Parasitic Diseases, Oswaldo Cruz Institute, Fiocruz, Ministry of Health, Rio de Janeiro, Brazil

Emeritus Professor Federal University of Rio de Janeiro, Rio de Janeiro, Brazil
e-mail: coura@ioc.fiocruz.br

M. Bentivoglio et al. (eds.), *Neglected Tropical Diseases and Conditions of the Nervous System*,
DOI 10.1007/978-1-4614-8100-3_11,

1 Introduction

Chagas disease or American trypanosomiasis is the infection caused by the flagellate protozoan *Trypanosoma* (*Schizotrypanum*) *cruzi* (*T. cruzi*), which can be transmitted to humans by many triatomine species (hematophagous insects of the order *Hemiptera*, family *Reduviidae*). Besides vector-borne transmission, there are other mechanisms and routes of infection: orally through food contaminated with the feces and urine of triatomines or through scent secretion by marsupials; blood transfusion; congenitally, across the placenta or through the birth canal during delivery; laboratory accidents, or, more rarely, sexually through sperm or menstrual fluid, as well as breast milk from mothers infected by *T. cruzi*.

Chagas disease was originally, and in nature continues to be, an enzootic disease of wild mammals transmitted through cannibalism or wild vectors. Over time, vectors have become adapted to peridomestic and domestic human environments. In the last three centuries, in the areas in which Chagas disease is endemic deforestation, to obtain land for agriculture and livestock breeding, has gradually modified the ecology. As a consequence, the food sources for triatomines, i.e. wild mammals, moved away, and triatomines have sought new food sources among domestic and peridomestic animals (dogs, cats, pigs, goats) and even humans, and they have become adapted to human homes. Due to this, Chagas disease presents three cycles. The first is a wild cycle, which is the most wide-ranging cycle, involving more than 130 species of wild triatomines and 100 different species of wild animals. The second is a peridomestic cycle, involving approximately ten species of triatomines adapted to the peridomicile, that feed on peridomestic and domestic animals. The third cycle involves about ten species of triatomines that can invade the domicile and become domiciled later on; they feed on human blood in areas of South America, Central America, Mexico and part of the USA (Coura and Dias 2009; Coura and Albajar 2010; Zeledon et al. 2012).

The *T. cruzi* parasites exist in nature since millions of years, as determined by DNA analyses of mummified bodies (Guhl et al. 1999; Afderheide et al. 2004). The primitive trypanosomes were monogenetic parasites of non blood-sucking insects. When these insects acquired the habit of blood-sucking, trypanosomes underwent morphological and functional changes, developing, for example, a flagellum and an undulating membrane enabling them to circulate in the bloodstream of mammals (Hoare 1972).

On the other hand, although triatomines have been known since the sixteenth century (Lent and Wygodzinsky 1979), they were only described in the eighteenth century by De Geer as *Triatoma rubrofasciata* in India, where Chagas disease does not exist. This disease was actually discovered in Brazil by Carlos Chagas in 1909, in a unique manner in the history of medicine since a single researcher discovered a disease, its etiological agent, its vector and reservoirs, as well as its domestic and wild cycles (Chagas 1909, 1912).

Enzootic infection by *T. cruzi* (named after the leading parasitologist Oswaldo Cruz, Chagas' mentor) is currently found from latitude 42° N (northern California, in the USA) to 43° S (southern Argentina and Chile). Over this vast area, it is estimated that about 15 million people are infected by *T. cruzi*, while many more millions of individuals are exposed to the infection in endemic areas extending from Mexico to the southern extremity of South America (Coura and Dias 2009). Of high relevance is the estimate that 500,000—one million people infected with *T. cruzi* have migrated from endemic zones to non-endemic countries like the USA, Canada, European and Asian countries. All these countries should now engage in the control of disease transmission in blood banks and bear the costs of treating a disease that represents for these countries a novel threat (Bisoffi et al. 2014; Coura and Albajar 2010).

2 Overview of Clinical Phases and Forms of Chagas Disease and Nervous System Involvement

Although Carlos Chagas (1911, 1913) initially described a "nervous form" of American trypanosomiasis based on clinical observations, current knowledge does not point to neurological impairment in Chagas disease as a distinct nosological entity but rather as a manifestation of occasional involvement of the central nervous system (CNS) during the acute phase of the disease, and of the peripheral nervous system (PNS) and autonomic nervous system (ANS) in the pathogenesis of the cardiac and digestive forms in the chronic phase. Given the limited knowledge at the time of the discovery of Chagas disease, the so-called "nervous form" could have been confounded with neurological manifestations of other diseases, such as endemic goiter and beriberi, very common at that time in the region where Carlos Chagas described American trypanosomiasis, or with other neurological and psychiatric diseases in co-morbidity with *T. cruzi* infection.

Chagas disease is characterized by acute and chronic phases. The acute phase can be asymptomatic or with nonspecific signs and symptoms: fever, lymphadenopathy, mild hepatosplenomegaly. A swelling can occur at the site of inoculation ("chagoma"). In the most severe cases, acute myocarditis and/or meningoencephalitis can occur, and are frequently fatal in infants. The acute phase is followed by a chronic phase, which in most cases is represented by an indeterminate (asymptomatic) form. This can last years or even throughout life. In a proportion of patients (about 30 %), chronic Chagas disease is symptomatic, with a cardiac form, a digestive form (with megaesophagus and/or megacolon), and a mixed form combining the cardiac and digestive forms.

As mentioned above, the CNS can be involved in the acute phase of Chagas disease. In addition, as it will be discussed further, reactivation of the acute form, with high parasite proliferation and CNS involvement can occur in immunosuppressed patients. The PNS and ANS are involved in the chronic cardiac and gastrointestinal forms. The ANS involvement affects especially the parasympathetic and enteric nervous systems, but targets also the sympathetic nervous system, with imbalance of parasympathetic and sympathetic regulatory mechanisms.

Chagas disease has even been considered "an illness of the nervous system" (Köberle 1961, 1968) starting during the acute phase of the disease. An excellent review of 31 studies on CNS involvement in Chagas disease was made 20 years ago by Pittella (1993). In the following year, the Pan-American Health Organization issued its Scientific Publication No 547 under the title "Chagas Disease and the Nervous System" (PAHO 1994) witnessing attention to this issue. Neurological impairment in Chagas diseases has been re-discussed in recent years (e.g. Pittella 2009; Py 2011; Carod-Artal 2013).

3 Pathogenesis and Evolution of Chagas Disease

Studies and debates on the pathogenesis of *T. cruzi* human infection and its evolution into acute and chronic phases have initiated since the disease discovery (Chagas 1909, 1916; Chagas and Villela 1922; see also Coura 1988, 2007; Coura and Borges-Pereira 2010, 2012). After the pioneering investigations by Gaspar Vianna (1911) on the pathology of the "Malady of Carlos Chagas", which included descriptions of nervous system damage, many studies and debates on this topic have followed. However, seminal findings have been described by Vianna (1911) himself. Namely, *T. cruzi* invade the blood, penetrate the host's cells, multiply through binary division of amastigote forms to produce "pseudocysts", which rupture and give rise to an inflammatory reaction that heals with fibrosis of the parasitized tissue (Fig. 1). Through rupturing of the "pseudocysts", new *T. cruzi* are released, circulate again in the blood and parasitize new cells, thus repeating the cycle for decades, while the patient continues to live in the chronic phase of the disease.

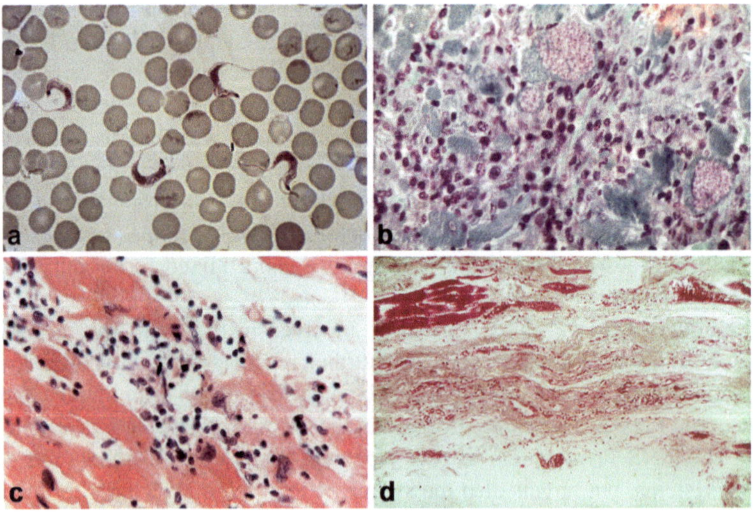

Fig. 1 The plate illustrates *T. cruzi* circulating in the blood (**a**); pseudocyst formation (**b**); lesions in cardiac fibers (**c**); healing with fibrosis (**d**)

Besides direct parasite infection, another main mechanism involved in the pathogenesis of Chagas disease is represented by autoimmunity induced by *T. cruzi* and its products. This mechanism, introduced by Santos-Buch and Teixeira (1974) and Teixeira (1975), and then investigated in several other studies, is complex and controversial. In recent years, the involvement of autoimmunity in Chagas disease has been critically reconsidered especially by Tarleton (2003), and even Teixeira and coworkers (2006), who have stated that this mechanism is still "open for investigation".

In a discussion on the evolution of chronic Chagas cardiopathy, Higushi (1999) took the view that CD4+ and CD8+ T lymphocytes are sensitized by *T. cruzi* and its antigens, with development of anti-myocardial T cells, macrophage activation and platelet aggregation. This may lead to ischemia, development of chronic Chagas myocarditis, impairment of the ANS, arrhythmia, cardiac dilatation and death due to ventricular fibrillation and/or heart failure.

In pathological studies on the digestive and cardiac forms of chronic Chagas disease, Köberle (1961, 1968) showed that megaesophagus and megacolon are consequences of denervation of the myenteric plexuses of these organs due to an inflammatory reaction induced directly by *T. cruzi*, and that impairment of the parasympathetic nervous system is involved in cardiopathy, as initially described by Vianna (1911). Consistently with this interpretation, denervation of the heart, esophagus, colon and other hollow viscera has been repeatedly considered responsible for chronic Chagas disease, produced directly by the inflammatory reaction caused by *T. cruzi* (see, for example, Prata 2001; Teixeira et al. 2006).

Molecular mechanisms potentially involved in neuronal damage and repair in *T. cruzi* infection, as well as the potential role of parasite-derived neurotrophic factors in the interplay between the parasite and the ANS have recently been reviewed (Chuenkova and Pereira-Perrin 2011). Concerning Chagasic cardiopathy, parasympathetic denervation of the heart, with loss of ganglion neurons and vagal nerve terminals, is an early event in *T. cruzi* infection. As the infection progresses, damage of the sympathetic nervous system also occurs and targets post-ganglionic nerve terminals in the heart rather than neurons in the cervical and stellate ganglia.

Neurodegeneration in the gastrointestinal tract targets the extrinsic innervation (neurons of the parasympathetic ganglia) and intrinsic innervation. Of special interest is the involvement of the enteric nervous system. Composed of millions of neurons, this system subserves the extensive intrinsic innervation of the gut comprising the myenteric plexus (of Auberbach) and the submucosal plexus (of Meissner). Enteric neurons comprise a variety of phenotypes in terms of neurotransmitters and neuromodulators, and control gastrointestinal motility, secretory processes and nutrients, as well as local blood flow (Furness 2012). Loss of enteric neurons in Chagas disease targets especially inhibitory neurons containing nitric oxide synthase and vasoactive intestinal peptide, possibly due to molecular mimicry between *T. cruzi* and enteric proteins (Rivera et al. 2011).

Puzzling issues are also represented by the fact that about 70 % of *T. cruzi*-infected individuals are asymptomatic despite retaining pathogenic parasites, and by the long latency (20–30 years) of the onset of chronic disease after the initial acute infection (Chuenkova and Pereira-Perrin 2011). Parasite-host interactions, including neuroregenerative responses of PNS and/or ANS fibers to the infection, diversity in virulence and tissue tropism of *T. cruzi* strains have been implicated in these enigmatic phenomena.

3.1 Biological Diversity of T. cruzi *Strains and Organ Tropism*

The etiological agent of Chagas disease exhibits great genotypic and phenotypic diversity with different biological properties, virulence and geographical distribution. The same host may be simultaneously infected by different *T. cruzi* strains (Devera et al. 2003). Classifications of the parasite heterogeneity have been proposed since decades. In the 1960s, when the present molecular tools were not available, Coura et al. (1966) proposed the designation of "cruzi complex" on the basis of morphological, immunological and pathogenic diversity (see also Devera et al. 2003). Subsequently, *T. cruzi* have been classified into "biodemes" according to biological variations (Andrade and Magalhães 1977), "zymodemes" according to isoenzyme variability, (Miles et al. 1980), and "schizodemes" according to molecular diversity (Morel et al. 1980). More recently, on the basis of genetic diversity *T. cruzi* have been partitioned into six groups, renamed by consensus as TcI- TcVI (Zingales et al. 2009, 2012), integrating the above-mentioned variations.

Heterogeneity of *T. cruzi* strains has been repeatedly implicated in the different presentations of Chagas disease and in the infection targeting of different organs. For example, the classification of parasite strains into three "biodemes" (types I, II, III) has taken into account tissue tropism together with strain morphology, virulence and pathogenicity (Andrade 1976; Andrade and Magalhães 1977). After *T. cruzi* inoculation in mice, type I shows a tropism for macrophages in the initial phase of the infection, with high virulence and parasitemia peaking 7–12 days after inoculation, predominance of slender forms of *T. cruzi*, and 100 % mortality of the mice on the 12th day after infection. Infection with *T. cruzi* type II involves myotropism, predominantly for the myocardium, during the acute phase, and a greater number of parasite wide forms, with parasitemia peaking from 12 to 20 days after infection. Type III infection involves tropism for skeletal muscles, with predominance of the parasite wide forms, parasitemia peaking 25–30 days after infection, and low mortality among the infected animals.

The genetic variation of strains and clones of *T. cruzi* has been correlated with specific receptors in the host's organs or systems, and it has been hypothesized that this could account for the parasite tropism for a given organ (Macedo and Pena 1998). According to this interpretation, different clonal-histotropic strains of *T. cruzi* would have affinity for different receptors expressed in cardiac or gastrointestinal tract tissue, so that Chagas disease pathology would be determined by host-parasite genetics. This theory, however, has not received experimental confirmation.

4 Involvement of the Nervous System in the Acute Phase of Chagas Disease

The involvement of the CNS in the acute phase of Chagas disease has been reported since the initial descriptions of the disease (Chagas 1911, 1913, 1916), especially meningoencephalitis, which is severe and fatal in infants (2–4 years of age). This is characterized by headache, vomiting, mental confusion, convulsions, meningeal signs and presence of *T. cruzi* in the cerebrospinal fluid. These patients present cranial hypertension, with elevated proteins and predominance of lymphocytes in the cerebrospinal fluid. In fatal cases, lesions disseminated in the leptomeninges and CNS have been reported, especially in the grey matter, with nodular encephalitis in multiple foci, diffuse vasculitis and nests of parasites in astrocytes and microglia. These lesions, originally described by Vianna (1911), have been repeatedly documented in subsequent studies (see, for example, the reviews by Sica 1994; Pittella 2009). Brain lesions with cognitive and behavioral disturbances as sequels have been reported in children of 6–12 years of age who had suffered acute meningoencephalitis due to *T. cruzi* infection (Sica 1994).

The involvement of the heart in the acute phase of Chagas disease affects not only cardiac myocytes but, as mentioned above, also the ANS. Damage to the parasympathetic and sympathetic innervation of the heart leads to atrio-ventricular (AV) block, supraventricular and ventricular extrasystoles and branch block of the bundle of His and Purkinje fibers. In cases of acute Chagas disease, ventricular and supraventricular extrasystoles have been reported in 38.5 % of the patients, and first and second-degree AV block in 30.8 % of the patients (Pinto et al. 2008).

Two other conditions that may lead to involvement of the CNS in acute Chagas infection are congenital transplacental transmission and immunosuppression. Transplacental transmission of *T. cruzi* may lead to abortion, prematurity and organ lesions, including premature death due to CNS involvement, or may lead to mental retardation of the survivors. Congenital Chagas infection was first recorded by Carlos Chagas himself (1911, 1913). He observed two newborns with episodes of convulsions who died on the 6th and 8th days of life and whose necropsies revealed the presence of *T. cruzi* in the CNS. Although infrequent, cases of congenital Chagas infection have been described in all endemic countries in South America (Freilij et al. 1994). Seminal pathological studies on the congenital form and on Chagas placentitis have been conducted by Bittencourt and coworkers (Bittencourt 1963; Bittencourt et al. 1972).

Immunosuppression (especially since the emergence of HIV-AIDS) makes the CNS a preferential target for Chagas infection, in the form of meningoencephalitis and focal lesions, similar to those produced by toxoplasmosis. In immunosuppressed patients Chagasic encephalitis can also occur in multiple foci with necrotizing features, or in tumoral or psedutotumoral forms (brain "chagoma"). Besides HIV-AIDS, other causes of immunosuppression and reactivation of acute Chagas disease include malignant tumors and chemotherapeutic treatment, as well as organ transplantation, which entail serological and parasitological investigations.

Immunosuppressive drugs include azathioprine, corticosteroids, cyclosporins or immunosuppressive monoclonal antibodies and anti-lymphocytic serum, Concerning organ transplantation, a protocol for the control of organ donors and recipient patients before and after transplantation has been established taking into account the following points (1) assessment on live or dead donor; (2) assessment on Chagas disease recipient; (3) post-transplantation control; (4) follow-up for presence of parasites after transplantation; and (5) treatment and follow-up of patients 30, 60 and 90 days after the treatment to assess the parasitological cure (Freilij and Storino 1994).

5 Nervous System Involvement in the Chronic Phase and Forms of Chagas Disease

The chronic phase of Chagas disease begins by definition 60–90 days after the acute phase, when parasitemia has greatly reduced due to the action of circulating antibodies. Acute manifestations of the disease are no longer apparent and *T. cruzi* are no longer detectable in the bloodstream by means of direct methods (fresh blood examination and concentration methods such as microhematocrit).

The chronic phase of Chagas disease may present in four distinct clinical forms mostly due to different degrees of PNS and ANS impairment (a) indeterminate form, without evident signs and symptoms, and with normal clinical examination and electrocardiographic and radiological examinations of the heart, esophagus and colon; (b) cardiac form, with damage of various severity of the heart and its autonomic innervation; this form may also lead to sudden death; (c) digestive form, with damage of various severity of the gastrointestinal tract, which may lead to dysperistalsis of the esophagus with dilatation and functional alterations of different severity, and megacolon; (d) mixed cardiac and digestive forms (Coura 2007). These forms are dealt with below.

Impairment of the PNS, with sensory disturbances, has also been reported in more than 10 % of the cases of chronic Chagas disease (Genovese et al. 1996). Electromyographic studies on such cases have shown reductions in motor and sensory conduction, including in cases of chronic indeterminate form of the disease.

Controlled electromyographic studies on different muscles, conducted in Chagas and non-Chagas disease patients in Brazil (De Faria et al. 1977) and Argentina (Sanz et al. 1978), have demonstrated muscle motor denervation in patients in the chronic phase of Chagas disease. This has been proven by histopathological and histochemical findings (Sica 1994).

5.1 Chronic Indeterminate Form with Mild Involvement of the Autonomic Nervous System

The chronic indeterminate form of Chagas disease is asymptomatic and initiates, by definition, just after the acute phase. It is characterized by low parasitemia and high antibody levels, and no clinical manifestations are demonstrable

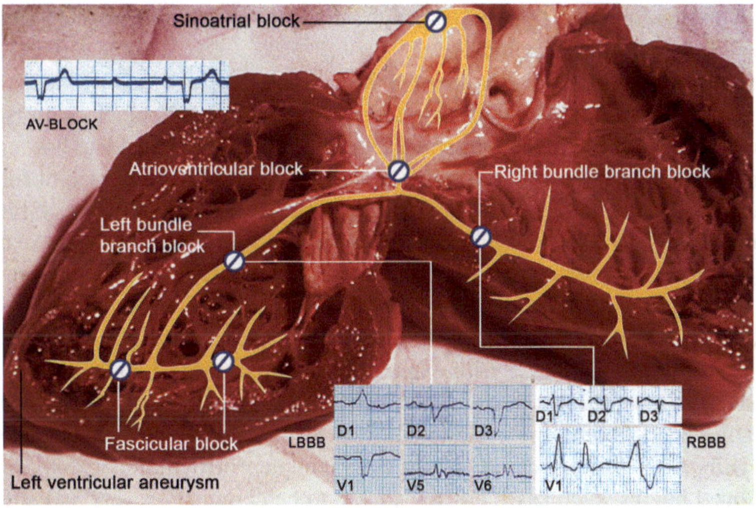

Fig. 2 Autoptic specimen of the heart of a patient affected by chronic Chagas cardiomyopathy, who died due to heart failure. The superimposed drawing and electrocardiographic charts show: Sinoatrial block; complete atrioventricular (AV) block; right bundle branch block; left bundle branch block; fascicular block and left ventricular aneurysm

through routine clinical examinations. Electrocardiographic and radiographic examinations of the heart, and radiological features of the esophagus and colon are normal. After the acute phase, the majority of the patients enter this form of the disease, and they may remain affected for one to two decades or for the rest of their lives. However, about 30 % of the cases may evolve to the chronic cardiac or digestive form. Although asymptomatic, the chronic indeterminate form may show ventricular extrasystoles and monometric alterations at examinations with sensitive methods, such as dynamic electrocardiography (Holter). These abnormalities result from sparse inflammatory foci, including localized neuronal lesions (Dias and Macedo 2005). Pathological studies on patients with the indeterminate form of Chagas disease who suffered unexpected or violent sudden death have shown myocardial abnormalities, including damage of the intracardiac autonomic innervation (Lopes et al. 1975, 1981) (Fig. 2).

5.2 *Chronic Cardiac Form with Severe Involvement of the Autonomic Nervous System*

The chronic cardiac form is the most severe and most frequent form of Chagas disease. The patients present A-V block, ventricular and supraventricular extrasystoles, frequent right-branch block and occasional left-branch block, fascicular block of the conduction system and complex arrhythmia, which leads to heart failure and sudden death in most cases (Fig. 2). The severity of the cardiac form is directly related to the severity of ANS damage, myocardial inflammatory lesions and residual fibrosis of these lesions.

Impairment of the heart autonomic innervation can be assessed by means of the electrocardiogram. Patients with first and second-degree A-V block, with isolated monomorphic ventricular and auricular extrasystoles can be considered of medium severity. Cases of third-degree A-V block (complete block), atrial fibrillation, branch block of the bundle of His with left anterior hemiblock, left-branch block of the bundle of His, fascicular block and frequent polymorphic ventricular extrasystoles in salvos (two or more sequential extrasystoles) are considered severe cases with the risk of sudden death due to ventricular fibrillation, even when there is no marked cardiac dilatation (Fig. 2).

On the other hand, cases with extensive inflammatory lesions and myocardial fibrosis, with a tendency towards dilated cardiopathy and ejection fraction <50 (thus indicating heart failure) are considered to be severe. When these signs are associated with ANS lesions of the heart with complex arrhythmia and the above-mentioned blocks, such cases are considered to be extremely severe and at risk of death due to heart failure and thromboembolism, or sudden death due to ventricular fibrillation.

5.3 Chronic Digestive Form of Chagas Disease with Dysperistalsis and Dilatation of the Esophagus and Colon

As mentioned above, the digestive form of Chagas disease is characterized by progressive development of megaesophagus and megacolon. Megaesophagus begins with altered motor coordination of deglutition, with dysphagia and odynophagia, initially due to dysperistalsis and food pressure on the walls of the organ, which progressively keeps dilating. Rezende et al. (1960) classified Chagas disease esophagopathy into the following groups according to viscera dilatation and the patients' capacity for swallowing, as assessed by means of radiography just before ingestion of contrast medium and 1 min after ingestion. *Group I* includes cases in which the esophagus presents a normal diameter, but without the capacity for complete evacuation. At the radiological examination, the contrast medium is retained inside the lower esophagus, thus forming a small residual column. The upper extremity of this column forms a flat surface, perpendicular to the walls of the esophagus. Above this column, the esophagus remains open, containing air, which confers a cylindrical shape (Fig. 3a). *Group II* includes cases showing already a moderate dilatation of the esophagus and considerable retention of contrast medium at the radiological examination, thus forming a residual column of varying height (Fig. 3b). The hallmark of this group is the uncoordinated motor activity of the esophagus, with the appearance of tertiary waves. Hypertonia of the lower esophagus is frequently observed. *Group III*: In this group, the esophagus presents a large increase in diameter and is hypotonic, with little contractile activity of its walls (Fig. 3c).

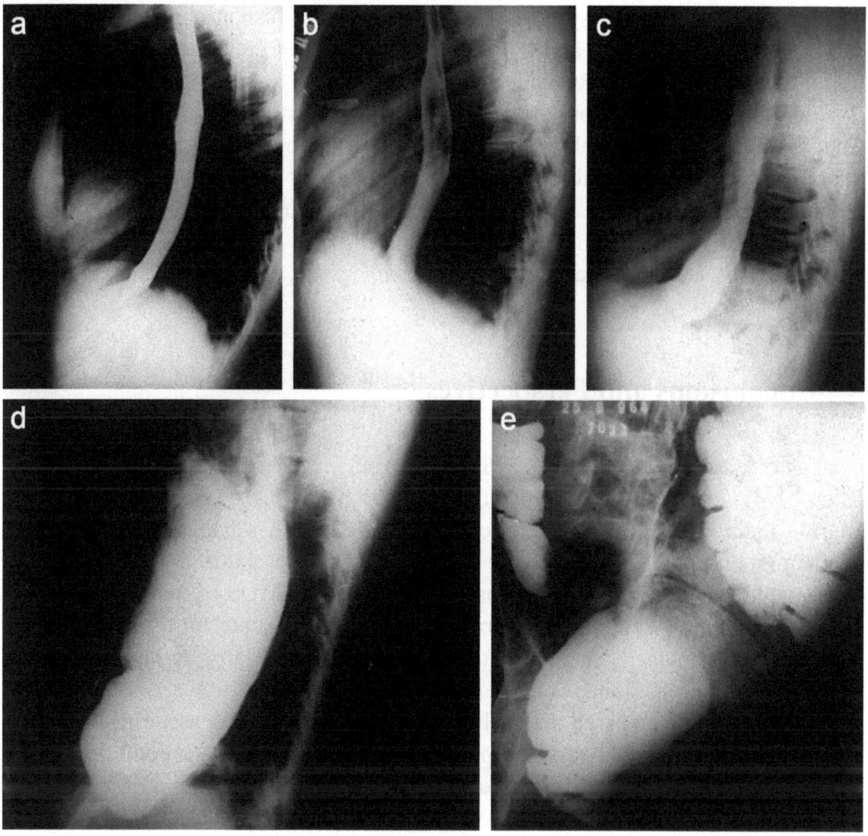

Fig. 3 Different degrees of severity of megaesophagus (**a–d**) and megacolon (**e**) at the radiological examination (see text). For megaesophagues severity: group I (**a**); group II (**b**); group III (**c**); group IV (**d**)

The degree of contrast retention is greater than in groups I and II. *Group IV* includes cases of dolichomegaesophagus. The esophagus acquires a large volume and is elongated and atonic. It is folded over the diaphragmatic cupola and thus produces a right paracardiac shadow on simple chest radiographs (Fig. 3d).

In approximately 5 % of the cases, the radiographic image presents intermediate characteristics between two groups, thus causing difficulty in allocating the case to a specific group. In such situations, it is advisable to classify the patients in the more severely affected group (Rezende-Filho et al. 2005).

Electromanometry is an important method for the study of motor alterations of the esophagus. It simultaneously records the pressure variations that occur at different levels of the esophagus and allows to ascertain esophageal motility under different conditions. On the other hand, pharmacological denervation tests, such as the use of methacholine at a dose of 0.05 mg/kg of body weight (Godoy and Vieira 1963), has been a useful method for proving the existence of the parasympathetic denervation that occurs in Chagas disease. The intensity of the motor response in megaesophagus cases is greater in groups I and II and less in the cases of greater esophageal dilatation in groups III and IV (Rezende-Filho et al. 2005).

Radiological examinations are of fundamental importance in diagnosing cases of Chagas colopathy (megacolon; Fig. 3e). The morphological characteristics related to colon diameter and distal length can be assessed, but this approach does not provide information on functional changes. Thus, the diagnosis of non-dilated Chagas colopathy requires other methods such as manometry and pharmacological tests, which are beyond the scope of the present chapter.

6 Conclusions and Perspectives

Chagas disease can affect the CNS, PNS, ANS in its acute and chronic phases (Sica 1994; Genovese et al. 1996; Junqueira 2012).

The main neurological alterations of the CNS in Chagas disease consist of meningoencephalitis in the acute phase, especially in children under the age of 4 years and in immunosuppressed patients due to HIV-AIDS or malignant tumors, irradiation, immunosuppressive therapies, in whom the acute form of Chagas disease can be reactivated. Such condition may be due especially to therapy to avoid organ transplant rejection.

The PNS can be involved in the chronic phase of Chagas disease, especially with sensory disturbances. However, impairment of the ANS in the chronic phase is a dominating feature, of key importance in the evolution and prognosis of the disease. In the cardiac form, alterations to the conduction system (sinoatrial node, AV node, right and left branches of the bundle of His and Purkinje fibers) are responsible for blocking cardiac stimuli, for arrhythmia and for sudden death in two-thirds of the cases. On the other hand, destruction of the myenteric plexuses is the determining factor for the development of megaesophagus and megacolon in Chagas disease. In the indeterminate form of the disease, ANS lesions are absent or mild, while they are severe in the cardiac, digestive and mixed forms. Depending on their severity, these lesions may result especially in autonomic parasympathetic dysfunction, but there may also be sympathetic dysfunction of varying intensity.

Severe acute phase cases of Chagas disease and cases exposed to re-infection may evolve with greater impairment of the ANS and therefore with greater severity of the chronic form of the disease, but this remains to be established. Novel means to predict disease evolution and which cases of the chronic indeterminate form will develop cardiac or digestive forms with more or less severe impairment of the ANS would be important for future studies.

Pathogenetic studies on mechanisms of viscera denervation in Chagas disease could also open novel perspectives on ANS regulation, vulnerability of ANS neurons to toxic stimuli and their regenerative capacity, ANS alterations in other diseases.

References

Afderheide AC, Salo W, Madden M, Streitz J, Bulkestra J, Guhl F, Arriaza B, Renier C, Wittmer LE Jr, Fornaciari G, Alisson A (2004) A 9000-years record of Chagas disease. Proc Natl Acad Sci USA 101:2034–2039

Andrade SG (1976) Tentative for grouping different *Trypanosoma cruzi* strains in some types. Rev Inst Med Trop Sao Paulo 18:140–141

Andrade SG, Magalhães JB (1977) Biodemes and zymodemes of *Trypanosoma cruzi* strains: correlations with clinical data and experimental pathology. Rev Soc Bras Med Trop 30:27–35

Bisoffi Z, Buonfrate D, Angheben A (2014) Travel, migration and neglected tropical diseases. Bentivoglio M, Cavalheiro EA, Kristensson K, Patel NB (eds) Neglected tropical diseases and conditions of the nervous system. Springer, New York

Bittencourt AL (1963) Placentite chagásica e transmissão congênita da doença de Chagas. Rev Inst Med Trop Sao Paulo 5:62–67

Bittencourt AC, Barbosa HS, Rocha T, Sodré I, Sodré A (1972) Incidência da transmissão congênita da doença de Chagas na maternidade Tsylla Balbina (Salvador, Bahia). Rev Inst Med Trop Sao Paulo 14:131–134

Carod-Artal FJ (2013) American trypanosomiasis. Handb Clin Neurol 114:103–123

Chagas C (1909) Nova Tripanozomiaze Humana. Estudo sobre a morfolojia e o ciclo evolutivo do Schizotrypanum cruzi n. gen.sp. ajente etiolojico de uma nova entidade mórbida do homem. Mem Inst Oswaldo Cruz 1:159–218

Chagas C (1911) Nova entidade mórbida do homem. Rezumo geral de estudos etiolójicos e clínicos. Mem Inst Oswaldo Cruz 3:219–275

Chagas C (1912) Sobre um trypanosoma do tatú, Tatusia novencincta, transmitido pelo Triatoma geniculata Latr. (1811). Possibilidade de ser o tatú um depositário do Trypanosoma cruzi no mundo exterior. Nota prévia. Brazil Med 26:305–306

Chagas C (1913) Les formes nerveuses d'une nouvelle trypanosomíase. Iconogr Salpetrière 26:1–9

Chagas C (1916) Processos patojênicos da tripanossomíase americana. Mem Inst Oswaldo Cruz 8:5–36

Chagas C, Villela E (1922) Forma cardíaca da Trypanosomiase Americana. Mem Inst Oswaldo Cruz 14:5–61

Chuenkova MV, Pereira-Perrin M (2011) Neurodegeneration and neuroregeneration in Chagas disease. Adv Parasitol 76:195–233

Coura JR (1988) Determinantes epidemiológicos da doença de Chagas no Brasil. A infecção e sua morbi-mortalidade. Mem Inst Oswaldo Cruz 93:392–402

Coura JR (2007) Chagas disease: what is known and what is needed. A background article. Mem Inst Oswaldo Cruz 102(Suppl I):113–122

Coura JR, Albajar PV (2010) Chagas disease: a new worldwide challenge. Nat Outlook 30(Suppl):56–57

Coura JR, Borges-Pereira J (2010) Chagas disease: 100 years after its discovery. A systemic review. Acta Trop 115:5–13

Coura JR, Borges-Pereira J (2012) Chagas disease. What is known and what should be improved: a systemic review. Rev Soc Bras Med Trop 45:286–296

Coura JR, Dias JCP (2009) Epidemiology, control and surveillance of Chagas disease—100 years after its discovery. Mem Inst Oswaldo Cruz 104(Suppl 1):31–40

Coura J, Ferreira LF, Rubens J, Pereira NC, Silva JR (1966) Tripanosoma do "complexo cruzi" em reservatório silvestre no Estado da Guanabara. Rev Inst Med Trop Sao Paulo 8:125–133

De Faria CR, Melo Souza SE, Rassi A (1977) Evidências eletromiográficas de desnervação motora em pacientes na fase crônica da doença de Chagas. Rev Goiana Med 23:125–130

Devera R, Fernandes O, Coura JR (2003) Should *Trypanosoma cruzi* be called "cruzi" complex? A review of the parasite diversity and the potential of selecting population after in vitro culturing and mice infection. Mem Inst Oswaldo Cruz 98:1–12

Dias JCP, Macedo VO (2005) Doença de Chagas. In: Coura JR (ed) Dinâmica das Doenças Infecciosas e Parasitárias. Editora Guanabara Koogan, Rio de Janeiro, pp 557–593

Freilij and Storino (1994) Chagas e Immunodepressión. In: Storino R, Milei J (eds) Enfermedad de Chagas. Morby Dogma, Argentina pp. 293–301

Freilij H, Altcheh J, Storino R (1994) Chagas congénito. In: Storino R, Milei J (eds) Enfermedad de Chagas. Morby Dogma, Argentina, pp 267–278

Furness JB (2012) The enteric nervous system and neurogastroenterology. Nat Rev Gastroenterol Hepatol 9:286–294

Genovese O, Ballario C, Storino R, Segura E, Sica RER (1996) Clinical manifestations of peripheral nervous system in human chronic Chagas disease. Arq Neuropsiquiatr 54:190–196

Godoy RA, Vieira CB (1963) Effects of cholinergic drugs on the megaesopagus of patients with Chagas disease. Acta Physiol Lat Am 11:107

Guhl F, Jamillo C, Vallejo GA, Yockteng R, Cardena- Arroyo Forniciari G (1999) Isolation of *Trypanosoma cruzi* DNA in 4000 years old mummified human tissue from northern Chile. Phys Anthropol 108:625–635

Higushi ML (1999) Human chronic chagasic cardiopathy: participation of parasites antigens subsets of lymphocytes and microvascular abnormalities. Mem Inst Oswaldo Cruz 94(Suppl 1): 263–267

Hoare CA (1972) The trypanosomes of mammals. A zoological monograph. Blackwell Scientific, Oxford, 749 p

Junqueira LF Jr (2012) Insights into the clinical and functional significance of cardiac autonomic dysfunction in Chagas disease. Rev Soc Bras Med Trop 45:243–252

Köberle F (1961) Patologia y anatomia patológica de la enfermedad de Chagas. Bol Ofi Sanit Panam 51:404–428

Köberle F (1968) Chagas disease and Chagas' syndrome: the pathology of American trypanosomiasis. Adv Parasitol 6:63–116

Lent H, Wygodzinsky P (1979) Revision of triatomines (*Hemiptera Reduviidae*) and their significance as vectors of Chagas disease. Bull Am Mus Nat Hist 163(art. 3):123–520

Lopes ER, Chapadeiro E, Almeida HO, Rocha H (1975) Contribuição ao estudo da anatomia patológica dos corações chagásicos falecidos subitamente. Rev Soc Bras Med Trop 9: 269–282

Lopes ER, Chapadeiro E, Andrade ZA, Almeida HO, Rocha A (1981) Anatomia patológica de corações chagásicos assintomáticos falecidos de modo violento. Rev Soc Bras Med Trop 76:189–197

Macedo AM, Pena SDJ (1998) Genetic variability of *Trypanosoma cruzi*: implications for the pathogeninesis of Chagas disease. Parasitol Today 14:119–124

Miles MA, Lanham SM, Souza AA, Póvoa M (1980) Further enzymatic characters of *Trypanosoma cruzi* and their evaluation for strain identification. Trans R Soc Trop Med Hyg 74:221–237

Morel C, Chiari E, Camargo EP, Mattei DM, Romanha AJ, Simpson L (1980) Strains and clones of *Trypanosoma cruzi* can be characterized by pattern of kinetoplast DNA minicircles. Proc Natl Acad Sci USA 77:6810–6814

PAHO-Pan America Health Organization (1994) Chagas disease and nervous system, Sci Publ no 547, Washington, 354 p

Pinto AY, Valente AS, Valente VC, Ferreira-Junior AG, Coura JR (2008) Fase aguda da doença de Chagas na Amazônia Brasileira. Estudos de 233 casos do Pará, Amapá e Maranhão observados entre 1988 e 2005. Rev Soc Bras Med Trop 1:602–614

Pittella JEH (1993) Central nervous system involvement in Chagas disease. An updating. Rev Inst Med Trop Sao Paulo 35:111–116

Pittella JE (2009) Central nervous system involvement in Chagas disease: a hundred-year-old history. Trans R Soc Trop Med Hyg 103:973–978

Prata A (2001) Clinical and epidemiological aspects of Chagas disease. Lancet Infect Dis 1:92–100

Py MO (2011) Neurologic manifestations of Chagas disease. Curr Neurol Neurosci Rep 11: 536–542

Rezende JM, Lauar KL, Oliveira AR (1960) Aspectos clínicos e radiológicos da aperistalse do esôfago. Rev Bras Gastroenterol 12:247–262

Rezende-Filho J, Moreira-Junior H, Rezende JM (2005) Métodos radiológicos e manométricos para diagnóstico da esofagopatia e da colopatia Chagásica. In: Coura JR (ed) Dinâmica das Doenças Infecciosas e Parasitárias. Editora Guanabara Koogan, Rio de Janeiro, pp 653–666

Rivera LR, Poole DP, Thacker M, Furness JB (2011) The involvement of nitric oxide synthase neurons in enteric neuropathies. Neurogastroenterol Motil 23:980–988

Santos-Buch CA, Teixeira ARL (1974) The immunology of experimental Chagas disease III. Rejection of allogenic heart cell in vitro. J Exp Med 140:38–53

Sanz OP, Ariustimufto G, Ratsunu AF, O'Neil E, Sica REP (1978) An electrophysiological investigation of skeletal muscle in human chronic Chagas disease. Arq Neuropsiquiatr 36:319–326

Sica R (1994) Compromisso del sistema nervioso. In: Storino R, Milei J (eds) Enfermedad de Chagas. Morby Dogma, Argentina, pp 303–320

Tarleton RL (2003) Chagas disease: a role for autoimmunity? Trends Parasitol 19:447–451

Teixeira ARL (1975) Autoimmune mechanism in Chagas disease. Pan Am Health Org Sci Publ 318:98–108

Teixeira ARL, Nascimento RL, Sturm NR (2006) Evolution and pathology in Chagas disease. A review. Mem Inst Oswaldo Cruz 101:463–491

Vianna G (1911) Contribuição para o estudo da anatomia patológica da "Moléstia de Carlos Chagas". Mem Inst Oswaldo Cruz 3:276–294

Zeledon R, Beard CB, Dias JCP, Leiby DA, Dorn PL, Coura JR (2012) An appraisal of the status of Chagas disease in the United States. Elsevier, New York

Zingales B, Andrade SG, Briones MRS, Campbell DA, Chiari E, Fernandes O, Guhl F, Lages Silva E, Macedo AM, Machado CR, Miles MA, Romanha AJ, Sturm NR, Tibarenc M, Schijman AG (2009) A new consensus for *Trypanosoma cruzi* intraspecific nomenclature: second revision. Mem Inst Oswaldo Cruz 104:1051–1054

Zingales B, Miles MA, Campbell DA, Tibayrenc M, Macedo AM, Teixeira MMG, Schijman AG, Llewellyn MS, Lages-Silva E, Machado CR, Andrade SG, Sturm SR (2012) The revised *Trypanosoma cruzi* subspecific nomenclature: rationale, epidemiological relevance and research applications. Infect Genet Evol 12:240–253

Neurodegeneration in Leprosy: Insights from Model Systems and Patients

Toshihiro Masaki and Anura Rambukkana

Abstract Despite the long history of leprosy as an infectious disease it remains a public health problem and millions are living with disability due to past or present leprosy. Disabilities due to nerve function loss, the pathologic hallmark of this infection, are associated directly with injury to the peripheral nervous system (PNS), which is primarily caused by the unique capacity of leprosy bacteria to invade Schwann cells, the glial cells of the PNS. Recent studies have suggested that *Mycobacterium leprae i*nterfere with Schwann cell signaling system and functions, which subsequently facilitates bacterial propagation due to concomitant Schwann cell de-differentiation. While these alterations during the long bacterial incubation period abrogate normal glial and neuronal functions, they also engender acute inflammatory responses that eventually destroy the peripheral nerves. Although the latter manifests clinically as sensorimotor loss and enable the diagnosis of patients with nerve damage, the underlying early events of nerve infection and inflammatory

T. Masaki
MRC Center for Regeneration Medicine, University of Edinburgh,
Little France Campus, Edinburgh EH16 4UU, UK

Center for Neuroregeneration, University of Edinburgh,
Little France Campus, Edinburgh EH16 4UU, UK

Teikyo University, 2-2-1 Senju-Sakuragi, Adachi-ku, Tokyo, Japan

A. Rambukkana (✉)
MRC Center for Regeneration Medicine, University of Edinburgh,
Little France Campus, Edinburgh EH16 4UU, UK

Center for Neuroregeneration, University of Edinburgh,
Little France Campus, Edinburgh EH16 4UU, UK

Center for Infectious Diseases, University of Edinburgh,
Little France Campus, Edinburgh EH16 4UU, UK
e-mail: a.rambuka@ed.ac.uk

M. Bentivoglio et al. (eds.), *Neglected Tropical Diseases and Conditions of the Nervous System*, 217
DOI 10.1007/978-1-4614-8100-3_12, © Springer Science+Business Media New York 2014

responses that cause the nerve injury are not well understood. These are the major challenges if we are to develop effective early diagnostics and new therapeutics for this devastating neglected disease. Here we highlight the recent knowledge on the early infectious process obtained in model systems and present those findings in the context of available clinical findings from leprosy patients.

Keywords Schwann cells • Mycobacterium leprae • Demyelination • Nerve regeneration • Neglected diseases • Peripheral Neuropathy • Nerve damage

1 Introduction

The involvement of the peripheral nervous system (PNS) and subsequent demyelination is the pathologic hallmark of human leprosy, and distinguishes leprosy from other diseases with skin lesions similar to leprosy. Although leprosy has been known to mankind since biblical times it remains a public health problem but unfortunately a neglected disease despite more than 200,000 new cases continue to be diagnosed each year. Also, an estimated two to three million people worldwide live with disability due to neurodegenerative conditions directly associated with the disease (Lockwood and Suneetha 2005; Rao 2005; Scollard et al. 2006). In spite of these startling statistics and long human history behind this disease, we know very little about the biology of the disease and pathogenesis. In particular, the large gaps in our understanding in neurobiology and neuropathogenesis of the disease have halted the progress towards developing new and effective diagnostics and therapeutics for the management of nerve damage. Nonetheless, it is encouraging that there is a new hope and enthusiasm for basic research because of the existence of many obvious parallels in the neurodegenerative processes between human leprosy and other demyelinating neurodegenerative diseases. Therefore, from the positive side, leprosy and its disease models could provide an exciting opportunity to gain new insights into the field of neurodegeneration.

2 Causative Organism and Its Neural Predilection

The distinctive nerve involvement in human leprosy is directly associated with the remarkable capacity of the causative organism, *Mycobacterium leprae (ML)* to invade the PNS. The genome sequence of *ML* is also unique, since this bacterial genome reveals a massive decay of functional genes, which causes *ML* to rely totally on host cell functions and possess the longest doubling time (~14 days) among bacteria (Cole et al. 2001). In the PNS, *ML* specifically targets the glial cells, the Schwann cells that enclose and insulate the axons. Schwann cells not only serve as a natural host but also a safe haven for the multiplication of *ML* since the blood-nerve barrier protects the organism from the host immune responses (Stoner 1979).

Therefore, Schwann cells serve as a primary non-immune source of infection, which not only causes nerve injury overtime but is also responsible for the continuous leakage of *ML* and ultimately facilitates dissemination of infection to other non-neural body tissues including skin, bone, smooth and skeletal muscles whose association is often seen with various stages of leprosy patients (Job et al. 1994; Scollard et al. 2006).

3 Pathogenesis

3.1 Early Events During Neural Infection and Bacterial Propagation Within the PNS

ML infection in humans initially presents with inflammation-mediated sensorimotor loss (Stoner 1979; Job 1989; Miko et al. 1993; Ooi and Srinivasan 2004). From the initial infection to the first symptoms of nerve involvement, we know nothing about the ongoing process in humans. This initial phase is therefore a completely 'black box', but it is critical for the propagation of ML and the establishment of bacterial niche within the PNS and subsequent nerve damage and disease progression. Because it is impossible to witness or diagnose the initial phase of nerve involvement, which can only be confirmed after neurological damage with functional loss (Job 1989; Lockwood et al. 2011), studies of early events of Schwann cell infection in humans cannot be performed, because it is unethical to obtain nerve biopsies without confirmed nerve involvement. Also, our lack of knowledge in this area is mainly due to high complexity of the infectious process with extremely long incubation and intractability of the system. Studies using a model of nine-banded armadillos, which are uniquely susceptible for *ML* infection and mimic the lepromatous form of human leprosy, are likely to fill the gaps in our knowledge in early infection and provide new insights into *ML*-induced changes. Other than nine-banded armadillos, there are no animal models that permit the observation or tracking early *ML* infection in Schwann cells and subsequent events continuously overtime. Nonetheless, combinations of well-defined primary neural tissue culture models and mouse models have provided valuable insights into early *ML* infection in Schwann cells (Rambukkana et al. 1997, 1998, 2002; Ng et al. 2000; Tapinos et al. 2006) (Fig. 1).

3.2 Why Do Leprosy Bacteria Target the Glial Cells of the PNS?

Schwann cells, the glial cells of PNS, which derive from neural crest precursors, comprise myelin-forming and non-myelin-forming phenotypes (Le Douarin and Dupin 2003; Jessen and Mirsky 2005). Although they can be considered as an

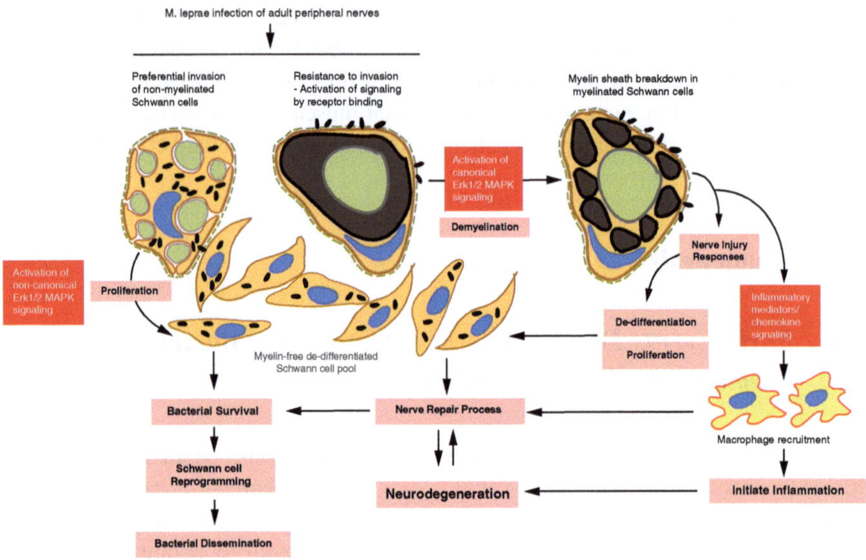

Fig. 1 Molecular, cellular and pathological as well as neuroregeneration events that follow the initial interaction of *M. leprae* with Schwann cell-axon units within the adult peripheral nerves. These events together with a cascade of complex inflammatory responses over a long incubation period subsequently cause neurodegeneration that manifests clinically in leprosy as the loss of sensory and motor neuronal functions

example of sophistication in cell differentiation they show remarkable plasticity, which contributes to the regeneration capacity of adult PNS even after severe injury (Radtke and Vogt 2009). By selecting Schwann cells *ML* have acquired several advantages (Cole et al. 2001; Rambukkana 2010). Recent studies have suggested that *ML* use the regenerative properties of PNS for the expansion of their bacterial niche (Rambukkana et al. 2002; Tapinos et al. 2006; Rambukkana 2010). Regeneration of damaged peripheral nerves has been documented in patients with advanced leprosy and this may also be associated with a part of *ML* efforts to maintain an intracellular niche during human infection (Miko et al. 1993). These and accumulating new findings led to the proposition that the high plasticity of non-myelinating Schwann cells that could easily be manipulated could be one of the reasons why *ML* have chosen Schwann cells as the primary residence for colonization and subsequent dissemination (Masaki et al. 2013) (Fig. 1). Indeed, high plasticity generally increases the susceptibility of somatic cells for manipulations when exposed to appropriate stimuli or microenvironment (Theise and Wilmut 2003). Likewise, during infection, bacterial efforts needed to manipulate host cells with high plasticity are relatively less demanding as compared to highly differentiated cells like myelin-forming Schwann cells. Indeed, myelin-forming Schwann cells have been shown to be much less preferred by *ML in vitro* and *in vivo* models as well as in leprosy patients (Rambukkana et al. 2002). Additionally, Schwann cells also serve as safe haven for *ML*, since they protect the organism from host immune assault (Stoner 1979; Job 1989). Thus, initial *ML* propagation in Schwann cells is likely to occur without hindrance from immune cells. Such favorable conditions, which are assisted with the non-toxic, non-cytopathic, and non-apoptotic nature of *ML*, permit bacterial residence within Schwann cells for a long period of time, a favorable condition that fits to extremely passive obligate *ML* life style (Lahiri et al. 2010; Rambukkana 2010).

3.3 Schwann Cell Infection, Demyelination and Regeneration

Under physiological and pathological conditions Schwann cells take decision to remain quiescent or to differentiate, to undergo programmed cell death or to proliferate and de-differentiate. *ML* appear to have evolved mechanisms to induce cell proliferation and de-differentiation after having caused demyelination of myelin-forming Schwann cells and invaded non-myelin-forming Schwann cells (Rambukkana 2010). Both demyelination and the invaded *ML* in non-myelin-forming Schwann cells in large numbers have been documented in lepromatous patients (Shetty et al. 1980, 1988). These findings suggest that *ML* multiply as long as Schwann cells, which lack anti-microbicidal machinery, can tolerate the bacterial load and then release bacteria and infect more Schwann cells (Rambukkana et al. 2002; Rambukkana 2004) (Fig. 1). These conditions generated by *ML* infection are somewhat similar to peripheral nerve injury *in vivo* where Schwann cells rapidly proliferate, an important event for the promotion of nerve regeneration (Fawcett and Keynes 1990). Once colonized, intracellular *ML* repress Schwann cell differentiation program and instead maintain infected Schwann cells in a de-differentiated stage that is favorable for intracellular bacterial propagation (Masaki et al. 2013). Although the latter conditions have significant advantage for the establishment of infection within the PNS, maintaining Schwann cells in de-differentiated stage will cause adverse effects on the functions of peripheral nerves. This could be an alternative mechanism for nerve dysfunctions after long incubation periods during human infections (Fawcett and Keynes 1990) (Fig. 1).

It is interesting to note that, in addition to neurodegeneration, regeneration of damaged axons has also been documented in patients with advanced leprosy, and this may also partly be associated with bacterial efforts to maintain an intracellular niche during human infections (Miko et al. 1993). Such studies on patients provide evidence that *ML* may assist the nerve regeneration process in order to maintain Schwann cell niche in active stage so that bacterial survival can be secured despite the immune responses. Using mouse models Masaki et al. (2013) have recently demonstrated that *ML* use the plasticity of adult Schwann cells to reprogram infected cells and thereby convert the infected cells to progenitor/stem cell-like cells which favour bacterial dissemination following colonization in Schwann cells. The capacity of these reprogrammed cells to secrete chemokines that recruit macrophages and transfer infection further assists the development of a secondary bacterial niche but also *ML* dissemination by granuloma formation, a pathogenic hallmark of human leprosy (Cosma et al. 2003). This new knowledge suggests that PNS not only serves as a primary niche for bacterial propagation but also promotes infection to other parts of the body particularly to skeletal muscles, bones and smooth muscles, which are known to harbour *ML* in leprosy patients. These accumulating evidences led us to the proposition that plasticity of non-myelinating Schwann cells could be one of the reasons why *ML* have chosen them as the primary non-immune residence for colonization, which subsequently lead to the peripheral nerve damage.

3.4 Contribution of Schwann Cell Infection to Early Neurodegeneration

A growing body of evidence now suggests that the lack of interaction between glial cells and neurons could be the cause for disease pathogenesis in the nervous system. Recent studies have shown that when glial-neuronal interaction is perturbed, neuronal functions are also abrogated. For example, knockdown of neuregulin 1, a major axonal ligand that communicates with Schwann cells via glial receptors ErbB3/2 complex, causes several abnormalities in neuronal functions (Nave and Salzer 2006). Interestingly, *ML* bind to neuregulin receptors and induce signaling that eventually causes demyelination (Tapinos et al. 2006). These findings suggest that *ML* interference with glial-neuronal communication could initiate neuronal dysfunctions. Furthermore, such non-immune mediated myelin damage could rapidly be repaired, as remyelination is known to occur spontaneously in adult peripheral nerve in response to injury (Fawcett and Keynes 1990). Nevertheless, *ML*-induced nerve injury conditions could be progressive during the course of an extremely slow infection. This continues to develop with Schwann cell propagation and bacterial replication that subsequently release immune factors and recruit inflammatory cells to the site of infected peripheral nerves (Masaki et al. 2013).

3.5 Activation of Signaling Pathways in Infected Schwann Cells

As noted above, homeostasis of glial cells and neurons and their reciprocal interactions play a significant role in maintaining a healthy PNS. Leprosy bacteria appear to take advantage of this intimate cell-cell communication system by infecting Schwann cells and altering or subverting critical Schwann cell functions (Tapinos et al. 2006). Induction of signaling in response to *ML* is one of the key events that occur very early after infection, when *ML* binds to ErbB2 receptor tyrosine kinase on Schwann cells (Tapinos et al. 2006). This leads to the sequential activation of Raf, Mek1/2 and Erk 1/2/MAPK signaling pathways in myelinated Schwann cells and is sufficient to induce demyelination (Tapinos et al. 2006). Such receptor-mediated MAP kinase signaling was the first to give evidence that the activation of Erk1/2 alone is sufficient to induce demyelination without immune responses or other lesions. Along with demyelinating changes, intracellular *ML* also can activate Erk1/2 signaling via PKC and LCK signaling without using canonical receptor-mediated activation, and induce Schwann cell proliferation (Rambukkana 2010). Thus, infection of Schwann cells can start not only demyelination but also cell proliferation (Fig. 1).

Interestingly, the lesson learned from leprosy bacteria infections on the activation of Schwann cell Erk1/2 signaling in demyelination has recently been reiterated in inducible transgenic mouse model. In this model, sustained activation of inducible Raf-kinase transgene that activates downstream Erk1/2 in myelinated Schwann cells is sufficient to cause demyelination (Napoli et al. 2012). Like in *ML* infection, transgene-induced Erk1/2 signal activation and subsequent demyelination also recruit inflammatory cells, which are usually required for regeneration following demyelination and injury. Thus, it is likely that initial *ML* infection in leprosy patients may also undergo similar series of events following *ML* interaction with Schwann cell-axon units—activation of Erk1/2 signaling, demyelination and subsequent inflammatory response. This recruits macrophages which in turn aggravate inflammatory conditions that eventually abolish the nerve functions. Nerve dysfunction due to inflammation at this stage seems to manifest clinically as sensory or sensorimotor loss (Job 1989).

3.6 *Association of Inflammation with Neurodegeneration in Human Leprosy*

Nerve involvement in leprosy patients can be identified following nerve function loss due to strong inflammatory responses, in the form of sensory or motor neuron damage (Minauchi and Igata 1987; Sabin et al. 1993). From this stage, the disease process is progressive and could lead to deformities and disabilities unless the patients are treated to reduce the bacterial load and inflammatory conditions. Although bacterial cure can be achieved by successful multidrug antibiotic therapy, neurological injury may continue to occur in these patients due to strong inflammation. This includes the delayed type hypersensitivity reactions producing acute inflammation known as type-1 reactions, which are episodes with enhanced T cell and macrophage responses in peripheral nerves leading to their destruction. However, we know little about the immune mechanisms involved in nerve damage in human leprosy.

4 Clinical Manifestations

4.1 *Phenotypes, Classification*

In humans, the estimated incubation period from initial infection to clinical manifestation of nerve injury is likely to be 5–7 years, although it could go up to 30–40 years, but no proper documented studies are available on this. The onset of nerve damage is insidious. However, if untreated, the neurological damage could be progressive, as in other neurodegenerative diseases, and result in permanent damage to the nerves and subsequent impairment in limbs and eyes. The bacilli favor the cooler areas of the body such as chin, malar areas of face, earlobes, buttocks, knees, and distal extremities. Disease phenotype of leprosy is thought to depend on host immune reactions to the bacteria, especially the T cell immunity (Ooi and Srinivasan 2004). When the host responds to the bacterial infection with little or no cellular immunity as in the lepromatous form of leprosy with high bacterial load, numerous skin lesions appear, and peripheral nerves are diffusely affected slowly and progressively. When the host responds with strong cellular immunity, as in the tuberculoid form of leprosy with low or no bacteria. The nerve is more markedly damaged due to aggressive immune attack. Thus, leprosy is classified according to the number of skin lesions and the number of bacilli found on slit-skin smear examination (Brown et al. 1996a, b; Chandi and Chacko 1986; Cole et al. 2001; Coruh and McDougall 1979). Paucibacillary disease *indeterminate* [tuberculoid tuberculoid (TT) and borderline tuberculoid (BT)] is defined as fewer than six skin lesions with no bacilli on slit-skin smear testing. Multibacillary disease [borderline borderline (BB), borderline lepromatous (BL) and lepromatous leprosy (LL)] is characterized by six or more lesions with or without positive skin smear results, mostly with high bacterial content.

4.2 *Neuropathies and Skin Lesions*

4.2.1 Diverse Clinical Signs and Symptoms of Neuropathies and Skin Lesions Across the Leprosy Spectrum

Regardless of host immune response, variability of nerve lesions occurs across the leprosy spectrum. Tuberculoid leprosy, particularly the TT type, is characterized by a well-defined uniformly circular or oval erythematous/hypo-pigmented skin plaque with maximal induration of the margins sloping towards the center. The surface is bald, dry and scaly, and completely anesthetic. In the vicinity of the skin lesions, thickened nerve is sometimes palpable. The involved dermal nerves are infiltrated with granulomas composed of epithelioid cells and lymphocytes. In peripheral neuropathy, a single nerve is enlarged, which results in a mononeuropathy such as wrist drop, foot drop or claw hand (Minauchi and Igata 1987; Ooi and Srinivasan 2004; Agrawal et al. 2005). Mononeuropathy multiplex is also seen. Commonly involved are greater auricular, radial cutaneous, ulnar, median, common peroneal, posterior tibial, and facial nerves (Minauchi and Igata 1987; Ooi and Srinivasan 2004; Agrawal et al. 2005).

Patients with the lepromatous form with high bacterial load, particularly LL patients, present with generalized disease with multi-system involvement, sparing only the central nervous system. Most strikingly involved tissues in addition to nerves are the skin, mucous membrane and reticuloendothelial systems (Sehgal 1994). Skin lesions are multiple, bilateral, and symmetric. Typically, the lesions are hypopigmented, shiny with ill-defined margins and merges imperceptibly with surrounding skin. Frequently involved are eye-brows, nose and lips, along with flattening of the bridge of the nose resulting in classical "leonine facies". Nerve involvement results in progressive bilateral symmetrical cutaneous sensory loss, and distal weakness, together with patch anesthesia and mononeuropathies (Minauchi and Igata 1987). The nerve trunks are bilaterally and symmetrically thickened and tender (Ridley and Jopling 1966; Sehgal 1994). Enlarged nerve may be functioning well, suggesting that bacilli invasion into nerves itself does not cause acute nerve damage, but it is the inflammatory reactions that are responsible for the nerve damage.

4.2.2 Bacterial Propagation in Cooler Tissues and Its Contribution to Sensory Loss

Interestingly, leprosy bacteria replicate most actively in the cooler tissues. This causes a unique pattern of neurologic deficits (Sabin et al. 1993). In LL patients, the distribution of neural deficits tends to expand symmetrically, which is presumed to be due to hematogenous spread of *ML* (Minauchi and Igata 1987; Scollard et al. 1999; Ooi and Srinivasan 2004). The relative coolness of the distal portions of the limbs results in "stocking-glove"-like distribution of the lesions. However, in all stages the evolution of the temperature-linked sensory loss can be distinguished from the distal symmetric sensory loss with fading proximal borders that is the hallmark of numerous sensory polyneuropathies (Sabin et al. 1993). There is ample evidence showing that sensory loss in LL is linked with low skin temperature. This feature directly correlates with bacterial preference to reside and replicate in non-myelinating Schwann cells, which enclose small sensory axons that extend to cooler body areas. *ML* is temperature-sensitive and prefers to live in cooler area in the body, and bacterial survival has been shown to be most effective below 37 °C *in vitro* (Truman and Krahenbuhl 2001).

Sensory loss in the skin appears first in the pinnae of the ears, the dorsal surfaces of the hands, the dorsomedial surfaces of the forearms, the dorsal surfaces of the feet, and the anterolateral aspects of the legs (Sabin et al. 1993). Then, sensory loss appears over the nose, malar areas, breasts, central abdomen, and buttocks. At this stage, sparing of the palms and soles may be apparent even when the dorsal areas of the hands and feet are affected. This is because palms and soles are not only warmer than the dorsal sides but also insulated by the thickened cuticle. In fact, it is noticeable when sensory loss of the medial arch terminates abruptly along the border of the cuticle of the forefoot, lateral arch, and heel (Sabin et al. 1993). Sensory loss extends to face and ears. When forehead is involved, there may be an abrupt return to normal sensation at the hairline, which is a highly diagnostic sign of leprosy, "hairline" sign. Finally, sensory loss expands to all over the body, leaving the islands of preserved sensation, such as the inguinal creases, the perineum, the axilla, the sternal area, and a stripe of up the center of the back from the intergluteal fold to the neck.

Motor dysfunction usually appears far later after the sensory loss expanded to significant skin area all over the body. Paralysis is caused by the destruction of the nerves located near the surface of the body. Ulnar nerve dysfunction usually appears first, leading to atrophy and weakness of intrinsic hand muscles, and an eventual clawhand deformity. Median nerve involvement may lead to further atrophy and weakness of intrinsic hand muscles. Facial nerve involvement may lead to complete lagophthalmos and lower lip ectropion. Peroneal nerve palsy may cause footdrop.

Wounds may occur in insensitive skin area, especially that of hands and feet, without recognition of leprosy, delaying the treatment. Wounds are often complicated with ulceration on the plantar surfaces and sides of the feet and toes and on the hands. Infection of the wounds is also a frequent complication (Sabin et al. 1993). Lacerations, burns, abrasions, and hematoma are quite common on the hands. Shortening of digits on feet and hands may occur through destructive osteo-myelitis. Amputations may occur traumatically or may be carried out surgically in an effort to prevent further tissue destruction due to severe sensory loss. Therefore, it is evident that sensorimotor neuropathy is a central feature for the variety of disabilities seen in human leprosy.

4.3 Neuritic Leprosy and Neuropathic Pain

This type of leprosy with neural signs and symptoms appears in patients without any skin involvement (Uplekar and Antia 1986; Jenkins et al. 1990; Ridley et al. 1994; Mahajan et al. 1996). Mononeuropathy or mononeuropathy multiplex are most frequently seen in leprosy patients. Even in mononeuropathy, multiple nerves can be enlarged. Commonly affected are the ulnar, radial, median, lateral popliteal, posterior tibial, facial, and sometimes trigeminal nerves. Sensory polyneuropathies showing only temperature and pain anesthesia are occasionally seen. Inflammatory reaction may be more extensive in nerves than in the skin. Nerve conduction studies show demyelinating features early in the disease, and axonal injury in later stages (McLeod et al. 1975; Brown et al. 1996a, b). Nerve biopsy studies showed epithelioid granulomas only in 14 % and acid-fast bacilli only in 16 % of the patients. This shows the difficulty of diagnosis of these patients. PCR analysis of *ML* was positive in 47 %, and may serve as a useful diagnostic tool (Jardim et al. 2003).

Neuropathic pain is a frequent complaint of leprous neuropathy during the disease, or even when the disease is arrested by bacteriological cure (Minauchi and Igata 1987; Haanpaa et al. 2004; Stump et al. 2004; Scollard et al. 2006). Severe pain usually occur during the stage of nerve swelling. Pain of variable severity may be felt even in the areas with complete loss of sensation or in artificial legs after amputation, resembling phantom limb pain. The pathogenesis of the neuropathic pain in human leprosy remains unclear.

4.4 *Autonomic Nervous System Involvement*

Autonomic nerve involvement is documented in lepromatous patients by sympathetic skin response and heart rate measurements to deep breathing (Brown et al. 1996a, b; Ulvi et al. 2003). Skin or muscle vasomotor reflexes, testing function of small unmyelinated autonomic nerves, may also be abnormal (Wilder-Smith et al. 2000). Orthostatic hypotension was frequently found in Japanese patients (Minauchi and Igata 1987). In addition, one-third of patients with LL in Japan had Raynaud's phenomenon (Minauchi and Igata 1987). Peripheral angiography revealed obstructions and/or narrowing and tortuosity of arteries, resulting in impaired blood supply. Similar results have been reported in other studies (Kaur et al. 1976; Chopra et al. 1981). The autonomic nervous system involvement may therefore also reflect invasion by *ML* of arterial smooth muscle cells (Coruh and McDougall 1979; Chopra et al. 1981).

5 Laboratory Findings

5.1 *Impaired Nerve Conduction*

Nerve injury in leprosy is thought to cause initially segmental demyelination followed by secondary axon loss. Accordingly, conduction studies in motor and sensory nerves have shown a prolongation of distal latencies and marked slowing in most patients with clinical evidence of nerve involvement, regardless of the type of leprosy (Verghese et al. 1970; McLeod et al. 1975; Minauchi and Igata 1987). Ulnar nerve frequently shows slowing of motor conduction velocity (MCV), especially in the elbow segment instead of the forearm, wrist, and humeral segments, reflecting the predilection of *ML* for cooler subcutaneous regions (Hackett et al. 1968; Verghese et al. 1970). Median nerve characteristically shows slowing in the forearm segment rather than in the carpal tunnel or distally. Conduction slowing is also seen in peroneal nerves in the segment between the popliteal fossa and the fibular head, and distally below the ankle (Swift et al. 1973), as well as in facial nerve (Dastur et al. 1966; Turkof et al. 2003). In the acute stage of motor impairment with focal nerve swelling, segmental conduction block may be seen.

5.2 *Nerve Pathology*

5.2.1 Demyelination in Early Leprosy

It is worth noting that, regardless of the form of leprosy, myelinated fibers predominantly show segmental demyelination in the earliest stage rather than Wallerian degeneration (Shetty et al. 1977, 1980). Nearly 10 % of nerve fibres show segmental demyelination in asymptomatic contacts exposed to leprosy patients (Shetty et al. 1977), suggesting demyelination as an early event in nerve involvement in human leprosy. Our experimental data showing that *ML* induce demyelinating Erk1/2 signaling pathway (Rambukkana et al. 2002; Tapinos et al. 2006) suggest that the bacterial-induced Erk1/2 signaling might play a key role in this early demyelination.

5.2.2 Diversity of Pathological Features Across the Leprosy Spectrum

In *indeterminate borderline leprosy* (BT) characterized by the earliest skin lesions of leprosy, bacteria are sometimes found in endoneurial macrophages or Schwann cells in dermal or subdermal nerves. Inflammatory changes are minimal and mononuclear cuffing of nerves may also be seen in these conditions (Chandi and Chacko 1986; Weng et al. 2000). In *tuberculoid leprosy* (TT), enlarged nerves and trunks are observed usually within the proximity of the skin lesion. Nerve biopsies reveal significant destruction of the nerve architecture with the presence of granulomas in the nerves comprised of epitheloid cells and giant cells with scattered mononuclear infiltration (Jopling and Morgan-Hughes 1965). The granulomas extend from the endoneurium to perineurium and they sometimes also involve the epineurium. Significant axonal loss may also be seen in these patients. However, these changes usually vary between axon fascicles (Minauchi and Igata 1987). *ML* are usually not easily identified and are never abundant in these patients. *ML* may be seen in Schwann cells, macrophages, or epithelioid cells. In the later stages, significant fibrosis and hyalinization of the endoneurium may occur with complete destruction of the nerve architecture (Miko et al. 1993).

In *lepromatous leprosy* (LL), despite high bacterial load, the nerve structure is more preserved than in tuberculoid type due to much less T cell-mediated immune responses (Job and Desikan 1968). The perineurium may be split into layers by edema or sheets of foamy cells, giving the appearance of an onion skin. The perineurium and endoneurium are infiltrated with abundant macrophages filled with masses of bacilli (globi) (Pereira et al. 1991). Shetty et al. (1980) reported that bacilli were mainly found in the non-myelinating Schwann cells. Indeed, sensory loss is the earliest symptom of human leprosy (Minauchi and Igata 1987; Ooi and Srinivasan 2004).

The inflammatory responses in lepromatous patients may bring increased numbers of lymphocytes, plasma cells, and mast cells to the endoneurium (Sabin et al. 1993). Although it is believed that granulomas are less frequently seen in lepromatous leprosy, extensive granulomatous infiltrations are reported in lepromatous patients (Haimanot et al. 1984). Fite stain usually reveals abundant bacilli in macrophages and Schwann cells and, to a lesser extent, in the endothelial cells of endoneurial and perineurial vessels (Job and Verghese 1975). As the disease progresses, fibrosis may supervene with nerve destruction. Abnormalities in endothelial cell structure may also occur due to inflammation, with resulting breaches in the blood-nerve barrier (Boddingius 1984).

6 Human Leprosy as a Model for Neurodegenerative Diseases

Although peripheral nerve damage is the characteristic feature in human leprosy, nerve impairment in this disease resembles many other non-infectious peripheral neuropathies and neurodegenerative conditions associated with sensory and motor neuron loss with known and unknown etiology (Sabin et al. 1993; Rambukkana 2010). Therefore, neurodegeneration of human leprosy could also serve as a model to investigate pathogenetic mechanisms for other neurodegenerative diseases with sensory nerve damage and demyelination, particularly those of still unknown etiologies. Because the causative agent of human leprosy is known, it would be possible to monitor how leprosy bacteria induce neurodegeneration in leprosy patients. However, the challenges of such studies should not be underestimated because of the immense obstacles researchers face not only for studies of patients but also for basic laboratory studies.

7 Conclusions and Future Perspectives

Despite the dramatic reduction in the prevalence of leprosy over the last 50 years largely due to the worldwide implementation of multiple-drug therapy (MDT) that produced bacterial cure nerve damage continues to be a problem in many of the treated patients. As in other neurodegenerative diseases, leprosy neuropathy could progress to later stages, and lack of treatment or failure to diagnose early it could certainly develop into debilitating conditions. Although MDT can cure bacterial load, it may cause unwanted leprosy reactions, and a variety of residual disabilities as described above require prolonged treatment or rehabilitation. Thus, we believe that discovery of new targets for diagnostics and preventing or halting the progression of the disease is critically important for thwarting nerve damage in the early stage of leprosy. Further understanding of how *ML* establish the productive infection within the PNS at the earliest stage, that eventually leads to neurodegeneration, dissemination and transmission of infection will bring new approaches to tackle this devastating neglected disease.

Finally, much is to be learned from the fascinating neurobiology, immunobiology, and cell biology associated with human leprosy. Therefore, both studies of patients and model system-related studies should be encouraged despite the obstacles researchers face under current challenging conditions. Such studies should provide entirely new insights into our understanding of human biology and how microbes, particular human neurotropic pathogens have evolved to manipulate our nervous system.

References

Agrawal A, Pandit L et al (2005) Neurological manifestations of Hansen's disease and their management. Clin Neurol Neurosurg 107:445–454

Boddingius J (1984) Ultrastructural and histophysiological studies on the blood-nerve barrier and perineurial barrier in leprosy neuropathy. Acta Neuropathol 64:282–296

Brown TR, Kovindha A et al (1996a) Leprosy neuropathy: correlation of clinical and electrophysiological tests. Indian J Lepr 68:1–14

Brown TR, Kovindha A et al (1996b) Abnormalities of the sympathetic skin response in lepromatous leprosy. Muscle Nerve 19:1357–1358

Chandi SM, Chacko CJ (1986) An ultrastructural study of the response of traumatized rabbit tibial nerve to epineurial infection with *Mycobacterium leprae*. Int J Lepr Other Mycobact Dis 54:79–83

Chopra JS, Kaur S et al (1981) Vascular changes in leprosy and its role in the pathogenesis of leprous neuritis. Lepr India 53:443–453

Cole ST, Eiglmeier K et al (2001) Massive gene decay in the leprosy bacillus. Nature 409:1007–1011

Coruh G, McDougall AC (1979) Untreated lepromatous leprosy: histopathological findings in cutaneous blood vessels. Int J Lepr Other Mycobact Dis 47(3):500–511

Cosma CL, Sherman DR et al (2003) The secret lives of the pathogenic mycobacteria. Annu Rev Microbiol 57:641–676

Dastur DK, Antia NH et al (1966) The facial nerve in leprosy. 2. Pathology, pathogenesis, electromyography and clinical correlations. Int J Lepr Other Mycobact Dis 34:118–138

Fawcett JW, Keynes RJ (1990) Peripheral nerve regeneration. Annu Rev Neurosci 13:43–60

Haanpaa M, Lockwood DN et al (2004) Neuropathic pain in leprosy. Lepr Rev 75:7–18

Hackett ER, Shipley DE et al (1968) Motor nerve conduction velocity studies of the ulnar nerve in patients with leprosy. Int J Lepr Other Mycobact Dis 36:282–287

Haimanot RT, Mshana RN et al (1984) Sural nerve biopsy in leprosy patients after varying periods of treatment: histopathological and bacteriological findings on light microscopy. Int J Lepr Other Mycobact Dis 52:163–170

Jardim MR, Antunes SL et al (2003) Criteria for diagnosis of pure neural leprosy. J Neurol 250: 806–809

Jenkins D, Papp K et al (1990) Leprotic involvement of peripheral nerves in the absence of skin lesions. Case report and literature review. J Am Acad Dermatol 23(5 Pt 2):1023–1026

Jessen KR, Mirsky R (2005) The origin and development of glial cells in peripheral nerves. Nat Rev Neurosci 6:671–682

Job CK (1989) Nerve damage in leprosy. Int J Lepr Other Mycobact Dis 57(2):532–539

Job CK, Desikan KV (1968) Pathologic changes and their distribution in peripheral nerves in lepromatous leprosy. Int J Lepr Other Mycobact Dis 36:257–270

Job CK, Verghese R (1975) Schwann cell changes in lepromatous leprosy–an electron microscope study. Indian J Med Res 63:897–901

Job CK, Chehl SK et al (1994) Transmission of leprosy in nude mice through thorn pricks. Int J Lepr Other Mycobact Dis 62:395–398

Jopling WH, Morgan-Hughes JA (1965) Pure neural tuberculoid leprosy. Br Med J 2: 799–800

Kaur S, Wahi PL et al (1976) Peripheral vascular deficit in leprosy. Int J Lepr Other Mycobact Dis 44:332–339

Lahiri R, Randhawa B et al (2010) Infection of mouse macrophages with viable *Mycobacterium leprae* does not induce apoptosis. J Infect Dis 201:1736–1742

Le Douarin NM, Dupin E (2003) Multipotentiality of the neural crest. Curr Opin Genet Dev 13:529–536

Lockwood DN, Suneetha S (2005) Leprosy: too complex a disease for a simple elimination paradigm. Bull World Health Organ 83:230–235

Lockwood DN, Suneetha L et al (2011) Cytokine and protein markers of leprosy reactions in skin and nerves: baseline results for the North Indian INFIR cohort. PLoS Negl Trop Dis 5(12):e1327
Mahajan PM, Jogaikar DG et al (1996) A study of pure neuritic leprosy: clinical experience. Indian J Lepr 68:137–141
Masaki T, Jinrong Q et al (2013) Reprogramming adult Schwann cells to stem cell-like cells by leprosy bacilli promotes dissemination of infection. Cell 152:51–67
McLeod JG, Hargrave JC et al (1975) Nerve conduction studies in leprosy. Int J Lepr Other Mycobact Dis 43:21–31
Miko TL, Le Maitre C et al (1993) Damage and regeneration of peripheral nerves in advanced treated leprosy. Lancet 342:521–525
Minauchi Y, Igata A (1987) Leprous neuritis. In: Matheus WB (ed) Handbook of clinical neurology. Elsevier, Amsterdam, pp 215–238
Napoli I, Noon LA et al (2012) A central role for the ERK-signaling pathway in controlling Schwann cell plasticity and peripheral nerve regeneration in vivo. Neuron 73(4):729–742
Nave KA, Salzer JL (2006) Axonal regulation of myelination by neuregulin 1. Curr Opin Neurobiol 16:492–500
Ng V, Zanazzi G et al (2000) Role of the cell wall phenolic glycolipid-1 in the peripheral nerve predilection of Mycobacterium leprae. Cell 103:511–524
Ooi WW, Srinivasan J (2004) Leprosy and the peripheral nervous system: basic and clinical aspects. Muscle Nerve 30:393–409
Pereira JH, Palande DD et al (1991) Mycobacteria in nerve trunks of long-term treated leprosy patients. Lepr Rev 62:134–142
Radtke C, Vogt PM (2009) Peripheral nerve regeneration: a current perspective. Eplasty 9:e47
Rambukkana A (2004) *Mycobacterium leprae*-induced demyelination: a model for early nerve degeneration. Curr Opin Immunol 16:511–518
Rambukkana A (2010) Usage of signaling in neurodegeneration and regeneration of peripheral nerves by leprosy bacteria. Prog Neurobiol 91:102–107
Rambukkana A, Salzer JL et al (1997) Neural targeting of *Mycobacterium leprae* mediated by the G domain of the laminin-alpha2 chain. Cell 88:811–821
Rambukkana A, Yamada H et al (1998) Role of alpha-dystroglycan as a Schwann cell receptor for *Mycobacterium leprae*. Science 282:2076–2079
Rambukkana A, Zanazzi G et al (2002) Contact-dependent demyelination by *Mycobacterium leprae* in the absence of immune cells. Science 296:927–931
Rao PN (2005) Leprosy program in India at the crossroads. Int J Lepr Other Mycobact Dis 73:211–215
Ridley DS, Jopling WH (1966) Classification of leprosy according to immunity. A five-group system. Int J Lepr Other Mycobact Dis 34:255–273
Ridley MJ, Waters MF et al (1994) Effect of *Mycobacterium leprae* in peripheral nerve trunk on the evolution of skin lesions. Int J Lepr Other Mycobact Dis 62:99–107
Sabin TD, Swift TR et al (1993) Leprosy. In: Dyck PJ, Thomas PK, Griffin JW, Low PA, Poduslo JF (eds) Peripheral neuropathy, 3rd edn. W.B.Saunders, Philadelphia, PA, pp 1354–1379
Scollard DM, McCormick G et al (1999) Localization of *Mycobacterium leprae* to endothelial cells of epineurial and perineurial blood vessels and lymphatics. Am J Pathol 154: 1611–1620
Scollard DM, Adams LB et al (2006) The continuing challenges of leprosy. Clin Microbiol Rev 19(2):338–381
Sehgal VN (1994) Leprosy. Dermatol Clin 12:629–644
Shetty VP, Mehta LN et al (1977) Teased fibre study of early nerve lesions in leprosy and in contacts, with electrophysiological correlates. J Neurol Neurosurg Psychiatry 40:708–711
Shetty VP, Mehta LN et al (1980) Study of the evolution of nerve damage in leprosy. Part I—lesions of the index branch of the radial cutaneous nerve in early leprosy. Lepr India 52:5–18
Shetty VP, Antia NH et al (1988) The pathology of early leprous neuropathy. J Neurol Sci 88:115–131

Stoner GL (1979) Importance of the neural predilection of *Mycobacterium leprae* in leprosy. Lancet 2:994–996

Stump PR, Baccarelli R et al (2004) Neuropathic pain in leprosy patients. Int J Lepr Other Mycobact Dis 72:134–138

Swift TR, Hackett ER et al (1973) The peroneal and tibial nerves in lepromatous leprosy. Clinical and electrophysiologic observations. Int J Lepr Other Mycobact Dis 41:25–34

Tapinos N, Ohnishi M et al (2006) ErbB2 receptor tyrosine kinase signaling mediates early demyelination induced by leprosy bacilli. Nat Med 12:961–966

Theise ND, Wilmut I (2003) Cell plasticity: flexible arrangement. Nature 425:21

Truman RW, Krahenbuhl JL (2001) Viable *M. leprae* as a research reagent. Int J Lepr Other Mycobact Dis 69:1–12

Turkof E, Richard B et al (2003) Leprosy affects facial nerves in a scattered distribution from the main trunk to all peripheral branches and neurolysis improves muscle function of the face. Am J Trop Med Hyg 68:81–88

Ulvi H, Yoldas T et al (2003) R-R interval variation and the sympathetic skin response in the assessment of the autonomic nervous system in leprosy patients. Acta Neurol Scand 107: 42–49

Uplekar MW, Antia NH (1986) Clinical and histopathological observations on pure neuritic leprosy. Indian J Lepr 58:513–521

Verghese M, Ittimani KV et al (1970) A study of the conduction velocity of the motor fibers of ulnar and median nerves in leprosy. Int J Lepr Other Mycobact Dis 38:271–277

Weng XM, Chen SY et al (2000) Immuno-histopathology in the diagnosis of early leprosy. Int J Lepr Other Mycobact Dis 68:426–433

Wilder-Smith EP, Wilder-Smith AJ et al (2000) Skin and muscle vasomotor reflexes in detecting autonomic dysfunction in leprosy. Muscle Nerve 23:1105–1112

Rabies: Neurology

Alan C. Jackson

Abstract Rabies is an ancient disease, but it remains an important problem and threat to humans in countries with endemic dog rabies, especially in Asia and Africa. Rabies in wildlife, particularly from bats, is the main threat to humans in North America. Human rabies often has distinctive clinical features reflecting the early brainstem involvement, including hydrophobia, but physicians in North America and Europe may not consider a diagnosis of rabies because the disease is rare and their lack of familiarity with the clinical manifestations. There is progressive neurological deterioration in rabies and the disease is virtually always fatal. With aggressive approaches there are a variety of medical complications, including multiple organ failure. The therapy of human rabies has proven to be disappointing. Entirely new approaches need to be considered in the future.

Keywords Encephalitis • Encephalomyelitis • Hydrophobia • Lyssavirus • Neglected diseases • Rabies virus

A.C. Jackson (✉)
Department of Internal Medicine (Neurology), University of Manitoba, Winnipeg, MB, Canada

Department of Medical Microbiology, University of Manitoba, Winnipeg, MB, Canada

Health Sciences Centre, GF-543, 820 Sherbrook Street, Winnipeg, MB, Canada R3A 1R9
e-mail: ajackson2@hsc.mb.ca

M. Bentivoglio et al. (eds.), *Neglected Tropical Diseases and Conditions of the Nervous System*, DOI 10.1007/978-1-4614-8100-3_13,

1 Introduction

Rabies encephalomyelitis has characteristic clinical features that usually distinguish it from other viral or other diseases of the central nervous system (CNS). Diagnosis may be especially challenging in the absence of a history of an animal bite, which is common in North America in association with bat rabies virus variants. There are two clinical forms of rabies: encephalitic and paralytic. In encephalitic rabies there is brainstem involvement resulting in hydrophobia with preservation of consciousness. This results is a very frightening disease that is virtually always fatal. Paralytic rabies may be misdiagnosed as Guillain-Barré syndrome or other peripheral nerve disorder. North American physicians are frequently not familiar with the disease and may fail to diagnose rabies even with typical clinical features. Currently rabies can be very effectively prevented after recognized exposures, but the therapy of human rabies remains an important challenge for the future.

Rabies is caused by caused by rabies virus infection that involves the nervous system. Rabies virus belongs in genotype 1 of the lyssaviruses and there are 10 other genotypes of lyssaviruses and 5 of these 10 genotypes have been recognized to very rarely cause human disease indistinguishable from rabies (Table 1). The genotype status of one additional case caused by Irkut virus (Leonova et al. 2009) is pending.

Table 1 Reported human rabies cases due to other *Lyssavirus* genotypes

Virus (Genotype)	Year	Location	Age of patient	Reference
Mokola (3)[a]	1968	Nigeria	3.5	Familusi and Moore (1972)
Mokola (3)	1971	Nigeria	6	Familusi et al. (1972)
Duvenhage (4)	1970	South Africa	31	Meredith et al. (1971)
Duvenhage (4)	2006	South Africa	77	Paweska et al. (2006)
Duvenhage (4)	2007	Kenya	34	van Thiel et al. (2009)
European Bat Lyssavirus 1 (5)	1985	Russia	11	Selimov et al. (1989)
European Bat Lyssavirus 1 (5)	2002	Ukraine	34	Botvinkin et al. (2005)
European Bat Lyssavirus 2 (6)	1985	Finland	30	Roine et al. (1988)
European Bat Lyssavirus 2 (6)	2002	Scotland	55	Johnson et al. (2012), Nathwani et al. (2003)
Australian Bat Lyssavirus (7)	1996	Australia	39	Samaratunga et al. (1998)
Australian Bat Lyssavirus (7)	1998	Australia	37	Hanna et al. (2000)
Australian Bat Lyssavirus (7)	2013	Australia	8	Anonymous (2013)
Irkut (pending)	2007	Russia	20	Leonova et al. (2009)

Adapted with permission from Jackson AC: Human disease, in Rabies: scientific basis of the disease and its management, Third Edition, edited by AC Jackson, 2013, Elsevier Academic Press, Oxford, UK, pp 269–298; Copyright Elsevier

[a]It is doubtful that this patient's clinical picture was actually caused by Mokola virus infection

2 Epidemiology

Globally about 55,000–75,000 humans die from rabies every year. Around half of the victims of rabies are children. Hence, rabies remains a very significant public health problem and is an important neglected disease. Surveillance for human rabies is poor, particularly in countries where the disease is prevalent. Laboratory confirmation diagnosis is infrequent and clinical diagnosis is inaccurate. Worldwide about 99 % of human cases of rabies are transmitted from dogs. Canine rabies continues to be a large problem in Asia and Africa. About three billion humans live in regions with endemic dog rabies, which puts a large proportion of the human population at risk of developing rabies (Knobel et al. 2007). In these regions there is dog-to-dog transmission of rabies virus. For a variety of economic, political, religious, and cultural reasons, canine rabies has not been controlled in many countries, although the methods of how to do this are very well developed and a "blueprint" for rabies control has recently been published (Lembo 2012) and is available on the Internet (http://www.rabiesblueprint.com/). Canine rabies has been well controlled in most of Latin America over the past few decades with dog control measures, particularly employing mass vaccination of dogs. There are about 30,000 human deaths from rabies each year in India. China recently had a resurgence of rabies, which is also related to endemic canine rabies, with a peak of over 3,300 human deaths in 2007 (Hu et al. 2009). Thailand markedly reduced the death rate due to human rabies from hundreds of cases to about ten cases per year by developing a very effective post-exposure rabies prophylaxis program without controlling dog rabies in the country. Outbreaks of rabies continue to occur in focal areas. For example, in 2008 an outbreak of human rabies deaths began on the island of Bali, Indonesia.

Rabies is also endemic in wildlife and typically a rabies virus variant becomes adapted to a particular species and can be readily transmitted within that species. However, when this variant is transmitted to another species, the new host is fully susceptible but cannot usually transmit the virus further and becomes a "dead-end" host. Rabies virus variants can be characterized from clinical specimens (e.g., saliva or skin biopsies) by using reverse transcription polymerase chain reaction (RT-PCR) amplification and sequencing or by characterization using monoclonal antibodies, which allows identification of the species of the vector that the variant is associated with and is also useful in identifying or confirming the animal source of human cases.

Bats are the most important wildlife vector. The majority of cases of human rabies occurring in the United States and Canada are caused by bat rabies virus variants. Many human cases of rabies due to these variants have no history of bat contact (38 %) or have either a house exposure with no direct contact (10 %) or direct contact with no recognized bite (18 %); only 38 % of cases are associated with a history of a bite or scratch (Table 2). The rabies virus variant associated with silver-haired bats (*Lasionycteris noctivagans*) and tricolored bats (previously called eastern pipistrelle bats) (*Perimyotis subflavus*) is responsible for most human rabies cases in North America. These bats are not normally found in homes, in contrast to little brown bats (*Myotis lucifugus*) and big brown bats (*Eptesicus fucus*), which are

Table 2 Indigenously acquired cases of human rabies from bats in the United States and Canada, 1950–2010[a]

Type of case	Number of cases
Bite or scratch	23 (37.7 %)
Direct contact with no recognized bite	11 (18.0 %)
House exposure, but no direct contact	6 (9.8 %)
No history of bat contact	21 (34.4 %)
Total	61

Adapted from: Update on rabies by Jackson AC in Research and Reports in Tropical Medicine 2:31–43, 2011, with permission from Dove Medical Press Ltd

[a]Data from De Serres et al. (2008), Pue et al. (2009), Blanton et al. (2010), and Blanton et al. (2011)

often found to have rabies, but variants associated with these bats are only infrequently responsible for human rabies cases. The second most common rabies virus variant causing human rabies in the United States is found in Brazilian (Mexican) free-tail bats (*Tadarida braziliensis*). Vampire bats also transmit rabies virus to humans and cattle in Latin America. In recent years there have been outbreaks in Brazil and Peru in indigenous populations in the Amazonian region (Schneider et al. 2009). Other genotypes of lyssaviruses (rabies-related viruses) are endemic in bats in Europe, Africa, Asia, and Australia and very rarely cause clinical illness identical to rabies (Table 1). Many human cases have likely gone unrecognized, particularly in Africa, because of the presence of endemic canine rabies.

Terrestrial wildlife rabies vectors include raccoons, skunks, foxes, and coyotes (Fig. 1). Raccoon rabies was first identified in Florida in the 1940s and has gradually spread north and currently has a distribution involving the entire eastern coast of the United States. Raccoon rabies has had multiple incursions into Canada from the United States beginning in 1999 (Ontario), but has been very effectively controlled and eradicated by Canadian wildlife control operations. In 2012 there were 1,953 laboratory-confirmed cases of raccoon rabies by passive surveillance in the United States (Dyer et al. 2013). Only two cases of human rabies due to the raccoon rabies virus variant have been recognized to date (Silverstein et al. 2003; Vora et al. 2013). Skunk rabies occurs mainly in the midwestern United States, California, and in the prairie provinces of Canada. Human cases have not occurred from skunk virus variants in recent decades. Fox rabies has been well controlled in Europe, Ontario, and Texas with oral immunization methods, and is not currently an important problem (Rosatte 2013). Coyote rabies was also controlled in Texas with oral immunization. There are a variety of other terrestrial vectors of rabies, including mongooses, wolves, and jackals. Lagomorphs (e.g., rabbits and hares) and rodents are not considered rabies vectors and exposures from small rodents, including mice, rats, squirrels, chipmunks, gerbils, hamsters, and guinea pigs, have not been known to transmit infection to humans (Manning et al. 2008). Woodchucks are responsible for the majority of rabies in rodents that is reported in the United States (Childs et al. 1997). However, all mammals are considered potentially susceptible to rabies. Opossums are considered relatively resistant to rabies (Baer et al. 1990).

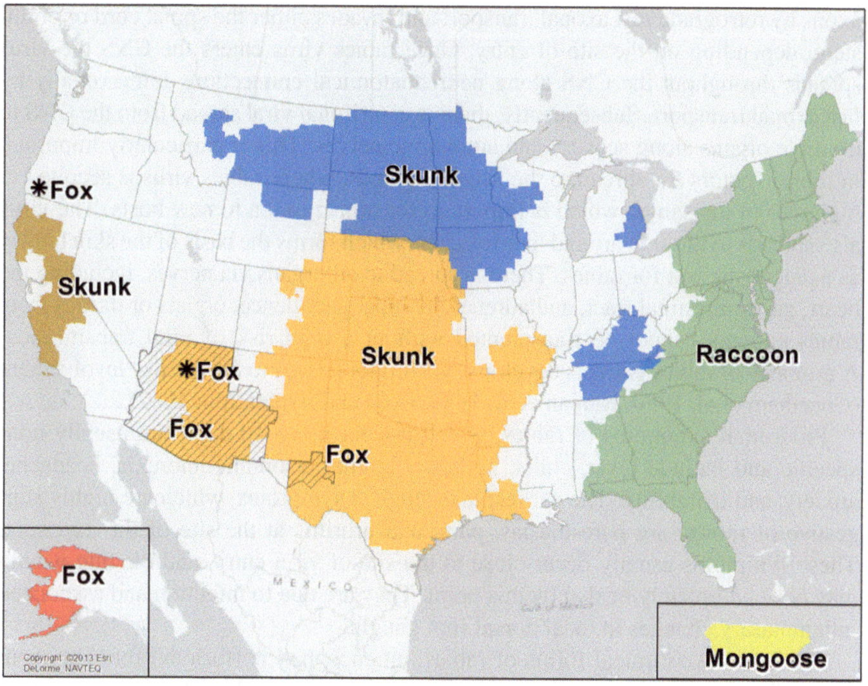

Fig. 1 Distribution of the major rabies virus variants among wild terrestrial reservoirs in the United States and Puerto Rico, 2008–2012. *Potential host shift event (Reproduced from JL Dyer et al., J Am Vet Med Assoc 243:805–815, 2013, Centers for Disease Control and Prevention)

3 Clinical Features of Human Disease

The vast majority of human rabies cases are associated with rabies virus transmission via an animal bite. A scratch contaminated with saliva or salivary contamination of an open wound may allow transmission, but this is much less common. Aerosol transmission in laboratory accidents and in caves containing millions of bats is a rare route of transmission. Finally, transmission via transplantation of tissues, including cornea (eight cases) and a vascular conduit (one case), or organs (seven cases) is well documented (Burton et al. 2005; Srinivasan et al. 2005; Maier et al. 2010; Jackson 2013; Vora et al. 2013), but rare.

In comparison to many other infectious diseases, the incubation period is relatively long in rabies, typically lasting 20–90 days. However, there are well documented cases with incubation periods of up to 6 years (Smith et al. 1991). Current evidence indicates that during the vast majority of the incubation period the virus remains close to the site of entry (inoculation). The nicotinic acetylcholine receptor, which is present on the post-synaptic membrane of the neuromuscular junction, is a rabies virus receptor and may serve to concentrate the virus and facilitate invasion into the peripheral nervous system (Lentz et al. 1982). Rabies virus spreads within axons by retrograde fast axonal transport and invades either the spinal cord or brainstem, depending on the site of entry. Once rabies virus enters the CNS the virus spreads throughout the CNS along neuroanatomical connections intraaxonally by fast axonal transport. Subsequently, there is centrifugal viral spread from the CNS to multiple organs along sensory and autonomic nerves. This is particularly important in rabies vectors for spread to the salivary glands, where rabies virus is secreted in high titer in the saliva, which is important for transmission to new hosts. The virus also spreads to the skin around hair follicles, which forms the basis of the skin biopsy as a diagnostic test for rabies. There is spread to all organs via nerves, including the heart, gastrointestinal tract, and adrenal medulla, and, hence, organs or tissues from rabies patients cannot be transplanted without a high risk of viral transmission. A minority of rabies patients develop myocarditis related to the cardiac involvement (Cheetham et al. 1970; Raman et al. 1988; Ross and Armentrout 1962).

Prodromal symptoms of rabies, which last for up to 10 days, are usually nonspecific and include fever, chills, malaise, fatigue, insomnia, anorexia, headache, anxiety, and irritability. The earliest neurological symptoms, which are highly suggestive of rabies, are paresthesias, pain, and pruritus at the site of the exposure. These symptoms usually occur close to the site of viral entry, and the bite wound may have completely healed by this point. They are due to infection and associated inflammatory changes in local dorsal root ganglia.

There are two clinical forms of rabies: an encephalitic (furious) form in about 80 % of patients and a paralytic (dumb) form in about 20 %. The main burden of infection in encephalitic rabies is likely in the brain, whereas it likely mainly involves the spinal cord, spinal nerve roots, and nerve plexuses in paralytic rabies. Fever occurs in both clinical forms. In encephalitic rabies, patients characteristically have episodes of generalized arousal or hyperexcitability, which are separated by lucid periods (Warrell 1976). The patients may demonstrate aggressive behavior, confusion, and hallucinations. Signs of autonomic dysfunction are common and include hypersalivation, piloerection (gooseflesh), sweating, priapism, and cardiac arrhythymias. About half of patients with encephalitic rabies develop hydrophobia, which is the most characteristic manifestation of rabies. Prior to developing hydrophobia, patients may experience pain in the throat or have difficulty swallowing. On attempts to swallow, patients with hydrophobia experience contractions of the diaphragm and other inspiratory muscles that usually last for about 5–15 s. Subsequently, the sight, sound, or even mention of water (or of any liquids) may trigger the spasms. A draft of air on the skin may have the same effect, and this is called aerophobia. Clinically the disease progresses with the subsequent development of paralysis and coma and virtually always results in a fatal outcome.

In paralytic rabies flaccid muscle weakness of the lower motor neuron variety develops early in the course of the disease and typically begins in the bitten extremity and then spreads to involve the other extremities and also involves the facial muscles. Sphincter involvement, pain, and sensory disturbances also occur. Bulbar and respiratory muscles eventually develop weakness in paralytic rabies, but the development of hydrophobia is unusual. Patients with paralytic rabies also progress to coma and death, and they typically survive longer than patients with encephalitic rabies.

Medical complications occur very frequently in rabies patients treated aggressively in critical care units. Cardiac and respiratory complications are particularly common. Cardiac problems, including sinus tachycardia, heart failure, hypotension, a variety of cardiac arrhythmias, and cardiac arrest (Hattwick 1974; Warrell et al. 1976), occur frequently in rabies. The cardiac manifestations probably reflect infection involving the autonomic nervous system (e.g., cardiac ganglia) and myocardium (Jackson et al. 1999). Respiratory complications include hyperventilation, hypoxemia, respiratory depression with apnea, atelectasis, and aspiration pneumonia (Hattwick 1974). Hypothalamic involvement in rabies may produce either hyperthermia or hypothermia. Endocrine complications include inappropriate secretion of antidiuretic hormone and diabetes insipidus (Bhatt et al. 1974; Hattwick 1974). Multiple organ failure commonly develops in patients who are treated aggressively in critical care units.

4 Diagnostic Investigations

Routine blood tests are usually normal in rabies. Computed tomography (CT) imaging is usually normal. Magnetic resonance imaging may show non-specific signal abnormalities in brain, spinal cord, nerve roots and/or plexuses (Laothamatas et al. 2011). Cerebrospinal fluid (CSF) analysis usually shows a mild mononuclear cell CSF pleocytosis with a cell count less than 100 cells per μL, although pleocytosis may be absent. In a previously unvaccinated patient neutralizing serum anti-rabies virus antibodies may develop, but this may not occur until illness is present for days to weeks. The presence of neutralizing anti-rabies virus antibodies in CSF is thought to be diagnostic of rabies, whereas these CSF antibodies are absent in patients vaccinated against rabies.

A laboratory diagnosis of rabies depends on the detection of rabies virus antigen or RNA in a body fluid or a tissue. Rabies virus can be isolated with viral cultures from CNS tissues, but only rarely from other clinical specimens such as saliva, CSF, or urine. Tissue sections from a full thickness skin biopsy (usually 5–6 mm in diameter) containing hair follicles (minimum of 10), which can be obtained from the posterior region of the neck at the hairline, should be evaluated for the presence of rabies virus antigen using the direct fluorescent antibody technique (Warrell et al. 1988). Corneal impression smears have relatively low sensitivity for the detection of rabies virus antigen (Warrell et al. 1988) and are now rarely used. The important recent advance in the laboratory diagnosis of rabies is an assay for the presence of rabies virus RNA in saliva using reverse transcription polymerase chain reaction (RT–PCR) amplification. RT–PCR can also be used on skin biopsy specimens (Dacheux et al. 2008) and for CSF, but the sensitivity in CSF is much lower. A negative test for the detection of rabies virus antigen or RNA (except on brain tissues) never excludes rabies. Repeat testing may be necessary to confirm a diagnosis of rabies. Biopsy specimens of brain tissues or post-mortem brain tissues can be evaluated for the detection of rabies virus antigen and for rabies virus isolation using culture techniques.

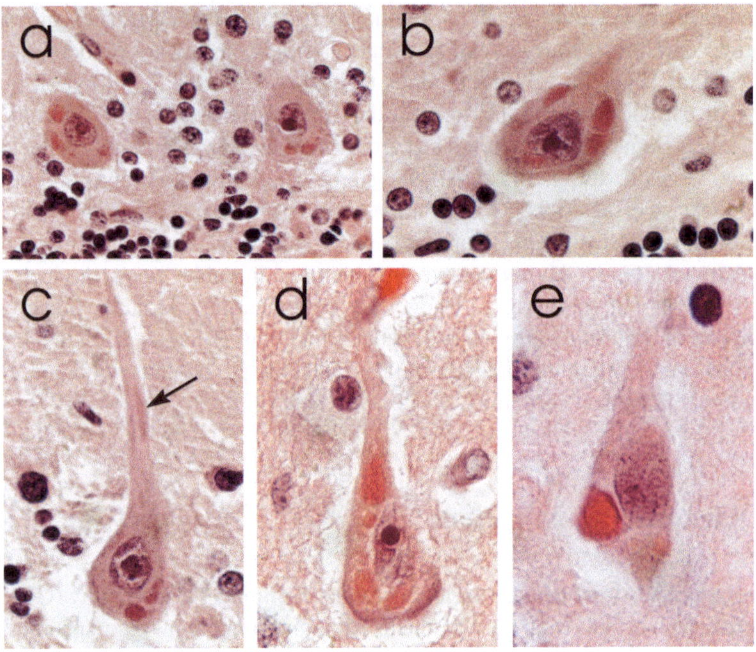

Fig. 2 Hematoxylin and eosin stained sections showing Negri bodies in the perikarya of (**a–c**) cerebellar Purkinje cells and (**d, e**) pyramidal neurons in the cerebral cortex of human rabies cases. The *arrow* in (**c**) indicates a Negri body in an apical dendrite. (Magnifications: A, ×280, B, ×445, C, ×535, D, ×710, E, ×840) (Adapted with permission from Rossiter JP and Jackson AC: Pathology, in Rabies: scientific basis of the disease and its management, Third Edition, edited by AC Jackson, 2013, Elsevier Academic Press, Oxford, UK, pp 351–386; Copyright Elsevier)

Pathologically, inflammatory changes are observed in the nervous system and characteristic eosinophilic inclusion bodies called Negri bodies (Fig. 2) are seen in the cytoplasm of neurons.

5 Human Rabies Due to Other Lyssaviruses

Rabies virus (genotype 1) is in the genus *Lyssavirus* that includes 10 other genotypes. Five of these 10 genotypes have been recognized to very rarely cause human disease in which the clinical and laboratory features are indistinguishable from rabies (Table 1). Cases due to other lyssaviruses have occurred in Africa (5, but one doubtful) and Europe (4) and less commonly in Australia (3) and Asia (1). Interestingly, no cases have been reported from the Americas.

6 Prognosis in Human Rabies

Rabies is virtually always fatal. Most surviving cases have received rabies vaccine prior to the onset of their illness (Table 3). One case did not receive rabies vaccine, but had the presence of neutralizing anti-rabies virus antibodies at the time of clinical presentation (Willoughby et al. 2005). A 17-year-old female from Texas (Holzmann-Pazgal et al. 2010) and an 8-year old female from California (Wiedeman et al. 2012) had atypical clinical courses, with the former patient not even requiring critical care, and both survived their illnesses. Rabies virus antibodies were detected

Table 3 Cases of human rabies with recovery[a]

Location	Year	Age of patient	Transmission	Immunization prior to onset	Outcome	Reference
United States	1970	6	Bat bite	Duck embryo vaccine	Complete recovery	Hattwick et al. (1972)
Argentina	1972	45	Dog bites	Suckling mouse brain vaccine	Mild sequelae	Porras et al. (1976)
United States	1977	32	Laboratory (vaccine strain)	Pre-exposure vaccination	Sequelae	Tillotson et al. (1977a, b)
Mexico	1992	9	Dog bites	Postexposure vaccination (combination)	Severe sequelae[b]	Alvarez et al. (1994)
India	2000	6	Dog bites	Postexposure vaccination (combination)	Severe sequelae[c]	Madhusudana et al. (2002)
United States	2004	15	Bat bite	No postexposure therapy	Mild sequelae	Hu et al. (2007); Willoughby et al. (2005)
Brazil	2008	15	Vampire bat bite	Postexposure vaccination	Severe sequelae	Ministerio da Saude in Brazil (2008)
Turkey	2008	17	Dog bites	Post-exposure vaccination (one dose)	Complete recovery	Karahocagil et al. (2013)
Chile	2013	25	Dog bite(s)	Postexposure vaccination (one dose)	Moderate sequelae	Galvez et al. (2013)

Adapted from Jackson AC: Therapy of human rabies, in Rabies: scientific basis of the disease and its management, Third Edition, edited by AC Jackson, 2013, Elsevier Academic Press, Oxford, UK, pp 573–587; Copyright Elsevier

[a]Recovery of cases with atypical features of rabies without the development of rabies virus neutralizing antibodies have not been not included because they are likely not cases of rabies (Blanton et al. 2011; Holzmann-Pazgal et al. 2010)

[b]Patient died less than 4 years after developing rabies with marked neurological sequelae (L. Alvarez, personal communication)

[c]Patient died about 2 years after developing rabies with marked neurological sequelae (S. Mahusudana, personal communication)

in these two patients, but they had only a low titer or absence of rabies virus neutralizing antibodies in sera and no detectable neutralizing anti-rabies virus antibodies in CSF and all other diagnostic tests for rabies were negative. The absence or minimal development of rabies virus neutralizing antibodies indicates that it is unlikely that either of these patients actually developed rabies and recovered. These two patients should not be considered rabies survivors.

7 Management of Human Rabies

All health care workers should initiate body substance precautions as soon a diagnosis of rabies is seriously considered and wear gowns, gloves, masks, and eye protection. They may also require post-exposure prophylaxis after high-risk contact with a patient with rabies.

Most survivors of rabies have received doses of rabies vaccine prior to the onset of their illnesses. An expert committee published a detailed document in 2003 on considerations for potential therapies for patients with rabies (Jackson et al. 2003). The best potential candidates for aggressive therapy were felt to be young and previously healthy patients with an early clinical diagnosis of rabies (Jackson et al. 2003). Therapies suggested for consideration include rabies vaccine, human rabies immune globulin, monoclonal antibodies (for the future), ribavirin, interferon-α, and ketamine. Previous animal work indicated that ketamine therapy might be useful (Lockhart et al. 1991). It was felt that combination therapies might be useful in situations in which specific therapies used individually had failed in the past; this approach has proved to be effective for both other infectious and non-infectious diseases,

In 2004 a 15-year-old female patient, who had bitten by a bat and not received rabies vaccine or other post-exposure therapy prior to the onset of clinical disease, survived rabies (Willoughby et al. 2005). After an incubation period of about a month she developed typical clinical features of rabies encephalitis. Five days after the onset of neurological symptoms she was transferred to a tertiary care hospital in Milwaukee, Wisconsin, and neutralizing anti-rabies virus antibodies were detected in both her sera and CSF. Nuchal skin biopsies were negative for rabies virus antigen. Rabies virus RNA was not detected in the skin biopsies or in saliva by RT - PCR, and viral isolation on saliva was negative. She was intubated, and put into a drug-induced coma, which included the non-competitive *N*-methyl-d-aspartate (NMDA) antagonist ketamine at 48 mg/kg/day as a continuous infusion and intravenous midazolam for 7 days. A burst-suppression pattern on her electroencephalogram was maintained with supplemental phenobarbital given as needed. Antiviral therapy included intravenous ribavirin and amantadine 200 mg per day administered enterally. She improved and was discharged from hospital with neurologic deficits and she subsequently demonstrated showed further neurologic improvement (Hu et al. 2007).

It is unknown if therapy with one or more specific agents played a significant role at all in her favorable outcome (Jackson 2005). Since that time there have been at least 26 cases in which the main components of this approach (the "Milwaukee Protocol") have been used with fatal outcomes (Table 4), and there have no documented survivors. The induction of coma per se has no established benefit for the management of infectious diseases of the nervous system, and to date there is no evidence supporting this approach in rabies or other viral encephalitides. There is no real justification for therapeutic coma becoming a routine therapy for the management of rabies. Excitotoxicity is not supported as an important basic mechanism in the pathogenesis of rabies and further studies on ketamine therapy both in vitro and in vivo have indicated a lack of efficacy (Weli et al. 2006). It is very likely that the Milwaukee patient would have recovered if she had received only supportive therapy.

Neutralizing anti-rabies virus antibodies are an important marker of an adaptive immune response that is essential for viral clearance (Lafon 2007) and recovery from rabies. The early presence of serum and CSF neutralizing anti-rabies virus antibodies was likely an important factor contributing to the favorable outcome of this patient. There have been seven survivors of rabies who received rabies vaccine prior to the onset of their disease and only one without vaccine (Table 3). This indicates that an early immune response is associated with a positive outcome. Therapy of encephalitic (furious) rabies with massive doses of human rabies immune globulin resulted in the rapid development of quadriplegia and bilateral facial paralysis (Hemachudha et al. 2003), indicating an immunopathological mechanism of neuronal injury. Bat rabies virus variants may be less neurovirulent than canine virus variants or other variants that are responsible for most human cases of rabies (Lafon 2005), and rabies due to canine rabies virus variants likely has a less favorable outcome than cases caused by bat rabies virus variants. Another patient infected with a bat rabies virus variant who received rabies vaccine prior to the onset of disease survived with a good neurological recovery (Hattwick et al. 1972). It is unknown if the causative bat rabies virus variant in the Milwaukee case was attenuated or had different biological properties than previously isolated variants because there was no viral isolation with this case. Most survivors of rabies to date have shown neutralizing anti-rabies virus antibodies in sera and CSF with other diagnostic laboratory tests negative for rabies virus antigen and RNA.

In 2007 a case of rabies in Canada was treated with the Milwaukee Protocol and after treatment he remained in a brain death-like state for about 4 weeks (McDermid et al. 2008; Reinke et al. 2013). At autopsy there was complete loss of neurons in the cerebral cortex and positive staining for rabies virus antigen was observed in both brainstem and cerebellar neurons (McDermid et al. 2008), indicating a failure of clearance of the viral infection from the brain and also failure of protection against neuronal injury and loss. In Germany lung and kidney/pancreas recipients from a rabies virus-infected donor developed rabies and were treated with major components of the Milwaukee Protocol, including intravenous midazolam, ketamine, and phenobarbital (in one) (Maier et al. 2010). One patient died within 2 days whereas the other survived 64 days after the onset of clinical rabies. At autopsy

Table 4 Cases of human rabies with treatment failures that used the main components of the "Milwaukee Protocol"

Case no.	Year of death	Age and sex of patient	Virus source	Country	Reference
1	2005	47 male	Kidney and pancreas transplant (dog)	Germany	Maier et al. (2010)
2	2005	46 female	Lung transplant (dog)	Germany	Maier et al. (2010)
3	2005	72 male	Kidney transplant (dog)	Germany	Maier et al. (2010)
4	2005	unknown	Dog	India	Bagchi (2005)
5	2005	7 male	Vampire bat	Brazil	[a]
6	2005	20–30 female	Vampire bat	Brazil	[a]
7	2006	33 male	Dog	Thailand	Hemachudha et al. (2006)
8	2006	16 male	Bat	USA (Texas)	Houston Chronicle (2006)
9	2006	10 female	Bat	USA (Indiana)	Christenson et al. (2007)
10	2006	11 male	Dog (Philippines)	USA (California)	Aramburo et al. (2011), Christenson et al. (2007)
11	2007	73 male	Bat	Canada (Alberta)	McDermid et al. (2008)
12	2007	55 male	Dog (Morocco)	Germany	Drosten (2007)
13	2007	34 female	Bat (Kenya)	The Netherlands	van Thiel et al. (2009)
14	2008	5 male	dog	Equatorial Guinea	Rubin et al. (2009)
15	2008	55 male	Bat	USA (Missouri)	Pue et al. (2009), Turabelidze et al. (2009)
16	2008	8 female	Cat	Colombia	Juncosa (2008)
17	2008	15 male	Vampire bat	Colombia	Badillo et al. (2009)
18	2009	37 female	Dog (South Africa)	Northern Ireland	Hunter et al. (2010)
19	2009	42 male	Dog (India)	USA (Virginia)	Blanton et al. (2010)
20	2010	11 female	Cat	Romania	Luminos et al. (2011)
21	2011	41 female	Dog (Guinea-Bissau)	Portugal	Santos et al. (2012)
22	2011	25 male	Dog (Afghanistan)	USA (Massachusetts)	Javaid et al. (2012)
23	2012	63 male	Brown bat	USA (Massachusetts)	Greer et al. (2013)
24	2012	9 male	Marmoset	Brazil	NE 10 (2012)
25	2012	41 male	Dog (Dominican Republic)	Canada (Ontario)	Branswell (2012)
26	2012	29 male	Dog (Mozambique)	South Africa	IAfrica.com (2012), Times Live (2012)

Updated from Jackson AC: Therapy in human rabies, in Research Advances in Rabies, Alan C. Jackson (ed), Advances in Virus Research 79:365–375, 2011; Copyright Elsevier

[a]Personal communication from Dr. Rita Medeiros, University of Para, Belem, Brazil

the two patients had 1.2–2.3 × 10^9 RNA copies/mg of CNS tissue, indicating ineffective viral clearance with the therapy. The longer surviving patient did show viral clearance from systemic organs and peripheral nerve. Viral clearance had also not occurred at the time of autopsy in a case from Belfast who was exposed in South Africa (Hunter et al. 2010). Hence, Milwaukee Protocol therapy has proved ineffective in promoting viral clearance from the CNS in rabies. It is very doubtful that the Milwaukee Protocol will prove to be useful in the therapy of human rabies. Unfortunately, promotion and repetition of this flawed approach has already impeded progress in the development of new and effective therapies for rabies. We need a better understanding of basic mechanisms underlying rabies pathogenesis in humans and animals in order to develop novel therapeutic approaches for the treatment of rabies.

8 Conclusions and Future Perspectives

A diagnosis of rabies must be considered clinically in order for the appropriate specimens to be collected and submitted for laboratory confirmation of the diagnosis. The absence of a history of an animal bite or other exposure never excludes a diagnosis of rabies and is relatively common in North America, where most indigenous rabies cases are caused by bat rabies virus variants.

Rabies in humans is virtually always a fatal disease, despite aggressive attempts at therapy. Most rabies survivors have received doses of rabies vaccine prior to the onset of clinical disease. A young survivor from Wisconsin developed neutralizing anti-rabies virus antibodies in serum and CSF relatively early in her clinical course. Her survival was likely due to excellent supportive critical care rather than due to any single component or combination of therapies that she received as part of the Milwaukee Protocol. The Milwaukee Protocol has failed in at least 26 cases without a well-documented subsequent survivor. Repetition of this approach has impeded progress in the rabies field by inhibiting the development of more novel therapies. More basic research is needed to provide a better understanding of exactly why patients die with rabies. Hopefully this information will prove to be helpful in designing new therapeutic approaches to the management of this dreaded disease.

References

Alvarez L, Fajardo R, Lopez E, Pedroza R, Hemachudha T, Kamolvarin N, Cortes G, Baer GM (1994) Partial recovery from rabies in a nine-year-old boy. Pediatr Infect Dis J 13:1154–1155

Anonymous (2013) Lyssavirus story breaks. Whitsunday Times. June 6

Aramburo A, Willoughby RE, Bollen AW, Glaser CA, Hsieh CJ, Davis SL, Martin KW, Roy-Burman A (2011) Failure of the Milwaukee protocol in a child with rabies. Clin Infect Dis 53:572–574. doi:10.1093/cid/cir483

Badillo R, Mantilla JC, Pradilla G (2009) Human rabies encephalitis by a vampire bat bite in an urban area of Colombia (Spanish). Biomédica 29:191–203

Baer GM, Shaddock JH, Quirion R, Dam TV, Lentz TL (1990) Rabies susceptibility and acetylcholine receptor (Letter). Lancet 335:664–665
Bagchi S (2005) Coma therapy. Calcutta, The Telegraph, 4 July 2005
Bhatt DR, Hattwick MAW, Gerdsen R, Emmons RW, Johnson HN (1974) Human rabies: diagnosis, complications, and management. Am J Dis Child 127:862–869
Blanton JD, Palmer D, Rupprecht CE (2010) Rabies surveillance in the United States during 2009. J Am Vet Med Assoc 237:646–657. doi:10.2460/javma.237.6.646
Blanton JD, Palmer D, Dyer J, Rupprecht CE (2011) Rabies surveillance in the United States during 2010. J Am Vet Med Assoc 239:773–783. doi:10.2460/javma.239.6.773
Botvinkin AD, Selnikova OP, Antonova LA, Moiseeva AB, Nesterenko EY, Gromashevsky LV (2005) Human rabies case caused from a bat bite in Ukraine. Rabies Bull Europe 29:5–7
Branswell H (2012) Testing suggests Toronto rabies case infected in Dominican Republic. Toronto, The Globe and Mail, 19 April 2012
Burton EC, Burns DK, Opatowsky MJ, El-Feky WH, Fischbach B, Melton L, Sanchez E, Randall H, Watkins DL, Chang J, Klintmalm G (2005) Rabies encephalomyelitis: clinical, neuroradiological, and pathological findings in 4 transplant recipients. Arch Neurol 62:873–882. doi:10.1001/archneur.62.6.873
Cheetham HD, Hart J, Coghill NF, Fox B (1970) Rabies with myocarditis: two cases in England. Lancet 1:921–922
Childs JE, Colby L, Krebs JW, Strine T, Feller M, Noah D, Drenzek C, Smith JS, Rupprecht CE (1997) Surveillance and spatiotemporal associations of rabies in rodents and lagomorphs in the United States, 1985–1994. J Wildl Dis 33:20–27
Christenson JC, Holm BM, Lechlitner S, Howell JF, Wenger M, Roy-Burman A, Hsieh CJ, LaBar L, Petru A, Davis S, Simons N, Apolinario P, Furst K, Benjamin B, Christie LJ, Schnurr D, Glaser CA, Rosenberg J, Sun B, Willoughby RE, Rupprecht CE (2007) Human rabies–Indiana and California, 2006. MMWR 56:361–365
Dacheux L, Reynes JM, Buchy P, Sivuth O, Diop BM, Rousset D, Rathat C, Jolly N, Dufourcq JB, Nareth C, Diop S, Iehle C, Rajerison R, Sadorge C, Bourhy H (2008) A reliable diagnosis of human rabies based on analysis of skin biopsy specimens. Clin Infect Dis 47:1410–1417. doi:10.1086/592969
De Serres G, Dallaire F, Cote M, Skowronski DM (2008) Bat rabies in the United States and Canada from 1950 through 2007: human cases with and without bat contact. Clin Infect Dis 46:1329–1337
Drosten C (2007) Rabies–Germany (Hamburg) ex Morocco. ProMED-mail 20070419.1287. Available at http://www.promedmail.org/ Accessed 5 Nov 2013
Dyer JL, Wallace R, Orciari L, Hightower D, Yager P, Blanton JD (2013) Rabies surveillance in the United States during 2012. J Am Vet Med Assoc 243:805–815
Familusi JB, Moore DL (1972) Isolation of a rabies related virus from the cerebrospinal fluid of a child with "aseptic meningitis." Afr J Med Sci 3:93–96
Familusi JB, Osunkoya BO, Moore DL, Kemp GE, Fabiyi A (1972) A fatal human infection with Mokola virus. Am J Trop Med Hyg 21:959–963
Galvez S, Basque M, Contreras L, Merino C, Ahumada R, Jamett J, Stefan M, Yung V, Favi M, Fernandez J, Mora J, Velasco-Villa A, Willoughby R, Pidal P (2013) Survivor of rabies encephalitis in Chile. Platform presentation at the XXIVth International Meeting on Research Advances and Rabies Control in the Americas in Toronto, Ontario, Canada on 27 Oct 2013
Greer DM, Robbins GK, Lijewski V, Gonzales RG, McGuone D (2013) Case records of the Massachusetts General Hospital: Case 1-2013: a 63-year-old man with paresthesias and difficulty swallowing. N Engl J Med 368:172–180
Hanna JN, Carney IK, Smith GA, Tannenberg AEG, Deverill JE, Botha JA, Serafin IL, Harrower BJ, Fitzpatrick PF, Searle JW (2000) Australian bat lyssavius infection: a second human case, with a long incubation period. Med J Aust 172:597–599
Hattwick MAW (1974) Human rabies. Public Health Rev 3:229–274
Hattwick MAW, Weis TT, Stechschulte CJ, Baer GM, Gregg MB (1972) Recovery from rabies: a case report. Ann Intern Med 76:931–942

Hemachudha T, Sunsaneewitayakul B, Desudchit T, Suankratay C, Sittipunt C, Wacharapluesadee S, Khawplod P, Wilde H, Jackson AC (2006) Failure of therapeutic coma and ketamine for therapy of human rabies. J Neurovirol 12:407–409. doi:10.1080/13550280600902295

Hemachudha T, Sunsaneewitayakul B, Mitrabhakdi E, Suankratay C, Laothamathas J, Wacharapluesadee S, Khawplod P, Wilde H (2003) Paralytic complications following intravenous rabies immune globulin treatment in a patient with furious rabies (Letter). International Journal of Infectious Diseases 7:76–77

Holzmann-Pazgal G, Wanger A, Degaffe G, Rose C, Heresi G, Amaya R, Eshofonie A, Lee-Han H, Awosika-Olumo A, Kuzmin I, Rupprecht CE (2010) Presumptive abortive human rabies—Texas, 2009. MMWR 59:185–190

Houston Chronicle (2006) Rabies, human—USA (Texas). ProMED-mail 20060513.1360. Available at http://www.promedmail.org/ Accessed 5 Nov 2013

Hu WT, Willoughby RE Jr, Dhonau H, Mack KJ (2007) Long-term follow-up after treatment of rabies by induction of coma (Letter). N Engl J Med 357:945–946

Hu R, Tang Q, Tang J, Fooks AR (2009) Rabies in China: an update. Vector-Borne Zoonotic Dis 9:1–12

Hunter M, Johnson N, Hedderwick S, McCaughey C, Lowry K, McConville J, Herron B, McQuaid S, Marston D, Goddard T, Harkess G, Goharriz H, Voller K, Solomon T, Willoughby RE, Fooks AR (2010) Immunovirological correlates in human rabies treated with therapeutic coma. J Med Virol 82:1255–1265. doi:10.1002/jmv.21785

IAfrica.com (2012) Rabies—South Africa (03): (Kwazulu-Natal), human ex Mozambique. ProMED-mail 20120608.1160328. Available at http://www.promedmail.org/ Accessed 5 Nov 2013

Jackson AC (2005) Recovery from rabies (Editorial). N Engl J Med 352:2549–2550. doi:10.1056/NEJMe058092

Jackson AC (2013) Human disease. In: Jackson, AC (ed) Rabies: scientific basis of the disease and its management, Third Edition. pp. 269–298. Elsevier Academic Press, Oxford, UK

Jackson AC, Ye H, Phelan CC, Ridaura-Sanz C, Zheng Q, Li Z, Wan X, Lopez-Corella E (1999) Extraneural organ involvement in human rabies. Lab Invest 79:945–951

Jackson AC, Warrell MJ, Rupprecht CE, Ertl HCJ, Dietzschold B, O'Reilly M, Leach RP, Fu ZF, Wunner WH, Bleck TP, Wilde H (2003) Management of rabies in humans. Clin Infect Dis 36:60–63. doi:10.1086/344905

Javaid W, Amzuta IG, Nat A, Johnson T, Grant D, Rudd RJ, Cherry B, Newman A, Blog D, Willoughby R, Fix C, Baty S, Cooper M, Ellis K, Alves DA, Cersovsky S, Rupprecht CE, Maxted AM, Lankau EW (2012) Imported human rabies in a U.S. Army soldier - New York, 2011. MMWR Morb Mortal Wkly Rep 61:302–305

Johnson N, Brookes SM, Healy DM, Spencer Y, Hicks D, Nunez A, Wells G, Fooks AR (2012) Pathology associated with a human case of rabies in the United Kingdom caused by European bat lyssavirus type-2. Intervirology 55(5):391–394

Juncosa B (2008) Hope for rabies victims: unorthodox coma therapy shows promise. First a U.S. girl—and now two South American kids survive onset of the deadly virus. Available at http://www.scientificamerican.com/ Accessed 5 Nov 2013

Karahocagil MK, Akdeniz H, Aylan O, Sünnetçioglu M, Ün H, Yapici K, Baran AI (2013) Complete recovery from clinical rabies: case report. Turkiye Klinikleri J Med Sci 33:547–552

Knobel DL, Kaare M, Fèvre E, Cleaveland S (2007) Dog rabies and its control. In: Jackson AC, Wunner WH (eds) Rabies, 2nd edn. Elsevier Academic, London, pp 573–594

Lafon M (2005) Bat rabies—the Achilles heel of a viral killer? Lancet 366:876–877

Lafon M (2007) Immunology. In: Jackson AC, Wunner WH (eds) Rabies, 2nd edn. Elsevier Academic, London, pp 489–504

Laothamatas J, Sungkarat W, Hemachudha T (2011) Neuroimaging in rabies. Adv Virus Res 79:309–327. doi:10.1016/B978-0-12-387040-7.00014-7

Lembo T (2012) The blueprint for rabies prevention and control: a novel operational toolkit for rabies elimination. PLoS Negl Trop Dis 6:e1388

Lentz TL, Burrage TG, Smith AL, Crick J, Tignor GH (1982) Is the acetylcholine receptor a rabies virus receptor? Science 215:182–184

Leonova GN, Belikov SI, Kondratov IG, Krylova NV, Pavlenko EV, Romanova EV, Chentsova IV, Petukhova SA (2009) A fatal case of bat lyssavirus infection in Primorye Territory of the Russian Far East. Rabies Bull Europe 33:5–8

Lockhart BP, Tsiang H, Ceccaldi PE, Guillemer S (1991) Ketamine-mediated inhibition of rabies virus infection *in vitro* and in rat brain. Antivir Chem Chemother 2:9–15

Luminos M, Barboi G, Draganescu A, Streinu Cercel A, Staniceanu F, Jugulete G, Visan A, Negulescu C, Turcitu MA (2011) Human rabies in a Romanian boy - an ante mortem case study. Rabies Bull Europe 35:5–10

Madhusudana SN, Nagaraj D, Uday M, Ratnavalli E, Kumar MV (2002) Partial recovery from rabies in a six-year-old girl (Letter). Int J Infect Dis 6:85–86

Maier T, Schwarting A, Mauer D, Ross RS, Martens A, Kliem V, Wahl J, Panning M, Baumgarte S, Muller T, Pfefferle S, Ebel H, Schmidt J, Tenner-Racz K, Racz P, Schmid M, Struber M, Wolters B, Gotthardt D, Bitz F, Frisch L, Pfeiffer N, Fickenscher H, Sauer P, Rupprecht CE, Roggendorf M, Haverich A, Galle P, Hoyer J, Drosten C (2010) Management and outcomes after multiple corneal and solid organ transplantations from a donor infected with rabies virus. Clin Infect Dis 50:1112–1119. doi:10.1086/651267

Manning SE, Rupprecht CE, Fishbein D, Hanlon CA, Lumlertdacha B, Guerra M, Meltzer MI, Dhankhar P, Vaidya SA, Jenkins SR, Sun B, Hull HF (2008) Human rabies prevention—United States, 2008: recommendations of the Advisory Committee on Immunization Practices. MMWR Recomm Rep 57((RR-3)):1–28

McDermid RC, Saxinger L, Lee B, Johnstone J, Noel Gibney RT, Johnson M, Bagshaw SM (2008) Human rabies encephalitis following bat exposure: failure of therapeutic coma. Can Med Assoc J 178:557–561. doi:10.1503/cmaj.071326

Meredith CD, Rossouw AP, HvP K (1971) An unusual case of human rabies thought to be of chiropteran origin. S Afr Med J 45:767–769

Ministerio da Saude in Brazil (2008) Rabies, human survival, bat—Brazil: (Pernambuco). ProMED-mail 20081114.3599. Available at http://www.promedmail.org/ Accessed 5 Nov 2013

Nathwani D, McIntyre PG, White K, Shearer AJ, Reynolds N, Walker D, Orange GV, Fooks AR (2003) Fatal human rabies caused by European bat lyssavirus type 2a infection in Scotland. Clin Infect Dis 37:598–601

NE 10 (2012) Rabies, human - Brazil (03): (Ceara). ProMED-mail 20120314.1070531. Available at http://www.promedmail.org/ Accessed 5 Nov 2013

Paweska JT, Blumberg LH, Liebenberg C, Hewlett RH, Grobbelaar AA, Leman PA, Croft JE, Nel LH, Nutt L, Swanepoel R (2006) Fatal human infection with rabies-related Duvenhage virus, South Africa. Emerg Infect Dis 12:1965–1967

Porras C, Barboza JJ, Fuenzalida E, Adaros HL, Oviedo AM, Furst J (1976) Recovery from rabies in man. Ann Intern Med 85:44–48

Pue HL, Turabelidze G, Patrick S, Grim A, Bell C, Reese V, Basilan R, Rupprecht C, Robertson K (2009) Human rabies - Missouri, 2008. MMWR 58:1207–1209

Raman GV, Prosser A, Spreadbury PL, Cockcroft PM, Okubadejo OA (1988) Rabies presenting with myocarditis and encephalitis. J Infect 17:155–158

Reinke SN, Resch L, Maingut F, Branton W, Jackson AC, Holt R, Slupsky C, Marrie T, Sykes BD, Power C (2013) Metagenomic and metabolomic characterization of rabies encephalitis: new insights into the treatment of an ancient disease. J Inf Dis 207:1451–1456

Roine RO, Hillbom M, Valle M, Haltia M, Ketonen L, Neuvonen E, Lumio J, Lahdevirta J (1988) Fatal encephalitis caused by a bat-borne rabies-related virus: clinical findings. Brain 111:1505–1516

Rosatte RC (2013) Rabies control in wild carnivores. In: Jackson AC (ed) Rabies: scientific basis of the disease and its management, 3rd edn. Elsevier Academic, Oxford, pp 617–670

Ross E, Armentrout SA (1962) Myocarditis associated with rabies: report of a case. N Engl J Med 266:1087–1089

Rubin J, David D, Willoughby RE Jr, Rupprecht CE, Garcia C, Guarda DC, Zohar Z, Stamler A (2009) Applying the Milwaukee rotocol to treat canine rabies in Equatorial Guinea. Scand J Infect Dis 41:372–375. doi:10.1080/00365540902798333

Samaratunga H, Searle JW, Hudson N (1998) Non-rabies lyssavirus human encephalitis from fruit bats: Australian bat lyssavirus (pteropid lyssavirus) infection. Neuropathol Appl Neurobiol 24:331–335

Santos A, Cale E, Dacheux L, Bourhy H, Gouveia J, Vasconcelos P (2012) Fatal case of imported human rabies in Amadora, Portugal, August 2011. Eurosurveillance 17:20130

Schneider MC, Romijn PC, Uieda W, Tamayo H, da Silva DF, Belotto A, da Silva JB, Leanes LF (2009) Rabies transmitted by vampire bats to humans: an emerging zoonotic disease in Latin America? Revista Panamericana de Salud Pública 25:260–269

Selimov MA, Tatarov AG, Botvinkin AD, Klueva EV, Kulikova LG, Khismatullina NA (1989) Rabies-related Yuli virus; Identification with a panel of monoclonal antibodies. Acta Virol 33:542–546

Silverstein MA, Salgado CD, Bassin S, Bleck TP, Lopes MB, Farr BM, Jenkins SR, Sockwell DC, Marr JS, Miller GB (2003) First human death associated with raccoon rabies—Virginia, 2003. MMWR 52:1102–1103

Smith JS, Fishbein DB, Rupprecht CE, Clark K (1991) Unexplained rabies in three immigrants in the United States: a virologic investigation. N Engl J Med 324:205–211

Srinivasan A, Burton EC, Kuehnert MJ, Rupprecht C, Sutker WL, Ksiazek TG, Paddock CD, Guarner J, Shieh WJ, Goldsmith C, Hanlon CA, Zoretic J, Fischbach B, Niezgoda M, El-Feky WH, Orciari L, Sanchez EQ, Likos A, Klintmalm GB, Cardo D, LeDuc J, Chamberland ME, Jernigan DB, Zaki SR (2005) Transmission of rabies virus from an organ donor to four transplant recipients. N Engl J Med 352:1103–1111. doi:10.1056/NEJMoa043018

Tillotson JR, Axelrod D, Lyman DO (1977a) Follow-up on rabies—New York. MMWR 26:249–250

Tillotson JR, Axelrod D, Lyman DO (1977b) Rabies in a laboratory worker—New York. MMWR 26:183–184

Times Live (2012) Rabies—South Africa (02): (Kwazulu-Natal), human ex Mozambique. ProMED-mail 20120528.1147931. Available at http://www.promedmail.org/ Accessed 5 Nov 2013

Turabelidze G, Pue H, Grim A, Patrick S (2009) First human rabies case in Missouri in 50 years causes death in outdoorsman. Mo Med 106:417–419

van Thiel PP, de Bie RM, Eftimov F, Tepaske R, Zaaijer HL, van Doornum GJ, Schutten M, Osterhaus AD, Majoie CB, Aronica E, Fehlner-Gardiner C, Wandeler AI, Kager PA (2009) Fatal human rabies due to Duvenhage virus from a bat in Kenya: failure of treatment with coma-induction, ketamine, and antiviral drugs. PLoS Negl Trop Dis 3:e428. doi:10.1371/journal.pntd.0000428

Vora NM, Basavaraju SV, Feldman KA, Paddock CD, Orciari L, Gitterman S, Griese S, Wallace RM, Said M, Blau DM, Selvaggi G, Velasco-Villa A, Ritter J, Yager P, Kresch A, Niezgoda M, Blanton J, Stosor V, Falta EM, Lyon GM III, Zembower T, Kuzmina N, Rohatgi PK, Recuenco S, Zaki S, Damon I, Franka R, Kuehnert MJ (2013) Raccoon rabies virus variant transmission through solid organ transplantation. JAMA 310:398–407

Warrell DA (1976) The clinical picture of rabies in man. Trans R Soc Trop Med Hyg 70:188–195

Warrell DA, Davidson NM, Pope HM, Bailie WE, Lawrie JH, Ormerod LD, Kertesz A, Lewis P (1976) Pathophysiologic studies in human rabies. Am J Med 60:180–190

Warrell MJ, Looareesuwan S, Manatsathit S, White NJ, Phuapradit P, Vejjajiva A, Hoke CH, Burke DS, Warrell DA (1988) Rapid diagnosis of rabies and post-vaccinal encephalitides. Clin Exp Immunol 71:229–234

Weli SC, Scott CA, Ward CA, Jackson AC (2006) Rabies virus infection of primary neuronal cultures and adult mice: failure to demonstrate evidence of excitotoxicity. J Virol 80:10270–10273. doi:10.1128/JVI.01272-06

Wiedeman J, Plant J, Glaser C, Messenger S, Wadford D, Sheriff H, Fritz C, Lindsay A, McKenzie M, Hammond C, Gordon E, Rupprecht CE, Petersen BW (2012) Recovery of a patient from clinical rabies—California, 2011. MMWR 61:61–65

Willoughby RE Jr, Tieves KS, Hoffman GM, Ghanayem NS, Amlie-Lefond CM, Schwabe MJ, Chusid MJ, Rupprecht CE (2005) Survival after treatment of rabies with induction of coma. N Engl J Med 352:2508–2514

Rabies: Neurobiology

Monique Lafon

Abstract Rabies virus is a neurotropic virus that replicates and propagates into the nervous system of the infected host. Successful achievement of the virus cycle from the site of entry (usually due to a bite) up to the site of exit (salivary glands) relies on the preservation of the neuronal network. Once the rabies virus has entered the nervous system, its progression is not interrupted by the host defence mechanisms. This virus has evolved sophisticated strategies to (1) disarm premature destruction of the infected neurons and prolong the life span of the infected neurons, (2) evade the innate immune response launched by the infected neurons, and (3) eliminate the protective T cells migrating into the nervous system. In addition, by targeting the nervous system that has the striking capacity to centrally control the immune response, the rabies virus infection benefits also from disarmed host defences. The successful adaptation of the virus to the mammalian nervous system may explain why rabies is fatal in almost all the cases.

Keywords Rabies virus • Neglected diseases • Neuroinflammatin • Innate immune response • IFN • PD-L1 • B7-H1 • Evasive strategies • T cells • Apoptosis

1 Introduction

Rabies is fatal encephalitis caused by the infection of the nervous system (NS) with rabies virus (RABV). This neurotropic rhabdovirus is transmitted through bites or scratches of inflected, mainly, dogs but also bats and other wild animals whose saliva contains viral particles (see Jackson 2014). After entry at the neuromuscular

M. Lafon (✉)
Unité de Neuroimmunologie Virale, Département de Virologie,
Institut Pasteur, 25, rue du Dr Roux, 75015 Paris, France
e-mail: Monique.lafon@pasteur.fr

M. Bentivoglio et al. (eds.), *Neglected Tropical Diseases and Conditions of the Nervous System*,
DOI 10.1007/978-1-4614-8100-3_14, © Springer Science+Business Media New York 2014

junctions or sensory muscle spindles through its interaction with different cellular receptors (Lafon 2005b), RABV particles propagate towards the cell body of neuron by retrograde transport using axonal vesicles (Klingen et al. 2008). Virus replication occurs in the cell bodies and dendrites (Ugolini 1995, 2010) from which newly formed viral particles are released to infect the next order neuron. RABV infects neurons almost exclusively and travels from one neuron to the next in the spinal cord to the brainstem, from where it reaches the salivary glands by centrifugal passage via cranial nerves. Once in the salivary glands, RABV is excreted in saliva and can be then transmitted to a new host (Dierks et al. 1969). Transmission is facilitated when RABV infection triggers a furious behaviour of the infected host resulting in a high occurrence of bites.

Viruses are obligatory parasites. Completion of virus cycle and successful transmission to a new host relies upon the evolution of strategies that allow the virus to hijack the cellular machinery, modulate host cell survival and circumvent host lines of defence. During RABV journey in the body, the virus faces the host immune defences at different steps. First, immediately after inoculation of RABV particles into the skin or muscles, the presence of viral elements is rapidly detected by the innate immune response which contributes to both eliminate microbes locally and to set up a specific immune response (B and T cells) in the periphery. Second, after entry of virus particles into nerves, the virus infection has to cope again with the innate immune response launched at that time by the infected neuron (Prehaud et al. 2005). So RABV infection can also be eliminated by the premature death of the infected neurons which can interrupt the neuronal network required for RABV progression into the NS (Thoulouze et al. 1997). The local innate immune response launched by the infected neurons can result in the production of antiviral, chemoattractive and inflammatory molecules that can trigger the infiltration of immune cells (monocytes, T and B cells) from the periphery (Vuaillat et al. 2008; Hooper et al. 2009). These cells have the ability to eliminate the infected neurons (Galelli et al. 2000). Finally, RABV will encounter the immune response once again at the end of its cycle when it is directed into the extra neural organs (including salivary glands) by centrifugal passage (Fig. 1).

However, despite these robust lines of defence, once a virulent RABV has entered the NS its progression is not interrupted. This might be the result of the intrinsic capacity of this virus to evade the innate immune response launched by the infected neurons (Rieder and Conzelmann 2009) to preserve the integrity of the infected neurons (Prehaud et al. 2010) but also to eliminate the protective T cells migrating into the NS (Lafon et al. 2008). In addition, RABV may also benefit from the intrinsic capacity of NS to control the immunological homeostasis of the body resulting in a down-regulation of the immune responsiveness in the periphery (Camelo et al. 2001) (Fig. 2).

The knowledge we have gained of the interactions of RABV with the immune responses has been obtained mainly from models of experimental rabies in mice using laboratory-adapted RABV strains after intramuscular or intraplantar

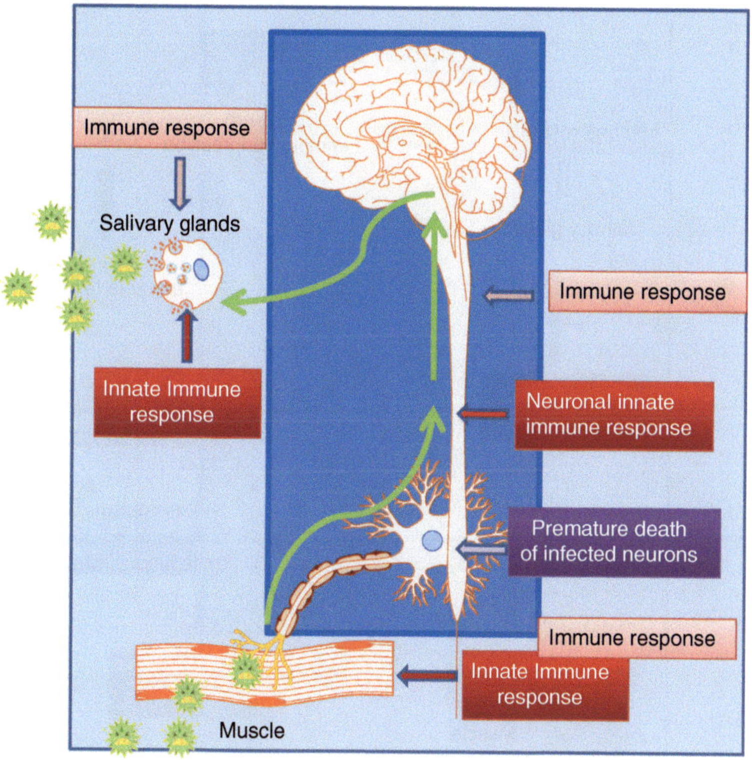

Fig. 1 Rabies virus (RABV) tricky travel. Propagation of RABV particles (*green dots*) from the site of entry (muscle) up to the brainstem and the exit site (salivary glands) is represented by the *green arrows*. RABV is a strictly neurotropic virus causing no viremia. In the periphery, innate (type 1 IFN) and adaptive (T, B neutralizing antibodies) immune responses are triggered by the detection of virus particles in the muscle. After entry into the nerve, RABV infection has to face two host's traps: the premature death of the infected neuron and the launch of an innate immune response by neurons and surrounding glial cells. This local innate immune response results in the production in the nervous system (NS) of antiviral, chemoattractive and inflammatory molecules which facilitates the infiltration of immune cells (monocytes, T and B cells) from the periphery. These cells have the potency to eliminate the infected neurons. Finally, RABV encounters the immune responses once again at the end of its cycle when it is directed into the salivary glands by centrifugal passage. The *dark blue frame* represents the immunological privileged sites of the NS. The *light blue frame* represents the extraneural peripheral tissues and organs (including the immune system in the periphery)

(footpad) route of administration to mimic natural transmission by a bite. Comparison of the features of pathogenic strains causing fatal encephalitis in mice and attenuated strains responsible of abortive non-fatal disease has allowed elucidating of which parameters control RABV pathogenicity.

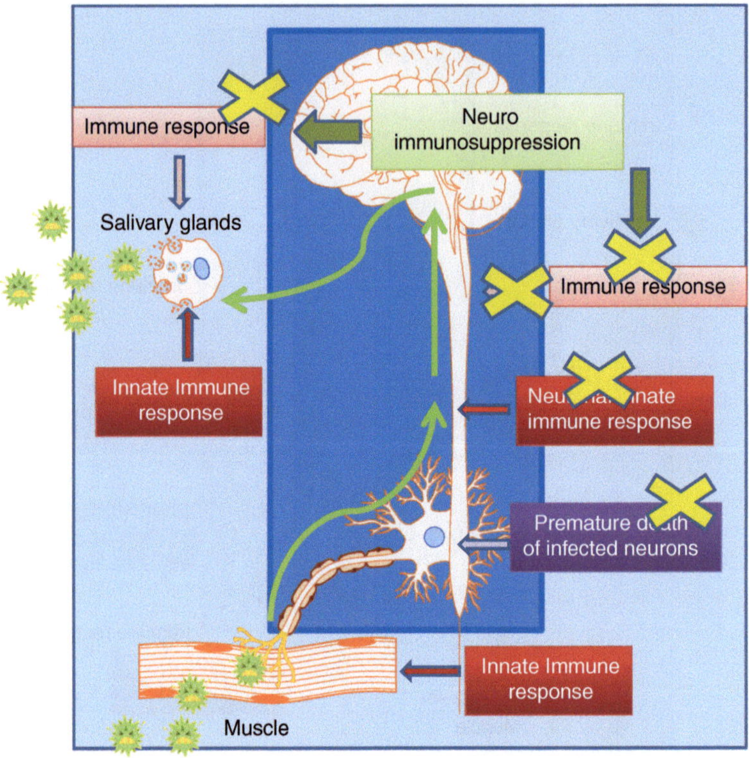

Fig. 2 Rabies virus (RABV) spoils host's plans to dampen the virus progression. RABV has evolved strategies to escape the host defenses and complete its cycle in the infected host. These host defenses can exert their protective functions at the site of entry and possibly also at the final stage of infection, in the salivary glands. However, they might be down-regulated by central neuro-immuno-suppression

2 RABV Preserves the Integrity of the Infected Neurons

The capacity to kill or injure infected cells is a widespread property of viruses. Nevertheless, a few viruses such as RABV are able to grow without inflicting major damage to the host cells. Analysis of autopsied rabies-infected patients indicates that despite heavy viral load, neuronal apoptosis is a rare event in natural rabies (Jackson et al. 2008). Experimentally, in a model of RABV infection of non-human primates 4 days after infection, infected motor neurons show no signs of degeneration with normal size, morphology and Nissl staining (Ugolini 2010). *In vitro*, rat spinal motoneurons never encounter death (Guigoni and Coulon 2002). Nevertheless, later in the infection, when the brain is heavily infected, degenerative changes are seen in dendrites and axons, and peripheral nerve dysfunction occurs (Juntrakul et al. 2005; Li et al. 2005; Scott et al. 2008; Kojima et al. 2009; Rossiter et al. 2009; Jackson et al. 2010).

Thus, it can be stated that successful RABV propagation in the NS requires that neuronal cell bodies are not damaged by premature apoptosis and that the integrity of axons and dendrites is preserved, at least during the period of time required for the virus to reach the brain.

Only attenuated RABV strains used for vaccines (such as the Evelyn Rotniki Abelseth (ERA) strain) or RABV related viruses of bat origin (such as Lagos bat virus) trigger apoptosis (Thoulouze et al. 1997; Kassis et al. 2004). Induction of apoptosis may contribute to the high immunogenicity of live rabies vaccines (Megret et al. 2005). M protein (Lagos bat virus) and the cytoplasmic domain of the G protein (in the case of the attenuated RABV ERA) have been found to control the commitment of infected cells toward death (Larrous et al. 2010; Prehaud et al. 2010; Babault et al. 2011).

Virulent RABV strains not only preserve the integrity of the infected neurons but also actively promote neuron survival (Prehaud et al. 2010). The capacity of RABV to promote neuronal survival or death depends upon the cellular partners recruited by the PDZ-binding site (PDZ-BS) of its envelope G protein. Neuronal survival requires selective association of the PDZ-BS with the PDZ domain of two closely related serine-threonine kinases (MAST1 and MAST2). However, a single amino acid change in the PDZ-BS that triggers neuronal death, allows G protein to recruit additional PDZ partners, notably a tyrosine-phosphatase, PTPN4. Silencing this phosphatase abrogates RABV-mediated apoptosis. PDZs are globular domains that play a central role in cell signaling by favoring spatial contacts between enzymes and their substrates. Disruption of the complexes formed by PDZ and the PDZ-BS encoded by their ligands may have profound effect on cellular signaling pathways—MAST2 functioning as an inhibitor of neuronal survival and PTPN4 as an inhibitor of cell death. It is likely that RABV G proteins antagonize PTPN4 and MAST2 functions by simply disrupting the interaction of PTPN4-PDZ and MAST2-PDZ with their respective cellular ligands (Prehaud et al. 2010; Babault et al. 2011; Terrien et al. 2012).

3 RABV Innate Immune Response

3.1 Innate Immune Response in the Periphery

The innate immune response is the first line of defence against infectious agents. It involves the release of type 1 interferons (IFN-α/β), inflammatory cytokines and chemokines, the activation of complement and the attraction of macrophages, neutrophils and NK cells into infected tissues. This innate immune response is triggered in the first hours following the entry of pathogens and is not pathogen specific. It contrasts with the adaptive immune response that is tailored to a specific pathogen and requires several days to develop. The innate immune system can sense the presence of micro-organisms through "pattern recognition receptors" (PRRs) that recognize danger signals and "pathogen associated molecular patterns" (PAMPs) expressed by microbes. Toll like receptors (TLRs) or retinoic acid inducible gene

(RIG) like receptors (RLRs) are important PRRs for the recognition of viral double-stranded RNAs and single-stranded RNAs (Liu et al. 2008; Hornung et al. 2006). Resulting signal transduction cascades involve TRIF, Myd88 or IPS-1 as adaptors of TLR3, or TLRs other than TLR3 and of RLRs, respectively.

Once RABV is inoculated in the skin or in muscle by bites or scratches, the presence of the virus is rapidly detected by host defence mechanisms. The IFN response triggered at the site of entry has an antiviral effect. This was demonstrated by comparing the viral load in the thigh muscle of two groups of mice, parental mice and mice lacking the type I IFN receptor (IFNAR) after injection in the hindlimb with the encephalitic RABV strain, CVS. The viral load measured by the accumulation of viral RNA was increased in mice lacking IFNAR compared to parental strain of mice (Chopy et al. 2011a). This observation suggests that some viral particles might be readily eliminated at this early step of infection.

3.2 Innate Immune Response in the NS

Like most tissues in the organism, the NS expresses different types of PPRs capable of sensing dangerous and pathogen signals (Boivin et al. 2002; Nguyen et al. 2002; Bottcher et al. 2003; Koedel et al. 2004; McKimmie et al. 2005). Central neurons express TLR1-4 as well as TLR 7 and 8 (Prehaud et al. 2005; Ma et al. 2006; Kim et al. 2007; Ma et al. 2007; Tang et al. 2007; Barajon et al. 2009). These express the RLRs (RIG-I and Mda-5) (Lafon et al. 2006; Menager et al. 2009; Peltier et al. 2010; Chopy et al. 2011b), but not the RLR LGP2, which seems to be actively degraded in neurons (Chopy et al. 2011b). Peripheral nerve plexuses and nerves (dorsal root ganglion sensory neurons and fibers of sciatic nerves) express TLRs (TLR3, 4 and 7) with prominent expression of TLR3 (Cameron et al. 2007; Barajon et al. 2009; Goethals et al. 2010). Neurons take an active part in the innate immune response in the brain being both responders to IFN and IFN producers, secreting type I IFN (predominantly IFN-β in the brain, no IFN-α and no Type III IFN λ) (Prehaud et al. 2005; Delhaye et al. 2006; Sommereyns et al. 2008).

RABV is known to trigger a RIG-I-mediated innate immune response in neurons (Hornung et al. 2006) by detecting the 5′tri-phosphate base pairing of the viral genome (Pichlmair et al. 2006). After infection, human neurons can mount a classical primary IFN response (activation of IRF3 and NF-kappa B), as well as a secondary IFN response, (activation of STATs and IRF7), leading to the production of cytokines (IL-6, TNF-alpha) and chemokines (CXCL10 and CCL5) (Prehaud et al. 2005; Chopy et al. 2011a, b).

3.2.1 RABV Dampens the IFN Response in the Infected Neurons

Viruses have evolved sophisticated strategies to escape the innate immune response (Randall and Goodbourn 2008; Versteeg and Garcia-Sastre 2010). This is the case for RABV (for review see Rieder and Conzelmann 2009). The N and the P protein of

RABV are multifunctional proteins involved in RNA synthesis and in counteracting the host innate immune response. The N protein limits RIG-I-signaling (Masatani et al. 2010, 2011), whereas the P protein inhibits IRF3 and IRF7 phosphorylation (Brzozka et al. 2005; Rieder et al. 2011), suppresses STAT1 nuclear translocation (Brzozka et al. 2006; Vidy et al. 2007) and sequesters an antiviral protein, the promyelocytic leukemia (PML) protein, in the cytoplasm (Blondel et al. 2010). As a result, down-regulation of the IFN response can be observed *in vitro*. For example, in RABV-infected human post-mitotic neurons (NT2-N), transcription of *IFN-β* gene is seen as early as 6h post infection, and IFN-β protein is produced during the first 24 h post infection, whereas, transcription and production decline thereafter (Prehaud et al. 2005).

It has been shown that virulence, at least for a Japanese vaccine strain (Nishigahara RABV strain), depends upon the capacity of this strain to evade the innate immune response and this process is controlled by the ability of the P and N protein to evade the innate immune response (Shimizu et al. 2006; Masatani et al. 2010). Thus, evasion of the IFN response in infected neurons may be critical for RABV progression in the NS through the neuronal network allowing the virus to reach the brainstem and the salivary glands.

3.2.2 RABV Limits the Inflammatory Response in the NS

Inflammation is a key component of host responses to cell damage or microbial entry leading to the production of inflammatory mediators including complement, adhesion molecules, cyclo-oxygenase enzymes and their products, as well as cytokines or chemokines. Release of these toxic factors has dramatic consequences when the site of inflammation is the NS, where severe dysfunction can lead to significant NS pathology with neuronal death (Brown and Neher 2010). In the brain, both neurons and glial cells can mount antiviral, inflammatory and chemokine responses. Astrocytes can respond to the presence of innate immune stimulus in the brain by producing proinflammatory cytokines and chemokines (Park et al. 2006). Nevertheless, an important role is taken by microglia in the induction of neuroinflammation, a feature, which may reflect the density or the sub cellular localization of the innate immune receptors (Bsibsi et al. 2006).

Transcriptome and proteomic analysis of the inflammatory response triggered in the NS of mice by various virulent strains of RABV showed that RABV infection stimulates the expression of chemokines (CCL5, CCL2, CCL9 and CXCL9) and inflammatory cytokines (IL-6, IL-12) (Camelo et al. 2000; Baloul et al. 2004; Wang et al. 2005; Chopy et al. 2011b; Sugiura et al. 2011). When compared with other encephalitic virus infections such as caused by Borna virus, RABV triggers only limited inflammation (Shankar et al. 1992; Fu et al. 1993). Moreover, the inflammatory reaction in the RABV-infected NS is transient with the expression of a majority of markers being rapidly down-regulated in the spinal cord and with a slight delay in the brain (Chopy et al. 2011b).

Comparison of the inflammatory reaction triggered by RABV strains of various degree of pathogenicity indicates that the more pathogenic strains trigger weaker inflammatory responses (Baloul and Lafon 2003; Wang et al. 2005; Laothamatas et al. 2008; Hicks et al. 2009).

It is likely that this low inflammatory reaction in the infected NS contributes to keeping intact the BBB, a condition that correlates with RABV pathogenicity, with non-pathogenic RABV strains triggering a transient opening of the BBB, but not pathogenic strains (Phares et al. 2006; Roy et al. 2007).

Limitation of neuroinflammation by virulent RABV occurs by several mechanisms. RABV infection avoids neuronal apoptosis (Lafon 2011) and rarely infects glial cells; two intrinsic features of the infection contributing to limit neuroinflammation. In addition, in the course of RABV infection the expression of anti-inflammatory molecules is up-regulated in the NS. This is the case for the anti-inflammatory soluble proteins TNFR1 and 2, which can interfere with the binding of TNF to its receptors (Chopy et al. 2011b). This is also the case for the suppressors of cytokine signaling (SOCS), a family of proteins that negatively control cytokine signal transduction, with SOCS-1 being up-regulated in the brain of RABV infected dogs in non infected cells in close vicinity of infected neurons (Nuovo et al. 2005). More importantly, RABV up-regulates the expression of HLA-G, a non-classical MHC molecule and B7-H1, the ligand of PD-1, (programmed death protein-1), in neurons, and also in the infected NS, in the case of B7-H1 (Lafon et al. 2005, 2008). Besides, their immune-tolerant properties, which are exploited by RABV, (see below Sect. 4), HLAG and B7-H1 molecules are now also considered as providing negative feedback that limits tissue inflammation (Carosella et al. 2001; Phares et al. 2010). This, in particular, is the case for B7-H1, which dampens the expression of pro-inflammatory molecules (such as iNos and TNF-α) during viral encephalitis (Phares et al. 2010), whereas, HLA-G influences the cytokine balance towards a Th2 pattern by promoting the secretion of IL-4, IL-3 and IL-10 and down-regulating the production of IFN-γ and TNF-α (Carosella et al. 2001).

Limitation of inflammation by RABV infection might result from (1) reducing the entry in the NS of mononuclear leukocytes, monocytes and macrophages, (2) maintaining the impermeability of the BBB, and (3) minimizing the release of neurotoxic molecules that can compromise NS function and host survival.

These conditions should preserve not only the integrity of the infected neuronal network, but also the life of the host, allowing the virus to reach the brainstem and the salivary glands before the premature death of the infected host.

4 RABV Adaptive Immune Response

Mounting an adaptive immune response against a microbe, even a neurotropic virus that rapidly enters the NS after its inoculation in muscle, always occurs in the periphery and never in the NS, which is devoid of lymphoid organs (Galea et al. 2007). The triggering of the adaptive immune response that takes place in the lymphoid organs such as the lymph nodes or spleen relies on the activation of plasmacytoid dendritic cells (DCs), pDCs and of type 1 IFN that they produce in a TLR7 and 9 dependent manner after encountering the microbe (Diebold et al. 2003; Steinman 1991). In natural rabies infection, the nature of cells producing type I IFN at the site of the injection is not known with certainty. Candidates are muscle cells, fibroblasts, keratinocytes, dendritic cells (DCs) and macrophages (Charlton and Casey 1979, 1981). There is experimental evidence that RABV can infect bone marrow derived conventional DCs (cDCs) and macrophages *in vitro*. Despite non-productive infection, RABV triggers the production of IFNs, cytokines and chemokines in these cells (Nakamichi et al. 2004; Faul et al. 2010). Infection of sentinel cells may be dispensable since macrophages activation was observed *in vitro* when inactivated RABV was added to the culture (Nakamichi et al. 2004). In cell culture, maturation of cDCs in the presence of RABV is controlled by type 1 IFN whose production might rely on the recognition of intra cytoplasmic RABV RNAs through RIG-I and mda-5 receptors and not TLR7 (Faul et al. 2010), a characteristic of cDCs (Eisenacher et al. 2007).

4.1 RABV Specific Immune Response in the Periphery

After the injection of the CVS strain of rabies in the hind limb of mice, the size of the draining popliteal lymph nodes and those of spleen increase. Draining lymph nodes are populated with activated T cells expressing the marker of activation CD69. Activation can also be observed among peripheral blood lymphocytes (Vuaillat et al. 2008).

A strong B cell response is mounted in the spleen. When mice were injected with a less pathogenic virus (the PV strain), similar activation of T cells was observed in lymph nodes and blood suggesting that adaptive immune response is independent of the virulence of the RABV strain. Indeed, when mice were injected with an encephalitic RABV bat strain (silver-haired bat rabies virus, SHBRV) or with a less pathogenic virus (CVS-F3, mutant of CVS encoding a mutation in the G protein), the resulting adaptive immune responses (neutralizing antibodies, $CD4^+$, $CD8^+$ T cells response) were not different (Roy and Hooper 2007). This suggests that adaptive immune response triggered by RABV strains in the periphery, an event that occurs late after the virus has entered the NS, is unrelated to RABV pathogenicity. This may explain why patients die of rabies despite having mounted an immune response in the periphery attested by the presence of neutralizing antibodies in the blood (Hunter et al. 2010). Nevertheless, in an experimental model of rabies in skunks, a cyclophosphamide-induced immunosuppression was found to reduce the infection of the salivary glands (Charlton et al. 1984), suggesting that the adaptive immune response may control to some extent the final stage of RABV infection. This control may also be exerted on viral particles at the site of entry if entry into the NS has been delayed.

4.2 RABV Provokes the Killing of Migratory T Cells

Infiltrating T cells control most of the infections of the NS. This is, for example, observed during the course of West Nile virus brain infection, where $CD8^+$ T cells attracted by the chemokines produced by inflammatory cells in the infected NS are a critical factor for controlling the infection (Klein et al. 2005; Zhang et al. 2008). In rabies, sterilisation of the infection by T cells is inefficient, and their activities are specifically inactivated by the virus (Lafon 2008). Immunohistochemical studies performed on autopsy materials obtained from deceased rabies-infected patients revealed that the cells undergoing death were leukocytes and not neurons (Hemachudha et al. 2005; Tobiume et al. 2009). This observation was reproduced in mice infected with the encephalitic RABV strain CVS. Immunocytochemistry of brain and spinal cord sections revealed that despite a heavy load of viral antigens, infected neurons do not undergo apoptosis. In contrast, the migrating T cells (CD3+) were apoptotic (Baloul and Lafon 2003; Baloul et al. 2004; Lafon 2005a; Kojima et al. 2009; Rossiter et al. 2009). Moreover, pathogenicity of the CVS strain was similar in immunocompetent mice Balb/c mice and in Nu/Nu Balb/c mice, indicating that T cells do not control the outcome of encephalitic rabies (Lafon 2005a). In striking contrast, deprivation of T cells transformed an abortive infection into a encephalitic rabies similar to that caused by the encephalitic strain CVS infection, showing that T cells is a critical factor in the restriction of the NS infection caused by an abortive RABV strain. Indeed, when apoptosis was analyzed in the spinal cord of immunocompetent mice infected with the abortive RABV strain PV, killing of T cells was not observed; instead, infected neurons died (Galelli et al. 2000). Altogether, these observations indicate that T cells have a protective potential to control RABV infection in the NS, nevertheless their capacity to control RABV infection is impeded with the encephalitic RABV strain. The mechanisms by which the encephalitic RABV strain evades the host T cell response was further studied as described below.

Tumors evade immune surveillance by multiple mechanisms, including the inhibition of tumour-specific T cell immunity. In order to escape attack from protective T cells, tumor cells up-regulate expression of certain surface molecules such as B7-H1, Fas-L, and HLA-G, which triggers death signalling in activated T cells expressing the corresponding ligands PD-1 for B7-H1, Fas for FasL and CD8—among others—for HLA-G (Gratas et al. 1998; Dong et al. 2002; Rouas-Freiss et al. 2003). Studies evaluating whether RABV-infected neurons up-regulate immunosubversive molecules to kill activated T cells following an evasive strategy similar to that selected by tumors cells have been undertaken both *in vivo* and *in vitro. In vitro*, RABV infection was found to up-regulate the expression of HLA-G at the surface of human neurons (Lafon et al. 2005; Megret et al. 2007). *In vivo*, comparison of experimental rabies in mice caused by CVS, which kills T cells, or by PV, which does not kill T cells, led to the finding that the CVS-infected NS, but not the PV-infected NS, up-regulates the expression of FasL. In mice lacking a functional FasL, there was less T cell apoptosis in the NS than in control mice.

Remarkably, RABV morbidity and mortality were reduced in these mice. Destruction of T cells through the Fas/FasL pathway can be enhanced by indoleamine 2, 3 dioxygenase (IDO), the expression of which in the infected neurons and the brain is up-regulated by RABV (Prehaud et al. 2005; Zhao et al. 2011). The enzyme IDO converts extracellular tryptophan into kynurenine, thereby reducing its concentration in the microenvironment that in turn markedly enhances the sensitivity of any nearby T cell for Fas-ligand induced apoptosis (Kwidzinski et al. 2003).

In addition, RABV-infected brain up-regulates the expression of another immunosubversive molecule, B7-H1 (Lafon et al. 2008). Whereas non-infected NS was almost devoid of B7-H1 expression, RABV infection triggers neural B7-H1 expression that increases as the infection progresses. Infected neurons and also other non-infected neuroectodermal cells, including astrocyte-like cells, were found positive for B7-H1. RABV infection of B7-H1 deficient mice resulted in a drastic reduction in clinical signs and mortality. Reduction of RABV virulence in $B7H1^{-/-}$ mice was concomitant of a reduction of $CD8^+$ T cell apoptosis among the migratory T cells.

Thus, despite the triggering of a classical adaptive immune response in the periphery and the infiltration of the lymphocytes into the infected NS, the protection, which could have been conferred in the NS by this immune response, is drastically impeded by RABV infection.

5 RABV Infection Triggers a CNS Mediated Immune-Unresponsiveness

The dampening of immune protection already triggered by RABV is completed by a central immunosupression caused by the neuronal reflex control of immunity triggered by the NS facing an excess of inflammation in its attempt to restore general homeostasis.

RABV infection by a pathogenic strain induces an immune-unresponsiveness (Wiktor et al. 1977a, b; Torres-Anjel et al. 1988; Perry et al. 1990; Hirai et al. 1992; Tshikuka et al. 1992; Kasempimolporn et al. 1997, 2001; Camelo et al. 2001) characterized by the impairment of T cells functions with an alteration of cytokine pattern, an inhibition of T cells proliferation and the destruction of immune cells without modification of immune cells proportion (CD4/CD8 ratio constant) in the lymphoid organs (Perry et al. 1990). This leads to the atrophy of the spleen and the thymus of RABV infected mammals. TNF-α receptor has been found to play a role in RABV immune-unresponsiveness, since immune cells lacking the TNF-α p55 receptor were less immunosuppressed compared to the wild-type (Camelo et al. 2000). Importantly, infection of the brain is required since immune-unresponsivenesss does not occur after the infection of the NS with an abortive RABV strain, which infects the spinal cord only (Camelo et al. 2001). This suggests that the property of the NS that centrally controls the immune response in the periphery might be triggered (Tracey 2009). NS modulates the immune functions through two main immune-neuroendocrine pathways: the hypothalamo-pituitary-adrenal (HPA) axis and the autonomic NS (ANS) composed of sympathetic and parasympathetic nerves fibres (Johnston and Webster 2009). The homeostatic reflex is activated after the brain senses the presence of an excess of inflammatory cytokines such as TNF-alpha, IL-1β or IL-6 in the periphery, by neuronal (mainly through local afferent fibres of the vagus nerve) and by humoral pathway (Johnston and Webster 2009) processed by the NS in hypothalamic and brainstem centres.

This general immune-unresponsiveness controlled by the NS may be advantageous for RABV propagation since a mouse strain showing a less efficient HPA axis is less susceptible to rabies (Roy and Hooper 2007). This centrally mediated immunosupression may limit peripheral control of infection in the muscle or the salivary glands (see Fig. 1).

Thus, RABV infection not only actively inhibits the T cell response and inflammation in the NS by up-regulating B7-H1 and FasL molecules, but also benefits from the intrinsic capacity of NS to trigger a central immunosuppression in order to maintain whole body homeostasis.

6 Conclusion and Future Perspectives

RABV has selected a battery of mechanisms to escape the host immunosurveillance, an explanation for why, in the absence of post-exposure treatment, rabies is one of the very few human infections with a near 100 % mortality rate. Despite the well-adapted viral strategies to escape the immune response, RABV infections can be limited if vaccine is injected promptly after exposure suggesting that the viral-mediated paralysis of the host immune response requires some time, which can be exploited for therapy. However, the efficacy of rabies post exposure treatment requires public education, prompt wound cleansing, vaccination and availability of rabies immunoglobulins. As half of the victims are children, pre-exposure vaccination of young individuals and eradication of RABV from its main reservoirs, the dogs, should be considered in an attempt to improve the global health of mankind.

In addition, improved knowledge of the immune evasive mechanisms evolved by RABV to infect the NS, may help identify new therapeutic targets such as the central neural immune reflex or neuroinflammation.

References

Babault N et al (2011) Peptides targeting the PDZ domain of PTPN4 are efficient inducers of glioblastoma cell death. Structure 19:1518–1524

Baloul L, Lafon M (2003) Apoptosis and rabies virus neuroinvasion. Biochimie 85:777–788

Baloul L, Camelo S, Lafon M (2004) Up-regulation of Fas ligand (FasL) in the central nervous system: a mechanism of immune evasion by rabies virus. J Neurovirol 10:372–382

Barajon I et al (2009) Toll-like receptors 3, 4, and 7 are expressed in the enteric nervous system and dorsal root ganglia. J Histochem Cytochem 57:1013–1023
Blondel D, Kheddache S, Lahaye X, Dianoux L, Chelbi-Alix MK (2010) Resistance to rabies virus infection conferred by the PMLIV isoform. J Virol 84:10719–10726
Boivin G, Coulombe Z, Rivest S (2002) Intranasal herpes simplex virus type 2 inoculation causes a profound thymidine kinase dependent cerebral inflammatory response in the mouse hindbrain. Eur J Neurosci 16:29–43
Bottcher T et al (2003) Differential regulation of Toll-like receptor mRNAs in experimental murine central nervous system infections. Neurosci Lett 344:17–20
Brown GC, Neher JJ (2010) Inflammatory neurodegeneration and mechanisms of microglial killing of neurons. Mol Neurobiol 41:242–247
Brzozka K, Finke S, Conzelmann KK (2005) Identification of the rabies virus alpha/beta interferon antagonist: phosphoprotein P interferes with phosphorylation of interferon regulatory factor 3. J Virol 79:7673–7681
Brzozka K, Finke S, Conzelmann KK (2006) Inhibition of interferon signaling by rabies virus phosphoprotein P: activation-dependent binding of STAT1 and STAT2. J Virol 80:2675–2683
Bsibsi M, Persoon-Deen C, Verwer RW, Meeuwsen S, Ravid R, Van Noort JM (2006) Toll-like receptor 3 on adult human astrocytes triggers production of neuroprotective mediators. Glia 53:688–695
Camelo S, Lafage M, Lafon M (2000) Absence of the p55 Kd TNF-alpha receptor promotes survival in rabies virus acute encephalitis. J Neurovirol 6:507–518
Camelo S, Lafage M, Galelli A, Lafon M (2001) Selective role for the p55 Kd TNF-alpha receptor in immune unresponsiveness induced by an acute viral encephalitis. J Neuroimmunol 113:95–108
Cameron JS et al (2007) Toll-like receptor 3 is a potent negative regulator of axonal growth in mammals. J Neurosci 27:13033–13041
Carosella ED, Moreau P, Aractingi S, Rouas-Freiss N (2001) HLA-G: a shield against inflammatory aggression. Trends Immunol 22:553–555
Charlton KM, Casey GA (1979) Experimental rabies in skunks: immunofluorescence light and electron microscopic studies. Lab Invest 41:36–44
Charlton KM, Casey GA (1981) Experimental rabies in skunks: persistence of virus in denervated muscle at the inoculation site. Can J Comp Med 45:357–362
Charlton KM, Casey GA, Campbell JB (1984) Experimental rabies in skunks: effects of immunosuppression induced by cyclophosphamide. Can J Comp Med 48:72–77
Chopy D, Detje CN, Lafage M, Kalinke U, Lafon M (2011a) The type I interferon response bridles rabies virus infection and reduces pathogenicity. J Neurovirol 17:353–367
Chopy D et al (2011b) Ambivalent role of the innate immune response in rabies virus pathogenesis. J Virol 85:6657–6668
Delhaye S et al (2006) Neurons produce type I interferon during viral encephalitis. Proc Natl Acad Sci USA 103:7835–7840
Diebold SS et al (2003) Viral infection switches non-plasmacytoid dendritic cells into high interferon producers. Nature 424:324–328
Dierks RE, Murphy FA, Harrison AK (1969) Extraneural rabies virus infection. Virus development in fox salivary gland. Am J Pathol 54:251–273
Dong H et al (2002) Tumor-associated B7-H1 promotes T-cell apoptosis: a potential mechanism of immune evasion. Nat Med 8:793–800
Eisenacher K, Steinberg C, Reindl W, Krug A (2007) The role of viral nucleic acid recognition in dendritic cells for innate and adaptive antiviral immunity. Immunobiology 212:701–714
Faul EJ, Wanjalla CN, Suthar MS, Gale M, Wirblich C, Schnell MJ (2010) Rabies virus infection induces type I interferon production in an IPS-1 dependent manner while dendritic cell activation relies on IFNAR signaling. PLoS Pathog 6:e1001016
Fu ZF et al (1993) Differential effects of rabies and borna disease viruses on immediate-early- and late-response gene expression in brain tissues. J Virol 67:6674–6681

Galea I, Bechmann I, Perry VH (2007) What is immune privilege (not)? Trends Immunol 28:12–18
Galelli A, Baloul L, Lafon M (2000) Abortive rabies virus central nervous infection is controlled by T lymphocyte local recruitment and induction of apoptosis. J Neurovirol 6:359–372
Goethals S, Ydens E, Timmerman V, Janssens S (2010) Toll-like receptor expression in the peripheral nerve. Glia 58:1701–1709
Gratas C, Tohma Y, Barnas C, Taniere P, Hainaut P, Ohgaki H (1998) Up-regulation of Fas (APO-1/CD95) ligand and down-regulation of Fas expression in human esophageal cancer. Cancer Res 58:2057–2062
Guigoni C, Coulon P (2002) Rabies virus is not cytolytic for rat spinal motoneurons in vitro. J Neurovirol 8:306–317
Hemachudha T, Wacharapluesadee S, Mitrabhakdi E, Wilde H, Morimoto K, Lewis RA (2005) Pathophysiology of human paralytic rabies. J Neurovirol 11:93–100
Hicks DJ, Nunez A, Healy DM, Brookes SM, Johnson N, Fooks AR (2009) Comparative pathological study of the murine brain after experimental infection with classical rabies virus and European bat lyssaviruses. J Comp Pathol 140:113–126
Hirai K et al (1992) Suppression of cell-mediated immunity by street rabies virus infection. Microbiol Immunol 36:1277–1290
Hooper DC, Phares TW, Fabis MJ, Roy A (2009) The production of antibody by invading B cells is required for the clearance of rabies virus from the central nervous system. PLoS Negl Trop Dis 3:e535
Hornung V et al (2006) 5′-Triphosphate RNA is the ligand for RIG-I. Science 314:994–997
Hunter M et al (2010) Immunovirological correlates in human rabies treated with therapeutic coma. J Med Virol 82:1255–1265
Jackson AC (2014) Rabies: neurology. In: Bentivoglio M, Cavalheiro EA, Kristensson K, Patel N (eds) Neglected tropical diseases and conditions of the nervous system. Springer, New York
Jackson AC, Randle E, Lawrance G, Rossiter JP (2008) Neuronal apoptosis does not play an important role in human rabies encephalitis. J Neurovirol 14:368–375
Jackson AC, Kammouni W, Zherebitskaya E, Fernyhough P (2010) Role of oxidative stress in rabies virus infection of adult mouse dorsal root ganglion neurons. J Virol 84:4697–4705
Johnston GR, Webster NR (2009) Cytokines and the immunomodulatory function of the vagus nerve. Br J Anaesth 102:453–462
Juntrakul S, Ruangvejvorachai P, Shuangshoti S, Wacharapluesadee S, Hemachudha T (2005) Mechanisms of escape phenomenon of spinal cord and brainstem in human rabies. BMC Infect Dis 5:104
Kasempimolporn S, Saengseesom W, Mitmoonpitak C, Akesowan S, Sitprija V (1997) Cell-mediated immunosuppression in mice by street rabies virus not restored by calcium ionophore or PMA. Asian Pac J Allergy Immunol 15:127–132
Kasempimolporn S, Tirawatnapong T, Saengseesom W, Nookhai S, Sitprija V (2001) Immunosuppression in rabies virus infection mediated by lymphocyte apoptosis. Jpn J Infect Dis 54:144–147
Kassis R, Larrous F, Estaquier J, Bourhy H (2004) Lyssavirus matrix protein induces apoptosis by a TRAIL-dependent mechanism involving caspase-8 activation. J Virol 78:6543–6555
Kim D et al (2007) A critical role of toll-like receptor 2 in nerve injury-induced spinal cord glial cell activation and pain hypersensitivity. J Biol Chem 282:14975–14983
Klein RS et al (2005) Neuronal CXCL10 directs CD8+ T-cell recruitment and control of West Nile virus encephalitis. J Virol 79:11457–11466
Klingen Y, Conzelmann KK, Finke S (2008) Double-labeled rabies virus: live tracking of enveloped virus transport. J Virol 82:237–245
Koedel U et al (2004) MyD88 is required for mounting a robust host immune response to Streptococcus pneumoniae in the CNS. Brain 127:1437–1445
Kojima D, Park CH, Satoh Y, Inoue S, Noguchi A, Oyamada T (2009) Pathology of the spinal cord of C57BL/6J mice infected with rabies virus (CVS-11 strain). J Vet Med Sci 71:319–324

Kwidzinski E et al (2003) IDO (indolamine 2,3-dioxygenase) expression and function in the CNS. Adv Exp Med Biol 527:113–118
Lafon M (2005a) Modulation of the immune response in the nervous system by rabies virus. Curr Top Microbiol Immunol 289:239–258
Lafon M (2005b) Rabies virus receptors. J Neurovirol 11:82–87
Lafon M (2008) Immune evasion, a critical strategy for rabies virus. Dev Biol (Basel) 131:413–419
Lafon M (2011) Evasive strategies in rabies virus infection. Adv Virus Res 79:33–53
Lafon M et al (2005) Modulation of HLA-G expression in human neural cells after neurotropic viral infections. J Virol 79:15226–15237
Lafon M, Megret F, Lafage M, Prehaud C (2006) The innate immune facet of brain: human neurons express TLR-3 and sense viral dsRNA. J Mol Neurosci 29:185–194
Lafon M et al (2008) Detrimental contribution of the immuno-inhibitor b7-h1 to rabies virus encephalitis. J Immunol 180:7506–7515
Laothamatas J et al (2008) Furious and paralytic rabies of canine origin: neuroimaging with virological and cytokine studies. J Neurovirol 14:119–129
Larrous F, Gholami A, Mouhamad S, Estaquier J, Bourhy H (2010) Two overlapping domains of a lyssavirus matrix protein that acts on different cell death pathways. J Virol 84:9897–9906
Li XQ, Sarmento L, Fu ZF (2005) Degeneration of neuronal processes after infection with pathogenic, but not attenuated, rabies viruses. J Virol 79:10063–10068
Liu L et al (2008) Structural basis of toll-like receptor 3 signaling with double-stranded RNA. Science 320:379–381
Ma Y et al (2006) Toll-like receptor 8 functions as a negative regulator of neurite outgrowth and inducer of neuronal apoptosis. J Cell Biol 175:209–215
Ma Y, Haynes RL, Sidman RL, Vartanian T (2007) TLR8: an innate immune receptor in brain, neurons and axons. Cell Cycle 6:2859–2868
Masatani T et al (2010) Rabies virus nucleoprotein functions to evade activation of the RIG-I-mediated antiviral response. J Virol 84:4002–4012
Masatani T et al (2011) Amino acids at positions 273 and 394 in rabies virus nucleoprotein are important for both evasion of host RIG-I-mediated antiviral response and pathogenicity. Virus Res 155(1):168–174
McKimmie CS, Johnson N, Fooks AR, Fazakerley JK (2005) Viruses selectively upregulate Toll-like receptors in the central nervous system. Biochem Biophys Res Commun 336:925–933
Megret F et al (2005) Immunopotentiation of the antibody response against influenza HA with apoptotic bodies generated by rabies virus G-ERA protein-driven apoptosis. Vaccine 23:5342–5350
Megret F et al (2007) Modulation of HLA-G and HLA-E expression in human neuronal cells after rabies virus or herpes virus simplex type 1 infections. Hum Immunol 68:294–302
Menager P et al (2009) Toll-like receptor 3 (TLR3) plays a major role in the formation of rabies virus Negri Bodies. PLoS Pathog 5:e1000315
Nakamichi K, Inoue S, Takasaki T, Morimoto K, Kurane I (2004) Rabies virus stimulates nitric oxide production and CXC chemokine ligand 10 expression in macrophages through activation of extracellular signal-regulated kinases 1 and 2. J Virol 78:9376–9388
Nguyen MD, Julien JP, Rivest S (2002) Innate immunity: the missing link in neuroprotection and neurodegeneration? Nat Rev Neurosci 3:216–227
Nuovo GJ, DeFaria DL, Chanona-Vilchi JG, Zhang Y (2005) Molecular detection of rabies encephalitis and correlation with cytokine expression. Mod Pathol 18(1):62–67
Park C et al (2006) TLR3-mediated signal induces proinflammatory cytokine and chemokine gene expression in astrocytes: differential signaling mechanisms of TLR3-induced IP-10 and IL-8 gene expression. Glia 53:248–256
Peltier DC, Simms A, Farmer JR, Miller DJ (2010) Human neuronal cells possess functional cytoplasmic and TLR-mediated innate immune pathways influenced by phosphatidylinositol-3 kinase signaling. J Immunol 184:7010–7021

Perry LL, Hotchkiss JD, Lodmell DL (1990) Murine susceptibility to street rabies virus is unrelated to induction of host lymphoid depletion. J Immunol 144:3552–3557

Phares TW, Kean RB, Mikheeva T, Hooper DC (2006) Regional differences in blood-brain barrier permeability changes and inflammation in the apathogenic clearance of virus from the central nervous system. J Immunol 176:7666–7675

Phares TW, Stohlman SA, Hinton DR, Atkinson R, Bergmann CC (2010) Enhanced antiviral T cell function in the absence of B7-H1 is insufficient to prevent persistence but exacerbates axonal bystander damage during viral encephalomyelitis. J Immunol 185:5607–5618

Pichlmair A et al (2006) RIG-I-mediated antiviral responses to single-stranded RNA bearing 5′-phosphates. Science 314:997–1001

Prehaud C, Megret F, Lafage M, Lafon M (2005) Virus infection switches TLR-3-positive human neurons to become strong producers of beta interferon. J Virol 79:12893–12904

Prehaud C et al (2010) Attenuation of rabies virulence: takeover by the cytoplasmic domain of its envelope protein. Sci Signal 3:ra5

Randall RE, Goodbourn S (2008) Interferons and viruses: an interplay between induction, signalling, antiviral responses and virus countermeasures. J Gen Virol 89:1–47

Rieder M, Conzelmann KK (2009) Rhabdovirus evasion of the interferon system. J Interferon Cytokine Res 29:499–509

Rieder M, Brzozka K, Pfaller CK, Cox JH, Stitz L, Conzelmann KK (2011) Genetic dissection of interferon-antagonistic functions of rabies virus phosphoprotein: inhibition of interferon regulatory factor 3 activation is important for pathogenicity. J Virol 85:842–852

Rossiter JP, Hsu L, Jackson AC (2009) Selective vulnerability of dorsal root ganglia neurons in experimental rabies after peripheral inoculation of CVS-11 in adult mice. Acta Neuropathol 118:249–259

Rouas-Freiss N, Moreau P, Menier C, Carosella ED (2003) HLA-G in cancer: a way to turn off the immune system. Semin Cancer Biol 13:325–336

Roy A, Hooper DC (2007) Lethal silver-haired bat rabies virus infection can be prevented by opening the blood-brain barrier. J Virol 81:7993–7998

Roy A, Phares TW, Koprowski H, Hooper DC (2007) Failure to open the blood-brain barrier and deliver immune effectors to central nervous system tissues leads to the lethal outcome of silver-haired bat rabies virus infection. J Virol 81:1110–1118

Scott CA, Rossiter JP, Andrew RD, Jackson AC (2008) Structural abnormalities in neurons are sufficient to explain the clinical disease and fatal outcome of experimental rabies in yellow fluorescent protein-expressing transgenic mice. J Virol 82:513–521

Shankar V, Kao M, Hamir AN, Sheng H, Koprowski H, Dietzschold B (1992) Kinetics of virus spread and changes in levels of several cytokine mRNAs in the brain after intranasal infection of rats with Borna disease virus. J Virol 66:992–998

Shimizu K, Ito N, Sugiyama M, Minamoto N (2006) Sensitivity of rabies virus to type I interferon is determined by the phosphoprotein gene. Microbiol Immunol 50:975–978

Sommereyns C, Paul S, Staeheli P, Michiels T (2008) IFN-lambda (IFN-lambda) is expressed in a tissue-dependent fashion and primarily acts on epithelial cells in vivo. PLoS Pathog 4:e1000017

Steinman RM (1991) The dendritic cell system and its role in immunogenicity. Annu Rev Immunol 9:271–296

Sugiura N et al (2011) Gene expression analysis of host innate immune responses in the central nervous system following lethal CVS-11 infection in mice. Jpn J Infect Dis 64:463–472

Tang SC et al (2007) Pivotal role for neuronal Toll-like receptors in ischemic brain injury and functional deficits. Proc Natl Acad Sci USA 104:13798–13803

Terrien E et al (2012) Interference with the PTEN-MAST2 interaction by a viral protein leads to cellular relocalization of PTEN. Sci Signal 5:ra58

Thoulouze MI, Lafage M, Montano-Hirose JA, Lafon M (1997) Rabies virus infects mouse and human lymphocytes and induces apoptosis. J Virol 71:7372–7380

Tobiume M et al (2009) Rabies virus dissemination in neural tissues of autopsy cases due to rabies imported into Japan from the Philippines: immunohistochemistry. Pathol Int 59:555–566

Torres-Anjel MJ, Volz D, Torres MJ, Turk M, Tshikuka JG (1988) Failure to thrive, wasting syndrome, and immunodeficiency in rabies: a hypophyseal/hypothalamic/thymic axis effect of rabies virus. Rev Infect Dis 10(Suppl 4):S710–S725

Tracey KJ (2009) Reflex control of immunity. Nat Rev Immunol 9:418–428

Tshikuka JG, Torres-Anjel MJ, Blenden DC, Elliott SC (1992) The microepidemiology of wasting syndrome, a common link to diarrheal disease, cancer, rabies, animal models of AIDS, and HIV-AIDS YHAIDS). The feline leukemia virus and rabies virus models. Ann N Y Acad Sci 653:274–296

Ugolini G (1995) Specificity of rabies virus as a transneuronal tracer of motor networks: transfer from hypoglossal motoneurons to connected second-order and higher order central nervous system cell groups. J Comp Neurol 356:457–480

Ugolini G (2010) Advances in viral transneuronal tracing. J Neurosci Methods 194(1):2–20, Epub ahead of print

Versteeg GA, Garcia-Sastre A (2010) Viral tricks to grid-lock the type I interferon system. Curr Opin Microbiol 13:508–516

Vidy A, El Bougrini J, Chelbi-Alix MK, Blondel D (2007) The nucleocytoplasmic rabies virus P protein counteracts interferon signaling by inhibiting both nuclear accumulation and DNA binding of STAT1. J Virol 81:4255–4263

Vuaillat C et al (2008) High CRMP2 expression in peripheral T lymphocytes is associated with recruitment to the brain during virus-induced neuroinflammation. J Neuroimmunol 193: 38–51

Wang ZW et al (2005) Attenuated rabies virus activates, while pathogenic rabies virus evades, the host innate immune responses in the central nervous system. J Virol 79:12554–12565

Wiktor TJ, Doherty PC, Koprowski H (1977a) In vitro evidence of cell-mediated immunity after exposure of mice to both live and inactivated rabies virus. Proc Natl Acad Sci USA 74:334–338

Wiktor TJ, Doherty PC, Koprowski H (1977b) Suppression of cell-mediated immunity by street rabies virus. J Exp Med 145:1617–1622

Zhang B, Chan YK, Lu B, Diamond MS, Klein RS (2008) CXCR3 mediates region-specific antiviral T cell trafficking within the central nervous system during West Nile virus encephalitis. J Immunol 180:2641–2649

Zhao P et al (2011) Innate immune response gene expression profiles in central nervous system of mice infected with rabies virus. Comp Immunol Microbiol Infect Dis 34(6):503–512

Japanese Encephalitis and Dengue Virus Infection: Neurology and Neurobiology

Usha K. Misra and Jayantee Kalita

Abstract Japanese encephalitis (JE) and Dengue virus infections are the most important arboviral infections. JE spreads by *Culex* mosquitoes and causes encephalomyelitis. JE virus has affinity for thalamus, striatum, midbrain and anterior horn cells of spinal cord; however it does not produce any systemic manifestation. JE results in death of 40 % of the patients and half of the survivors have neurological sequelae. There is no antiviral therapy for JE and the management is symptomatic; however it can be prevented by vaccination, Dengue occasionally produces neurological manifestations, which include encephalopathy, encephalitis, immune-mediated syndromes and muscle dysfunctions. The underlying pathophysiology is increased vascular permeability and coagulopathy. The neurological complications of dengue are more common in severe dengue infection. Management of dengue is also supportive and vaccines are under evaluation.

Keywords Flavivirus • Neurotropism • Arbovirus • Neglected tropical diseases

1 Introduction

Arbovirus refers to a group of viruses requiring a blood sucking arthropod to complete its life cycle. These viruses include the genus *Flaviviridae* (Latin 'flavus' means yellow). The *flavivirus* has evolved from a common ancestor about 10,000–20,000 years back in Africa. The genus *Flaviviridae* comprises of about 70 viruses; 40 of these are mosquito-borne, 60 tick-borne and the vector is unknown in 18 of them.

U.K. Misra (✉) • J. Kalita
Department of Neurology, Sanjay Gandhi Post Graduate Institute of Medical Sciences, Raebareily Road, Lucknow 226014, Uttar Pradesh, India
e-mail: drukmisra@rediffmail.com; ukmisra@sgpgi.ac.in

M. Bentivoglio et al. (eds.), *Neglected Tropical Diseases and Conditions of the Nervous System*, 269
DOI 10.1007/978-1-4614-8100-3_15, © Springer Science+Business Media New York 2014

The mosquito-borne viruses are composed of three major serological groups, though overlapping:

1. Yellow fever serological group
2. Dengue serological group
3. Japanese encephalitis serological group

The clade of mosquito-borne viruses, which is represented by yellow fever virus are mostly found in the old world and comprises divergent viruses, which circulate between the *Aedes* mosquito and primates. A subsequent clade of the yellow fever virus, which is transmitted by *Aedes* mosquito, results in hemorrhagic manifestations and these viruses are grouped as dengue serological group. Another clade which is transmitted by *Culex* mosquito, causes encephalitis and is grouped as Japanese encephalitis (JE) serological group (Mackenzie et al. 2002; Solomon 2004). JE and dengue virus infections are important public health problems in South East and East Asia and are the subject of this review.

2 Japanese Encephalitis

JE was recognized in horses and humans in 1871. In 1934, Hyashi reproduced a disease in monkey by intracerebral inoculation of homogenate of fatal case and in 1935 JE virus (JEV) was isolated from human brain in Tokyo, Japan (Nakayama strain). JEV was also isolated from the brain of a sick horse in 1937. Mosquito transmission of JEV was suspected in 1930s by isolating JEV from *Culex tritaeniorhynchus* (Mitamura et al. 1936). In 1959, pigs and birds were recognized as reservoir of JEV. The term Japanese B encephalitis was used to distinguish it from summer epidemics of von Economo's *Encephalitis lethargica*, which is also known as type A encephalitis. Complete nucleotide sequence of JEV was reported in 1990 (McMinn 1997) and JE vaccine was introduced in the Second World War to protect American soldiers.

2.1 *Epidemiology*

The JE endemic area has been increasing over past 70 years (Fig. 1). JE though is primarily concentrated in South East Asia. The global burden of JE in 24 endemic countries is 67,900 cases annually and 75 % of them are children. The overall annual incidence of JE is 1/1,000,000 population which is 5.4/100,000 in children (Campbell et al. 2011). The increasing area of JE may be because of demographic, environmental and social factors. Rapid urbanization, deforestation, shrinkage of agricultural land, bird migration, human movement, climate change as well as viral virulence may be responsible. JEV transmission occurs round the year in the tropics with epidemics following rains when the mosquito density is the highest.

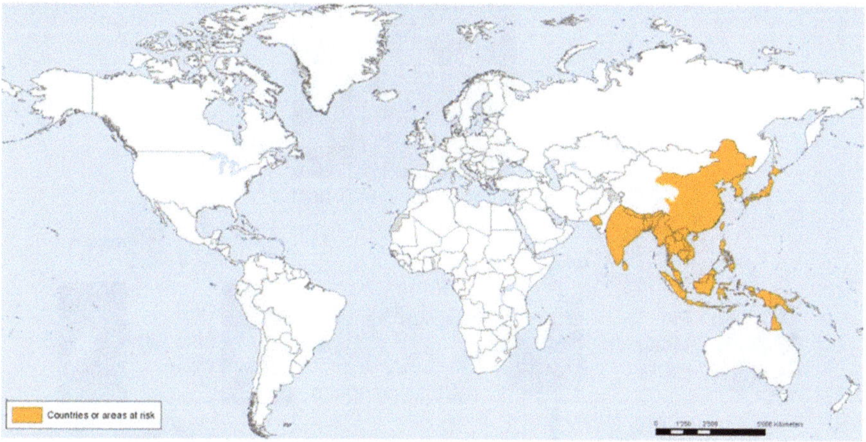

Fig. 1 World map shows distribution of Japanese encephalitis. http://gamapserver.who.int/mapLibrary/Files/Maps/Global_JE_ITHRiskMap.png

JEV affects children in the endemic areas because of herd immunity in the adults, whereas both adults and children are affected in newly invaded areas because of lack of immunity. The travelers to endemic areas may have severe illness because of absence of protective antibodies. The age specific attack rate is the highest between 3 and 6 years, which has been attributed to exposure of poorly clothed village children to mosquito bites. Most of the JEV infection in humans is subclinical. The estimated ratio of symptomatic to asymptomatic infection ranges between 1:25 and 1:1,000 (Solomon and Winter 2004).

2.2 *Virus*

JEV is a single stranded positive sense RNA virus wrapped in a nucleo-capsid which is surrounded by a 50 nm glycoprotein containing envelope. The viral RNA has 5′ untranslated region (UTR), a longer 3′ UTR and an open reading frame in between (Chambers et al. 1990). JEV genome RNA encodes three structural proteins—(1) Capsid protein (C), (2) Precursor to membrane protein (prM), (3) Envelope protein (E); and seven nonstructural proteins (NS) NS1, NS2a, NS2b, NS3, NS4a, NS4b and NS5. The E glycoprotein is the most immunogenic protein, composed of 500 amino acids, and the main component of the surface projection of the virion. E protein is responsible for induction of neutralizing antibodies and accounts for the protective response in the host. E protein is also regarded as the cell receptor binding protein and mediates cell fusion and cell entry. The *flavivirus* enters the cell by endocytosis after attaching to a heparan sulfate receptor. Subsequently the lipid membrane of virus fuses with endosome membrane. The viral RNA enters the cytoplasm of the host cell (Chambers et al. 1990). The virulence of JEV depends on E protein and change in any amino acid may alter its virulence (Cecilia and Gould 1991; Ni and Barrett 1996).

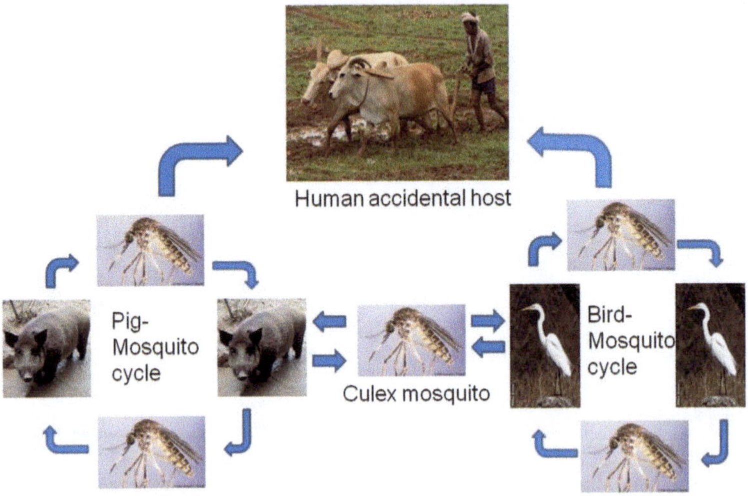

Fig. 2 Life cycle of Japanese encephalitis virus (with permission from http://www.encephalitisindia.org)

2.3 *Life Cycle of JEV*

JE is a zoonotic disease and is transmitted between mosquitoes, pigs and water birds (Fig. 2). Humans are accidentally infected by the bite of JEV infected mosquitoes and are dead end host. Humans do not participate in the spread of JEV because of short and low level viremia. Pigs and ardeid birds are the most important host for the maintenance, amplification and spread of JEV. The pigs are the main participants in the transmission cycle with respect to humans. JEV infected animals usually remain asymptomatic though fatal encephalitis has been reported in horses and fetal losses in sows (Burke and Leake 1988). The bird mosquito cycle is thought to be important for maintaining and amplifying JEV in the environment. *Culex* mosquitoes, especially *Culex tritaeniorhynchus*, is the most important vector for JEV transmission. *C. tritaeniorhynchus* breeds in the rice field with shallow water, can fly up to 1.5 km horizontally and up to 1.5 m vertically (Asahina and Noguchi 1968; Scherer et al. 1959).

2.4 *Pathogenesis and Pathology*

Following an infected mosquito bite, JEV multiplies in the dermis and then in lymph nodes. Subsequently there is a short viremia during which the virus enters the brain. JEV multiplies in the endothelium of cerebral vessels and thus crosses the blood-brain barrier (Dropulie and Masters 1990). JEV replicates in the neurons. There is perivascular cuffing and infiltration of inflammatory cells, and phagocytosis of neurons

(Johnson et al. 1985; Miyake 1964). In fulminant cases, evidence of inflammation may be lacking but viral antigen may be demonstrated in morphologically normal neurons. This may account for normal CSF in some patients. At autopsy, the lesions are mostly restricted to thalamus, basal ganglia, midbrain, cerebral cortex, cerebellum and anterior horn cells of spinal cord. In the acute stage, histopathological examination reveals areas of congestion, small hemorrhages, thrombi formations and neuronophagia. In the sub-acute stage, degeneration, loss of neurons and proliferation of glial cells are more prominent whereas inflammatory cell infiltration is not obvious. In the chronic stage, degeneration or loss of neurons, gliosis, and inflammatory cell infiltration are apparent. In the late stage, there may be calcification of basal ganglia, thalamus and cerebral cortex. The highest concentration of JEV RNA and antigen has been demonstrated in thalamus and brainstem (Srivastava et al. 2012a, b). In the cerebellum, JEV antigen is localized to neurons in the granular and molecular layers, but not to Purkinje cells (Johnson et al. 1985). Co-infection of cysticercosis with JE has been reported in up to 30 % of the patients (Shankar et al. 1983).

2.5 Immunology

Following primary JEV infection, a rapid IgM response occurs in serum within 7 days, and by 1 month, IgM level declines and IgG level increases. Both symptomatic and asymptomatic JEV infections have IgM and IgG response in serum, but only JE patients have this response in the CSF. In the patients who had an earlier flavivirus infection, there is an early rise of IgG and slow rise of IgM. These patients may have milder illness due to neutralization of extracellular viruses and lysis of infected cell through antibody dependent cytotoxicity (Carmenaga et al. 1974). Persistence of JEV in the nervous system has been reported in 5 % of laboratory confirmed cases of JE (Ravi et al. 1993).

The cellular immune response probably restricts the viral replication before it invades the central nervous system (CNS). Immunosuppression by cyclophosphamide results in rapidly progressive encephalitis following JEV infection (Nathanson and Cole 1970). It is therefore likely that cytotoxic lymphocytes help in clearance of JEV. Neuronal injury and host defense in JE have been attributed to a number of pro-inflammatory cytokines and chemokines, which are released by activated microglia and astrocytes in the brain. Cytokines such as tumor necrosis factor (TNF)-α and interferon (IFN)-α/β, -γ activate intracellular antiviral pathways after binding to specific receptors on the surface of infected and uninfected cells. Other cytokines also contribute to antiviral responses (Burke and Morrilnl 1987; Ghoshal et al. 2007). In clinical studies, however, pro-inflammatory cytokines and chemokines were not related to disease severity and mortality (Kalita et al. 2010; (Solomon and Winter 2004). In JE patients with movement disorders, CSF norepinephrine, dopamine, 3,5 dihydroxyphenyl acetic acid, serotonin and homovalinic acid were reduced (Kalita et al. 2007).

In an experimental study these neurotransmitters and metabolites were decreased in thalamus, striatum, midbrain and cortex (Misra et al. 2010a). In experimental studies, the maximum damage of thalamus and midbrain in JEV infection was demonstrated by documenting maximum number of JEV RNA copies, JEV antigen labeling and cytokine expression in these areas (Kumar et al. 2009a; Misra et al. 2010a; Srivastava et al. 2009, 2012b).

2.6 *Clinical Features*

The incubation period of JEV in humans is 5–15 days. The clinical picture of JEV infection ranges from febrile headache to aseptic meningitis to severe encephalitis. Following 2–4 days of nonspecific fever, headache, gastrointestinal disturbance, the encephalitic illness follows. The encephalitis manifests with fever, behavioral abnormality or altered sensorium. Seizures, focal neurological deficit, decerebration, and tremulous eye movements may be associated. Cranial nerve palsy is rare. Severe cases with JE may develop respiratory paralysis which needs artificial ventilation.

2.6.1 Seizures

Seizures may occur in 6.7–47.2 % of the patients with JE, especially children (Gourie Devi et al. 1995; Kalita et al. 2003; Misra and Kalita 2001; Misra et al. 2008; Solomon et al. 2002). Seizures are focal with secondary generalization and may rarely progress to status epilepticus. High frequency of seizures resulting in poor outcome has been reported from Vietnam (Solomon et al. 2002).

2.6.2 Acute Flaccid Paralysis

JE is an encephalomyelitis and the patient may have varying degree of anterior horn cell involvement ranging from patchy focal reflex loss, muscle weakness to wasting simulating polio-like illness (Misra and Kalita 1997; Solomon et al. 1998). Anterior horn cell involvement in JE patients has been documented by electromyography, nerve conduction, motor evoked potential (Misra and Kalita 1997) and autopsy studies (Zimmerman 1946). Polio-like illness due to JEV infection without encephalitis has been reported from Vietnam but in our experience all the patients with anterior horn cell involvement have associated encephalitis.

2.6.3 Movement Disorders

Movement disorders are common in JE and are noted as the patient recovers from coma (Dickerson et al. 1952; Misra and Kalita 1997, 2010). In a study, 74 out of 209 encephalitis patients developed movement disorders. Among the patients with

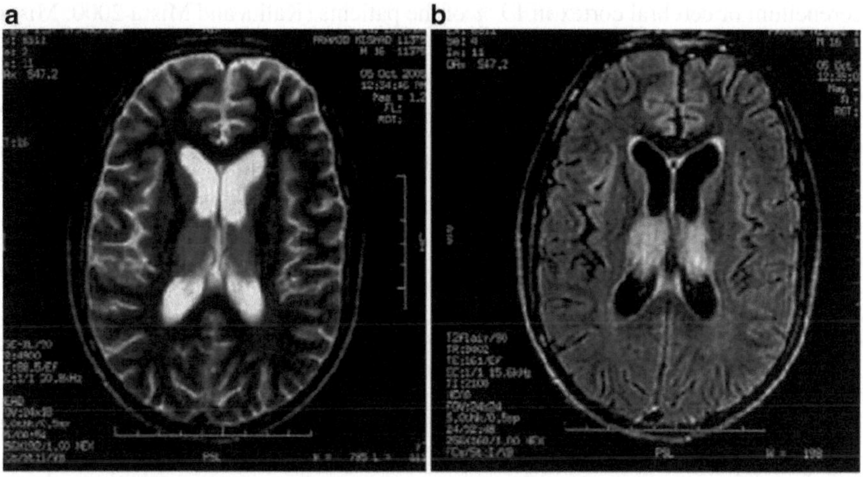

Fig. 3 Cranial MRI in axial section shows bilateral thalamic lesion on T2 sequence (**a**) which is more apparent on FLAIR sequence (**b**)

movement disorders 62.2 % had JE, 8.1 % had dengue, 1.4 % had miscellaneous and 28.3 % had nonspecific encephalitis (Misra and Kalita 2010). Two major types of movement disorders are noted in JE:

1. Parkinsonian features—characterized by mask-like face, abulia, akinesia, rigidity with or without tremor.
2. Dystonia: Nearly all types of dystonia occur in JE and include oromandibular, limb and axial dystonia (Kalita et al. 2011). The dystonia may be fixed or kinetic, mild to severe and sometime simulate status dystonicus (Kalita and Misra 2002a). The movement disorders in JE are usually refractory to medical treatment but regress spontaneously with time.

2.6.4 Laboratory Findings

There may be peripheral leucocytosis and hyponatremia. CSF reveals lymphocytic pleocytosis (10–100/mm^3), protein increase (50–200 mg/dl) and normal glucose. In early stage, CSF may reveal polymorphonuclear pleocytosis.

2.6.5 Imaging

Cranial CT scan is abnormal in about 55 % of the hospitalized patients with JE and reveals thalamic and basal ganglia hypodensity with evidence of hemorrhage occasionally. MRI, however, is more sensitive than CT scan and reveals changes in up to 94 % of the hospitalized patients with JE. The MRI findings are present in thalamus in 94 % (Fig. 3), basal ganglia in 35 %, midbrain in 58 %, pons in 26 % and cerebellum or cerebral cortex in 19 % of the patients (Kalita and Misra 2000; Misra et al. 1994). In addition to thalamus and midbrain lesions, temporal lobe involvement in JE has been reported in 17.7 % of the patients (Handique et al. 2006). Newer MRI sequences have revealed greater sensitivity of FLAIR sequence in demonstrating the pathological changes in JE (Fig. 2) (Misra et al. 2010b). In the acute stage, single photon emission computed tomography reveals hyperperfusion (Kimura et al. 1997) whereas in the sub-acute and chronic stage there is hypoperfusion of thalamus, basal ganglia and frontal area (Kalita et al. 1999; Misra et al. 2010b).

2.6.6 Electrophysiology

Electroencephalography (EEG) reveals non-specific slowing of delta to theta range. There may be lateralized epileptiform discharges and rarely alpha coma. The EEG changes improve within 1–3 months as the consciousness improves (Kalita and Misra 1998).

Nerve conduction studies are normal, except for compound muscle action potentials, which may be reduced due to loss of anterior horn cells. Electromyography reveals fibrillations and sharp waves with reduced interference. Anterior horn cell involvement on the basis of such findings was reported in 35.4 % of the patients (Misra and Kalita 1997). Pyramidal tract involvement results in abnormal central motor conduction in 73.9 % of the patients. Interestingly, the somatosensory evoked potentials are normal in spite of the high frequency of thalamic involvement in JE (Kalita and Misra 2002b).

2.6.7 Diagnosis

Thalamic lesions on CT or MRI are highly suggestive of JE in a patient with acute encephalitis in an JE endemic area. The confirmation of JE is based on detection of antibody, antigen, viral RNA, virus isolation and culture. Currently CSF IgM capture enzyme linked immunosorbent assay (ELISA) is the most acceptable method for the diagnosis of JE with 90 % specificity (Burke et al. 1985). The sensitivity of IgM capture ELISA is 75 % in the first few days of illness and increases to 95 % by the end of first week (Burke et al. 1985). In the early stage of disease, CSF JEV antigen (Ravi et al. 1989) and JEV RTPCR may be detected. Virus isolation from blood is difficult because of low level and short viremia. At autopsy, JEV can be isolated from the CSF and brain The yield of different laboratory tests of JE has been reported as 22.7–32.6 % for hemagglutination inhibition, 40–88.2 % for immunofluorescence, 47–68.2 % for MAC ELISA, 98.3 % for MACDOT, 86.5 % for PCR and 78 % for microneutralization (Tiroumourougane et al. 2002). This variability may be due to sampling time, type of specimen and clinical course of JE.

2.6.8 Differential Diagnosis

In the acute stage, JE may be difficult to differentiate clinically from other encephalitides. The absence of systemic manifestations helps to narrow down the possibilities. Herpes simplex virus (HSV) encephalitis has to be differentiated from JE as it has specific antiviral therapy. HSV encephalitis can be differentiated by characteristic fronto-temporal involvement on CT/MRI and positive PCR for HSV. Acute disseminated encephalitis may be differentiated by multiple white matter lesions. For details on differential diagnosis, see http://www.encephalitisindia.org.

2.6.9 Management

There is no specific antiviral treatment for JE. IFN-α and ribavirin have been found ineffective in clinical trials (Kumar et al. 2009b; Solomon et al. 2003). Corticosteroids do not reduce mortality or affect the outcome (Hoke et al. 1992). General care and symptomatic treatment are the same as in other encephalitides.

2.6.10 Outcome

About 20–40 % of the patients with JE die in the acute stage and 40–50 % of the survivors have sequelae, which include cognitive impairment, behavioral abnormality, focal weakness, movement disorders and seizures. Poor prognostic predictors are great age, high fever, deep coma, hypotonia, seizures, raised intracranial pressure and low level of IgG and IgM (Kumar et al. 1990; Misra et al. 1998; Solomon et al. 2002). The MRI changes, however, are not correlated with the outcome (Kalita and Misra 2000).

2.7 Prevention

There are two main strategies for prevention of JE

- Immunization
- Interfering with enzootic cycle

Immunization is the main way of prevention of JE. There are three types of JE vaccines (1) Formalin inactivated (2) Live attenuated and (3) Chimeric. Live attenuated vaccines are safe and effective and efficacy and safety of chimeric vaccine are under evaluation (Table 1).

Human exposure to mosquito bites can be reduced by wearing proper clothes, remaining indoor during dusk and dawn, using mosquito nets and distancing human habitat from rice fields and pigsties.

Table 1 Vaccines against JEV infection

Type	Dose	Seroconversion	Side effects
Formalin inactivated	0.5–1 ml SC on 0, 7 and 30 days then yearly till 10 years	91 %	2–20 % (urticaria, angioedema, myalgia, headache, encephalomyelitis, seizure, neuropathy)
Life attenuated			
SA14-14-2	0.5 ml SC booster every 5 years	80 % after single dose, 98 % after two doses	Safe
Vero cell	0.5 ml I/M, 0 and 28 days	High	Safe
Chimeric vaccine	Single dose	Under evaluation	

GI gastrointestinal

3 Dengue

3.1 Epidemiology

Dengue is regarded as the second most important mosquito-borne disease after malaria. The incidence of dengue has multiplied many folds in last 50 years. About four billion people live in dengue endemic areas of Asia, Africa, Australia, the Americas and Southern Europe (Fig. 4). Recently about 100 million patients with dengue fever (DF), 500,000 patients with dengue hemorrhagic fever (DHE)/dengue shock syndrome (DSS) and more than 25,000 deaths have been reported yearly (Brady et al. 2012; WHO 2009) (Fig. 5).

3.2 Virus

Dengue virus (DENV) is a positive-strand RNA virus, about 15 nm in diameter and 100 Kb in length. The RNA genome is composed of seven non-structural protein genes (NS) and three structural protein genes encoding the capsid, membrane and envelope protein, respectively. The dengue virus serological group belongs to the family *Flaviviridae* and has four serotypes: DEN1, DEN2, DEM 3 and DEN 4. Although there is cross reactivity in these serological groups, infection by one serotype does not provide immunity to other serotypes; hence an individual living in an endemic area has the possibility of four attacks of dengue virus infection.

The DENV may have originated in Africa about 1,000 years back and might have migrated to South East Asia as all four serotypes of dengue virus are isolated from this region. DEN I serotype was isolated during the Second World War in the Pacific region by American and Japanese investigators. DEN-2 was isolated later. DEN 3 and DEN 4 were isolated in 1950s during the outbreaks in The Philippines

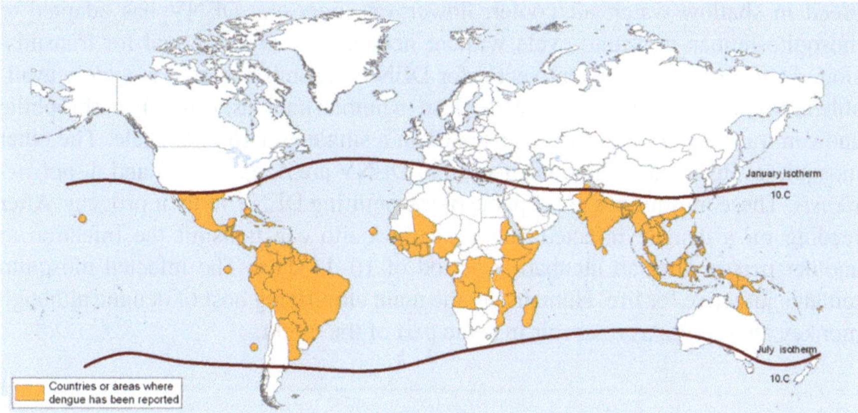

Fig. 4 World distribution of dengue virus infection (http://gamapserver.who.int/mapLibrary/Files/Maps/Global_DengueTransmission_ITHRiskMap.png)

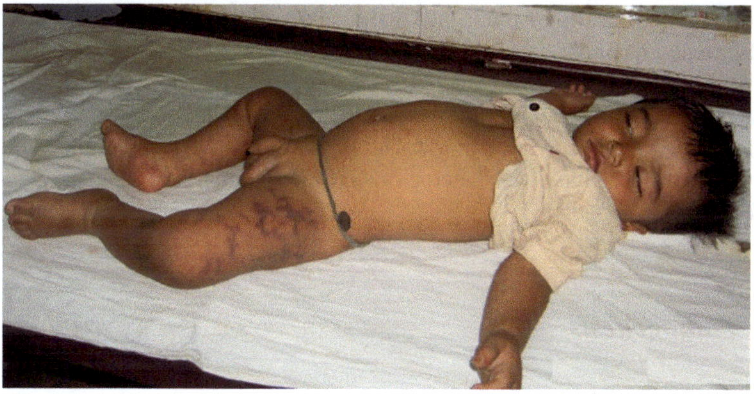

Fig. 5 Photograph of a child with dengue hemorrhagic fever showing ecchymosis

and Thailand. The dengue serotypes have been classified into different genotypes based on sequencing E gene or from the junction of E and N genes. DEN 4 has three genotypes and DEN 1 and 2 have five genotypes. The epidemic potential and virulence of DENV depends on the genotype (Lewis et al. 1993; Rico-Hesse 1990). The transmission of dengue from Asia to other regions of the world might have occurred due to migration of individuals and increase in travels to new areas.

3.3 *Life Cycle of DENV*

Initially DENV was maintained in canopy dwelling *Aedes* mosquito and lower primates. Following deforestation, the mosquitoes have migrated to urban areas as they breed in shallow water, air cooler, flower vas, tiers etc. DENV has adapted to mosquito-human-mosquito cycle without needing any other animal for transmission. *A. aegypti* is an efficient vector for DENV transmission as it is widely available in tropics and subtropics, breeds close to human habitats, is highly anthrophilic and can transmit to more than one person in a single gonotrophic cycle. The other mosquitoes involved in the transmission of DENV are *A. albopictus* and *A. polynesiensis*. These mosquitoes are capable of transmitting DENV to their progeny. After feeding on a dengue infected person, a mosquito can transmit the infection to another person after an incubation period of 10–14 days. The infected mosquito remains infective for life. Humans are the main amplifying host of dengue although monkey also serves as reservoir in some part of the world.

3.4 Virus Entry and Replication

The monocyte, macrophage and dendritic cells are the primary cellular targets of DENV (Balsitis et al. 2009). DENV is attached to the cells by binding to DC-SIGN, CD209, C type lactin and mannose receptor. It enters the cells through clathrin-mediated endocytosis. Inside the cell, the viral nucleocapsid is released into the cytoplasm primarily by the help of anionic lipid which is released from the endosomes (Laughlin et al. 2012). This is followed by protein translation, viral RNA replication, immature virus assembly and viral maturation. Numerous mature viruses are released into the circulation to infect more target cells.

3.5 Pathogenesis

Following DENV infection, three major changes determine the clinical findings (a) increased vascular permeability, (b) thrombocytopenia, and (c) coagulopathy. With abatement of fever, there is increased vascular permeability due to high concentration of soluble IL2, IFNγ, IL2 receptor, CD4, CD8, CCL2 and vascular endothelial growth factor. The RNA load and NSI antigen determine the disease severity (Wang et al. 2006). The increase in vascular permeability in dengue results in hypotension. Shock in dengue is due to venous pooling. If the hypotension is not corrected, the diastolic blood pressure rises and pulse pressure narrows. Finally both systolic and diastolic blood pressures fall abruptly.

Increase in activated partial thromboplastin time and reduction in fibrinogen level along with thrombocytopenia correlate with the severity of dengue infection. Reduced fibrinogen is due to direct interaction of virus with plasminogen. Release of heparin sulfate and chondroitin sulfate from glycocalyx may also contribute to coagulopathy. In DENV infection, coagulopathy is usually minor and resolves in a few days as the virus is cleared. In the patients with DSS, the coagulopathy may be compounded by prolonged hypotension and tissue hypoxia leading to hemorrhage possibly due to

Table 2 Clinical case classification of dengue (World Health Organization 2009)

Criteria for dengue with and without warning sign		
Probable dengue	Warning sign	Criteria for severe dengue
Live in/travel to dengue endemic area	1. Abdominal pain or tenderness	*Severe plasma linkage leading to*
	2. Persistent vomiting	(a) Shock (DSS)
Fever and two of the followings symptoms	3. Clinical fluid accumulation	(b) Fluid accumulation with respiratory distress
	4. Mucosal bleeding	
1. Nausea, vomiting	5. Lethargy, restlessness	*Severe bleeding*
2. Rash	6. Hepatomegaly > 2 cm	as evaluated by clinician
3. Aches and pain	7. ↑ hematocrit with rapid ↓platelet count	*Severe organ involvement*
(a) Positive tourniquet test		– Liver: AST, ALI ≥ 1,000U/l
(b) Leucopenia		– CNS: impaired consciousness
(c) Any warning sign		– Heart or other organs

AST aspartate transaminase, *ALT* alanine transaminase, *CNS* central nervous system

disseminated intravascular coagulation. DHF and DSS are more commonly seen in secondary DENV infection and are attributed to two mechanisms:

(a) Antibody-dependent enhancement (ADE): The neutralizing antibody-developed after the primary DENV infection declines over time. During the secondary DENV infection, binding of cross reactive neutralizing antibody at subneutralizing concentration enhances the infection of monocytes and dendritic cells via Fc receptors which is known as ADE and results in severe dengue (Laughlin et al. 2012).
(b) Memory T cells are reactivated by secondary DENV infection of different serotypes which express altered cytokine level resulting in plasma leakage and increased disease severity (Laughlin et al. 2012; Rothman 2010).

These postulations however do not explain the occurrence of DHF/DSS in primary DENV infection suggesting the role other host and viral factors.

3.6 Clinical Features

The incubation period of DENV infection is 2–7 days. Most of the DENV infections are asymptomatic and the symptomatic patients clinically present with undifferentiated fever, dengue fever (DF), dengue hemorrhagic fever (DHF) and dengue shock syndrome (DSS). The patient with dengue fever may manifest with muscle pain, anorexia, gastrointestinal symptoms, conjunctival congestion, lymphadenopathy and skin rash. Dengue hemorrhagic fever is associated with thrombocytopenia, evidence of plasma leakage in addition to fever and hemorrhagic manifestations. Association of DHF with shock is classified as DSS. In most of the patients with DSS consciousness is retained. The duration of shock is usually short, either the patient recovers following fluid replacement or dies within 12–24 h. Once the shock improves, the patient recovers in 2–3 days. The WHO guideline for case definition of dengue is presented in Table 2 (WHO 2009).

Table 3 Neurological complications associated with dengue

Dengue CNS involvement: At least one of the following
• Impaired consciousness (children ≤ 5 years Blantyre Coma Scale ≤ 4, >5 years GCS score ≤ 14)
• Neck stiffness
• Focal neurological sign
• Seizure
Types of CNS involvement
Dengue encephalopathy
Dengue CNS involvement and
Presence of the following dengue associated complication: hepatic failure, metabolic acidosis, renal failure, hyponatremia, prolong shock, disseminated intravascular coagulation, brain hemorrhage AND
Normal CSF except RBC in CSF in patients with brain hemorrhage
Dengue encephalitis: Dengue CNS involvement and one of the following
(a) CSF: presence of DENV RNA, IgM or NSI antigen
(b) CSF pleocytosis in absence of other neuroinvasive pathogen
Dengue immune-mediated neurological complications
Acute disseminated encephalopathy
Acute transverse myelitis
Acute optic neuritis
Guillain-Barré syndrome
Dengue-associated muscle dysfunction
Transient muscle dysfunction: CK rise with or without muscle weakness
Rhabdomyolysis

3.6.1 Neurological Manifestations of Dengue

Although dengue virus is regarded as non-neurotropic virus, recently a number of neurological manifestations have been reported especially from Southeast Asian countries (Cam et al. 2001; Despres et al. 1998; Domingues et al. 2008; Kalita et al. 2005; Misra et al. 2006; Solomon et al. 2000). The neurological complications therefore have been included in the case definition of severe dengue by the WHO (2009). It is defined by the presence of at least one of the following: impaired consciousness, neck stiffness, focal neurological signs or seizures. The frequency of neurological complications in hospitalized dengue patients ranges between 0.5 and 47 % (Campbell et al. 2011; Soares et al. 2006). The frequency of dengue-associated encephalitis like illness has been reported in Asia in 4.1–20 % of the cases and Brazil in 47 % and Puerto Rico in 26 %. CNS involvement occurs in all subgroups however more common in DHF/DSS. The neurological complications of dengue infection (Table 3) are as follows:

Dengue Encephalopathy

Dengue encephalopathy is the commonest neurological manifestation of DENV infection and is defined as altered sensorium in a patient with DENV infection in whom CSF is normal without evidence of CNS invasion by DENV (absent IgM antibody, NS1 antigen and RNA). Shock, hypoxia, hyponatremia, acute liver or kidney failure, intracerebral hemorrhage and cerebral edema may lead to dengue encephalopathy independently or in various combinations. In some patients the effects of metabolic alterations and DENV on the nervous system may coexist. In a retrospective study from Indonesia, 6.2 % of hospitalized DHF patients had encephalopathy (Hendarto and Hadinegoro 1992). In Thailand 2.7 % serologically confirmed dengue patients had encephalopathy (Pancharoen and Thisyakorn 2001).

Dengue Encephalitis

Dengue encephalitis is clinically similar to dengue encephalopathy and is characterized by presence of CSF pleocytosis, IgM antibodies, NS1 antigen, or DENV RNA. Dengue encephalitis manifests with impaired consciousness, headache, disorientation and behavioral abnormality. DENV specific antibodies in the CSF are found in 22–33 % of the patients with dengue encephalitis (Soares et al. 2006). Detection of DENV in the CSF may be limited by low sensitivity of RT-PCR compared to serum due to low CSF viral load (Varatharaj 2010). In view of these limitations, combination of clinical symptoms, signs and CSF should be considered for the diagnosis of dengue encephalitis.

Cranial MRI is usually normal in encephalitis and encephalopathy but may reveal nonspecific changes consistent with brain edema, subdural or intracerebral hemorrhage (Cam et al. 2001; Misra et al. 2006). EEG reveals delta slowing in 55 % of the patients, which correlates with level of consciousness but not with outcome (Kalita and Misra 2006).

Dengue-Associated Immune Mediated Syndromes

Dengue-associated immune mediated syndromes include disseminated encephalomyelitis, acute transverse myelitis, Guillain-Barré syndrome and occasionally optic neuritis and mononeuropathy (phrenic, long thoracic, oculomotor). In dengue associated transverse myelitis, intrathecal synthesis of DENV specific IgG has been demonstrated (Puccioni-Sohler et al. 2009).

Dengue-Associated Transient Muscle Dysfunction

Dengue-associated transient muscle dysfunction is characterized by increased creatine kinase (CK) with or without myalgia and muscle weakness. In severe muscle dysfunction there may be rhabdomyolysis. Dengue encephalitis represents a more severe illness with protracted course and worse outcome than the muscle dysfunction group. In 17 patients with DEN infection, 11 had encephalitis and 6 muscle dysfunction. All the patients with muscle dysfunction had elevated CK

Table 4 Sensitivity and specificity of laboratory diagnostic tests in dengue virus infection (modified from Tang and Ooi 2012)

Category	Methods	Sensitivity (%)	Specificity (%)
Virus detection	Virus isolation		
	– Mosquito	71.5–84.2	100
	– Cell Culture	40.5	100
	Viral RNA		
	– Conventional RTPCR	48.4–100	100
	– Real time PCR	58.9–100	100
	– NASBA	98.5	100
	Viral antigen		
	– NSI	54.2–93.4	92.5–100
Antibody detection	IgM detection	61.5–100	52–100
	IgG detection	46.3–99	80–100
	Rapid IgM detection (strip test)	20.5–97.7	76.6–90.6
Ag and Ab combined detection	NS1 and IgM	89.9–92.9	75.0–100
	NS1 and IgM and IgG	93.0	100

Ag antigen, *Ab* antibody, *NASBA* nucleic acid sequence based amplification, *RT-PCR* reverse transcriptase polymerase chain reaction

(Misra et al. 2006). In a study on 39 patients with dengue, 31 had muscle involvement; weakness and raised serum CK in 16 (clinical) and the remaining 15 had raised CK without weakness (subclinical). The severity of muscle weakness correlated with thrombocytopenia and serum CK. Electromyography of these patients revealed subtle myopathic changes without fibrillations, sharp waves or complex repetitive discharges suggesting absence of inflammation and necrosis. Muscle histology reveals interstitial hemorrhage with edema, myophagocytosis and occasional necrosis (Misra et al. 2012). These patients recovered in 2 weeks. Quantitative EMG studies have revealed improvement in duration of motor unit potential at 1 month follow up, which correlated with clinical recovery and serum CK (Kalita et al. 2012).

3.6.2 Diagnosis

The diagnosis of dengue in a clinically suspected patient is confirmed by detection of antigen, viral RNA, isolation of virus or antibody against DENV (Tang and Ooi 2012). The sensitivity and specificity of the test depends on the timing of sample collection. Within the first week of illness, dengue infection can be confirmed by virus isolation, PCR or NSI antigen detection in the blood. After 5–90 days, IgM antibodies can be detected by ELISA, plaque reduction or neutralization test. IgG in primary dengue is elevated after 10 days of illness limiting its clinical utility. In the early stage, if serological test is negative, it should be repeated in 2nd or 3rd week if the clinical suspicion is high. The sensitivity and specificity of different diagnostic tests are summarized in Table 4. The secondary DENV infection can be diagnosed by early (after 4 days of illness) and rapid rise of IgG and measuring the IgG and IgM ratio. The window period for detection of virus, RNA and NS1 is reduced in secondary DENV infection compared to primary. The other investigations such as hemoglobin, hematocrit, platelet count, prothrombin time, activated partial thromboplastin time, liver and kidney function tests and serum electrolytes should be carried out and repeated as indicated.

3.6.3 Differential Diagnosis

The DENV infection may clinically simulate malaria, leptospira, viral hepatitis, Chikungunya virus infections, septicemia, scrub typhus and many other conditions producing rash, thrombocytopenia and coagulopathy. Many of these conditions require specific therapy and should therefore be differentiated from dengue (Soares et al. 2011).

3.6.4 Management

Dengue fever is usually a self limiting disease; however it is difficult to predict which patient of DF is likely to develop DHF/DSS. Both DHF and DSS develop suddenly after abatement of fever; therefore, patients should be monitored for warning signs for at least 2 days. There is no specific antiviral drug for DENV infection. The mainstay of treatment is correction of fluid, electrolytes, thrombocytopenia and coagulopathy. WHO has provided guidelines for fluid management in DSS and DHF (WHO 2009). Blood transfusion is indicated if there is significant bleeding or drop in hematocrit. Fresh frozen plasma and platelets are needed if there is coagulopathy and thrombocytopenia with bleeding. Mere dip in platelet is not an indication for platelet transfusion unless it is below $10{,}000/mm^3$. The complications require specific management such as dialysis for renal failure, artificial ventilation for respiratory failure and anticonvulsants for seizures. Temperature should be corrected by paracetamol and cold springing. Aspirin and non-steroidal anti-inflammatory drugs are avoided due to thrombocytopenia.

3.6.5 Outcome

Outcome is generally good without sequelae. DHF and DSS have higher mortality if not treated adequately.

3.7 Prevention

Health education, hygiene and prevention of mosquito breeding and mosquito bite are the key for prevention of dengue. There are three major difficulties in developing a dengue vaccine; the vaccine has to induce immunity to all four serotypes, the lack of immunocompetent animal models for the human disease and the possibility

Table 5 Dengue vaccines in different phases of development (modified from Laughlin et al. 2012)

Vaccine type	Developers	Phases at Nov. 2011
Live attenuated	Sanofi Pasteur/Acambis	Phase 3
	GSK/WRAIR	Phase 2 (suspended)
	NIAID, NIH	Phase 1
	Iviragen/CDC	Phase 1
Subunit	Merck/Hawai Biotech	Phase 1
DNA	NMRC	Phase 1

of ADE in secondary DENV infections, which may be fatal. In spite of these challenges, tetravalent life attenuated and non-replicating dengue vaccines are in different phases of development (Laughlin et al. 2012). The details about dengue vaccines are summarized in Table 5. The preliminary trial of tetravalent vaccine in volunteers has reported seroconversion in 75–90 % and found to be safe (Durbin et al. 2011).

4 Future Perspectives

JEV and DENV infections are two important mosquito borne diseases which are of special concern in the developing countries. Both these diseases do not have specific antiviral drugs, and provides research opportunity. In JE, there is a short viremia, therefore the therapeutic options are only to minimize cell injury and enhance recovery. DENV infections have longer viremia and there is opportunity to intervene before severe complications such as DHF or DSS develop. Predicting the severe illness by evaluating biological markers may be important. The therapeutic options may be at the level of viral entry, cell fusion, replication and multiplication of DENV. Strategies to modify vascular permeability preventing thrombocytopenia may also be helpful. Search for safe and effective vaccine and novel strategies for vector control should continue for prevention of JE and dengue. From neuroscience perspective, the basis of tropism of JEV to thalamus, striatum and midbrain needs investigation. Whether the neurological manifestations of dengue reflect a neuroinvasion with DENV infection of neurons or are secondary to the severe systemic disease needs further evaluation. These answers will help in reducing brain damage in both these diseases.

References

Asahina S, Noguchi K (1968) Long distance flight of *Culex tritaeniorhynchus*. Jpn J Sanit Zool 19:110–112

Balsitis SJ, Coloma J, Castro G et al (2009) Tropism of dengue virus in mice and humans defined by viral nonstructural protein 3-specific immunostaining. Am J Trop Med Hyg 80:416–424

Brady OJ, Gething PW, Bhatt S, Messina JP, Brownstein JS, Hoen AG, Moyes CL, Farlow AW, Scott TW, Hay SI (2012) Refining the global spatial limits of dengue virus transmission by evidence-based consensus. PLoS Negl Trop Dis 6(8):e1760. doi:10.1371/journal.pntd.0001760, Epub 2012 Aug 7

Burke DS, Leake CJ (1988) Japanese encephalitis. In: Monath TP (ed) The arboviruses: epidemiology and ecology, vol III. CRC, Boca Raton, FL, pp 63–92

Burke DS, Morrilnl JC (1987) Levels of interferon in the plasma and cerebrospinal fluid of patients with acute Japanese encephalitis. J Infect Dis 155:797–799

Burke DS, Nisalak A, Ussery MA, Laorakpongse T, Chantavibul S (1985) Kinetics of IgM and IgG responses to Japanese encephalitis virus in human serum and cerebrospinal fluid. J Infect Dis 151:1093–1099

Cam BV, Fonsmark L, Hue NB, Phuong NT, Poulsen A, Heegaard ED (2001) Prospective case–control study of encephalopathy in children with dengue hemorrhagic fever. Am J Trop Med Hyg 65:848–851

Campbell GL, Hills SL, Fischer M, Jacobson JA, Hoke CH, Hombach JM, Marfin AA, Solomon T, Tsai TF, Tsu VD, Ginsburg AS (2011) Estimated global incidence of Japanese encephalitis: a systematic review. Bull World Health Organ 89:766–774

Carmenaga DL, Nathanson N, Cole GA (1974) Cyclophosphamide: potentiated West Nile encephalitis: relative influence of cellular and humoral factors. J Infect Dis 130:634–641

Cecilia D, Gould EA (1991) Nucleotide changes responsible for loss of neuroinvasiveness in Japanese encephalitis virus neutralization-resistant mutants. Virology 181:70–77

Chambers TJ, Hahn CS, Galler R, Rice CM (1990) Flavivirus genome organization, expression, and replication. Annu Rev Microbiol 44:649–688

Despres P, Frenkiel MP, Ceccaldi PE, Duarte Dos Santos C, Deubel V (1998) Apoptosis in the mouse central nervous system in response to infection with mouse neurovirulent dengue viruses. J Virol 72:823–829

Dickerson RB, Newton JR, Hansen JE (1952) Diagnosis and immediate prognosis of Japanese B encephalitis. Am J Med 12:277–288

Domingues RB, Kuster GW, Onuki-Castro FL, Souza VA, Levi JE, Pannuti CS (2008) Involvement of the central nervous system in patients with dengue virus infection. J Neurol Sci 267:36–40

Dropulie B, Masters CL (1990) Entry of neurotropic arboviruses into the central nervous system: an in vitro study using mouse brain endothelium. J Infect Dis 161:685–691

Durbin AP, Kirkpatrick BD, Pierce KK, Schmidt AC, Whitehead SS (2011) Development and clinical evaluation of multiple investigational monovalent DENV vaccines to identify components for inclusion in a live attenuated tetravalent DENV vaccine. Vaccine 29:7242–7250

Ghoshal A, Das S, Ghosh S, Mishra MK, Sharma V, Koli P, Sen E, Basu A (2007) Proinflammatory mediators released by activated microglia induces neuronal death in Japanese encephalitis. Glia 55:483–496

Gourie Devi M, Ravi V, Shankar SK (1995) Japanese encephalitis: an overview. In: Rose CF (ed) Recent advances in tropical neurology. Elsevier, Amsterdam, pp 221–235

Handique SK, Das RR, Barman K, Medhi N, Saharia B et al (2006) Temporal lobe involvement in Japanese encephalitis: problems in differential diagnosis. Am J Neuroradiol 27:1027–1031

Hendarto SK, Hadinegoro SR (1992) Dengue encephalopathy. Acta Paediatr Jpn 34:350–357, overseas edition

Hoke CH, Vaughn DW, Nisalak A, Intralawan P, Poolsuppasit S et al (1992) Effect of high dose dexamethasone on the outcome of acute encephalitis due to Japanese encephalitis virus. J Infect Dis 165:631–637

Johnson RT, Burke DS, Elwell M, Leake CJ, Nisalak A et al (1985) Japanese encephalitis: immunocytochemical studies of viral antigen and inflammatory cells in fatal cases. Ann Neurol 18:567–573

Kalita J, Misra UK (1998) EEG in Japanese encephalitis: a clinicoradiological correlation. Electroencephalogr Clin Neurophysiol 106:238–243

Kalita J, Misra UK (2000) Comparison of CT scan and MRI findings in the diagnosis of Japanese encephalitis. J Neurol Sci 174:3–8

Kalita J, Misra UK (2002a) Markedly severe dystonia in Japanese encephalitis. Mov Disord 15:1168–1172
Kalita J, Misra UK (2002b) Neurophysiological changes in Japanese encephalitis. Neurol India 50:262–266
Kalita J, Misra UK (2006) EEG changes in dengue virus infection with neurological manifestations: a CT MRI correlation. Clin Neurophysiol 117:2252–2256
Kalita J, Das BK, Misra UK (1999) SPECT studies of regional cerebral blood flow in 8 patients with Japanese encephalitis in subacute and chronic stage. Acta Neurol Scand 99:213–218
Kalita J, Misra UK, Pandey S, Dhole TN (2003) A comparison of clinical and radiological findings in adults and children with Japanese encephalitis. Arch Neurol 60:1760–1764
Kalita J, Misra UK, Mahadevan A, Shankar SK (2005) Acute pure motor quadriplegia: is it dengue myositis? Electromyogr Clin Neurophysiol 45:357–361
Kalita J, Kumar S, Vijaykumar K, Palit G, Misra UK (2007) A study of CSF catecholamine and its metabolites in acute and convalescent period of encephalitis. J Neurol Sci 252:62–66
Kalita J, Srivastava R, Mishra MK, Basu A, Misra UK (2010) Cytokines and chemokines in viral encephalitis: a clinicoradiological correlation. Neurosci Lett 473:48–51
Kalita J, Misra UK, Pradhan PK (2011) Oromandibular dystonia in encephalitis. J Neurol Sci 304:107–110
Kalita J, Misra UK, Maurya PK, Shankar SK, Mahadevan A (2012) Quantitative electromyography in dengue-associated muscle dysfunction. J Clin Neurophysiol 29:468–471
Kimura K, Dosaka A, Hashimoto Y, Yasunaga T, Uchino M et al (1997) Single-photon emission CT findings in acute Japanese. Am J Neuroradiol 18:465–469
Kumar R, Mathur A, Kumar A, Sharma S, Chakraborty S et al (1990) Clinical features and prognostic indicators of Japanese encephalitis in Lucknow (India). Indian J Med Res 91:321–327
Kumar S, Misra UK, Kalita J, Khanna VK, Khan MY (2009a) Imbalance in oxidant/antioxidant system in different brain regions of rat after the infection of Japanese encephalitis virus. Neurochem Int 55:648–654
Kumar R, Tripathi P, Baranwal M, Singh S, Tripathi S, Banerjee G (2009b) Randomized, controlled trial of oral ribavirin for Japanese encephalitis in children in Uttar Pradesh, India. Clin Infect Dis 48:400–406
Laughlin CA, Morins DM, Cassetti MC, Denis AC, Martin JS, Whitehead CS, Fauci AS (2012) Dengue research opportunities in Americas. J Infect Dis 206:1121–1127
Lewis JA, Chang GJ, Lanciotti RS, Kinney RM, Mayer LW, Trent DW (1993) Phylogenetic relationships of dengue-2 viruses. Virology 197:216–224
Mackenzie JS, Johansen CA, Ritchie SA, van den Hurk AF, Hall RA (2002) Japanese encephalitis as an emerging virus: the emergence and spread of Japanese encephalitis virus in Australasia. Curr Top Microbiol Immunol 267:49–73
McMinn PC (1997) The molecular basis of virulence of the encephalitogenic flaviviruses. J Gen Virol 78:2711–2722
Misra UK, Kalita J (1997) Anterior horn cells are also involved in Japanese encephalitis. Acta Neurol Scand 96:114–117
Misra UK, Kalita J (2001) Seizures in Japanese encephalitis. J Neurol Sci 190:57–60
Misra UK, Kalita J (2010) Spectrum of movement disorders in encephalitis. J Neurol 257:2052–2058
Misra UK, Kalita J, Kumar S, Mathur A (1994) Radiological and neurophysiological changes in Japanese encephalitis. J Neurol Neurosurg Psychiatry 57:1484–1487
Misra UK, Kalita J, Srivastava M (1998) Prognosis of Japanese encephalitis: a multivariate analysis. J Neurol Sci 161:143–147
Misra UK, Kalita J, Syam UK, Dhole TN (2006) Neurological manifestations 331 of dengue virus infection. J Neurol Sci 244:117–122
Misra UK, Tan CT, Kalita J (2008) Viral encephalitis and epilepsy. Epilepsia 49:13–18
Misra UK, Srivastava R, Kalita J, Khan MY (2010a) Sequential changes in serum cytokines and chemokines in a rat model of Japanese encephalitis. Neuroimmunomodulation 17:411–416

Misra UK, Kalita J, Phadke RV, Wadwekar V, Boruah DK, Srivastawa A, Maurya PK, Bhattacharya A (2010b) Usefulness of various MRI sequences in the diagnosis of viral encephalitis. Acta Trop 116:206–211
Misra UK, Kalita J, Maurya PK, Kumar P, Shankar SK, Mahadevan A (2012) Dengue-associated transient muscle dysfunction: clinical, electromyography and histopathological changes. Infection 40:125–130
Mitamura T, Kitoka M, Watanabe M et al (1936) Study of Japanese encephalitis virus :animal experiments and mosquito transmission experiments. Kansai Lji 1:260, quoted by Vaughn DW, Hoke CH (1992) The epidemiology of Japanese encephalitis: prospects for prevention. Epidemiol Rev 14:197–221
Miyake M (1964) The pathology of Japanese encephalitis. Bull World Health Organ 30:153–160
Nathanson N, Cole GA (1970) Fatal Japanese encephalitis virus infection in immunosuppressed spider monkeys. Clin Exp Immunol 6:161–166
Ni H, Barrett ADT (1996) Molecular differences between wild-type Japanese encephalitis virus strains of high and low mouse neuroinvasiveness. J Gen Virol 77:1449–1455
Pancharoen C, Thisyakorn U (2001) Neurological manifestations in dengue patients. Southeast Asian J Trop Med Public Health 32:341–345
Puccioni-Sohler M, Soares CN, Papaiz-Alvarenga R, Castro MJ, Faria LC, Peralta JM (2009) Neurologic dengue manifestations associated with intrathecal specific immune response. Neurology 73:1413–1417
Ravi V, Vanajakshi S, Gowda A, Chandramukhi A (1989) Laboratory diagnosis of Japanese encephalitis using monoclonal antibodies and correlation of findings with the outcome. J Med Virol 29:221–223
Ravi V, Desai AS, Shenoy PK, Satishchandra P, Chandramuki A, Gourie-Devi M (1993) Persistence of Japanese encephalitis virus in the human nervous system. J Med Virol 40:326–329
Rico-Hesse R (1990) Molecular evolution and distribution of dengue viruses type 1 and 2 in nature. Virology 174:479–493
Rothman AL (2010) Cellular immunology of sequential dengue virus infection and its role in disease pathogenesis. Curr Top Microbiol Immunol 338:83–98
Scherer WF, Buescher EL, Flemings MB, Noguchi A, Scanlon J (1959) Ecologic studies of Japanese encephalitis virus in Japan. III. Mosquito factors. Zootropism and vertical flight of *Culex tritaeniorhynchus* with observations on variations in collections from animal-baited traps in different habitats. Am J Trop Med Hyg 8:665–677
Shankar SK, Rao TV, Mruthyunjayanna BP, Devi MG, Deshpande DH (1983) Autopsy study of brains during an epidemic of Japanese encephalitis in Karnataka. Indian J Med Res 78:431–440
Soares CN, Faria LC, Peralta JM, de Freitas MR, Puccioni-Sohler M (2006) Dengue infection: neurological manifestations and cerebrospinal fluid (CSF) analysis. J Neurol Sci 249:19–24
Soares CN, Cabral-Castro MJ, Peralta JM, de Freitas MR, Zalis M, Puccioni-Sohler M (2011) Review of the etiologies of viral meningitis and encephalitis in a dengue endemic region. J Neurol Sci 303:75–79
Solomon T (2004) Flavivirus encephalitis. N Engl J Med 351:370–378
Solomon T, Winter PM (2004) Neurovirulence and host factors in flavivirus encephalitis evidence from clinical epidemiology. Arch Virol Suppl 18:161–170
Solomon T, Kneen R, Dung NM, Khanh VC, Thuy TT et al (1998) Poliomyelitis-like illness due to Japanese encephalitis virus. Lancet 351:1094–1097
Solomon T, Dung NM, Vaughn DW, Kneen R, Thao LT et al (2000) Neurological manifestations of dengue infection. Lancet 355:1053–1059
Solomon T, Dung NM, Kneen R, le Thao TT, Gainsborough M, Nisalak A, Day NP, Kirkham FJ, Vaughn DW, Smith S, White NJ (2002) Seizures and raised intracranial pressure in Vietnamese patients with Japanese encephalitis. Brain 125(Pt 5):1084–1093
Solomon T, Ni H, Beasley DW, Ekkelenkamp M, Cardosa MJ, Barrett AD (2003) Origin and evolution of Japanese encephalitis virus in southeast Asia. J Virol 77:3091–3098
Srivastava R, Kalita J, Khan MY, Misra UK (2009) Free radical generation by neurons in rat model of Japanese encephalitis. Neurochem Res 34:2141–2146

Srivastava R, Kalita J, Khan MY, Gore MM, Bondre VP, Misra UK (2012a) Temporal changes of Japanese encephalitis virus in different brain regions of rat. Indian J Med Res 138:219–223

Srivastava R, Kalita J, Khan MY, Misra UK (2012b) Status of proinflammatory and anti-inflammatory cytokines in different brain regions of a rat model of Japanese encephalitis. Inflamm Res 61:381–389

Tang KF, Ooi EE (2012) Diagnosis of dengue and update. Expert Rev Anti Infect Ther 10:895–907

Tiroumourougane SV, Raghava P, Srinivasan S (2002) Japanese encephalitis. Postgrad Med J 78:205–215

Varatharaj A (2010) Encephalitis in the clinical spectrum of dengue infection. Neurol India 58:585–591

Wang WK, Chen HL, Yang CF et al (2006) Slower rates of clearance of viral load and virus-containing immune complexes in patients with dengue hemorrhagic fever. Clin Infect Dis 43:1023–1030

World Health Organization (2009) Dengue guidelines for diagnosis, treatment, prevention and control. WHO, Geneva

Zimmerman HM (1946) The pathology of Japanese B encephalitis. Am J Pathol 22:965–991

Part III
Neglected Toxic Disorders of the Nervous System

Mental Health Problems Associated with the Use and Abuse of Khat (*Catha edulis*)

Michael Odenwald

Abstract During the past decades khat use has become a frequent and frenetic, but unnoticed, phenomenon in a number of very poor countries in the world where its use is not controlled and where its economic importance grows. Khat research is still in its beginnings and lags behind the increasingly uninhibited pattern of khat use. In this chapter, potential khat-induced hazards are reviewed and discussed. The currently held view on khat dependence as rare phenomenon contrasts with newer empirical data that demonstrate the existence of a growing binge user group. More research is urgently needed on this topic. Khat can be related to psychosis in several ways, among which evidence is strongest for khat-induced brief psychotic states. Khat can also exacerbate pre-existing psychotic states. The question whether khat constitutes a risk factor for schizophrenia requires urgently further studies. Numerous somatic disorders have also been associated with khat use, among which evidence is strongest for coronary heart diseases. Research has failed so far to prove risks and causal associations based on strong research designs. However, there is enough evidence to state that khat may represent an important but neglected threat for the public health of countries around the Horn of Africa, especially related to substance dependence and psychosis.

Keywords Khat • Qat • Miraa • *Catha edulis* • Dependence • Addiction • Psychosis • Schizophrenia

---

M. Odenwald (✉)
Fachbereich Psychologie, Fach 27, University of Konstanz, 78457 Konstanz, Germany
e-mail: michael.odenwald@uni-konstanz.de

M. Bentivoglio et al. (eds.), *Neglected Tropical Diseases and Conditions of the Nervous System*, 293
DOI 10.1007/978-1-4614-8100-3_16, © Springer Science+Business Media New York 2014

1 The Development from a Niche to a Cash Crop

Khat refers to the young and tender leaves and shoots of the khat tree (*Catha edulis*). It is an evergreen tree of the *Celastraceae* family, normally reaching 6 m in height, but in an equatorial climate it may grow to 25 m. *Catha edulis* can be found in the Abyssinian highlands, the Horn of Africa, Eastern and Southern Africa, the Arab peninsula, and Afghanistan. Khat has many names including *qat* (Yemen), *jad/chaad* (Ethiopia, Somalia), *miraa* (Kenya) or *marungi* (Uganda, Rwanda). The khat leaves have been consumed for centuries for their mildly stimulating properties caused by several alkaloids contained in the leaves. Cathinone, [(−)-α-aminopropiophenene], is considered to be the main psychoactive compound, but it is unstable and swiftly decomposes after harvesting (Feyissa and Kelly 2008). Thus, chewing the fresh leaves is the preferred use form (Al-Motarreb et al. 2002; Krikorian 1984). More stable stimulant compounds are cathine [(+) norpseudoephedrine] and (−)-norephedrine. There are still compounds that have not been studied intensively, such as the cathedulins [for a detailed description of the khat alkaloids and their pharmacological properties see Patel (2014)]. For centuries, khat has been used in certain regions and single ethnic groups in and around the Horn of Africa. Since the end of the nineteenth century with successive transport innovations, the railway network, road haulage, and air cargo, khat has made its way from a niche crop to a cash crop, representing today the economic backbone for Ethiopia and Yemen and to a smaller extent for Kenya (Anderson et al. 2007). Additionally, there is evidence that khat imports from neighboring countries drain Somalia's economy (Hansen 2010).

2 Prevalence of Khat Use

The estimate of the exact prevalence of khat use is difficult, as the consumption still largely depends on socio-economic, ethnic and geographic factors and as its use is constantly growing and changing. The current literature is a patchwork of different snapshots produced with different methods.

In Yemen, the habit of chewing khat was restricted to the northern highlands as well as some urban centers in the south and has expanded since 1991 to the whole of the country (Gatter 2012). Different prevalence estimates vary between 60–90 % of adult males and 10–50 % of adult women (Ali et al. 2004; Numan 2004). Based on the Household Budget Survey 1998, Yemenite households spent around 9–10 % of their income on buying khat (World Bank 2001).

In Ethiopia, khat chewing has traditionally been a habit in the southern and eastern part of the country, especially in Harerghe and among the Muslim populations. Different regional studies found varying prevalence estimates. Current prevalence rates of 75 % of adult males and 35 % of adult women in a traditional khat growing area with homogeneous ethnic composition (Alem et al. 1999) are in contrast with prevalence of 40 % of men and 18 % of women in another rural area with mixed ethnic background (Belew et al. 2000). A representative national assessment of 16,606 adolescents and young adults (15–24 years of age) has reported a 4-weeks prevalence of 16 % (Kebede et al. 2005). Other studies among Ethiopian high school or university students revealed current prevalence rates between 10 and 65 % (Adugna et al. 1994; Kebede 2002; Negussie 2012; Reda et al. 2012; Zein 1988) and great differences between urban and rural areas as well as between ethnic groups (Kassaye et al. 1999; Reda et al. 2012).

In Kenya, khat chewing is traditionally practiced by some Meru tribes (Nyambene Hills) and Muslims, mostly of Somali origin, in the northern part of the country. No prevalence data for the general population are available. Some studies of patients of general health services in different parts of the country disclosed a lifetime prevalence rate of 10.7 % in a region without khat production (Othieno et al. 2000) and a current prevalence rate of approximately 30 % in a khat-producing region (Ndetei et al. 2009; Omolo and Dhadphale 1987b).

Thirty years ago, a cross-sectional assessment of khat use in Somalia reported that in the north of the country 64 % of adult males from the general population regularly consume khat compared to 21 % in the south (Elmi 1983). More recently, in Hargeisa, North-Western Somalia, 31 % of males above the age of 12 chewed khat in the week before the interview (Odenwald et al. 2005). Among active combatants, self-reported khat use in the past week was 26.2 % in the north and 50.7 % in the south of Somalia, and excessive khat use patterns were seven times higher in the south (Odenwald et al. 2007a, b).

Other countries where khat is traditionally used include Djibouti, Saudi-Arabia, Tanzania, Madagascar and South Africa. The general trend in traditional use countries is that khat production and use spreads to other regions and to segments of the population traditionally not in contact with the leaves, e.g. in Kenya (Roba 2009), Ethiopia (Feyisa and Aune 2003), Somalia (Elmi 1983), the eastern parts of Yemen (Gatter 2012). In recent years, khat production and use have also been spreading to African countries where it has been unknown previously, e.g., Uganda and Rwanda (Beckerleg 2008). In sum, khat has become a large regional market without the involvement of multinationals. Today it is a source of income for millions of farmer, pickers, packers, traders (Anderson et al. 2007; Gatter 2012) and an every-day and leisure drug for up to ten million users on any day (Balint et al. 1991; Kalix 1990).

Because of mass migration, khat use spread to high-income countries, e.g., countries of the European Community (Anderson et al. 2007; Griffiths et al. 2010), North America (Gegax 2002), Australia (Douglas and Pedder 2010) and Israel (Litman et al. 1986). Khat use in western countries is still almost exclusively limited to immigrant groups. Some studies have investigated the prevalence of khat use among adult immigrants in Western countries. Among Somalis in the UK, a lifetime history of khat use was found among 38–78 % (58–79 % of males, 16–76 % of females) and current use was found among 34–67 % of males and 17 % of females (Bhui et al. 2003; Griffiths et al. 1997; Patel et al. 2005).

In addition to the growing number of users, khat chewing patterns have also qualitatively changed in the course of the last decades. What one century ago has been a formalized and strongly regulated social habit, around the year 2000 carries features of excessiveness, informality and decoupling from normative control among large user groups (Klein and Beckerleg 2007; Nabuzoka and Badhadhe 2000). This is manifest in the consumption of higher khat quantities in one "session" (Dhadphale et al. 1981; Griffiths 1998; Nabuzoka and Badhadhe 2000; Odenwald et al. 2007a, b; Patel et al. 2005), longer consumption time and binge use sessions of more than 24 h in a row (Nabuzoka and Badhadhe 2000; Odenwald et al. 2007a, b), use in the morning as eye opener and parallel use of other drugs, e.g. benzodiazepines or alcohol (Nabuzoka and Badhadhe 2000; Odenwald et al. 2007a, b; Omolo and Dhadphale 1987a; Selassie and Gebre 1996; Zein 1988). While khat chewers used to be traditionally 'initiated' at around 20 years of age, nowadays, they start using the drug as early as 8 years in life and consumption has become part of the youth culture (Carrier 2005; Nabuzoka and Badhadhe 2000; Patel et al. 2005). Furthermore, the formerly adult male habit is now practiced more and more by women (Alem et al. 1999; Griffiths 1998; Nabuzoka and Badhadhe 2000; Patel et al. 2005) and is also reported in pregnant and breastfeeding women (Belew et al. 2000; Eriksson et al. 1991; Khawaja et al. 2008).

3 Khat: Associated Health Problems—A Neglected Topic

Over the past decades, numerous somatic and mental health problems have been associated with khat use. However, khat research is in its infancy and robust information on the subject is very limited. Moderate khat use is not generally considered noxious. Serious adverse effects, such as khat-induced psychotic states, are usually associated with excessive and prolonged use. However it is often difficult to determine the relative impact of the drug itself in relation to other risk factors that may also be associated with consumption, such as tobacco smoking, poor diet or the residues from pesticides (al'Absi and Grabowski 2012; Date et al. 2004). The following paragraphs will mainly focus on the potential psychiatric consequences and the state of evidence.

3.1 Addiction

The dependence potential of the khat leaves remains poorly understood. The current opinion on withdrawal symptoms upon discontinuation is that they are expected to be mild, brief and only after prolonged use (Giannini et al. 1986; Kalix 1991). These symptoms include profound lassitude, anergia, difficulty in initiating normal activity, slight trembling, several days after ceasing, nightmares, often paranoid in nature for example being attacked, strangled or followed (Kennedy et al. 1980). However, the existing empirical base is meager and not related to current excessive use patterns.

In the same token, today it is generally believed that khat use does not induce tolerance (Giannini et al. 1986; Kalix 1991). It has been argued that the chewing mode of ingestion limits the possible amount to consume in a certain time and tolerance development is thus prevented (Halbach 1972; Kennedy et al. 1980). This view can be criticized based on recent definitions of tolerance. Among stimulant users, tolerance development, the upward shift in the set point for reward and the subsequent dysphoria ('opponent process') are closely related to the development of 'binge' consumption patterns. Users need to increase the dose and the frequency of drug administration in order to experience the desired psychological effects (Koob and Le Moal 2005). Thus, khat tolerance development might not only include increases in the amount of consumption per time unit but rather to the extension of the time spent for consuming it. Recent studies indicate that a growing group of binge users consume khat for more than 24 h in a row (Nabuzoka and Badhadhe 2000; Patel et al. 2005; Widmann 2012) in such large quantities a novice would never manage (Griffiths 1998; Luqman and Danowski 1976; Nabuzoka and Badhadhe 2000; Nencini et al. 1984; Patel et al. 2005; Dhadphale et al. 1981; Mion and Oberti 1998). In addition, the development of tolerance to physiological effects of khat was reported several times (Nabuzoka and Badhadhe 2000; Nencini et al. 1984). No study has ever directly targeted the topic of khat tolerance to desired psychological effects, e.g. euphoria.

While the data on physical aspects of khat dependency are scarce, researchers early on recognized the potential of khat to induce psychological dependency (Eddy et al. 1965; Kennedy et al. 1980). This is best illustrated in descriptions of typical scenes of inner city khat markets at the Horn of Africa or in Eastern Africa at the time just before a khat delivery arrives (Hansen 2010). At these hours of the day, users speed to khat markets frequently causing traffic accidents. An aggressive and nervous atmosphere prevails until the khat trucks arrive. Additionally, observational data confirm the existence of a specific 'drug language' among Somali khat users (Odenwald et al. 2010).

From the scientific point of view, the potential of khat to induce psychological dependence is confirmed by a number of studies, using the Severity of Dependence Scale (Gossop et al. 1995), a five-item instrument measuring the psychological component of dependence. This instrument has recently been adapted and validated for the study of khat addiction (Kassim et al. 2010). It has been shown that khat chewers scoring high on this instrument show more khat-related behaviors and have higher khat alkaloid levels in their saliva (Kassim et al. 2012). About 10 % of a sample of Somali khat users scored at a level comparable with a clinical population with severe heroin dependence in need for treatment (Griffiths 1998). In a more recent study with Yemenite immigrants living in the UK this figure was 39 % (Kassim and Croucher 2006).

Little information is available on the prevalence of khat dependence as defined by ICD or DSM. According to the World Drug Reports, khat is considered the main drug of abuse besides alcohol in Ethiopia and for other countries of the region (UNODC 2004, 2009). As for any other drug, not every khat user will develop a full-blown dependence syndrome (Anthony et al. 1994). Assuming a parallel between khat and amphetamine, the high prevalence of khat chewing in countries like Yemen would produce khat dependence as high as 5–10 % of the adult male population. An Ethiopian study using the Composite International Diagnostic Interview (CIDI, World Health Organization 1997) found a prevalence of khat dependence according to ICD-10 of 5 % among males (among females 1.3 %) in a representative sample from a traditional khat producing area (Awas et al. 1999). The simple application of the ICD-10 dependence criteria to a group of 25 chronic psychotic patients in Somalia revealed a percentage of 84 % (Odenwald et al. 2012) and 100 % among 33 khat-using male Somali refugees in Nairobi (Widmann 2012).

A common characteristic of chronic central stimulant abuse are marked neurocognitive deficits (Ersche and Sahakian 2007). A recent review found that there are virtually no studies and the authors developed suggestions how to study these effects among khat users (Hoffman and Al'Absi 2010). To date, only very few empirical studies have reported on neuro-cognitive effects of khat use. Aircrew members of an Arabic Airline who were daily khat chewers (25), occasional chewers (39) and non-chewers (24) presented for the Standard Aviation Medical Examination and participated in a standardized neuropsychological test battery (Khattab and Amer 1995). Daily khat chewers performed worst in subtests for perceptual speed, long-term memory, visual memory and visual perception and had a faster electroencephalographic background activity compared occasional khat chewers and non-chewers. Duration and amount of khat use were negatively correlated with performance. Recently, several studies found poorer working memory among severe khat users compared to controls (Colzato et al. 2011; Hoffman and Al'Absi 2012; Mikulica 2012). Other studies reported different other executive functions being impaired among khat users: inhibitory control (Colzato et al. 2010), problem solving (Mikulica 2012), cognitive flexibility (Colzato et al. 2011), cognitive control (Colzato et al. 2012) and processing speed (Hoffman and Al'Absi 2012).

While khat users traditionally were mono-substance users, more recently concomitant use is increasingly reported indicating that khat users functionally modulate their physiological and psychological state with a set of other substances (Negussie 2012; Tulloch et al. 2012; Zein 1988). Most obvious is tobacco use which seems to be strongly linked to khat use (al'Absi and Grabowski 2012). In a recent study, khat chewers showed greater tobacco use and khat dependence correlated with tobacco dependence but, most importantly, there were clear signs of an enhancement effect (Kassim et al. 2011).

In sum, the topic of khat addiction urgently needs further empirical studies. Current evidence supports the hypothesis that excessive and prolonged khat use can produce a dependency and neurocognitive deficit syndrome qualitatively similar to that produced by amphetamine. The questions of khat-related withdrawal and tolerance need further studying.

3.2 Psychosis and Other Psychiatric Disorders

Besides the acute psychological effects and withdrawal effects, khat has also been discussed in the context of other psychiatric disorders.

First, several studies demonstrated increased khat use among individuals with mental disorders (Bhui et al. 2006) or with mental distress (Belew et al. 2000). In many of them, it can be seen as functional substance use to counteract symptoms of depression, anxiety and Posttraumatic Stress Disorder (Odenwald et al. 2009; Odenwald et al. 2007a, b) as well as medication side effects (Teferra et al. 2011). Khat helps individuals to feel better and forget stressful war events. But in these burdened individuals, khat use may be also related to suicidal thoughts (Bhui et al. 2003) and might be a risk factor for psychotic developments (Odenwald et al. 2009).

Among otherwise healthy chewers, khat-induced depressive symptoms occur frequently (Hassan et al. 2002). Also psychotic symptoms are frequently reported by khat users, e.g. among Somali khat chewers in the UK 19–20 % reported paranoia after khat use (ever in life) and 14–15 % hallucinations (Griffiths 1998; Griffiths et al. 1997; Patel et al. 2005). Among Somali combatants, paranoid symptoms were also more frequent in khat users (2.9 % vs. 8.9 %, $p<0.001$) and risk of paranoia was increasing with higher khat use (Odenwald et al. 2009).

More than 20 descriptions of cases with khat-induced brief psychotic episodes are available today (Odenwald 2007; Warfa et al. 2007). Three sub-types with different symptom clusters were identified (Pantelis et al. 1989): a paranoid psychotic state with prominent delusions of persecution often associated with auditory hallucinations in a setting of clear consciousness; a manic syndrome with grandiose delusions; and more seldom, a depressive syndrome, possibly related to cessation after a period of excessive use. All cases were native of countries in which khat is used traditionally and most were detected in Western countries where the patients had immigrated to. Most reported excessive khat use before the onset of psychotic symptoms and violent behavior in the course of the acute psychiatric development. Most of them had completely remitted after 2–4 weeks, given abstinence is maintained even without medication. However, most of these cases had repeated such episodes.

Very few studies have addressed the topic of khat effects on pre-existing psychotic disorders. The exacerbation of psychotic symptoms in patients with pre-existing psychotic disorders is mentioned and described in case-reports (Granek et al. 1988; Mion et al. 1997), qualitative studies (Bimerew et al. 2007; Teferra et al. 2011) and quantitative studies (Odenwald et al. 2005, 2012). However, it becomes clear that the effects of khat use on psychotic exacerbation largely depends on specific factors; it is not so much the question of khat use *per se* but on khat chewing patterns, amount of chewing and whether a patient discontinues the antipsychotic medication or not (Odenwald et al. 2012; Teferra et al. 2011).

A few empirical studies ever addressed the question of whether khat use might be a causal factor for the development of chronic psychotic disorders. In the UK, severe schizophrenia and other psychotic disorders were more prevalent among Somali khat users that non-users drawn from a large NHS data base (Tulloch et al. 2012). In the US, a large study among different refugee groups found an unexpected high prevalence of psychotic disorders among young Somali males the group who misuses khat most (Kroll et al. 2010). In contrast to international studies which have proven the even distribution of schizophrenia between sexes (Jablensky et al. 1992) a higher prevalence of male patients with severe mental disorders was reported in Somalia (8 % vs. 2 %; $p<0.001$) where khat use is more prevalent among males (Odenwald et al. 2005). In this study, the onset of lifetime khat use preceded the first psychotic episode in 31 of 38 randomly selected chronic psychotic patients for more than a year. The average age of onset of khat intake was found to be about 4 years earlier among cases who later developed a chronic psychotic disorder (16.6 ± 4.8 years) compared to matched healthy controls (20.7 ± 7.0 years, $p=0.010$). The lapse of time between first khat use and onset of positive symptoms was on average 8.6 ± 6.6 years (median 7 years). In the weeks preceding the onset of psychotic symptoms, patients had chewed mostly excessively (78 %>two bundles/day) compared 4 % of controls ($p<0.001$).

Taken together, the empirical base is thin and all of these questions need to be further studied. It is unclear whether already moderate use or only excessive, chronic or prolonged chewing might have these adverse mental health effects. Of special importance in this context seems to be the question as to whether sensitization towards khat effects might be associated with the development of long-lasting psychotic disorders, relapses or deterioration of psychotic symptoms in vulnerable individuals as discussed for methamphetamine (Yui et al. 2002).

3.3 *Physiological Disorders*

Besides psychiatric sequelae, numerous somatic health problems have been associated with khat use (Al-Habori 2005; Al-Motarreb et al. 2010). As with mental health, adverse effects are commonly linked with prolonged or excessive use. However, the major shortcoming of this literature is that other explanations have not been systematically ruled out, for example tobacco smoking which is frequently combined with khat use (al'Absi and Grabowski 2012), and pesticide content in the leaves (Daba et al. 2011; Date et al. 2004). Observed negative somatic consequences associated to khat use include mucosal problems (Halboub et al. 2009), hypertension (Getahun et al. 2010), cardiovascular complications (Ali et al. 2010), duodenal ulcers (Al-Motarreb et al. 2010), sexual dysfunction (Mohammed and Engidawork 2011), hepatoxicity (Chapman et al. 2010) and reduced birth weight of infants born to *khat* chewing mothers (Mwenda et al. 2003) just to mention a few. By the same token, the argument for possible medicinal uses has only been touched upon (Al-Hebshi et al. 2010; Bredholt et al. 2009).

4 Conclusions and Future Perspectives

Khat is a Janus-faced and contested topic, far away from international attention. As here reviewed, during the past decades khat use has become a frequent and frenetic phenomenon in a number of very poor countries in the world where it is not controlled. While khat research is still in its infancy and lags behind the more and more uninhibited khat use patterns, it has drawn attention to potential khat-induced hazards (such as addiction, psychosis, heart and liver diseases) with cross-sectional and correlational methods. But research has failed so far to prove risks and causal associations with strong research designs. Despite the scarcity of data and the methodological shortcomings, there is enough evidence to state that khat might be an important but neglected threat for the public health of countries around the Horn of Africa, which cannot develop a khat research program by their own means. Thus it is mandatory that international resources need to be invested to build up an interdisciplinary khat research field that should help to observe and understand the current developments and provide decision-makers with a knowledge base. The current knowledge justifies intelligent khat prevention and control measures at national levels as suggested by the WHO (World Health Organization Eastern Mediterranean Regional Office 2007), that could balance culture and economy *vs.* potential dangers.

References

Adugna F, Jira C, Molla T (1994) Khat chewing among Agaro secondary school students, Agaro, southwestern Ethiopia. Ethiop Med J 32:161–166

al'Absi M, Grabowski J (2012) Concurrent use of tobacco and khat: added burden on chronic disease epidemic. [Letter Research Support, N.I.H., Extramural Research Support, Non-U.S. Gov't]. Addiction 107:451–452. doi:10.1111/j.1360-0443.2011.03684.x

Alem A, Kebede D, Kullgren G (1999) The prevalence and socio-demographic correlates of khat chewing in Butajira, Ethiopia. Acta Psychiatr Scand Suppl 397:84–91

Al-Habori M (2005) The potential adverse effects of habitual use of *Catha edulis* (khat). Expert Opin Drug Saf 4:1145–1154

Al-Hebshi NN, Al-Sharabi AK, Shuga-Aldin HM, Al-Haroni M, Ghandour I (2010) Effect of khat chewing on periodontal pathogens in subgingival biofilm from chronic periodontitis patients. J Ethnopharmacol 132:564–569

Ali AA, Al-Sharabi AK, Aguirre JM, Nahas R (2004) A study of 342 oral keratotic white lesions induced by qat chewing among 2500 Yemeni. J Oral Pathol Med 33:368–372

Ali WM, Zubaid M, Al-Motarreb A, Singh R, Al-Shereiqi SZ, Shehab A, Al Suwaidi J (2010) Association of khat chewing with increased risk of stroke and death in patients presenting with acute coronary syndrome. Mayo Clin Proc 85:974–980

Al-Motarreb A, Baker K, Broadley KJ (2002) Khat: pharmacological and medical aspects and its social use in Yemen. Phytother Res 16:403–413

Al-Motarreb A, Al-Habori M, Broadley KJ (2010) Khat chewing, cardiovascular diseases and other internal medical problems: the current situation and directions for future research. J Ethnopharmacol 132:540–548

Anderson D, Beckerleg S, Hailu D, Klein A (2007) The khat controversy: stimulating the debate on drugs. Berg, Oxford

Anthony JC, Warner LA, Kesseler RC (1994) Comparative epidemiology of dependence on tobacco, alcohol, controlled substances, and inhalants: basic findings from the National Comorbidity Survey. Exp Clin Psychopharmacol 2:244–268

Awas M, Kebede D, Alem A (1999) Major mental disorders in Butajira, southern Ethiopia. Acta Psychiatr Scand Suppl 397:56–64

Balint GA, Ghebrekidan H, Balint EE (1991) *Catha edulis*, an international socio-medical problem with considerable pharmacological implications. East Afr Med J 68:555–561

Beckerleg S (2008) Khat special edition introduction. Subst Use Misuse 43:749–761
Belew M, Kebede D, Kassaye M, Enquoselassie F (2000) The magnitude of khat use and its association with health, nutrition and socio-economic status. Ethiop Med J 38:11–26
Bhui K, Abdi A, Abdi M, Pereira S, Dualeh M, Robertson D, Ismail H (2003) Traumatic events, migration characteristics and psychiatric symptoms among Somali refugees–preliminary communication. Soc Psychiatry Psychiatr Epidemiol 38:35–43
Bhui K, Craig T, Mohamud S, Warfa N, Stansfeld SA, Thornicroft G, McCrone P (2006) Mental disorders among Somali refugees—developing culturally appropriate measures and assessing socio-cultural risk factors. Soc Psychiatry Psychiatr Epidemiol 41:400–408
Bimerew MS, Sonn FCT, Korlenbout WP (2007) Substance abuse and the risk of readmission of people with schizophrenia at Ammanual Psychiatric Hospital, Ethiopia. Curationis 30:74–81
Bredholt T, Dimba EA, Hagland HR, Wergeland L, Skavland J, Fossan KO, Gjertsen BT (2009) Camptothecin and khat (*Catha edulis* Forsk.) induced distinct cell death phenotypes involving modulation of c-FLIPL, Mcl-1, procaspase-8 and mitochondrial function in acute myeloid leukemia cell lines. Mol Cancer 8:101
Carrier N (2005) 'Miraa is cool': the cultural importance of miraa (khat) for Tigania and Igembe youth in Kenya. J Afr Cult Stud 17:201–218
Chapman MH, Kajihara M, Borges G, O'Beirne J, Patch D, Dhillon AP, Morgan MY (2010) Severe, acute liver injury and khat leaves. N Engl J Med 362:1642–1644
Colzato LS, Ruiz M, van den Wildenberg WP, Bajo M, Hommel B (2010) Long-term effects of chronic khat use: impaired inhibitory control. Front Psychol 1:129. doi:10.3389/fpsyg.2010.00219
Colzato LS, Ruiz MJ, van den Wildenberg WP, Hommel B (2011) Khat use is associated with impaired working memory and cognitive flexibility. PLoS One 6:e20602
Colzato LS, Ruiz MJ, van den Wildenberg WP, Hommel B (2012) Khat use is associated with increased response conflict in humans. Hum Psychopharmacol 27:315–321. doi:10.1002/hup.2229
Daba D, Hymete A, Bekhit AA, Mohamed AM, Bekhit Ael D (2011) Multi residue analysis of pesticides in wheat and khat collected from different regions of Ethiopia. Bull Environ Contam Toxicol 86:336–341
Date J, Tanida N, Hobara T (2004) Qat chewing and pesticides: a study of adverse health effects in people of the mountainous areas of Yemen. Int J Environ Health Res 14:405–414
Dhadphale M, Mengech A, Chege SW (1981) Miraa (*catha edulis*) as a cause of psychosis. East Afr Med J 58:130–135
Douglas H, Pedder M (2010) Legal regulation of the drug khat in Australia. J Law Med 18: 284–301
Eddy NB, Halbach H, Isbell H, Seevers MH (1965) Drug dependence: its significance and characteristics. Bull World Health Organ 32:721–733
Elmi AS (1983) The chewing of khat in Somalia. J Ethnopharmacol 8:163–176
Eriksson M, Ghani NA, Kristiansson B (1991) Khat-chewing during pregnancy-effect upon the off-spring and some characteristics of the chewers. East Afr Med J 68:106–111
Ersche KD, Sahakian BJ (2007) The neuropsychology of amphetamine and opiate dependence: implications for treatment. Neuropsychol Rev 17:317–336
Feyisa TH, Aune JB (2003) Khat expansion in the Ethiopian highlands. Effects on the farming system in Habro district. MT Res Dev 23:186–190
Feyissa AM, Kelly JP (2008) A review of the neuropharmacological properties of khat. Prog Neuropsychopharmacol Biol Psychiatry 32:1147–1166. doi:10.1016/j.pnpbp.2007.12.033
Gatter P (2012) Politics of qat. The role of a drug in ruling Yemen. Reichert, Wiesbaden
Gegax TT (2002) Meet the khat-heads. Newsweek 140:35
Getahun W, Gedif T, Tesfaye F (2010) Regular Khat (*Catha edulis*) chewing is associated with elevated diastolic blood pressure among adults in Butajira, Ethiopia: a comparative study. BMC Public Health 10:390
Giannini AJ, Burge H, Shaheen JM, Price WA (1986) Khat: another drug of abuse? J Psychoactive Drugs 18:155–158

Gossop M, Darke S, Griffiths P, Hando J, Powis B, Hall W, Strang J (1995) The Severity of Dependence Scale (SDS): psychometric properties of the SDS in English and Australian samples of heroin, cocaine and amphetamine users. Addiction 90:607–614

Granek M, Shalev A, Weingarten AM (1988) Khat-induced hypnagogic hallucinations. Acta Psychiatr Scand 78:458–461

Griffiths P (1998) Qat use in London: a study of qat use among a sample of Somalis living in London. Drugs Prevention Initiative, Paper 26. Home Office, London

Griffiths P, Gossop M, Wickenden S, Dunworth J, Harris K, Lloyd C (1997) A transcultural pattern of drug use: qat (khat) in the UK. Br J Psychiatry 170:281–284

Griffiths P, Lopez D, Sedefov R, Gallegos A, Hughes B, Noor A, Royuela L (2010) Khat use and monitoring drug use in Europe: the current situation and issues for the future. J Ethnopharmacol 132:578–583

Halbach H (1972) Medical aspects of the chewing of khat leaves. Bull World Health Organ 47:21–29

Halboub E, Dhaifullah E, Abdulhuq M (2009) Khat chewing and smoking effect on oral mucosa: a clinical study. Acta Med (Hradec Kralove) 52:155–158

Hansen P (2010) The ambiguity of khat in Somaliland. J Ethnopharmacol 132:590–599

Hassan NA, Gunaid AA, El-Khally FM, Murray-Lyon IM (2002) The effect of chewing Khat leaves on human mood. Saudi Med J 23:850–853

Hoffman R, Al'Absi M (2010) Khat use and neurobehavioral functions: suggestions for future studies. J Ethnopharmacol 132:554–563

Hoffman R, Al'Absi M (2012) Working memory and speed of information processing in chronic khat users: preliminary findings. Eur Addict Res 19:1–6. doi:10.1159/000338285

Jablensky A, Sartorius N, Ernberg G, Anker M, Korten A, Cooper JE, Bertelsen A (1992) Schizophrenia: manifestations, incidence and course in different cultures. A World Health Organization ten-country study. Psychol Med Monogr Suppl 20:1–97

Kalix P (1990) Khat—a dangerous drug? Harefuah 118:555–556

Kalix P (1991) The pharmacology of psychoactive alkaloids from ephedra and *catha*. J Ethnopharmacol 32:201–208

Kassaye M, Sherief HT, Fissehaye G, Teklu T (1999) "Drug" use among high school students in Addis Ababa and Butajira. Ethiop J Health Dev 13:101–106

Kassim S, Croucher R (2006) Khat chewing amongst UK resident male Yemeni adults: an exploratory study. Int Dent J 56:97–101

Kassim S, Islam S, Croucher R (2010) Validity and reliability of a severity of dependence scale for khat (SDS-khat). J Ethnopharmacol 132:570–577

Kassim S, Islam S, Croucher RE (2011) Correlates of nicotine dependence in U.K. resident Yemeni khat chewers: a cross-sectional study. Nicotine Tob Res 13:1240–1249. doi:10.1093/ntr/ntr180

Kassim S, Hawash A, Johnston A, Croucher R (2012) Validation of self-reported khat chewing amongst khat chewers: an exploratory study. J Ethnopharmacol 140:193–196. doi:10.1016/j.jep.2011.12.040

Kebede Y (2002) Cigarette smoking and khat chewing among college students in North West Ethiopia. Ethiop J Health Dev 16:7–17

Kebede D, Alem A, Mitike G, Enquselassie F, Berhane F, Abebe Y, Gebremichael T (2005) Khat and alcohol use and risky sex behaviour among in-school and out-of-school youth in Ethiopia. BMC Public Health 5:109

Kennedy JG, Teague J, Fairbanks L (1980) Qat use in North Yemen and the problem of addiction: a study in medical anthropology. Cult Med Psychiatry 4:311–344

Khattab NY, Amer G (1995) Undetected neuropsychophysiological sequelae of khat chewing in standard aviation medical examination. Aviat Space Environ Med 66:739–744

Khawaja M, Al-Nsour M, Saad G (2008) Khat (*Catha edulis*) chewing during pregnancy in Yemen: findings from a national population survey. Matern Child Health J 12:308–312

Klein A, Beckerleg S (2007) Building castles of spit—the role of khat chewing in worship, work and leisure. In: Goodman J, Lovejoy P, Sherrat A (eds) Consuming habits. Routledge, London

Koob GF, Le Moal M (2005) Neurobiology of addiction. Elsevier, New York
Krikorian AD (1984) Kat and its use: an historical perspective. J Ethnopharmacol 12:115–178
Kroll J, Yusuf AI, Fujiwara K (2010) Psychoses, PTSD, and depression in Somali refugees in Minnesota. Soc Psychiatry Psychiatr Epidemiol 46:481–493
Litman A, Levav I, Saltz-Rennert H, Maoz B (1986) The use of khat. An epidemiological study in two Yemenite villages in Israel. Cult Med Psychiatry 10:389–396
Luqman W, Danowski TS (1976) The use of khat (*Catha edulis*) in Yemen. Social and medical observations. Ann Intern Med 85:246–249
Mikulica J (2012) Cambridge gambling task, spatial span and tower of Hanoi performance among khat chewers. Master, University of Konstant, Konstanz
Mion G, Oberti M (1998) Epidemiologic study of qat use in the National Army of Djibouti. Med Trop (Mars) 58:161–164
Mion G, Ruttimann M, Oberti M, Aversenq C (1997) Acute Khat-induced psychotic crisis. Ann Fr Anesth Reanim 16:201–202
Mohammed A, Engidawork E (2011) Reproductive parameters are differentially altered following subchronic administration of *Catha edulis* Forsk (Khat, Celastraceae) extract and cathinone in male rats. J Ethnopharmacol 134:977–983
Mwenda JM, Arimi MM, Kyama MC, Langat DK (2003) Effects of khat (*Catha edulis*) consumption on reproductive functions: a review. East Afr Med J 80:318–323
Nabuzoka D, Badhadhe FA (2000) Use and perception of khat among young Somalis in a UK city. Addict Res 8:5–26
Ndetei DM, Khasakhala LI, Ongecha-Owuor FA, Kuria MW, Mutiso V, Kokonya DA (2009) Prevalence of substance abuse among patients in general medical facilities in Kenya. Subst Abus 30:182–190
Negussie B (2012) Substance use among high school students in Dire Dawa, Ethiopia. Harar Bull Health Sci 4:38–52
Nencini P, Ahmed AM, Amiconi G, Elmi AS (1984) Tolerance develops to sympathetic effects of khat in humans. Pharmacology 28:150–154
Numan N (2004) Exploration of adverse psychological symptoms in Yemeni khat users by the symptoms checklist-90 (SCL-90). Addiction 99:61–65
Odenwald M (2007) Chronic khat use and psychotic disorders: a review of the literature and future prospects. Sucht 53:9–22
Odenwald M, Neuner F, Schauer M, Elbert TR, Catani C, Lingenfelder B, Rockstroh B (2005) Khat use as risk factor for psychotic disorders: a cross-sectional and case-control study in Somalia. BMC Med 3:5
Odenwald M, Hinkel H, Schauer E, Neuner F, Schauer M, Elbert T, Rockstroh B (2007a) The consumption of khat and other drugs in Somali combatants: a cross-sectional study. PLoS Med 4:e341. doi:10.1371/journal.pmed.0040341
Odenwald M, Lingenfelder B, Schauer M, Neuner F, Rockstroh B, Hinkel H, Elbert T (2007b) Screening for posttraumatic stress disorder among Somali ex-combatants: a validation study. Confl Health 1:10. doi:10.1186/1752-1505-1181-1110
Odenwald M, Hinkel H, Schauer E, Schauer M, Elbert T, Neuner F, Rockstroh B (2009) Use of khat and posttraumatic stress disorder as risk factors for psychotic symptoms: a study of Somali combatants. Soc Sci Med 69:1040–1048
Odenwald WN, Bhui K, Elbert T (2010) The stimulant khat–another door in the wall? A call for overcoming the barriers. J Ethnopharmacol 132:615–619
Odenwald M, Lingenfelder B, Peschel W, Haibe FA, Warsame AM, Omer A, Elbert T (2012) A pilot study on community-based outpatient treatment for patients with chronic psychotic disorders in Somalia: change in symptoms, functioning and co-morbid khat use. Int J Ment Health Syst 6:8. doi:10.1186/1752-4458-6-8
Omolo OE, Dhadphale M (1987a) Alcohol use among khat (*Catha*) chewers in Kenya. Br J Addict 82:97–99
Omolo OE, Dhadphale M (1987b) Prevalence of khat chewers among primary health clinic attenders in Kenya. Acta Psychiatr Scand 75:318–320

Othieno CJ, Kathuku DM, Ndetei DM (2000) Substance abuse in outpatients attending rural and urban health centres in Kenya. East Afr Med J 77:592–595

Pantelis C, Hindler CG, Taylor JC (1989) Use and abuse of khat (*Catha edulis*): a review of the distribution, pharmacology, side effects and a description of psychosis attributed to khat chewing. Psychol Med 19:657–668

Patel NB (2014) Neurobiology of khat (*Catha edulis* forsk). In: Bentivoglio M, Cavalheiro EA, Kristensson K, Patel N (eds) Neglected tropical diseases and conditions of the nervous system. Springer, New York

Patel SL, Wright S, Gammampila A (2005) Khat use among Somalis in four English cities. Home Office Online Report 47/05. Home Office, London. http://www.homeoffice.gov.uk/rds/pdfs05/rdsolr4705.pdf

Reda AA, Moges A, Biadgilign S, Wondmagegn BY (2012) Prevalence and determinants of khat (*Catha edulis*) chewing among high school students in eastern Ethiopia: a cross-sectional study. PLoS One 7:e33946

Roba AW (2009) The ecology, economics of production and trade of Miraa around Masabit Mountain in Kenya. Paper presented at the changing use and misuse of *Catha edulis* (khat) in a changing world: tradition, trade and tragedy, Linköping. http://www.esf.org/index.php?eID=tx_nawsecuredl&u=0&file=fileadmin/be_user/activities/research_conferences/Docs_NEW/2009/274_adano_pres.pdf&t=1354839740&hash=9b35235171d3d20d569913156e43729177d285fc

Selassie SG, Gebre A (1996) Rapid assessment of drug abuse in Ethiopia. Bull Narc 48:53–63

Teferra S, Hanlon C, Alem A, Jacobsson L, Shibre T (2011) Khat chewing in persons with severe mental illness in Ethiopia: a qualitative study exploring perspectives of patients and caregivers. Transcult Psychiatry 48:455–472. doi:10.1177/1363461511408494

Tulloch AD, Frayn E, Craig TK, Nicholson TR (2012) Khat use among Somali mental health service users in South London. Soc Psychiatry Psychiatr Epidemiol 47:1649–1656. doi:10.1007/s00127-011-0471-8

UNODC (2004) World Drug Report 2004. United Nations, New York

UNODC (2009) World Drug Report 2009. United Nations, New York

Warfa N, Klein A, Bhui K, Leavey G, Craig T, Stansfeld SA (2007) Khat use and mental illness: a critical review. Soc Sci Med 65:309–318

Widmann M (2012) Trauma and psychotic symptoms among khat users—a pilot study with Somali refugees. Master, University of Konstanz, Konstanz

World Bank (2001) Yemen economic update, Issue 6. The World Bank Group Sana'a Office, Sana'a

World Health Organization (1997) Composite international diagnostic interview (CIDI): Core version 2.1 (Version 1.1. ed.). WHO, Geneva

World Health Organization Eastern Mediterranean Regional Office (2007) Report of the sixth meeting of the Regional Advisory Panel on Impact of Drug Abuse (RAPID 6); key messages on Khat; Cairo, 16–19 October 2007. WHO EMRO, Cairo

Yui K, Ikemoto S, Goto K, Nishijima K, Yoshino T, Ishiguro T (2002) Spontaneous recurrence of methamphetamine-induced paranoid-hallucinatory states in female subjects: susceptibility to psychotic states and implications for relapse of schizophrenia. Pharmacopsychiatry 35:62–71

Zein AZ (1988) Polydrug abuse among Ethiopian university students with particular reference to khat (*Catha edulis*). J Trop Med Hyg 91:71–75

Neurobiology of Khat (*Catha edulis* Forsk)

Nilesh B. Patel

Abstract Around 20 million individuals in eastern Africa and the Arabian Peninsula chew the fresh leaves and twigs of *Catha edulis* Forsk (khat) for its pyschostimulatory effect, a practice deeply rooted in their traditions and cultures. In 1975, the main active ingredient of khat, cathinone, was identified, and found to be structurally related to and with effects similar to amphetamines and other psychostimulants. Animal studies on the neurobiology of khat are sparse and sporadic, being a neglected area of research in the field of drugs of abuse, and most work has focused on the action of cathinone rather than on khat extracts. Like other psychostimulants, the target of khat and cathinone action on the central nervous system is the dopaminergic system involving the nucleus accumbens. Studies on peripheral tissue also show it effects on the serotoninergic system. In animal self-administration studies, cathinone exhibits an addictive and abuse potential and produces psychomotor sensitization. However, there is little information from either human or animal studies on the short- and long-term effect on brain function of daily or frequent khat use with different patterns of consumption; nor is there information on pre-natal and adolescent exposure to khat or its neurotoxic potential. More research on the effects of khat use is needed as it contains a cocktail of alkaloids which is consumed by the user.

Keywords Drugs of abuse • Cathinone • Psychostimulants • Miraa • Qat • Amphetamine

N.B. Patel (✉)
Department of Physiology, School of Medicine, College of Health Science,
University of Nairobi, P. O. Box 30197-00100, Nairobi, Kenya
e-mail: npatel@uonbi.ac.ke

M. Bentivoglio et al. (eds.), *Neglected Tropical Diseases and Conditions of the Nervous System*, 307
DOI 10.1007/978-1-4614-8100-3_17, © Springer Science+Business Media New York 2014

1 Introduction

Khat, *Catha edulis* Forsk, is, either daily or frequently, consumed by an estimated 20 million individuals for its pyschostimulatory effect attributed mainly to the presence of cathinone in its fresh leaves and twigs. Cathinone is structurally related to amphetamine and produces many similar effects in humans (see Odenwald 2014) though in some respects khat/cathinone effects may differ from known amphetamines. This chapter starts by a brief review of the history of khat use to show why it has not attracted the international exposure or research importance as other drugs of abuse, followed by a review of studies on its abuse potential, which mainly involve its main active ingredient, cathinone, and then possible mechanisms of action. The literature is sparse and sporadic given that it is a neglected area of research despite its potential adverse impact on the millions of individuals, families, society and economy where khat is grown and consumed routinely.

2 History of Khat Use

This hardy perennial dicotyledonous evergreen shrub is indigenous to and cultivated along eastern Africa, from Madagascar to the Horn of Africa, and the Arabian Peninsula, especially Yemen. Its fresh young leaves and twigs commonly called khat or by local traditional name (Fig. 1) are chewed for their psychostimulant effect by around 20 million inhabitants in these regions (Al-Motarreb et al. 2002; Saha and Dollery 2006; Magdum 2011). Figure 2 shows the distribution of khat use along eastern Africa and Arabian Peninsula with percentage of users and local names. Peter Forsskål, who died in an ill-fated 1761 Danish expedition to the Arabian Peninsula, gave the plant's botanical classification—*Catha edulis* belonging to the Celastraceae family. Carsten Niebuhr, who edited and published Forsskål's botanical collection from this expedition in *Flora-Aegytiaco-Arabia* (1775), named the plant *Catha edulis* Forsk in memory of his friend and colleague. When and

Fig. 1 (**a**) "Kilo" of khat. "Kilo" is not a weight reference but a term used for a bundle. Three to four bundles can be consumed per day or more when there is khat party or binging. (**b**) The leaves and the bark of the twigs are chewed and tucked into the cheek. (**c**) Muguka, leaves from different part of *Catha edulis*, cheaper than miraa and the use of which has increased in recent times

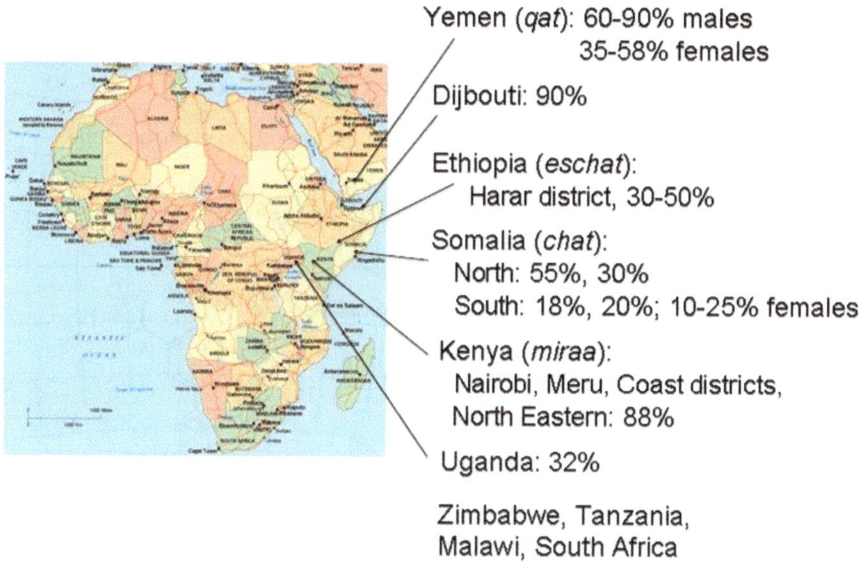

Fig. 2 Major areas of khat use along the eastern Africa and Arabian Peninsula with local names for khat and available estimated percentage of users. (Compiled from various sources)

where the chewing of the young fresh leaves and twigs of *Catha edulis* (khat) started is not known. According to some its use can be traced to the Ancient Egyptians who considered the plant a divine food, which was capable of releasing humanity's divinity for transcendence into "apotheosis" making the person god-like (Giannini et al. 1986). Whether the use originated in Ethiopia or Yemen is a debated issue.

Several early travellers to the Arabian Peninsula commented this habit.

In the 18th century, Neibuhr wrote "…never saw the Arabian use opium like the Turks and Persians. Instead of taking this gratification, they chew kaad [khat]. These are buds of a certain tree, which are brought in small boxes from the hills of Yemen." In early 19th century, Abdullah bin Adbul Kadir (1854), a traveller from Malay, described the prevalence of khat chewing in Al Hudaydah, Yemen: "You observed a new peculiarity in this city—everyone chewed leaves as goats chew the cud. There is a type of leaf, rather wide and about two fingers in length, which is widely sold, as people would consume these leaves just as they are; unlike betel leaves, which need certain condiments to go with them, these leaves were just stuffed fully into the mouth and munched. Thus when people gathered around, the remnants from these leaves would pile up in front of them. When they spat, their saliva was green. I then queried them on this matter: 'What benefits are there to be gained from eating these leaves?' To which they replied, 'None whatsoever, it's just another expense for us as we've grown accustomed to it'. Those who consume these leaves have to eat lots of ghee [clarified butter] and honey, for they would fall ill otherwise. The leaves are known as *Kad* [khat]." These and other observations are similar of the use of coca leaves among the people of western South America.

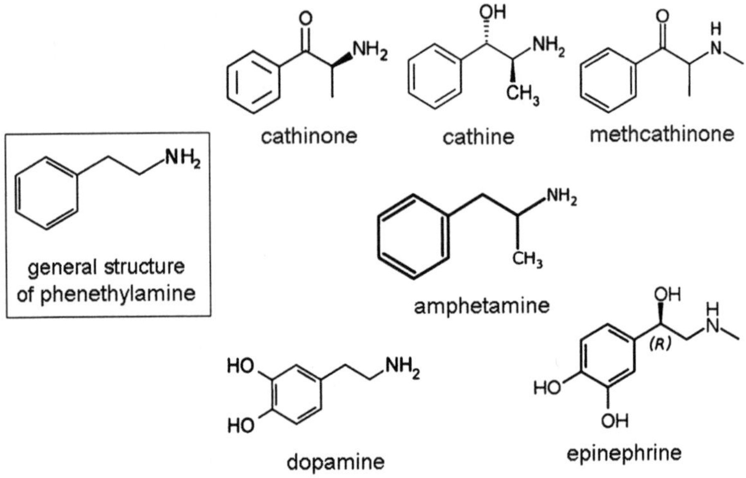

Fig. 3 Phenethylamine group of molecules: note that the addition or substitution on the phenyl ring, ethyl chain or amino group of the molecule produces a large group of psychoactive and bioactive molecules including cathinone and cathine, which are active ingredients found in khat

In 1909, the epidemic of opium use in China resulted in the Shanghai conference, which led to the 1912 International Opium Convention of The Hague, the start of the international regulations in the trafficking, trading and controlling access to drugs of abuse. This convention was subsequently taken up by the League of Nations, and later by the United Nations. In 1935, the matter of khat use and its regulation was first discussed, and debated on and off for a number of years, ultimately cumulating in the resolution that khat use was a regional issue and as confined to a few countries (WHO 1964), possibly not an issue of international regulation, but one that needed to be studied further.

In the deliberations on whether to enter khat on the list of internationally controlled substances, it was recognized that its active ingredient had to be isolated, identified, and its abuse potential studied before a decision could be made. In the 1930s, Wolfes had isolated cathine (d-norpseudoephedrine), but subsequent studies showed that cathine could not fully account for the activity of khat in the user. The United Nations Narcotics Laboratory, after studying fresh khat samples from different sources, isolated 20 components, which included a large group of alkaloids (cathedulins), and in particular, isolated cathinone (alpha-aminopropiophenone), a liable substance, which was a more active ingredient of khat than cathine (UNODC 1980). Once cathinone has been isolated, synthesized and its molecular structure determined (Schorno and Steinegger 1978; Braendan 1979) it was found to be similar to amphetamine, both belonging to the phenethylamine group (Fig. 3). In 1980, the World Health Organisation (WHO) classified khat as a drug of abuse that can produce mild to moderate psychological dependence. Currently, khat and cathinone are on the list of banned or controlled substances in several countries and khat has a higher international profile due to the spread of its use to parts of the world where it was not originally used.

3 Pharmacological Effects on the Dopamine, Serotonin, and Norepinephrine Systems

Two widely used techniques to determine where in the brain a drug could affect neurotransmitter release are *in vitro* tissue and *in vivo* microdialysis methods. Of the number of neurotransmitters systems in the brain, studies on drugs of abuse have highlighted the role of three: dopamine, serotonin, and norepinephrine (noradrenaline). In an *in vitro* study of rabbit striatal and rat nucleus accumbens tissue prelabeled with ^{3}H-dopamine, there was increase in the release of radioactivity on cathinone application (Kalix 1986; Kalix et al. 1987). However, without further analysis, it is not possible to conclude if the increased radioactivity released from these tissues was due only to dopamine or a mixture of dopamine and its metabolites. In *in vivo* microdialysis studies of the anterior caudate-putamen and nucleus accumbens, (−)-cathinone and (+)-amphetamine increased dopamine levels in a dose-dependent manner with amphetamine having a higher effect at the largest doses used, but at lower doses the amphetamine difference was only seen in the nucleus accumbens (Pehek et al. 1990). In synaptosomal preparation, d,l-cathinone like d-amphetamine, released and blocked uptake of ^{3}H-dopamine, and with repeated high doses of d,l-cathinone there was long-lasting dopamine depletion in various rat brain regions and decreased number of dopamine uptake sites similar to amphetamines, but regional brain levels of norepinephrine and serotonin were not altered (Wagner et al. 1982; Fleckenstein et al. 1999). Cathinone also inhibited the firing of dopaminergic neurons (reversible with haloperidol) in the substantia nigra pars compacta, with potency similar to amphetamine (Mereu et al. 1983). These studies, along with the drug discrimination studies (described below), further emphasized the role of dopamine in the action of cathinone, but this is probably only a part of its mechanism. Dopamine systems are involved in motor function through the striatal tissue and "mental" reward system involving the nucleus accumbens. Hence, increase in dopamine could affect both motor and mental function, though cathinone appears to have a less disruptive effect on motor behavior compared to amphetamine.

Like amphetamine, but a third less potent, cathinone induces release of ^{3}H-serotonin (5-HT) from rat striatal preparations (Kalix 1984) and has four times higher affinity for serotonin receptor in an isolated rat stomach fundus preparation (Glennon and Liebowitz 1982). Repeated high doses of cathinone do not alter regional brain levels of serotonin (Wagner et al. 1982). However, high-dose administrations of cathinone to striatal synaptosomes obtained from drug-treated rats rapidly decreased serotonin transporter function, which should result in higher serotonin levels *in vivo* (Fleckenstein et al. 1999). Thus, while, *in vitro* studies show that cathinone causes release of serotonin and has higher binding affinity to peripheral serotonin receptors compared to amphetamine, serotonin is not found to be involved in studies of cathinone's discriminative mechanism (discussed in later section), but this does not rule out serotonin's involvement in other effects of cathinone.

While regional brain levels of norepinephrine (noradrenaline) or serotonin are not altered on a long-term basis by repeated administration of d,l-cathinone

(Wagner et al. 1982), there is increased release of radioactivity from rabbit heart atria pre-labeled with ^{3}H-norepinephrine (Kalix 1983). The effect on rat right ventricle may involve competitive blockade of norepinephrine transporter rather than simple displacement of norepinephrine (Cleary et al. 2002). These changes in peripheral neurotransmitters are not unexpected as khat, cathinone, and cathine produce sympathomimetic effects in the user, but whether the central neural mechanism involve norepinephrine is not known.

4 Experimental Studies on Behavior and Potential for Addiction

4.1 Behavioral Studies

Like amphetamine, cathinone or khat extract produce psychomotor sensitization -hyperlocomotion (Kalix 1980a; Calcagnetti and Schechter 1992a; Banjaw et al. 2005), behavioral sensitization (Banjaw and Schmidt 2005)—as well as pre-pulse inhibition (Banjaw et al. 2005), and increased isolation induced aggression in rats (Banjaw et al. 2006). The psychomotor sensitization reflects nucleus accumbens involvement (Wise and Bozarth 1987) and the psychostimulant-induced activity is blocked by the dopamine release inhibitors CGS 10746B and isradipine (Calcagnetti and Schechter 1992b).

Like other psychostimulants, cathinone produces a hyperthermic response (Kalix 1980b), which is linked to the neurotoxic effect of amphetamines. Treatment of rats with khat extract produces seizures and decreases seizure threshold elicited by pentylenetetrazol (Oyungu et al. 2007, 2009). In addition, cathinone induces head-twitch response (Connor et al. 2002), which is a behavioral proxy for serotonin receptor 5-HT$_{2A}$ activation (Schmid and Bohn 2010). This is one of the serotonin receptors postulated to be involved in the mechanism of activity of hallucinogens (Vollenweider 2001). Taken together, these findings suggest that khat activity could have some other different and subtle effects compared to amphetamines.

4.2 Addictive and Abuse Potentials

4.2.1 Self-administration and Self-reinforcing Experimental Studies

A routinely used experimental approach to assess whether a drug has addictive and abuse potential is to avail the drug to laboratory animals and observe whether they will self-administer the drug. The frequency and amount of the self-administration gives an indicator of the reinforcing property and abuse potential of the drug.

Using a self –administration continuous set-up, Yanagita (1979, 1986) found in rhesus monkeys a spree type usage of cathinone, as seen with cocaine, with spree periods of 6–59 h. Compared to amphetamine and cocaine, both d,l-and l-cathinone maintained significantly higher rates of responding with l-cathinone being more potent (Schuster and Johanson 1979). Woolverton and Johanson (1984) found the reinforcing effects of d,l-cathinone were comparable to cocaine. In mice (Kuz'min and Evartan 1991), comparison of the pattern of intravenous self-administration of morphine, cocaine, amphetamine, cathinone, and ephedrine produced similar bell-shaped concentration curves typical of compounds with addictive potential. Rats also demonstrated cathinone self-administration, which was increased by block of the dopamine D1 receptor with SCH 2390 but not with D2 receptor antagonist, spiperone, suggesting that dopamine D1 receptors are involved in cathinone's reinforcing effects (Gosnell et al. 1996).

4.2.2 Conditioned Place Preference

In conditioned place preference (CPP), an experimental approach to assess the abuse potential of a drug, rats given cathinone either by the intravenous (iv) or intracerebroventricular (ICV) route showed preference for the environment to which they were exposed to with cathinone (Schechter 1991a) and this effect was dose dependent (Schechter and Meehan 1993). Interestingly, no tolerance to CPP was found with repeated cathinone administration unlike that observed in drug discriminative behavior (Schechter and McBurney 1991). CPP was attenuated or blocked by pretreatment with dopamine release inhibitor, CGS 10746B (Schechter 1991a; Calcagnetti and Schechter 1993).

4.2.3 Drug Discrimination Experimental Studies

While self-reinforcing and CPP studies provided evidence that cathinone has addictive and abuse potential, these experimental approaches cannot answer the question of whether (1) the feeling produced in an animal (interoceptive cue) with cathinone is similar to other drugs, especially other psychostimulants, and (2) if mechanism of action of cathinone is similar to that of other known psychostimulants. In the drug discrimination experimental paradigm, the animal is trained to perform a particular task under the influence of a drug and another in the absence of that drug. In a two-lever food-motivated operant task, animals press one lever when in the drug induced mental state (interoceptive cue) and the other when not. The drug used for the training is the training drug, and once the animal has learnt the task it is given other test drugs, and if the animal presses the levers correctly, it is assumed that the interoceptive cue induced in the animal by the test drug is similar to that produced by the training drug.

Table 1 summarizes the drugs that have been found to be able to replace or generalize for the cathinone cue and those that could not. The drugs that can substitute for

Table 1 Drugs that can or cannot substitute for cathinone in the drug discrimination experimental paradigm

Can	Cannot
d-amphetamine[a,b,c]	Apomorphine[a,d,c]
Cocaine[a, e]	Fenfluramine[a]
Pripradol[a]	Fentyamyl[a]
Cathine[a]	Phydroxyamphetamine[a,f]
Methamphetamine[e]	Phenylethylamine[a,d]
Methylphenidate[a]	Deuterated phenylehylamine[e,d,g]
	Chlorodiazepoxide[a]

[a]Goudie et al. (1986)
[b]Rosecrans et al. (1979)
[c]Schechter et al. (1984)
[d]Apomorphine, phenethylamine and deuterated phenethylamine produced 29 % and 60 % generalization (substitution) for the cathinone cue, respectively
[e]Schechter and Gennon (1985)
[f]A polar conger of amphetamine
[g]A long lasting derivative of phenylethylamine which is resistant to metabolism by monoamine oxidase

cathinone are known psychostimulants, and cathinone can also substitute when amphetamine or cocaine is used as the training drug. These results support the view that cathinone is a psychostimulant and represents a "natural" amphetamine (Kalix 1992). When cathinone was used to substitute for amphetamine as the test drug, it was found to be twice as potent as amphetamine (Rosecrans et al. 1979). In rats, injecting cathinone into the nucleus accumbens produced discriminative behavior at a much lower concentration than either by intraperitoneal or ICV route, indicating that one site of action of cathinone in the brain is the nucleus accumbens (Schechter et al. 1992).

In the drug discrimination studies, the time-course for cathinone interoceptive cue behavior in rats was earlier than amphetamine (5 vs. 15–30 min) and effective for about 1 h (Schechter 1989).

Positive results were found in all drug discrimination studies in which the role of dopamine as part of cathinone's mechanism of action was tested, i.e. the dopaminergic system is involved as reported for other psychostimulants. However, depending on the agent used, the assessment of the involvement of the dopaminergic system in cathinone's effect differs. Unlike amphetamine, with haloperidol, a dopamine antagonist, Goudie et al. (1986) found at most 50 % reduction in the cathinone cue, and Rosecrans et al. (1979) found that it did not affect the generalization of amphetamine stimulus to cathinone. Pretreatment with haloperidol failed to alter the stimulant properties of cathinone but did partially antagonize those of amphetamine and cocaine (Huang and Wilson 1986) and attenuated cathinone discrimination (Schechter 1986c). However, co-administration with CGS 1074613, a dopamine release inhibitor, totally antagonized cathinone's generalization to amphetamine (Schechter and Bojaw 1988) and pretreatment blocked cathinone discrimination (Schechter 1992). Use of serotonin antagonist, BC 105/B (Rosecrans et al. 1979), 5-HT receptor blocker, pirenperone (Schechter 1986a), 5-HT_3 receptor antagonist, MDL 72222 (Schechter 1992) or inhibiting serotonin synthesis with p-chlorophenylalanine had no effect on the cathinone discrimination in rats (Schechter 1991b). Phenoxybenzamine, an alpha-adrenergic antagonist, also had no effect (Rosecrans et al. 1979). Thus, at least from drug discrimination behavior studies, alteration of the dopaminergic neurotransmitter system appears to play a major role in the mechanism of action of cathinone.

5 Tolerance

It is widely assumed khat does not produce tolerance in humans but, as pointed out by Halbach (1979), this could be simply due to self-limiting process of its ingestion. And, in drug discrimination studies with cathinone and cathine development of tolerance is observed. With chronic administration of cathinone over 8 days, tolerance developed as shown by the decrease in the discriminative ability of rats and lasted up to 15 days after cessation of the chronic cathinone treatment (Schechter 1986b). With the same experimental approach, acute tolerance to cathine was found to develop when rats were tested 24 h after cathinone treatment (Schechter 1990a), and this tolerance to cathine was also seen with consecutive administration of cathine, but not with cathinone (Schechter 1990a), and involved the dopaminergic system (Schechter 1990b).

While cathinone shows similarity with amphetamine in its effect and mechanism of action, there are several studies that suggest that cathinone differs in an important aspect from amphetamine that could make it a more desirable drug of use. As Rosecrans et al. (1979) commented, "…dl-cathinone is self-administered at lower doses than d-amphetamine, but was less effective than d-amphetamine in disrupting the operant behavior of primates. Thus, dl-cathinone appears less disruptive to behavior, which might explain why this drug was more potent than d-amphetamine in our discrimination study and in self-administration research. From this, one might also predict that individuals using this drug would be less disrupted and better able to function under its effects. In addition, this drug might produce fewer long term d- amphetamine-like behavioral problems because of its apparent lack of a dopamine agonist action." However, studies show that cathinone increases levels of dopamine, but as pointed out there are at least two functions dopamine is involved in—motor function and reward function. Hence it is possible that cathinone has more effect on the reward function than on the motor function.

6 Possible Differences in the Overall Effect of Khat and Cathinone

Most of the studies on khat focus on the effect of cathinone and not on khat (*Catha edulis*) *per se*. There may be functional differences between cathinone and the cocktail of active substances (cathinone, cathine, norephedrine, cathedulins and other alkaloids) ingested by chewing khat. In post-mortem neurotransmitter analysis done 5 days after 9 consecutive days of S-(−)-cathinone, d-(−)-amphetamine or *Catha edulis* extract treatment in rats, for study of behavioral sensitization, Banjaw and

Schmidt (2005) found only *Catha edulis* extract treated rats had reduced levels of dopamine, 3,4-dihydroxyphenylacetic acid (DOPAC) and homovanillic acid (HVA) in the anterior caudate-putamen. In a study of isolation induced aggression (Banjaw et al. 2006), compared to cathinone, rats treated with *Catha edulis* extract had higher elevation of dopamine in the nucleus accumbens with both showing similar depletion of serotonin and 5-hydroindoleacetic acid in the anterior and posterior striatum. Hence khat, as a whole, may cause changes that are different compared to cathinone and these changes may also differ depending on the pattern of khat use, i.e., daily, frequently or binging. In addition, khat contains 62 types of cathedulins (Kite et al. 2003) most of which have not been studied for their biological effects. For example, study of cathedulin fractions showed increased release of dopamine from striatal tissue and binding to both D1 and D2 receptors (Houghton et al. 2011).

7 Conclusions and Future Perspectives

Studies on neural basis of the action of khat and its most active ingredient, cathinone, are few and sporadic. Until recently khat was hardly known outside the region of its traditional and culture use. After the isolation of cathinone, its structural determination and effects, it was found to be similar to amphetamine and other psychostimulants. Khat can be considered a natural source of amphetamine, as coca leaves are the natural source of cocaine. While cathinone shows a clear abuse potential, the effects of khat in users are not clear, as the intake is limited due to the method of consumption as well as the different patterns of consumption. Cathinone, like other psychostimulants, acts in part via the dopaminergic system affecting the reward and motor circuitry of the brain. However, cathinone appears to be less disruptive to motor function compared to amphetamine and this property could make it a more desirable drug of use. Analysis by Nutt et al. (2007) put khat dependence and physical harm lower than tobacco, alcohol, and cannabis, but given the paucity of studies, khat may be a more potent natural drug of abuse than the current literature suggests. How much cathinone the khat user consumes is not known, but a number of socioeconomic studies show that regular khat consumption has a negative impact on the user, family, society, and national economy. There are, in addition, several important questions that need urgent study (1) the short- and long-term effects on executive and cognitive brain function with routine khat use, especially as khat use often starts during the adolescent period, (2) as the traditional and culture control have eroded, khat use is also reported to have increased among women, and pre-natal and postnatal exposure to khat needs to assessed, and (3) there is a complete lack of information on its neurotoxic effects in terms of brain regions and neurotransmitter systems. Despite the historical paucity of studies on khat and cathinone, understanding of action of cathinone-like substances will increase in coming years due to increasing recreational use of synthetic cathinones as designer drugs of abuse—mephedrone, methylone, methcathinone, “bath salts”, “plant food”—in high-income countries. However, studies on khat (*Catha edulis* Forsk) itself cannot be neglected as it is khat with its of cocktail of alkaloids that is consumed by a large number of persons.

References

Al-Motarreb A, Baker K, Broadley KJ (2002) Khat: pharmacological and medical aspects and its social use in Yemen. Phytother Res 16:403–413

Banjaw MY, Schmidt WJ (2005) Behavioural sensitization following repeated intermittent oral administration of *Catha edulis* in rats. Behav Brain Res 156:181–189

Banjaw MY, Fendt M, Schmidt WJ (2005) Clozapine attenuates the locomotor sensitisation and the prepulse inhibition deficit induced by a repeated oral administration of *Catha edulis* extract and cathinone in rats. Behav Brain Res 160:365–373

Banjaw MY, Miczek K, Schmidt WJ (2006) Repeated *Catha edulis* oral administration enhances the baseline aggressive behavior in isolated rats. J Neural Transm 113(5):543–556

Braendan OJ (1979) Research on the chemical composition of khat. NIDA Res Monogr 27:320–321

Calcagnetti DJ, Schechter MD (1992a) Increases in the locomotor activity of rats after intracerebral administration of cathinone. Brain Res Bull 29:843–846

Calcagnetti DJ, Schechter MD (1992b) Psychostimulant-induced activity is attenuated by two putative dopamine release inhibitors. Pharmacol Biochem Behav 43(4):1023–1031

Calcagnetti DJ, Schechter MD (1993) Place preference for the psychostimulant cathinone is blocked by pretreatment with a dopamine release inhibitor. Prog Neuropsychopharmacol Biol Psychiatry 17:637–649

Cleary L, Buber R, Docherty JR (2002) Effects of amphetamine derivatives and cathinone on noradrenaline-evoked contractions of rat right ventricle. Eur J Pharmacol 451(3):303–308

Connor JD, Rostom A, Makonnen E (2002) Comparison of effects khat extract and amphetamine on motor behaviors in mice. J Ethnopharmacol 8:65–71

Fleckenstein AE, Haughey HM, Metzger RR et al (1999) Differential effects of psychostimulants and related agents on dopaminergic and serotonergic transporter function. Eur J Pharmacol 382:45–49

Giannini AJ, Burge H, Shakeen M, Price WA (1986) Khat: another drug of abuse. J Psychoactive Drugs 18:155–158

Glennon RA, Liebowitz SM (1982) Serotonin receptor affinity of cathinone and related analogues. J Med Chem 25:393–397

Gosnell BA, Yracheta JM, Bell SM, Lane KE (1996) Intravenous self-administration of cathinone by rats. Behav Pharmacol 7:526–531

Goudie AJ, Atkinson J, West CR (1986) Discriminative properties of the psychostimulant dl-cathinone in a two lever operant task. Lack of evidence for dopaminergic mediation. Neuropharmacol 25:85–94

Halbach H (1979) Khat-the problem today. NIDA Res Monogr 27:318–319

Houghton P, Ismail M, Salvage S (2011) Not cathinone alone – dopamine, khat constituents and brain tissue. http://darc-khat.middlesex.wikispaces.net/file/view/Not+cathinone+alone+_dopamine+khat+constituents+and+brain+tissue_+Houghton+et+al.pdf

Huang D, Wilson MC (1986) Comparative discriminative stimulus properties of dl-cathinone, d-amphetamine, and cocaine in rats. Pharmacol Biochem Behav 24(2):205–210

Kalix P (1980a) Hypermotility of the amphetamine type induced by a constituent of khat leaves. Br J Pharmacol 68:11–13

Kalix P (1980b) Hyperthermic response to (-)-cathinone, an alkaloid of *Catha edulis* (khat). J Pharm Pharmacol 32:662–663

Kalix P (1983) Effect of the alkaloid (-) cathinone on the release of radioactivityfrom rabbit atria prelabelled with ^{3}H-norepinephrine. Life Sci 32:801–807

Kalix P (1984) Effect of the alkaloid (-)-cathinone on the release of radioactivity from rat striatal tissue prelabelled with ^{3}H-serotonin. Neuropsychobiology 12(2–3):127–129

Kalix P (1986) A comparison of the effects of some phenethylamines on the release of radioactivity from isolated rat caudate nucleus prelabelled with ^{3}H-dopamine. Arzneimittelforschung 36:1019–10121

Kalix P (1992) Cathinone, a natural amphetamine. Pharmacol Toxicol 70:77–86

Kalix P, Geisshüsler S, Brenneisen R (1987) The effect of phenylpentenyl-khatamines on the release of radioactivity from rat striatal tissue prelabelled with [^{3}H]dopamine. J Pharm Pharmacol 39:135–137

Kite GC, Ismail M, Simmonds MS, Houghton PJ (2003) Use of doubly protonated molecules in the analysis of cathedulins in crude extracts of khat (*Catha edulis*) by liquid chromatography/serial mass spectrometry. Rapid Commun Mass Spectrom 17(14):1553–15564

Kuz'min AV, Evartan EE (1991) The intravenous self-administration of narcotics in mice. Zh Vyssh Nerv Deiat Im I P Pavlova 41:1253–1260

Magdum SS (2011) An overview of khat. Addict Disord Their Treat 10:72–83

Mereu GP, Pacitti C, Argiolas A (1983) Effect of (-)-cathinone, a khat leaf constituent, on dopaminergic firing and dopamine metabolism in the rat brain. Life Sci 32:1383–1389

Nutt D, King LA, Saulsbury W, Blakemore C (2007) Development of a rational scale to assess the harm of drugs to potential users. Lancet 369:1047–1053

Odenwald M (2014) Mental health problems associated with the use and abuse of khat (*Catha edulis*). In: Bentivoglio M, Cavalheiro EA, Kristensson K, Patel N (eds) Neglected tropical diseases and conditions of the nervous system. Springer, New York

Oyungu E, Kioy PG, Patel NB (2007) Effect of *Catha edulis* (khat) on behaviour and its potential to induce seizures in Sprague Dawley rats. East Afr Med J 84:219–225

Oyungu E, Kioy PG, Patel NB (2009) Proconvulsant effect of khat (*Catha edulis*) in Sprague-Dawley rats. J Ethnopharmacol 121:476–478

Pehek EA, Schechter MD, Yamamoto BK (1990) Effects of cathinone and amphetamine on the neurochemistry of dopamine in vivo. Neuropharmacology 29:1171–1176

Rosecrans JA, Campbell OL, Dewey WL, Harris LS (1979) Discriminative stimulus and neurochemical mechanism of cathinone: a preliminary study. NIDA Res Monogr 27:328–329

Saha S, Dollery C (2006) Severe ischaemic cardiomyopathy associated with khat chewing. J R Soc Med 99:316–318

Schechter MD (1986a) Discriminative properties of l-cathinone compared to dl- and d-cathinone. Pharmacol Biochem Behav 24:1161–1165

Schechter MD (1986b) Induction of and recovery from tolerance to the discriminative stimulus properties of l-cathinone. Pharmacol Biochem Behav 25:13–16

Schechter MD (1986c) Dopaminergic mediation of a behavioral effect of l-cathinone. Pharmacol Biochem Behav 25:337–340

Schechter MD (1989) Temporal parameters of cathinone, amphetamine and cocaine. Pharmacol Biochem Behav 34:289–292

Schechter MD (1990a) Rats become acutely tolerant to cathine after amphetamine or cathinone administration. Psychopharmacology (Berl) 101:126–131

Schechter MD (1990b) Dopaminergic nature of acute cathine tolerance. Pharmacol Biochem Behav 36:817–820

Schechter MD (1991a) Effect of learned behavior upon conditioned place preference to cathinone. Pharmacol Biochem Behav 38:7–11

Schechter MD (1991b) Effect of serotonin depletion by p- chlorophenylalanine upon discriminative behaviours. Gen Pharmacol 22:889–8893

Schechter MD (1992) Effect of altering dopamine or serotonin neurotransmitters upon cathinone discrimination. Pharmacol Biochem Behav 41:37–41

Schechter MD, Bojaw W (1988) CGS 10746B is able to attenuate the effects of amphetamine: further evidence for dopaminergic mediation. Pharmacol Biochem Behav 30:1089–1092

Schechter MD, Glennon RA (1985) Cathinone, cocaine and methamphetamine: similarity of behavioral effects. Pharmacol Biochem Behav 22(6):913–916

Schechter MD, McBurney D (1991) Effect of repeated administrations upon cathinone discrimination and conditioned place preference. Gen Pharmacol 22:779–782

Schechter MD, Meehan SM (1993) Conditioned place preference produced by the psychostimulant cathinone. Eur J Pharmacol 232:135–138

Schechter MD, Rosecrans JA, Glennon RA (1984) Comparison of behavioral effects of cathinone, amphetamine and apomorphine. Pharmacol Biochem Behav 20:181–184

Schechter MD, Schechter JB, Calcagnetti DJ (1992) Direct microinjection of cathinone into the rat brain produces discriminative stimuli. Pharmacol Biochem Behav 42:619–623

Schmid CL, Bohn LM (2010) Serotonin, but not N-Methyltryptamines, activates the serotonin 2A receptor via a β-arrestin2/Src/Akt signaling complex in vivo. J Neurosci 30:13513–13524

Schorno X, Steinegger E (1978) The phyenylalkylamines of *Catha edulis* forsk: the absolute configuration of cathinone. United Nations Document MNAR/7/1978

Schuster CR, Johanson CE (1979) Behavioral studies of cathinone in monkeys and rats. NIDA Res Monogr 27:324–325

UNODC United Nations Office on Drugs and Crime (1980) Bulletin Narcotics (3)

Vollenweider FX (2001) Brain mechanism of hallucinogens and entactogens. Dialogues Clin Neurosci 3:265–279

Wagner GC, Preston K, Ricaurte GA, Schuster CR, Seiden LS (1982) Neurochemical similarities between d, l-cathinone and d-amphetamine. Drug Alcohol Depend 9:279–284

WHO Expert Committee on Addiction-Producing Drugs (1964) World Health Organization Technical Reprot Series No 273: pp 10

Wise RA, Bozarth MA (1987) A psychomotor stimulate theory of addiction. Psychol Rev 94:469–492

Woolverton WL, Johanson CE (1984) Preference in rhesus monkeys given a choice between cocaine and d, l-cathinone. J Exp Anal Behav 41:35–43

Yanagita T (1979) Studies on cathinones: cardiovascular and behavioral effects in rats and self-administration experiment in rhesus monkeys. NIDA Res Monogr 27:326–327

Yanagita T (1986) Intravenous self-administration of (-)-cathinone and 2-amino-1-(2,5-dimethoxy-4-methyl)phenylpropane in rhesus monkeys. Drug Alcohol Depend 17:135–141

Konzo: Neurology of a Permanent and Non-progressive Motor Neuron Disorder Associated with Food (Cassava) Toxicity

Daniel Okitundu, Dieudonné Mumba, and Desiré Tshala-Katumbay

Abstract Epidemics of neurodegenerative diseases putatively caused by food toxins have been reported under the tropics with no clear understanding of their pathogenetic mechanisms. These diseases include the disease named konzo, which has been well documented in sub-Saharan Africa, mostly among children and women of childbearing age. Konzo is highly prevalent in Congo-Kinshasa, Mozambique, Tanzania, Central African Republic, Angola, and Cameroun. The main clinical picture consists of a symmetrical, permanent and irreversible spastic paraparesis (motor neuron disease) with no signs of sensory or genitourinary impairments. Impaired cognition is possible but yet to be elucidated. The exact pathogenetic mechanisms of the disease remain unknown. Serological studies rule out the role of retroviruses such as the human lymphotropic viruses HIV-I/II or HTLV-I/II. Epidemiological studies consistently show an association between outbreaks of the disease and chronic dietary reliance on foodstuffs derived from insufficiently processed cyanogenic cassava (aka manioc or tapioca). Biochemical and toxicological studies suggest that the metabolites of linamarin (α-hydroxyisobutyronitrile β-D-glucopyranoside, the main cassava cyanogen), notably cyanide (mitochondrial toxin), thiocyanate (AMPA chaotropic agent), and cyanate (protein carbamoylating agent) may be important players in the pathogenesis of konzo. Experimental data

D. Okitundu
Department of Neurology, University of Kinshasa, Kinshasa, DR Congo

D. Mumba
Department of Tropical Medicine, University of Kinshasa, Kinshasa, DR Congo

Institut National de Recherches Biomédicales, Kinshasa, DR Congo

D. Tshala-Katumbay, M.D., Ph.D. (✉)
Department of Neurology, School of Medicine, Oregon Health & Science University, Portland, OR, USA

Center for Research on Occupational and Environmental Toxicology, Oregon Health & Science University, 3181 SW Sam Jackson Park Road, L606, Portland, OR 97239, USA
e-mail: tshalad@ohsu.edu

M. Bentivoglio et al. (eds.), *Neglected Tropical Diseases and Conditions of the Nervous System*, DOI 10.1007/978-1-4614-8100-3_18,

suggest that thiol-redox and protein-folding mechanisms may also be perturbed. Factors of susceptibility including genetics, poor nutrition, and environmental exposures, or their interaction, have yet to be elucidated. There is no cure for konzo. Prevention of the disease remains of paramount importance.

Keywords Cyanogens • Motor neuron disease • Neurodegeneration • Neglected tropical diseases • Neurotoxicity

1 Epidemiology

Cassava toxicity has been incriminated in the pathogenesis of a number of neuroendocrine disorders including konzo, a permanent and clinically distinct disease of the upper motor neuron system (Fig. 1). The name "konzo" derives from a *Yaka* designation of a fetish used by hunters and means tied legs in the *yaka* language (Trolli 1938; van den Abeele and Vandenput 1956; van der Beken 1993). The literature suggests that konzo was known from the *Yaka* indigenous population of the Bandundu province in Zaïre, presently known as the Democratic Republic of Congo (DRC), since the end of the nineteenth century. However, the syndrome was first documented in the medical literature only decades later (Trolli 1938). Since then outbreaks of konzo have been reported in many other sub-Saharan African countries including Mozambique (where it is called *mantakassa*), Tanzania, Central African Republic (CAR), Cameroon, Angola, Uganda, and DRC (Chabwine et al. 2011;

Fig. 1 Spastic stance and fixed joint (ankle ankylosis) in a young boy moderately affected by konzo. Subjects moderately affected by the disease use sticks to walk while those mildly affected may walk with no support. Severely affected subjects may not be able to walk with no support

Ciglenečki et al. 2011; Cliff et al. 1985; Diasolua Ngudi et al. 2011; Howlett et al. 1992; Mbelesso et al. 2009; Ministry of Health Mozambique 1984a, b; Mlingi et al. 2011; Rosling and Tylleskär 2000; Tshala-Katumbay et al. 2001a; Tylleskär et al. 1991, 1994). While isolated cases of the disease may be found, the disease mostly occurs in an outbreak manner that suggests that the disease may be triggered by environmental changes. Most epidemiological studies show a link between outbreaks of konzo, agro-ecological collapse, and consumption of foodstuffs derived from insufficiently processed bitter cassava (*Manihot esculenta Crantz*), a staple for millions of people dwelling under the tropics (Banea et al. 1992; Cliff et al. 2011; Mlingi et al. 2011; Rosling and Tylleskär 2000; Tylleskär et al. 1995a, b). Children and women of childbearing age seem to be more affected for reasons that are yet to be elucidated. The total number of cases of konzo has been estimated reaching hundreds of thousands with most of the cases occurring in the DRC. Accurate regional prevalence rates (perhaps as high as 5 % in certain rural areas) have been difficult to obtain because of unreliable demographic data and poorly functioning surveillance systems (Rosling and Tylleskär 2000).

2 Neurological Signs

The onset and clinical picture of konzo are so distinct that the disease is recognizable by lay people within the affected regions. The disease has a sudden onset often preceded by physical exertion such as a long walk. During the initial phase, subjects affected by konzo experience trembling in their legs, a sensation of leg weakness, heaviness, or stiffness; and muscle cramping usually confined to calf musculature. Acutely reversible somatosensory symptoms are often reported and may include paresthesia, numbness, muscle aching, and a sensation of electrical discharge in the back and legs. Blurred vision and difficulties in swallowing have been occasionally reported. Clinical deficits are usually more severe at the onset of the disease confining the affected subjects to beds. Within a few days, the course of the disease stabilizes and the deficits are mostly confined to the motor system. However, a "second attack" remains possible. The most visible feature is the cross-legged (scissoring) gait of affected subjects who may able to walk and/or run. The observed scissoring gait is a good reflection of the *Yaka* designation of the disease, i.e. konzo (tied legs). Once stabilized, the most prominent sign is a symmetrical postural abnormality with a spastic (cross-legged or scissoring) gait during ambulation. In case of subjects mildly affected by the disease, the spasticity of legs is revealed only when the subject is asked to run (Howlett et al. 1990; Tshala-Katumbay et al. 2001a, b; Tylleskär et al. 1995a). The World Health Organization (WHO) has adopted the following definition and epidemiological criteria for the disease: (1) a heavy reliance on cassava as staple food, (2) abrupt onset (<1 week) of leg weakness and a non-progressive course of the disease in a formerly healthy person, (3) a symmetric spastic abnormality when walking and/or running, (4) bilaterally exaggerated knee and/or ankle jerks without signs of disease of the spine. Based on the ability to walk, the following WHO classification of severity has been proposed: (1) mild form = subject is able to walk without support, (2) moderate form = subject has to use one or two sticks, and (3) severe form = subject is unable to walk (WHO 1996).

On neurological examination, the main clinical picture of konzo consists of an isolated symmetric spastic paraparesis. Deep tendon reflexes of the lower limbs are exaggerated and extensor plantar responses can be elicited in most cases when patients are tested in the recumbent position. Ankle clonus is frequently found. Upper extremities also show pathological reflexes in severely affected subjects, with a clearly noticeable palmomental reflex. Severely affected subjects may show a tetraparesis associated with weakness in their trunk and pseudobulbar signs in the form of speech and swallowing difficulties (Cliff et al. 1999; Cliff and Nicala 1997; Howlett et al. 1990; Tshala-Katumbay et al. 2001a, b; Tylleskär et al. 1995a). A bilateral optic neuropathy may be seen in subjects affected by konzo. This condition encompasses visual impairment, temporal pallor of the optic discs, and defect of visual fields. A pendular nystagmus has also been reported in few cases. The presence of visual symptoms at disease onset and/or optic neuropathy on subsequent examination do not correlate with the severity of konzo (Mwanza et al. 2003a, b). Hearing and sensory function, as well as urinary, bowel and sexual functions, appear to be normal. Subclinical forms and cognitive deficits have been suggested but still need to be confirmed in their nature and origin (Tshala-Katumbay et al. 2001a, b). Physically, stunting and goitre are commonly found (Rosling and Tylleskär 2000).

3 Ancillary Investigations

Biochemical markers of disease susceptibility include low serum levels of prealbumin, albumin, and inorganic urinary sulphate. Markers of toxicant cyanogenic exposure include high blood levels of linamarin (α-hydroxyisobutyronitrile β-D-glucopyranoside), cyanide, thiocyanate, and urinary thiocyanate, the main cyanide detoxification metabolite (Banea-Mayambu et al. 1997; Mlingi et al. 1993; Tylleskär et al. 1991, 1995a, b). Peripheral and central nerve conduction studies show a prominent dysfunction of the pyramidal system with evidence of subclinical involvement of sensory pathways (Table 1). Non-epileptic electroencephalographic abnormalities are found, while magnetic resonance imaging from two subjects has remained unremarkable (Tylleskär et al. 1993; Tshala-Katumbay et al. 2000, 2002a, b). Levels of cyanate, carbamoylated proteins, and markers of oxidative damage may be elevated in subjects severely intoxicated by insufficiently processed cassava (Kassa et al. 2011; Lundquist et al. 1979, 1983, 1995; Rosling 1994; Spencer 1999; Tor-Agbidye et al. 1999).

Table 1 Neuroepidemiology and clinical electrophysiology of konzo (Tshala-Katumbay and Spencer 2007)

Explorations	Abnormalities
Epidemiology (putative causal factors)	Heavy and chronic dietary reliance on insufficiently processed bitter (toxic) cassava
Neurology	Spastic para/tetraparesis Pseudobulbar signs and optic neuropathy
Motor evoked potentials (MEP)	Frequent inability to elicit MEP[a]. When present, central motor conduction time is often increased[b]
Peripheral nerve conduction studies	Normal motor and sensory nerve conduction. Increased amplitude of F-waves
Somatosensory evoked potentials (SEP)	Cortical responses following tibial stimulation frequently absent. If present, the latency is prolonged. Median SEP often normal
Visual evoked potentials (VEP)	Frequent delay and decreased amplitude of P100
Electroencephalography (EEG)	Frequent generalized slowing of background activity and non-specific paroxysmal activities

[a]Consistent with reduction of the upper motor neuron pool
[b]Consistent with loss of pyramidal conductivity from spinal tract (axonal) damage

4 Differential Diagnosis

The diagnosis of konzo is relatively straightforward when the disease occurs in its epidemic form as several families within a community are affected within a common timeframe. The association with poor nutrition and overconsumption of foodstuffs derived from insufficiently processed bitter cassava is needed for the diagnosis of konzo. The disease must be differentiated from lathyrism, another spastic paraparesis associated with poor nutrition and overconsumption of the grass pea *Lathyrus sativus* (Bradbury and Lambein 2011); and tropical spastic paraparesis (TSP), a neurological entity endemic to tropical Africa, Latin America, and the Seychelles and Japan islands (Gessain and Gout 1992; Proietti et al. 2005). In certain parts of the world, for example the Bandundu province of the DRC, clusters of TSP coexists with konzo (Carton et al. 1986; Kayembe et al. 1990). Whereas konzo appears to be a toxiconutritional disease, the etiology of TSP is linked to the infection by the human T-cell lymphotropic virus type I (HTLV-I) (Gessain and Gout 1992). Because of this association, TSP has been named HTLV-I Associated Myelopathy (HAM). The differential diagnosis of konzo, lathyrism and TSP/HAM may be difficult when (a) either konzo or lathyrism coexist with TSP/HAM or (b) a TSP/HAM subject tests negative to HTLV-I while residing in a konzo- or lathyrism-affected area. In these cases, the differential diagnosis is made by the history of the disease obtained after a carefully conducted structured interview, the dietary habits, and findings at physical examination. TSP/HAM is a slowly progressive spastic paraparesis whereas lathyrism and konzo are non-progressive conditions usually of acute or subacute onset. In addition, clinical signs of sensory and sphincter involvement may be evident in the extremities of TSP/HAM subjects. In the absence of co-morbidity, subjects affected by konzo or lathyrism usually test negative for HTLV-I antibodies or protein immunoblots (Carton et al. 1986; Kayembe et al. 1990; Tshala-Katumbay et al. 2001a; Tylleskär et al. 1996).

The process of identifying the cause of a spastic paraparesis under the tropics may be challenging when the physician is faced with an isolated case. In this situation, the differential diagnosis should be made against other causes of non-compressive myelopathy including but not limited to the subacute myelo-optic neuropathy (SMON) due to clioquinol (5-chloro-7-iodo-8-quinolinol; iodochlorhydroxyquin) intoxication (Benvenisti-Zarom et al. 2005; Konagaya et al. 2004), infections or liver failure (Berger and Sabet 2002; McArthur et al. 2005; Utku et al. 2005), hereditary spastic paraplegia (HSP), primary lateral sclerosis (PLS), or amyotrophic lateral sclerosis (ALS) (Strong and Gordon 2005). In many cases, the history of the illness, the presence of signs indicating a systemic disease, the genetic and serum and/or cerebrospinal fluid laboratory analyses, the virological testing against other viruses such as the human immunodeficiency viruses type I (HIV-I) and II (HIV-I-II) as well as the neuroimaging findings may help refine the diagnosis. An earlier detection of treatable causes of myelopathy and spastic paraparesis such as tuberculosis remains of paramount importance (Table 2).

5 On the Pathogenesis of Konzo

The molecular mechanisms of konzo remain unknown, and the absence of an animal model of konzo is a major drawback to study this question. Since many subjects affected by konzo report intense physical activity prior to onset of the disease, failure in cellular energy production, in light of the putative role of cyanide toxicity, has been suggested but not proven. Nevertheless, epidemiological studies have consistently shown an association between the occurrence of konzo, a diet dominated by insufficiently processed bitter (toxic) cyanogenic cassava, and a low protein intake (Rosling and Tylleskär 2000). Bitter (poisonous) varieties of cassava contain large amounts of cyanogenic glucosides namely linamarin (~90 %) and lotaustralin (~10 %). Levels of cyanogenic glucosides in cassava—the plant's chemical defence system against predators—depend on environmental conditions, including season, soil fertility and moisture (Dixon et al. 1994; Mahungu 1994; Sundaresan et al. 1987). The above-mentioned glucosides are stored in the plant cell vacuoles, while a cyanogen-cleavage enzyme (β-glucosidase, syn. linamarase) is present in the cell wall. Once the physical integrity of the cassava root tissue is disrupted, as in cassava processing for food preparation, the cyanogenic glucosides come into contact with linamarase and are hydrolysed. This leads to the formation of glucose and cyanohydrins (Du et al. 1995; Joachim and Pandittesekere 1991; Mkpong et al. 1990). At $pH>5$, the cyanohydrins spontaneously breakdown into ketones and hydrogen cyanide (HCN) gas escapes (O'Brien et al. 1992). Lower pH would lead to persistence of cyanohydrins in the finished food product, with the result that HCN may be released by bacterial enzymatic cleavage in the human gut (O'Brien et al. 1992; Rosling 1994).

Table 2 Tropical non-compressive spastic/tetra paraparesis (Tshala-Katumbay and Spencer 2007). See text for abbreviations

	Toxiconutrional				Infectious			Neurodegenerative
Aetiology/lesion	Konzo	Lathyrism	Combined degeneration of the spinal cord	SMON	TSP/HAM	HIV	Spinal cord TB/cysticercosis/schistosomiasis/syphilis	ALS/PLS/HSP
Dietary/toxic factors	Cassava (*Manihot esculenta*)	Grass pea (*Lathyrus sativus*)	B 12 deficiency	Clioquinol intoxication	No	No	No	Controversial in sporadic ALS cases
Onset	Acute	Acute/subacute	Subacute	Subacute	Subacute	Subacute	Subacute	Subacute
Upper motor neuron disorder	Yes	Yes	No	No	No	No	No	Yes (PLS, HSP) No (ALS)
Lower motor neuron involvement	No	No	Possible	No	Possible	Possible	Possible	Yes (ALS) No (PLS, HSP)
Sensory/sphincter involvement[a]	No	No	Yes	Yes	Yes	Yes	Possible	No
Cranial nerve involvement	Optic neuropathy[a]	No	No	Optic neuropathy	Possible	Possible	Possible	Bulbar palsy but rare in HSP
Clinical course	Non-progressive	Non-progressive	Progressive	Progressive	Progressive	Progressive	Progressive	Progressive
Virological testing	Negative	Negative	Negative	Negative	Positive	Positive	Negative	Negative
Bacterial or parasitic	No	No	No	No	No	No	Yes	No
Genetic susceptibility	Postulated	No	No	No	No	No	No	Documented

TB tuberculosis

[a]More common in cassava-associated ataxic myeloneuropathy

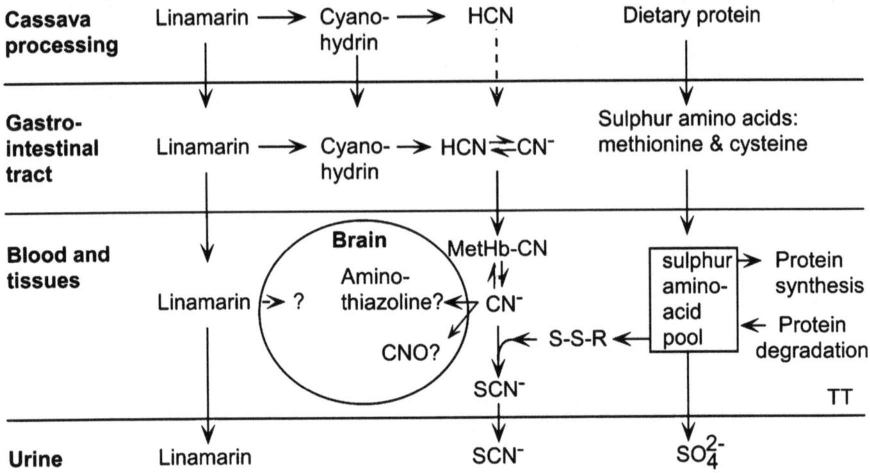

Fig. 2 Metabolism of linamarin in humans. Urinary SCN is considered a good marker of exposure to cassava cyanogens. See text for abbreviations

Several traditional processing methods have been developed to remove cyanogenic glucosides and their degradation products from cassava prior to its consumption. These include soaking or grating followed by drying or heating and fermentation (Bokanga 1995; Nweke 1994; Oke 1994). Toxic cyanogenic exposure arises when adherence to established effective processing techniques is no longer possible, e.g. in times of food shortages induced for example by armed conflict or drought. In some circumstances, a minor change in the sequence of the different processing steps can lead to up to hundred-fold increase levels of cyanogenic compounds in the final food product. The exact toxic agent responsible for konzo remains unknown. Whereas linamarin is excreted intact in urine, a portion breaks down into cyanohydrins, ketone bodies, and HCN, which is normally sequestered by a saturable mechanism involving methemoglobin (MetHB). One major metabolic pathway of free HCN (high acute toxicity) leads to the conversion of cyanide (CN) to thiocyanate (SCN, low acute toxicity). This pathway is mostly regulated by the enzyme rhodanese (thiosulfate sulphur transferase), which is reliant on the dietary intake of sulfur amino acids (SAA) notably methionine and cystein. The toxicity of CN may therefore be exacerbated in a context of SAA deficiency. Under normal conditions, CN is also metabolized to cyanate (OCN) and trace amounts of iminothiazolidine carboxylic acid. However, under conditions of SAA deficiency (low protein intake), the production of neurotoxic OCN is amplified several fold (Spencer 1999; Tor-Agbidye et al. 1999; Rosling and Tylleskär 2000) (Fig. 2). It is known that weeks or months of dependency on incompletely detoxified cassava combined with low intake of proteins—the source of SAA providing the sulphur substrate for the conversion of CN to SCN—leads to outbreaks of konzo (Rosling and Tylleskär 2000). Cyanide has been suggested to play a role in the pathogenesis of konzo because of its potential inhibition effect on mitochondrial energy production. However, this proposal is not supported by rodent studies of chronic cyanide intoxication (Spencer 1999). Other neurotoxic candidates for the causation of konzo include (1) OCN, an established neurotoxic compound that induces axonopathy in humans and animals and (2) SCN, a chaotropic agent that preferentially increases glutamate binding to neuronal AMPA-binding sites resulting in an effect comparable to the AMPA (α-amino-3-hydroxy-5-methyl-4-isoxazolepropionic acid) receptor action of the neuroexcitatory amino acid β-*N*-oxalyl-amino-L-alanine (BOAA), the pulative toxic agent the konzo lookalike lathyrism (Tshala-Katumbay and Spencer 2007; Spencer 1999). OCN probably has a significant role in cassava-related neurodegeneration, because of its ability to carbamoylate proteins and induce neurological disease (Cerami et al. 1973; Farias and Vial 1993; Kassa et al. 2011; Mellado et al. 1982; Ohnishi et al. 1975; Tellez et al. 1979). Experimental data suggest that thiol redox and protein folding mechanisms may be also perturbed in konzo (Kassa et al. 2011). Because of the complexity of the putative pathogenetic mechanisms, levels of exposure to cassava cyanogens, deficiency in essential nutrients, serum markers of protein modifications including oxidation and carbamoylation, and metagenomic variations, should be further explored among human populations affected by the konzo (Bonmarin et al. 2002; Kassa et al. 2011; Tshala-Katumbay and Spencer 2007).

6 Treatment and Prevention

There is no effective treatment for konzo. Once the disease process has stabilized, the disability remains unchanged and irreversible. Attempts with physical therapy have been made to reduce muscle spasm and contractures. Centrally acting spasmolytics, dorsal rhizotomy or partial surgical transection of the thigh adductor muscles, the latter tried with some occasional success in subjects with lathyrism; or intramuscular injection of botulinum toxin, used with success to reduce adductor spasticity in patients with cerebral palsy (Mall et al. 2006; Rosling and Tylleskär 2000; Spencer 1994), could be tried in konzo. So far, prevention is of paramount importance. Efforts should include (1) promotion of efficient detoxification methods of cassava prior to its consumption, and perhaps, the development and distribution of low-toxin cassava strains, (2) adoption of balanced diets, and (3) securing food and farming systems in areas at risk for konzo (Banea et al. 2012; Boivin 1997; Bradbury et al. 2011; Nzwalo and Cliff 2011).

7 Future Perspectives

Future research will elucidate toxico-dietary and/or genetic interactions that modulate the risk for konzo and determine whether the neurodegenerative process extends into cognition. Emerging functional metagenomics (microbiome) may help clarify the role of the gut flora, which may possibly key in regulating the susceptibility to the disease konzo through differential regulation of cassava (linamarin) metabolism. An experimental model of konzo is needed to elucidate the exact molecular pathways that trigger neuronal death, mostly within the motor system, under circumstances of dietary cyanogenic exposure. Focus on prevention remains of paramount importance since the condition appears to be irreversible.

References

Banea M, Bikangi N, Nahimana G, Nunga M, Tylleskär T, Rosling H (1992) High prevalence of konzo associated with a food shortage crisis in the Bandundu region of Zaire. Ann Soc Belg Med Trop 172:295–309

Banea JP, Nahimana G, Mandombi C, Bradbury JH, Denton IC, Kuwa N (2012) Control of konzo in DRC using the wetting method on cassava flour. Food Chem Toxicol 50:1517–1523

Banea-Mayambu JP, Tylleskär T, Gitebo N, Matadi N, Gebre-Medhin M, Rosling H (1997) Geographical and seasonal association between linamarin and cyanide exposure from cassava and the upper motor neurone disease konzo in former Zaire. Trop Med Int Health 2:1143–1151

Benvenisti-Zarom L, Chen J, Regan RF (2005) The oxidative neurotoxicity of clioquinol. Neuropharmacology 49:687–694

Berger JR, Sabet A (2002) Infectious myelopathies. Semin Neurol 22:133–142

Boivin MJ (1997) An ecological paradigm for a health behavior analysis of "konzo", a paralytic disease of Zaire from toxic cassava. Soc Sci Med 45:1853–1862

Bokanga M (1995) Biotechnology and cassava processing in Africa. Food Technol 49:86–90

Bonmarin I, Nunga M, Perea WA (2002) Konzo outbreak, in the south-west of the Democratic Republic of Congo. J Trop Pediatr 48:234–238

Bradbury JH, Lambein F (2011) Konzo and neurolathyrism: similarities and dissimilarities between these crippling neurodegenerative diseases of the poor. Food Chem Toxicol 49:537–538

Bradbury JH, Cliff J, Denton IC (2011) Uptake of wetting method in Africa to reduce cyanide poisoning and konzo from cassava. Food Chem Toxicol 249:539–542

Carton H, Kazadi K, Kabeya, Odio W, Billiau A, Maertens K (1986) Epidemic spastic paraparesis in Bandundu (Zaire). J Neurol Neurosurg Psychiatry 49:620–627

Cerami A, Allen TA, Graziano JH, DeFuria FG, Manning JM, Gillette PN (1973) Pharmacology of cyanate I. General effects on experimental animals. J Pharmacol Exp Ther 185:653–666

Chabwine JN, Masheka C, Balol'ebwami Z, Maheshe B, Balegamire S, Rutega B, Wa Lola M, Mutendela K, Bonnet MJ, Shangalume O, Balegamire JM, Nemery B (2011) Appearance of konzo in South-Kivu, a wartorn area in the Democratic Republic of Congo. Food Chem Toxicol 49:644–649

Ciglenečki I, Eyema R, Kabanda C, Taafo F, Mekaoui H, Urbaniak V (2011) Konzo outbreak among refugees from Central African Republic in Eastern region, Cameroon. Food Chem Toxicol 49:579–582

Cliff J, Nicala D (1997) Long-term follow-up of konzo patients. Trans R Soc Trop Med Hyg 91:447–449

Cliff J, Lundquist P, Mårtensson J, Rosling H, Sörbo B (1985) Association of high cyanide and low sulphur intake in cassava-induced spastic paraparesis. Lancet 2(8466):1211–1213

Cliff J, Nicala D, Saute F, Givragy R, Azambuja G, Taela A, Chavane L, Gani A (1999) Ankle clonus and thiocyanate, linamarin, and inorganic sulphate excretion in school children in communities with Konzo, Mozambique. J Trop Pediatr 45:139–142

Cliff J, Muquingue H, Nhassico D, Nzwalo H, Bradbury JH (2011) Konzo and continuing cyanide intoxication from cassava in Mozambique. Food Chem Toxicol 249:631–635
Diasolua Ngudi D, Banea-Mayambu JP, Lambein F, Kolsteren P (2011) Konzo and dietary pattern in cassava-consuming populations of Popokabaka, Democratic Republic of Congo. Food Chem Toxicol 49:613–619
Dixon A, Aseidu R, Bokanga M (1994) Breeding of cassava for low cyanogenic potential: problems, progress and prospects. Acta Hortic 375:153–161
Du L, Bokanga M, Lindberg-Møller B, Halkier B (1995) The biosynthesis of cyanogenic glucosides in roots of cassava. Phytochemistry 39:323–326
Farias G, Vial C (1993) Functional domains on chemically modified tau protein. Cell Mol Neurobiol 13:173–182
Gessain A, Gout O (1992) Chronic myelopathy associated with human T-lymphotropic virus type I (HTLV-I). Ann Intern Med 117:933–946
Howlett WP, Brubaker GR, Mlingi N, Rosling H (1990) Konzo, an epidemic upper motor neuron disease studied in Tanzania. Brain 113:223–235
Howlett W, Brubaker G, Mlingi N, Rosling H (1992) A geographical cluster of konzo in Tanzania. J Trop Geogr Neurol 2:102–108
Joachim A, Pandittesekere D (1991) Investigations of the hydrocyanic acid content of manioc (*Manihot utillissima*). Trop Agriculturist 100:150–163
Kassa RM, Kasensa NL, Monterroso VH, Kayton RJ, Klimek JE, David LL, Lunganza KR, Kayembe KT, Bentivoglio M, Juliano SL, Tshala-Katumbay DD (2011) On the biomarkers and mechanisms of konzo, a distinct upper motor neuron disease associated with food (cassava) cyanogenic exposure. Food Chem Toxicol 49(3):571–578
Kayembe K, Goubau P, Desmyter J, Vlietinck R, Carton H (1990) A cluster of HTLV-1 associated tropical spastic paraparesis in Equateur (Zaire): ethnic and familial distribution. J Neurol Neurosurg Psychiatry 53:4–10
Konagaya M, Matsumoto A, Takase S, Mizutani T, Sobue G, Konishi T, Hayabara T, Iwashita H, Ujihira T, Miyata K, Matsuoka Y (2004) Clinical analysis of longstanding subacute myelo-optico-neuropathy: sequelae of clioquinol at 32 years after its ban. J Neurol Sci 218:85–90
Lundquist P, Mårtensson J, Sörbo B, Öhman S (1979) Method for determining thiocyanate in serum and urine (Addendum in Clin Chem 1983, 29:403). Clin Chem 25:678–681
Lundquist P, Mårtensson J, Sörbo B, Öhman S (1983) Adsorption of thiocyanate by anion-exchange resins and its analytical application. Clin Chem 29:403
Lundquist P, Kågedal B, Nilsson L, Rosling H (1995) Analysis of the cyanide metabolite 2-aminothiazoline-4-carboxylic acid in urine by High-Performance Liquid Chromatography. Anal Biochem 228:27–34
Mahungu N (1994) Relationship between cyanogenic potential of cassava and other agronomic traits. Acta Hortic 375:125–129
Mall V, Heinen F, Siebel A, Bertram C, Hafkemeyer U, Wissel J, Berweck S, Haverkamp F, Nass G, Doderlein L, Breitbach-Faller N, Schulte-Mattler W, Korinthenberg R (2006) Treatment of adductor spasticity with BTX-A in children with CP: a randomized, double-blind, placebo-controlled study. Dev Med Child Neurol 48:10–13
Mbelesso P, Yogo ML, Yangatimbi E, Paul-Sénékian V, Nali NM, Preux PM (2009) Outbreak of konzo disease in health region No. 2 of the Central African Republic. Rev Neurol (Paris) 165:466–470
McArthur JC, Brew BJ, Nath A (2005) Neurological complications of HIV infection. Lancet Neurol 4:543–555
Mellado W, Slebe JC, Maccioni RB (1982) Tubulin carbamoylation. Functional amino groups in microtubule assembly. Biochem J 203:675–681
Ministry of Health Mozambique (1984a) Mantakassa: an epidemic of spastic paraparesis associated with chronic cyanide intoxication in a cassava staple area in Mozambique. 1. Epidemiology and clinical and laboratory findings in patients. Bull World Health Organ 62:477–484
Ministry of Health Mozambique (1984b) Mantakassa: an epidemic of spastic paraparesis associated with chronic cyanide intoxication in a cassava staple area in Mozambique. 2. Nutritional factors and hydrocyanic content of cassava products. Bull World Health Organ 62:485–492

Mkpong O, Yan H, Chism G, Sayre R (1990) Purification, characterization, and localization of linamarase in cassava. Plant Physiol 93:176–181

Mlingi N, Assey V, Swai A, McLarty D, Karlén H, Rosling H (1993) Determinants of cyanide exposure from cassava in a konzo-affected population in northern Tanzania. Int J Food Sci Nutr 44:137–144

Mlingi NL, Nkya S, Tatala SR, Rashid S, Bradbury JH (2011) Recurrence of konzo in southern Tanzania: rehabilitation and prevention using the wetting method. Food Chem Toxicol 49:673–677

Mwanza JC, Lysebo DE, Kayembe DL, Tshala-Katumbay D, Nyamabo LK, Tylleskar T, Plant GT (2003a) Visual evoked potentials in konzo, a spastic paraparesis of acute onset in Africa. Ophthalmologica 217:381–386

Mwanza JC, Tshala-Katumbay D, Kayembe DL, Eeg-Olofsson KE, Tylleskar T (2003b) Neuro-ophthalmologic findings in konzo, an upper motor neuron disorder in Africa. Eur J Ophthalmol 13:383–389

Nweke F (1994) Cassava processing in Sub-Saharan Africa: implications for expanding cassava production. Outlook Agric 23:197–205

Nzwalo H, Cliff J (2011) Konzo: from poverty, cassava, and cyanogen intake to toxico-nutritional neurological disease. PLoS Negl Trop Dis 5(6):e1051. doi:10.1371/journal.pntd.0001051

O'Brien G, Mbome L, Taylor A, Poulter N (1992) Variations in cyanogen content of cassava during village processing in Cameroon. Food Chem 44:131–136

Ohnishi A, Peterson CM, Dyck PJ (1975) Axonal degeneration in sodium cyanate-induced neuropathy. Arch Neurol 32:530–534

Oke O (1994) Eliminating cyanogens from cassava through processing: technology and tradition. Acta Hortic 375:163–174

Proietti FA, Carneiro-Proietti AB, Catalan-Soares BC, Murphy EL (2005) Global epidemiology of HTLV-I infection and associated diseases. Oncogene 24:6058–6068

Rosling H (1994) Measuring effects in humans of dietary cyanide exposure from cassava. Acta Hortic 375:271–283

Rosling H, Tylleskär T (2000) Cassava. In: Spencer PS, Schaumburg HH, Ludolph A (eds) Experimental and clinical neurotoxicology, 2nd edn. Oxford University Press, New York, pp 338–343

Spencer P (1994) Lathyrism. In: Vinken PJ (ed) Handbook of clinical neurology. Intoxications of the nervous system, part II. Elsevier, Amsterdam, pp 1–20

Spencer P (1999) Food toxins, AMPA receptors, and motor neuron diseases. Drug Metab Rev 31:561–587

Strong MJ, Gordon PH (2005) Primary lateral sclerosis, hereditary spastic paraplegia and amyotrophic lateral sclerosis: discrete entities or spectrum? Amyotroph Lateral Scler Other Motor Neuron Disord 6:8–16

Sundaresan S, Nambisan B, Amma CSE (1987) Bitterness in cassava in relation to cyanoglucoside content. Indian J Agric Sci 57:37–40

Tellez I, Johnson D, Nagel RL, Cerami A (1979) Neurotoxicity of sodium cyanate. New pathological and ultrastructural observations in *Macaca nemestrina*. Acta Neuropathol 47:75–79

Tor-Agbidye J, Palmer V, Laserev M, Craig A, Blythe L, Sabri M, Spencer P (1999) Bioactivation of cyanide to cyanate in sulfur amino acid deficiency: relevance to neurological disease in humans subsisting on cassava. Toxicol Sci 50:228–235

Trolli G (1938) Paraplégie spastique épidémique, "Konzo" des indigènes du Kwango. In: Trolli G (ed) Résumé des observations réunies, au Kwango, au sujet de deux affections d' origine indéterminée. Fonds Reine Elisabeth, Brussels, pp 1–36

Tshala-Katumbay DD, Spencer PS (2007) Toxic disorders of the upper motor neuron system. In: Eisen A and Shaw P (eds) Handb Clin Neurol, Elsevier, New York, pp 353–372

Tshala-Katumbay D, Mutombo Lukusa V, Edebol Eeg-Olofsson K (2000) EEG findings in konzo: a spastic para/tetraparesis of acute onset. Clin Electroencephalogr 31:196–200

Tshala-Katumbay D, Banea Mayambu JP, Kazadi-Kayembe T, Nunga-Matadi R, Bikangi-Nkiabungu F, Edebol Eeg-Olofsson K, Tylleskar T (2001a) Neuroepidemiology of konzo – a spastic para/tetraparesis of acute onset – in a new area of the Democratic Republic of Congo. Afr J Neurol Sci 20:8–12

Tshala-Katumbay D, Eeg-Olofsson KE, Tylleskar T, Kazadi-Kayembe T (2001b) Impairments, disabilities and handicap pattern in konzo-a non-progressive spastic para/tetraparesis of acute onset. Disabil Rehabil 23:731–736

Tshala-Katumbay D, Edebol Eeg-Olofsson K, Kazadi-Kayembe T, Fallmar P, Tylleskar T, Kayembe-Kalula T (2002a) Abnormalities of somatosensory evoked potentials in konzo – an upper motor neuron disorder. Clin Neurophysiol 113:10–15

Tshala-Katumbay D, Eeg-Olofsson KE, Kazadi-Kayembe T, Tylleskar T, Fallmar P (2002b) Analysis of motor pathway involvement in konzo using transcranial electrical and magnetic stimulation. Muscle Nerve 25:230–235

Tylleskär T, Banea M, Bikangi N, Fresco L, Persson LÅ, Rosling H (1991) Epidemiological evidence from Zaire for a dietary aetiology of konzo, an upper motor neuron disease. Bull World Health Organ 69:581–590

Tylleskär T, Howlett WP, Rwiza HT, Aquilonius SM, Stålberg E, Lindén B, Mandahl A, Larsen HC, Brubaker GR, Rosling H (1993) Konzo: a distinct disease entity with selective upper motor neuron damage. J Neurol Neurosurg Psychiatry 56:638–643

Tylleskär T, Légué F, Peterson S, Kpizingui E, Stecker P (1994) Konzo in the Central African Republic. Neurology 44:959–961

Tylleskär T, Banea M, Rosling H (1995a) Konzo, an upper motoneuron disease associated to cassava and agro-ecological collapse in Africa. In: Clifford Rose F (ed) Recent advances in tropical neurology. Elsevier, Amsterdam, pp 355–366

Tylleskär T, Banea M, Bikangi N, Nahimana G, Persson LA, Rosling H (1995b) Dietary determinants of a non-progressive spastic paraparesis (Konzo): a case-referent study in a high incidence area of Zaire. Int J Epidemiol 24:949–956

Tylleskär T, Banea M, Böttiger B, Biberfeld G, Rosling H (1996) Konzo, an epidemic spastic paraparesis in Africa, is not associated with antibodies to HTLV-I, HIV or HIV gag-encoded proteins. J Acquir Immune Defic Syndr Hum Retrovirol 12:317–318

Utku U, Asil T, Balci K, Uzunca I, Celik Y (2005) Hepatic myelopathy with spastic paraparesis. Clin Neurol Neurosurg 107:514–516

Van den Abeele M, Vandenput R (1956) Les principales cultures du Congo Belge. Ministère des Colonies. Direction de l'Agriculture, des Forêts et de l'Elevage, Bruxelles

Van der Beken A (1993) Proverbes yaka du Zaire. Harmattan, Paris

WHO (1996) Konzo, a distinct type of upper motoneuron disease. Wkly Epidemiol Rec 71: 225–232

Food Preservation, Snake Venoms and Stroke in the Tropics

Albert K. Akpalu

Abstract This chapter reviews how do non-traditional risk factors, like envenomation and food preservation, influence or add on to the burden of stroke in sub-Saharan Africa and the tropics. The major means of food preservation in the tropics, namely sub-Saharan Africa, has been by processing of food with salt. A possible direct effect between stroke and salt intake has been evaluated, with data consistent with the hypothesis that a high intake of salt may increase the risk of stroke, independent of effects on blood pressure. Hypertension continues to be the major risk factor for stroke in the tropics. WHO regards population-wide strategies as an integral part of the overall approach to the prevention of cardiovascular disease worldwide, especially those that include dietary modification. There is good evidence that a reduction in salt intake reduces blood pressure and strokes. WHO recently recognized snake bites as a neglected tropical disease. Little is known about the prevalence of stroke among snake bite victims. Much of the available information on the occurrence of stroke following a venomous snake bite is derived from case reports or small cohorts. This chapter examines the mechanisms related to how snake venom causes stroke and the importance of health education and provision of appropriate antivenins for prompt treatment.

Keywords Neglected diseases • Cerebrovascular disease • Diet • Salt • Nervous system • Envenomation • Snake bite

Stroke is the second leading cause of death worldwide, with the traditional stroke risk factors of hypertension, obesity, diabetes and smoking playing an important role (Connor et al. 2007). How do non-traditional risk factors, like envenomation and food preservation, influence or add on to the burden of stroke in sub-Saharan Africa and the tropics? This chapter reviews the available literature and discusses the potential role of neuroscience research in bridging knowledge gap in this field.

A.K. Akpalu (✉)
Neurology Unit, Department of Medicine, Korle bu Teaching Hospital, Accra, Ghana
e-mail: a_akpalu@yahoo.com

M. Bentivoglio et al. (eds.), *Neglected Tropical Diseases and Conditions of the Nervous System*,
DOI 10.1007/978-1-4614-8100-3_19,

1 Food Preservation

Industrial food production and processing is necessarily connected with the use of salt. Salt (sodium chloride) is used as a preservative, spice, agent for color maintenance, texture, and to regulate fermentation by stopping the growth of bacteria, yeast and mold (Doka et al. 2010).

Besides kitchen salt, other types of salt that also contain sodium are used in various technological processes in the industry of food preparation. Most of the "hidden" salt (70–75 %) is ingested with industrial food, which, unfortunately, has been increasingly consumed due to the modern way of life. Bread and bakery products, meat products, various sauces, dried fish, various types of cheese, fast food, conserved vegetables, ready-made soups and food additives are the most common industrial foods rich in sodium (Doka et al. 2010).

Sodium chloride is a widely used food additive due to its low cost, its ability to increase liking of foods via flavor modification, and other functional abilities in a food matrix (Hooper et al. 2002; Kilcast and den Ridder 2007). Sodium chloride is required in food processing for some technological functions such as dough development in bread, and water binding and preservation in meats. However, it is also believed that the sodium content of many food products exceeds technological requirements and primarily acts to enhance sensory effects (Hutton 2002).

Sodium reduces water activity in foods and is therefore able to limit the growth of pathogens and spoilage organism in a variety of food systems (Doyle and Glass 2010). For example, in processed meats and cheeses sodium chloride limits the growth and production of toxin by *Clostridium botulinum* (Taormina 2010). Similarly, sodium diacetate and sodium lactate limit the growth of *Listeria monocytogenes* and lactic acid bacteria in ready-to-eat meats, which makes it safe for human consumption (Seman et al. 2002). Sodium reduction may therefore pose a risk for unwanted bacteria growth and shorten shelf life (Liem et al. 2011).

Many actions have been taken all over the world to restrict salt consumption. The World Health Organization (WHO) recommends the upper limit of salt input of 5 g/day (WHO 2002, 2010). These actions appeal to food industry to reduce the proportion of salt in their products. Besides lower salt addition during manufacture, food industry can use salt substitutes, in particular potassium chloride, in combination with additives that can mask the absence of salt, and flavor intensifiers that also enhance the product salinity. However, food industry is still quite resistant to reducing salt in their products for fear from losing profits (Doka et al. 2010).

The major means of food preservation in the tropics, in sub-Saharan Africa in particular, has been by processing of food with salt. Typical preserved foods include: salted fish; varieties include tilapia considered a delicacy: "koobi" (Fig. 1), momoni, kako; salted pigs' feet and salted beef (Kerry et al. 2005; Cappuccio et al. 2006). Salt is commonly added to food as stock cubes or monosodium glutamate. Processed foods are also not as common in high-income countries (Kerry et al. 2005).

Fig. 1 Salted Tilapia ("koobi") on sale at a Ghanaian market stall

Dietary salt intake plays a critical role in blood pressure regulation. However the question as to whether high dietary salt intake increases risk of stroke, either indirectly via effects on blood pressure or directly via alternative mechanisms, has received limited attention (Perry 2000).

Hypertension is the major risk factor for stroke in the tropics with rates in excess of 70 % (Wiredu and Nyame 2001; Cappuccio et al. 2000; Akpalu and Nyame 2009). A positive correlation between salt intake and hypertension has been shown consistently (Cappuccio et al. 2006). Diet is a major contributor to variation in the occurrence of hypertension and cardiovascular disease, including stroke, worldwide (Goldstein et al. 2011). A possible direct effect between stroke and salt intake has been evaluated, with data consistent with the hypothesis that a high intake of salt may increase the risk of stroke, independent of effects on blood pressure (Perry and Beevers 1992).

It is well known that there is a direct link between reduction in salt intake and drop in both the systolic and diastolic blood pressure (Cappuccio et al. 2006; Sacks et al. 2001). In West Africa, the lower the salt intake, the lower the blood pressure: a reduction in the average salt intake in the whole community may lead to a small but significant reduction in population systolic pressure (Cappuccio et al. 2006; Adeyemo et al. 2002).

On the basis of the overall evidence from relevant studies, a modest reduction in salt intake of ~3 g/day (50 mmol of sodium) in non-hypertensive individuals would reduce blood pressure by 2 to 4/1 to 2 mmHg (He and MacGregor 2003).

In an international study of salt and blood pressure, it was estimated that with a 100 mmol lower daily sodium intake the average decrease in blood pressure from age 25 to 55 would be by 9.0 mmHg for systolic and 4.5 mmHg for diastolic pressure (Intersalt 1988).This reduction, if sustained long-term, would be responsible for a reduction in stroke events of 12–14 % per year (He and MacGregor 2002, 2003). The efficacy of salt reduction in lowering blood pressure levels has been confirmed recently in two small short-term clinical trials in Nigeria (Adeyemo et al. 2002) and Jamaica (Forrester et al. 2005), respectively.

From prospective studies, randomized clinical trials of blood pressure lowering treatment with drugs and short term trials of salt restriction, it is estimated that stroke rates fall by at least 10 % in non-hypertensive individuals whose salt intake was reduced by 50 mmol/day and systolic blood pressure by 1.3 mmHg (He and MacGregor 2003). In hypertensive individuals, the benefit would be much greater. There is no evidence so far to guide to extrapolate these benefits to African populations. It can be argued, however, that the presumptive preventive action could be greater, given the less competing cardiovascular risk in African populations (Cappuccio et al. 2006).

WHO regards population-wide strategies as an integral part of the overall approach to the prevention of cardiovascular disease worldwide, especially those that include dietary modification (WHO 2002; Sacks et al. 2001). There is good evidence that a reduction in salt intake reduces blood pressure (Sacks et al. 2001) and that people of black African origin living in Africa respond well (Cappuccio et al. 2000).

In the western world, it is very difficult to implement successful salt reduction strategies in the population since most of the salt is ingested in processed food. Therefore, interventions should involve the participation of the local indigenous food industry and an attempt by food research institutes in the tropics to look at strategies of desalinating these tasty ingredients in the local cuisine, by soaking salty fish overnight or boiling to remove excess salt (Campbell et al. 2012; Webster et al. 2011).

A recent Cochrane review concludes that despite collating more event data than previous systematic reviews of "Random Controlled Trials" (665 deaths in about 6,250 participants), there is still insufficient power to exclude clinically important effects of reduced dietary salt on mortality or cardiovascular disease morbidity. Estimates of benefits from dietary salt restriction are consistent with the predicted small effects on clinical events attributable to the small blood pressure reduction achieved (Taylor et al. 2011).

Many national and international agencies have acknowledged the role of lifestyle and diet, in particular sodium intake, on blood pressure levels. Diets high in salt are now recognized as one of the leading risks to cardiovascular health in the world as they increase blood pressure in both children and adults (Sacks et al. 2001). Furthermore, a recent meta-analysis of randomized trials has demonstrated that modest reductions in dietary sodium intake are associated with significant reductions in blood pressure in both normotensive and hypertensive subjects and a 20 % reduction in cardiovascular accidents (He and MacGregor 2002, 2011a, b; Asaria et al. 2007).

Furthermore, sodium reduction is noted as one of the most cost-effective and most easily implemented strategies to improve population health (WHO 2007; He and MacGregor 2011a, b; Bibbins-Domingo et al. 2010). Reducing dietary salt is recommended by WHO and many national governmental and nongovernmental health organizations. Some agencies, however, do not promote a reduction in dietary sodium, namely, nongovernmental or commercial organizations such as the Salt Institute, as they are sponsored by either the food or salt industries (WHO 2007, 2010; Webster et al. 2011).

Exposure to excess salt *in utero* may increase blood pressure in offspring or sodium may increase vascular and cardiac disease in the absence of changes in blood pressure (Bibbins-Domingo et al. 2010; He and MacGregor 2010). The burden of disease associated with excess dietary salt is not only high, but may also be underestimated (He and MacGregor 2003; Campbell et al. 2012). WHO currently recommends a daily consumption of less than 5 g of salt, although some agencies recommend that no more than 1,500 mg of sodium should be consumed per day calculated as 2/3 tsp of table salt (CDC 2009; WHO 2010; Goldstein et al. 2011). In most populations, sodium intake is 5.7 g or more/day after age 5, with many populations consuming and average of over 10 g/day (Brown et al. 2009; Anderson et al. 2010).

Excess sodium intake results in adverse effects beyond blood pressure increase. For example, in a population of overweight adults, a daily intake of sodium greater than 2,300 mg/day was associated with 61 % increase in coronary heart disease mortality, 89 % increase in stroke mortality, and 39 % increase in all-cause mortality over a 19-year period (He et al. 1999). Along with the other sodium-related illnesses (i.e., gastric cancers, kidney stones, etc.), it is clear that the economic costs associated with such illnesses can be substantial. This suggests that even a relatively affluent, well-educated population may have difficulty identifying and avoiding high-salt foods even if they perceive it is a health issue and have chosen to follow a low-salt diet (Campbell et al. 2012).

One factor that may increase hypertension in urban areas may be changes in diet, particularly an increase in salt intake (Poulter et al. 1984, 1990; Simmons et al. 1986). Africans are more sensitive to salt than other ethnic groups, with higher consumption in urban areas (Wilson 1986). Understanding patterns and sources of salt intake may help develop strategies of health promotion that aim to reduce salt intake in sub-Saharan Africa. This strategy is now one of the priorities recognized by WHO to reduce cardiovascular disease in developing countries (WHO 2002).

A universal reduction in salt additives during the manufacturing process has a strong potential to reduce health disparities in vulnerable populations while improving overall population health (Campbell et al. 2012).

2 Snake Venom and Stroke

Most snake bite victims are among the poorest people in the world, so this condition poses an economic threat to populations whose misery is already stretched to the limits. WHO recently recognized snake bites as a neglected tropical disease (Williams et al. 2010).This will hopefully allow a better recognition of cases and reduce the impact burden of this human threat (Williams et al. 2010).

Fig. 2 (**a**) *Naja nigricollis nigricollis*; (**b**) Carpet viper. Reproduced with permission from "Amphibians & Reptiles of Morocco and Western Sahara", available from https://www.moroccoherps.com (accessed January 03, 2013)

Venomous snake species are distinguished by characteristic morphology, including the size, color patterns, and shape of their head, tail, and fangs. However, misidentification of the offending snake by the victim or even by experts is common. When available, antigen detection in the serum of the victim by immunodiagnostic tests may be of value for indirect identification of the snake (Habib et al. 2001; Del Brutto and Del Brutto 2012).

Venomous snakes are classified into four families: Viperidae, Atractaspididae, Elapidae, and Colubridae. In Africa, the most dangerous snakes are the mambas (*Dendroaspis* sp.), the Egyptian cobra (*Naja haje annulifera*), the puff adder (*Bitis arietans*), and some carpet vipers (*Echis* sp.) (Enwere et al. 2000; Habib et al. 2001; Michael et al. 2011) (Fig. 2).

In resource-poor countries, snake envenoming has been considered an occupational disease mainly affecting farmers and hunters. Most bites occur when snakes are inadvertently stepped on and they feel threatened, as few snake species are aggressive to humans (Njoku et al. 2008; Michael et al. 2011; Del Brutto and Del Brutto 2012).

2.1 Snake Venom Composition

Snake venom is a mixture of proteins, carbohydrates, and other substances (Doley and Kini 2009). The venom is produced in the venom apparatus, consisting of glands, muscles, ducts, and fangs, which are located behind and below the eyes of the snake (Warrell 2010). Most venom components are primarily directed to immobilize and kill the prey. Relevant components of the venom have cytotoxic, hypotensive, neurotoxic, or anticoagulant effects (Del Brutto and Del Brutto 2012). The composition of the venom is species-specific, i.e., neurotoxins most often predominate in the venom of elapids, while cytotoxic and anticoagulant/procoagulant substances are most often found in the venom of vipers and colubrids (Doley and Kini 2009).

Cytotoxic enzymes (phospholipases A2, metalloproteinases) activate pro-inflammatory mechanisms that cause edema, blister formation, and local tissue necrosis, which are common occurrences at the site of a venomous snake bite. These enzymes also favor the release of bradykinin, prostaglandin, cytokines, and sympathomimetic amines, which are also responsible for pain (Vaiyapuri et al. 2010; Teixeira et al. 2009).

Other venom toxins, like aminopeptidases, can affect physiological functions causing, for example, systemic hypotension (Vaiyapuri et al. 2010). Some peptides of the venom act as angiotensin-converting enzyme inhibitors, causing a drop in arterial blood pressure (Joseph et al. 2004). Other toxins, such as safarotoxins and endothelins, are potent vasoconstrictors of coronary arteries and may cause myocardial ischemia or cardiac arrhythmias (Kochva et al. 1993).

Neurotoxins are major components of some snake venoms, particularly elapids. These toxins do not cross the blood–brain barrier, but cause paralysis by affecting the neuromuscular transmission at either presynaptic or postsynaptic levels (Lewis and Gutmann 2004). Presynaptic neurotoxins are phospholipase A2 complexes, called β-neurotoxins, which inhibit the release of acetylcholine from the presynaptic terminal. Such inhibition may be irreversible as these toxins interfere with the formation of new acetylcholine-containing synaptic vesicles (Lewis and Gutmann 2004).

Examples of β-neurotoxins include taipoxin, paradoxyn, trimucrotoxin, viperotoxin, pseudocerastes, textilotoxin, and crotoxin (Doley and Kini 2009). On the other hand, postsynaptic neurotoxins are three-finger protein complexes, called α-neurotoxins, which have a curare-like mechanism of action, causing a reversible blockade of acetylcholine receptors. The best characterized α-neurotoxins are irditoxin. Some venoms contains both α- and β-neurotoxins, producing complex blockade of neuromuscular transmission (Doley and Kini 2009; Del Brutto and Del Brutto 2012).

Snake venom toxins may also interfere with blood coagulation and cause hemorrhages or thrombosis. Metalloproteinases activate factor X and serine proteases are potent prothrombin activators. In addition, a number of non-enzymatic proteins, called snake venom C-type lectins and some of the three-finger toxins, have anticoagulant or procoagulant activity and may be either agonists or antagonists of platelet aggregation (Doley and Kini 2009).

Paradoxically, some of the components of the venom of snakes may also have immunomodulatory, anti-inflammatory and anti-tumoral effects and are currently under investigation as potential therapeutic agents for human diseases (Koh et al. 2006).

Another toxin, ancrod, a serine protease derived from the venom of the Malayan pit viper (*Calloselasma rhodostoma*), has been used for years for therapy of patients with acute ischemic stroke because of its defibrinogenating properties *in vivo* by converting plasma fibrinogen into a soluble, non cross-linked form of fibrin (Levy et al. 2009; Olinger et al. 1988). Because of this property, the toxin ancrod has been used in clinical trials for the treatment of thrombotic stroke. Some other snake venom toxins are used in clinical laboratories for the assay of hemostatic parameters (Marsh and Fyffe 1996).

2.2 Local Symptoms and Signs of Snake Bite

Considering the systemic manifestations of snake bite, and specifically stroke, an understanding of the local features of envenomation is important. Pain is the earliest symptom after a viper's bite, but may be minor or absent after envenoming by many elapids (Teixeira et al. 2009). This is followed by swelling, blister formation, and necrosis of the skin and subcutaneous tissue. These manifestations may be complicated by tourniquet applications. The entire limb may become edematous, and a compartmental syndrome may develop (Vigasio et al. 1991). Pain and local necrotic symptoms are especially severe after viper and colubrid bites (Warrell 2010; Njoku et al. 2008). Venom ophthalmia is a particular syndrome characterized by ocular pain, hyperemia, blepharitis, and corneal erosions, which may occur when the venom of spitting cobras enters the eye of the victims (Chu et al. 2010).

2.3 Systemic Manifestations of Snake Bite

Some venom components cause vasodilatation and capillary leakage, which, by themselves or together with the hypovolemia resulting from acute bleeding, may cause arterial hypotension and shock (Otero-Patiño 2009). Cardiac ischemia or arrhythmias because of constriction of coronary arteries may occur after the bite of burrowing vipers due to safarotoxins in their venom (Kochva et al. 1993). Acute renal failure may also occur in severely envenomed snake bite victims and may be related to hypovolemic shock, consumption coagulopathy, rhabdomyolysis, and direct nephrotoxicity causing tubular necrosis (Sitprija 2008).

Pituitary hemorrhages, causing acute hypopituitarism, may occur after the bite of Russell's vipers due to the presence of hemorrhagins in their venom (Antonypillai et al. 2011).

Clinical manifestations related to thrombotic and hemorrhagic complications are common in envenomed snake bite victims, particularly in those bitten by vipers and colubrids. The venom of these snakes contains toxins that alter the coagulation system and the function of platelets in different ways, representing the basis for thrombosis and hemorrhages at different locations, including the upper and lower digestive tract, the lungs, the retroperitoneal space, the genitourinary tract, and the nervous system (Markland 1998).

2.4 Neurological Complications of Snake Bite

Neurological signs and symptoms after a venomous snake bite are most often related to the toxic effects of the venom, i.e., anticoagulant/procoagulant activity or neurotoxicity. As noted previously, both effects may be combined in the same venom causing complex and severe neurological damage (Doley and Kini 2009).

Some patients develop neurological complications related to cerebral hypoxia, which, in turn, is related to the hypotensive shock that may accompany some snake bite envenoming (Del Brutto and Del Brutto 2012). The neurological symptoms include drowsiness, confusion and convulsions (Bashir and Jinkins 1985). Subarachnoid hemorrhage has been reported in three of 115 patients from Nigeria, two of whom died (Warrell et al. 1977). A flaccid dysarthria has been reported after snake bite amongst other rare manifestations (Vir et al. 2010).

2.5 Cerebrovascular Accidents Following Snake Bite

Little is known about the prevalence of stroke among snake bite victims. Much of the available information on the occurrence of stroke following a venomous snake bite is derived from case reports or small cohorts. Most reports of stroke cases following snake bites derive from rural areas of developing countries where neuroimaging is not available (Bashir and Jinkins 1985; Del Brutto and Del Brutto 2012; Warrell et al. 1975, 1977).

Seven patients with cerebral infarction have been reported among 109 snake bite victims (6.4 %) from Martinique, but there is no information on the computerized tomography (CT) findings of these patients (Thomas et al. 1998). Fifteen patients with intracranial hemorrhages among 294 cases of snake bites (5.1 %) have been reported from Ecuador; though the diagnosis was only clinical (Kerrigan 1991).

A more recent hospital-based study showed a 2.6 % prevalence of cerebrovascular complications among 309 snake bite victims attending a general hospital in Ecuador (Mosquera et al. 2003). This report may be relatively accurate because neuroimaging studies were performed in every patient presenting with deterioration of consciousness or focal neurological signs. Brazilian studies including 322 patients bitten by venomous snakes reported only one stroke case (0.3 %) (Bucaretchi et al. 2001).

In Africa, data is limited and numbers of stroke event following venomous bites are few but lethal. It is possible that cases of cerebrovascular accidents die before getting the necessary help (Warrell et al. 1977; Habib et al. 2001; Njoku et al. 2008; Michael et al. 2011) and data is lost as traditional norms are followed by burying the victims without a post-mortem examination.

2.6 Cerebral Infarction

Well documented cases (25) of cerebral infarctions following snake bites have been reviewed (Del Brutto and Del Brutto 2012). The offending snake was identified as a viper in 23 of these cases, and only one patient was bitten by an elapid. All but two patients had cortical infarcts which were located in multiple cerebral arterial territories in 16; only two patients had hemorrhagic transformation of the infarcts. All patients had pain and local signs at the bite site, but only five developed systemic manifestations (hypotensive shock, renal failure, rhabdomyolysis). Laboratory tests were abnormal in 13 of 16 cases. Most common abnormalities included low platelet counts, increased fibrinogen levels, and prolongation of prothrombin and thromboplastin times. Snake antivenom was administered to 24 of the 25 patients. Only three patients died, 16 were left with mild to severe sequelae, and six recovered completely (Lee et al. 2001; Numeric et al. 2002; Thomas et al. 2006; Mosquera et al. 2003; Cole 1996; Hung et al. 2002; Narang et al. 2009; Bashir and Jinkins 1985; Boviatsis et al. 2003; Merle et al. 2005).

The cause of ischemic stroke in snake bite victims is controversial. Proposed aetiopathogenesis include the presence of venom toxins causing hypercoagulability and endothelial damage, immune-mediated vasculitis, systemic hypotension (Hoskote et al. 2009; Mugundhan et al. 2008) and arterial thrombosis (Bashir and Jinkins 1985).

That infarctions are most likely related to prothrombotic effects of the venom and to endothelial damage is supported by the finding of multiple cerebral infarctions in more than 60 % of cases. None of the patients had infarctions in watershed areas, which are common in hypotensive shock, and vasculitis seems unlikely as many patients developed the infarct within the first few hours after the envenoming (Gawarammana et al. 2009). Angiography, performed in some of these cases, has shown occlusion of large and medium size intracranial arteries, a finding that is uncommon in immune-mediated vasculitis or in hypotensive shock (Bashir and Jinkins 1985; Lee et al. 2001).

2.7 Cerebral Haemorrhage

Well-documented cases of snake bite causing bilateral intracranial haemorrhage have been reported (see for review Del Brutto and Del Brutto 2012; Bartholdi et al. 2004; Kitchens and Eskin 2008; Kouyoumdjian et al. 1991; Midyett 1998; Yap and Ihle 2003).

Nine patients had lobar haemorrhages, which were multiple in five cases. Some of the lobar bleedings extended into the ventricular system or the subarachnoid and subdural spaces. One patient had a cerebellar haemorrhage, one an epidural hematoma, and the other a subarachnoid haemorrhage. Of interest, in CT scans the attenuation coefficient of many of these bleedings was rather low, and haemorrhages appeared almost isodense to the brain parenchyma on CT, a finding that could be explained by the associated severe anemia (Del Brutto and Del Brutto 2012).

The offending snake was a viper in ten patients and an elapid in the remaining two. Eleven patients had abnormal laboratory tests, including low platelet counts, increased clotting times, increased prothrombin and thromboplastin times, and decrease fibrinogen levels. Antivenin was given to all but one patient. Despite therapy, nine patients died and the other three were left with irreversible sequelae. In all cases, intracranial haemorrhages were related to abnormalities in hemostatic factors ranging from decreased platelets to a severe consumption coagulopathy (Del Brutto and Del Brutto 2012).

2.8 Treatment of Cerebrovascular Complications Following Snake Bite

The use of first aid measures in the management of snake bite by patients in rural communities in Africa is a popular practice. Common first aid measures employed include tourniquet (ropes, pieces of cloth), use of the "black stone" (se further), and application of traditional medicine and incision of site of bites (Madaki et al. 2005; Njoku et al. 2008; Michael et al. 2011). The use of first aid measure did not prevent spread of the venom. However, the use of the tourniquet, traditional herbs and the "black stone" appears to have beneficial effects by reducing the average antivenom requirement of patients (Madaki et al. 2005). On the other hand, current evidence suggests that, in addition to the general measures, ancient practices such as the use of tight tourniquets, as well as incision, sustained suction, or the use of herbal medicines to the wound should be discouraged (Warrell 2010; Gold et al. 2002; Michael et al. 2011).

The severity of envenomation must be evaluated and strokes will fall under moderate to severe envenoming (Gold et al. 2002). Severe anemia must be corrected with transfusion of packed red blood cells, and venom-induced hemostatic abnormalities (including those causing ischemic or hemorrhagic strokes) require the administration of antivenin. In the case of bleeding, the use of fresh frozen plasma, cryoprecipitates, and human fibrinogen concentrates is indicated (Gold et al. 2002).

2.9 The Use of the "Black Stone"

The "black stone" also known as a viper's stones, "Belgian Stone", *schwarze Steine*, *pierre noire*, *piedritas negras* or serpent-stones, are animal bones, which are widely used and promoted as a treatment for snake bite in Africa, South America and Asia (Chippaux et al. 2007; Madaki et al. 2005; Michael et al. 2011). Since no actual clinical trial has ever been performed, an experimental approach has been used to evaluate its efficacy against the venoms of *Bitis arietans*, *Echis ocellatus* and *Naja nigricollis* (Chippaux et al. 2007). Local application of the "black stone" after intramuscular venom injection had no demonstrable effect on the outcome of envenomation and did not change the LD50 of *B. arietans* venom. The results show that, contrary to widespread belief, no efficacy of the "black stone" in the treatment of envenomation may be expected (Chippaux et al. 2007).

2.10 Antivenin

Antivenin is obtained from the serum of animals that have been immunized with the venom of a snake. Introduced more than one century ago, antivenin has significantly reduced the mortality of snake bite envenoming (Gold et al. 2002; Juckett and Hancox 2002). Antivenin may correct hemostatic abnormalities and improve the clinical manifestations of the patient from minutes to a few hours after administration, but this effect seems to be less remarkable when administered several hours or days after the bite, particularly when cerebrovascular complications have appeared (Narvencar 2006).

Recurrent envenoming may occur because of delayed absorption of venom remnants from the wound, after the effect of the antivenin has been cleared. In these cases, repeated doses may be needed (Warrell 2010).

Development of antivenins containing neutralizing antibodies that can persist in the blood for prolonged times may reduce the risk of delayed envenoming (Gutierrez et al. 2003). The use of antivenin is not free of adverse reactions. Therefore, it should only be given to patients who develop evidence of envenoming. The use of corticosteroids or epinephrine before antivenin administration may help to reduce the occurrence of hypersensitivity reactions, although these drugs must be used on individual basis (Gold et al. 2002).

3 Conclusions and Future Perspectives

Diet in the context of reduction of salt intake is one of the important lifestyle factors for the prevention of stroke (Goldstein et al. 2011; Kokubo 2012). Saltiness is an important sensory attribute of many foods, and sodium chloride contributes more than just saltiness to the characteristic flavor of many food types. Whilst ensuring adequate dietary sodium intake is vital to health, intake of excessive amounts of sodium has been linked to development of hypertension and subsequent pathologies (Karppanen et al. 2005). Changing the dietary sodium content of a population that has adapted to a high sodium diet will not be easy, and will entail a number of strategies (Kilcast and den Ridder 2007).

Part of the reason why previous attempts to reduce sodium in processed foods were not successful can be ascribed to loss of palatability of foods (Mattes 1997). One strategy to reduce sodium is to replace sodium with potassium salts. However, while potassium chloride elicits weak saltiness, at higher concentrations it also elicits metallic and bitter taste limiting its utility in food.

The "stealth" approach of gradual sodium reduction in processed foods (Girgis et al. 2003), thereby modifying consumers' salt taste experience over time is recognized as arguably the best current strategy to reduce sodium in foods. Initiatives carried out in several countries will provide much needed data about whether this approach is successful in reducing sodium intake and blood pressure at population level (Liem et al. 2011).

There are potential barriers such as the technological function of sodium in foods, and decreased consumer liking, that could be put forward by the food industry to stop sodium reduction. However, mandating sodium reduction in certain food categories would urge food industry to devise creative solutions, which will enable sodium reduction without compromising on food safety and consumer acceptance. There is no doubt that sodium reduction in foods will be difficult, but equally there is no doubt that reducing the level of sodium in foods is essential for population health (Liem et al. 2011).

In addition to safety risks and processing challenges involved in producing low sodium foods, there is also an economic consideration. Sodium chloride is relatively cheap and any substitute used will increase the cost of the product. Furthermore, in order to find suitable salt replacers, bitter blockers and or flavor enhancers substantial effort and funds need to be invested in research, development and consumer testing. The food industry will need to take all functions of sodium in foods into account when addressing sodium reduction (Liem et al. 2011).

Snake bite envenoming is a health challenge in the developing world. Implementation of public campaigns directed to populations at risk will undoubtedly reduce the number of snake bites. Such campaigns must focus on simple preventive measures such as the use of flashlights when walking on footpaths at night, or to avoid walking barefoot or wearing only sandals and short pants. Instead, long trousers and boots must be worn during agricultural or fishing activities (Del Brutto and Del Brutto 2012).

The data reviewed above indicate that cerebral haemorrhage has a greater mortality than cerebral infarction following snake bite. The mechanism of action and predilection of viperidae to cause complications and death is a matter for future research and bench to clinic translation.

Health education and provision of appropriate antivenins for prompt treatment is essential. Development of a new generation of more innocuous antivenins presenting less severe adverse reactions in humans holds a lot of promise (Winkel et al. 2006).

Defibrogenation with snake venom-derived enzymes as the toxin ancrod still holds a lot of promise for stroke (Levy et al. 2009), and may be more cost effective than the standard thrombolytic therapy still unavailable to the majority of people living in the tropics. Herein lies a huge gap of knowledge and a bright prospect for future neuroscience research and application from bench to clinic.

References

Adeyemo AA, Prewitt TE, Luke A, Omotade OO, Rotimi CN, Brieger WR, Cooper RS (2002) The feasibility of implementing a dietary sodium reduction intervention among free-living normotensive individuals in south West Nigeria. Ethn Dis 2002(12):207–212

Akpalu AK, Nyame PK (2009) Plasma homocysteine as a risk factor for strokes in Ghanaian adults. Ghana Med J 43(4):157–163

Anderson CAM, Appel LJ, Okuda N et al (2010) Dietary sources of sodium in China, Japan, the United Kingdom, and the United States, women and men aged 40 to 59 years: the INTERMAP Study. J Am Diet Assoc 110:736–745

Antonypillai CN, Wass JA, Warrell DA, Rajaratnam HM (2011) Hypopituitarism following envenoming by Russell's viper (*Daboia siamensis* and *D. russelii*) resembling Sheehan's syndrome: first case report from Sri-Lanka, a review of the literature and recommendations for endocrine management. QJM 104:97–108

Asaria P, Chisholm D, Mathers C, Ezzati M, Beaglehole R (2007) Chronic disease prevention: health effects and financial costs of strategies to reduce salt intake and control tobacco use. Lancet 370:2044–2053

Bartholdi D, Selic C, Meier J, Jung HH (2004) Viper snakebite causing symptomatic intracerebral hemorrhage. J Neurol 251:889–891

Bashir R, Jinkins J (1985) Cerebral infarction in a young female following snake bite. Stroke 16:328–330. doi:10.1161/01.STR.16.2.328

Bibbins-Domingo K, Chertow GM, Coxson PG et al (2010) Projected effect of dietary salt reductions on future cardiovascular disease. N Engl J Med 362(7):590–599

Boviatsis EJ, Kouyialis AT, Papatheodorou G, Gavra M, Korfias S, Sakas DE (2003) Multiple hemorrhagic brain infarcts after viper envenomation. Am J Trop Med Hyg 68:253–257

Brown IJ, Tzoulaki I, Candeias V, Elliott P (2009) Salt intakes around the world: implications for public health. Int J Epidemiol 38(3):791–813

Bucaretchi F, Herrera SRF, Hyslop S, Baracat ECE, Vieira RJ (2001) Snakebites by Bothrops spp in children in Campina, Sao Paulo, Brazil. Rev Inst Med Trop Sao Paulo 43:329–333

Campbell NRC, Johnson JA, Campbell TS (2012) Sodium consumption: an individual's choice? Int J Hypertens. doi:10.1155/2012/860954

Cappuccio FP, Plange-Rhule J, Phillips RO, Eastwood JB (2000) Prevention of hypertension and stroke in Africa. Lancet 356:677–678

Cappuccio FP, Kerry SM, Micah FB, Plange-Rhule J, Eastwood JB (2006) A community programme to reduce salt intake and blood pressure in Ghana. BMC Public Health 6:13

Centers for Disease Control and Prevention (CDC) (2009) Application of lower sodium intake recommendations to adults—United States, 1999–2006. Morb Mortal Wkly Rep 58:281–283

Chippaux JP, Ramos-Cerrillo B, Stock RP (2007) Study of the efficacy of the black stone on envenomation by snake bite in the murine model. Toxicon 49:717–720

Chu ER, Weinstein SA, White J, Warrell DA (2010) Venom ophthalmia caused by venom of the spitting elapids and other snakes: report of ten cases with review of epidemiology, clinical features, pathophysiology and management. Toxicon 56:259–272

Cole M (1996) Cerebral infarct after rattlesnake bite. Arch Neurol 53:957–958

Connor MD, Walker R, Modi G, Warlow CP (2007) Burden of stroke in black populations in sub-Saharan Africa. Lancet Neurol 6:269–278

Del Brutto OH, Del Brutto VJ (2012) Neurological complications of venomous snake bites: a review. Acta Neurol Scand 125:363–372

Doka JJ, Nola IA, Andabaka D (2010) Food processing industry – the salt shock to the consumers. Acta Med Croatica 64:97–103

Doley R, Kini RM (2009) Protein complexes in snake venom. Cell Mol Life Sci 66:2851–2871

Doyle ME, Glass KA (2010) Sodium reduction and its effect on food safety, food quality, and human health. Compr Rev Food Sci Food Saf 9:44–56

Enwere GC, Obu HA, Jobarteh A (2000) Snake bites in children in The Gambia. Ann Trop Pediatr 20:121–124

Forrester T, Adeyemo A, Soarres-Wynter S, Sargent L, Bennett F et al (2005) A randomized trial on sodium reduction in two developing countries. J Hum Hypertens 19:55–60

Gawarammana I, Mendis S, Jeganathan K (2009) Acute ischemic strokes due to bites by *Daboia russelli* in Sri Lanka – first authenticated case series. Toxicon 54:421–428

Girgis S, Neal B, Prescott J, Prendergast J, Dumbrell S, Turner C (2003) A one-quarter reduction in the salt content of bread can be made without detection. Eur J Clin Nutr 57:616–620

Gold BS, Dart RC, Barish RA (2002) Bites of venomous snakes. N Engl J Med 347:347–356

Goldstein LB, Bushnell CD, Adams RJ et al (2011) Guidelines for the primary prevention of stroke: a guidelines for healthcare professionals from the American Heart Association/American Stroke Association. Stroke 42:517–584

Gutierrez JM, Leon G, Lomonte B (2003) Pharmacokinetic pharmacodynamic relationships of immunoglobulin therapy for envenomation. Clin Pharmacokinet 42:721–741

Habib AG, Gebi UI, Onyemelukwe GC (2001) Snake bite in Nigeria. Afr J Med Sci 30:171–178

He FJ, MacGregor GA (2002) Effect of modest salt reduction on blood pressure: a meta-analysis of randomized trials. Implications for public health. J Hum Hypertens 16:761–770

He FJ, MacGregor GA (2003) How far should salt intake be reduced? Hypertension 42: 1093–1099

He FJ, MacGregor GA (2010) Reducing population salt intake worldwide: from evidence to implementation. Prog Cardiovasc Dis 52:363–382

He FJ, MacGregor GA (2011a) Salt and cardiovascular disease mortality. Lancet 377(9778):1626

He FJ, MacGregor GA (2011b) Salt reduction lowers cardiovascular risk: meta-analysis of outcome trials. Lancet 378(9789):380–382

He J, Ogden LG, Vupputuri S, Bazzano LA, Loria C, Whelton PK (1999) Dietary sodium intake and subsequent risk of cardiovascular disease in overweight adults. J Am Med Assoc 282: 2027–2034

Hooper L, Bartlett C, Davey Smith G, Ebrahim S (2002) Systematic review of long term effects of advice to reduce dietary salt in adults. Br Med J 325:628

Hoskote SS, Iyer VR, Kothari VM, Sanghvi DA (2009) Bilateral anterior cerebral artery infarction following viper bite. J Assoc Physicians India 57:67–69

Hung DZ, Wu ML, Deng JF, Yang DY, Lin-Shiau SY (2002) Multiple thrombotic occlusions of vessels after Russells's viper envenoming. Pharmacol Toxicol 91:106–110

Hutton T (2002) Sodium technological functions of salt in the manufacturing of food and drink products. Br Food J 104:126–152

Intersalt (1988) An international study of the electrolyte excretion and blood pressure. Results for 24 h urinary sodium and potassium excretion. Intersalt Cooperative Research Group. Br Med J 297:319–328

Joseph R, Pahari S, Hodgson WC, Kini RM (2004) Hypotensive agents from snake venoms. Curr Drug Targets Cardiovasc Haematol Disord 4:437–459

Juckett G, Hancox JG (2002) Venomous snakebites in the United States: management review and update. Am Fam Physician 65:1367–1374

Karppanen H, Karppanen P, Mervaala E (2005) Why and how to implement sodium, potassium, calcium, and magnesium changes in food items and diets? J Hum Hypertens 19:S10–S19

Kerrigan KR (1991) Venomous snakebite in eastern Ecuador. Am J Trop Med Hyg 44:93–99

Kerry SM, Emmett L, Micah FB, Martin-Peprah R, Antwi S, Phillips RO, Plange-Rhule J, Eastwood JB, Cappuccio FP (2005) Rural and semi-urban differences in salt intake, and its dietary sources, in Ashanti, West Africa. Ethn Dis 15:33–39

Kilcast D, den Ridder C (2007) Sensory issues in reducing salt in food products. In: Kilcast D, Angus F (eds) Reducing salt in foods – practical strategies. Woodhead, Cambridge, UK, pp 195–199

Kitchens C, Eskin T (2008) Fatality in a case of envenomation by *Crotalus adamanteus* initially successfully treated with polyvalent ovine antivenom followed by recurrence of defibrinogenation syndrome. J Med Toxicol 4:180–183

Kochva E, Bdolah A, Wollberg Z (1993) Safarotoxins and endothelins: evolution, structure and function. Toxicon 31:541–568

Koh DC, Armugam A, Jeyaseelan K (2006) Snake venom components and their applications in biomedicine. Cell Mol Life Sci 63:3030–3041

Kokubo Y (2012) Traditional risk factor management for stroke: a never ending challenge for health behaviours of diet and physical activity. Curr Opin Neurol 25:11–17

Kouyoumdjian JA, Polizelli C, Lobo SM, Guimares SM (1991) Fatal extradural haematoma after snake bite (*Bothrops moojeni*). Trans R Soc Trop Med Hyg 85:552

Lee BC, Hwang SH, Bae JC, Kwon SB (2001) Brainstem infarction following Korean viper bite. Neurology 56:1244–1245

Levy DE, del Zoppo GJ, Demaerschalk BM, Demchuk AM, Diener HC, Howard G, Kaste M, Pancioli AM, Ringelstein EB, Spatareanu C, Wasiewski WW (2009) Ancrod in acute ischemic stroke. Results of 500 subjects beginning treatment within 6 hours of stroke onset in the ancrod stroke program. Stroke 40:3796–3803

Lewis RL, Gutmann L (2004) Snake venoms and the neuromuscular junction. Semin Neurol 24:175–179

Liem DG, Miremadi F, Keast RS (2011) Reducing sodium in foods: the effect on flavor. Nutrients 3:694–711. doi:10.3390/nu3060694

Madaki J, Obilom R, Mandong B (2005) Pattern of first-aid measures used by snake-bite patients and clinical outcome at Zamko Comprehensive Health Centre, Langtang, Plateau State. Niger Med Pract 48:10–13

Markland FS (1998) Snake venom and the hemostatic system. Toxicon 36:1749–1800
Marsh NA, Fyffe TL (1996) Practical applications of snake venom toxins in haemostasis. Boll Soc Ital Biol Sper 72:263–278
Mattes RD (1997) Physiologic responses to sensory stimulation by food: nutritional implications. J Am Diet Assoc 97:406–413
Merle H, Donnio A, Ayeboua L, Plumelle Y, Smadja D, Thomas L (2005) Occipital infarction revealed by quadranopsia following snakebite by *Bothrops lanceolatus*. Am J Trop Med Hyg 73:583–585
Michael GC, Thacher TD, Shehu MIL (2011) The effect of pre-hospital care for venomous snake bite on outcome in Nigeria. Trans R Soc Trop Med Hyg 105:95–101
Midyett FA (1998) Neuroradiologic findings in brown snake envenomation: computed tomography demonstration. Australas Radiol 42:248–249
Mosquera A, Idrovo LA, Tafur A, Del Brutto OH (2003) Stroke following *Bothrops* spp. snakebite. Neurology 60:1577–1580
Mugundhan K, Thruvarutchelvan K, Sivakumar S (2008) Posterior circulating stroke in a young male following snake bite. J Assoc Physicians India 56:713–714
Narang SK, Paleti S, Azeez Asad MA, Samina T (2009) Acute ischemic infarct in the middle cerebral artery territory following a Russell's viper bite. Neurol India 57:479–480
Narvencar K (2006) Correlation between timing of ASV administration and complications in snake bites. J Assoc Physicians India 54:717–719
Njoku CH, Isezuo SA, Makusidi MA (2008) An audit of snake bite injuries seen at the Usmanu Danfodiyo University Teaching Hospital Sokoto. Niger Postgrad Med J 15:112–115
Numeric P, Moravie V, Didier M, Chatot-Henry D, Cirille S, Bucher B, Thomas L (2002) Multiple cerebral infarctions following a snakebite by *Bothrops caribbaeus*. Am J Trop Med Hyg 67:287–288
Olinger CP, Brott TG, Barsan WG et al (1988) Use of ancrod in acute or progressive ischemic cerebral infarction. Ann Emerg Med 17:1208–1209
Otero-Patiño R (2009) Epidemiological, clinical and therapeutic aspects of *Bothrops asper* bites. Toxicon 54:998–1011
Perry IJ (2000) Dietary salt intake and cerebrovascular damage. Nutr Metab Cardiovasc Dis 4:229–235
Perry IJ, Beevers DG (1992) Salt intake and stroke: a possible direct effect. J Hum Hypertens 1:23–25
Poulter N, Khaw KT, Hopwood BE, Mugambi M, Peart WS, Rose G, Sever PS (1984) Blood pressure and its correlates in an African tribe in urban and rural environments. J Epidemiol Community Health 38:181–185
Poulter NR, Khaw KT, Hopwood BEC et al (1990) The Kenyan Luo migration study: observations on the initiation of a rise in blood pressure. BMJ 300:967–972
Sacks FM, Svetkey LP, Vollmer WM, Appel LJ, Bray GA, Harsha D, Obarzanek E, Conlin PR, Miller ER III, Simons-Morton DG, Karanja N, Lin P-H, for the DASH-Sodium Collaborative Research Group (2001) Effects on blood pressure of reduced dietary sodium and the Dietary Approaches to Stop Hypertension (DASH) diet. N Engl J Med 344:3–10
Seman DL, Borger AC, Meyer JD, Hall PA, Milkowski AL (2002) Modeling the growth of *Listeria monocytogenes* in cured ready-to-eat processed meat products by manipulation of sodium chloride, sodium diacetate, potassium lactate, and product moisture content. J Food Prot 65:651–658
Simmons D, Barbour G, Congleton J, Levy J, Meacher P, Saul H, Sowerby T (1986) Blood pressure and salt intake in Malawi: an urban rural study. J Epidemiol Community Health 40:188–192
Sitprija V (2008) Animal toxins and the kidney. Nat Clin Pract Nephrol 4:616–627
Taormina PJ (2010) Implications of salt and sodium reduction on microbial food safety. Crit Rev Food Sci Nutr 50:209–227

Taylor RS, Ashton KE, Moxham T, Hooper L, Ebrahim S (2011) Reduced dietary salt for the prevention of cardiovascular disease: a meta-analysis of randomized controlled trials (Cochrane review). Am J Hypertens 8:843–853. doi:10.1038/ajh.2011.1 15

Teixeira C, Cury Y, Moreira V, Picolob G, Chaves F (2009) Inflammation induced by *Bothrops asper* venom. Toxicon 54:988–997

Thomas L, Tyburn B, Ketterle J et al (1998) Prognostic significance of clinical grading of patients envenomed by *Bothrops lanceolatus* in Martinique. Trans R Soc Trop Med Hyg 92:542–545

Thomas L, Chausson N, Uzan J, Kaidomar S, Vignes R, Plumelle Y, Bucher B, Smadja D (2006) Thrombotic stroke following snake bites by the "Fer-de-Lance" Bothrops lanceolatus in Martinique despite antivenom treatment: a report of three recent cases. Toxicon 48:23–28

Vaiyapuri S, Wagstaff SC, Watosn KA, Harrison RA, Gibbins JM, Hutchinson EG (2010) Purification and functional characterization of rhiminopeptidase A, a novel aminopeptidase from the venom of Bitis gabonica rhinoceros. PLoS Negl Trop Dis 4:e796

Vigasio A, Battiston B, De Filippo G, Brunelli G, Calabrese S (1991) Compartmental syndrome due to viper bite. Arch Orthop Trauma Surg 110:175–177

Vir D, Sachin, Gupta D, Modi M, Panda N (2010) Neurological manifestations in speech after snake bite: a rare case. Pan Afr Med J 4:13

Warrell DA (2010) Snake bite. Lancet 375:77–88

Warrell DA, Ormerod LDL, Davidson NMD (1975) Bites by puff-adder *(Bitis arietans)* in Nigeria, and value of antivenom. Br Med J 4:697–704

Warrell DA, Davidson NMcD, Greenwood BM, Ormerod LD, Pope HM, Watkins BJ, Prentice CR (1977) Poisoning by bites of the saw-scaled or carpet viper *(Echis carinatus)* in Nigeria. Q J Med 46:33–62

Webster JL, Dunford EK, Hawkes C, Neal BC (2011) Salt reduction initiatives around the world. J Hypertens 29:1043–1050

Williams D, Gutiérrez JM, Harrison R, Warrell DA, White J, Winkel KD, Gopalakrishnakone P; Global Snake Bite Initiative Working Group; International Society on Toxinology (2010) The global snake initiative: an antidote for snake bite. Lancet 375:89–91

Wilson TW (1986) History of salt supplies in West Africa and blood pressures today. Lancet 84:784–785

Winkel KD, Mirtschin P, Pearn J (2006) Twentieth century toxinology and antivenom development in Australia. Toxicon 48:738–754

Wiredu EK, Nyame PK (2001) Stroke mortality in Accra: a study of risk factors. Ghana Med J 35(4):151–161

World Health Organization (2002) The World Health Report 2002: reducing risks, promoting healthy life. World Health Organization, Geneva, pp 1–248

World Health Organization (2007) Reducing salt intake in populations. In: Report of a WHO Forum and Technical Meeting, Paris, France, 5–7 Oct 2006, WHO, pp 1–65

World Health Organization (2010) Creating an enabling environment for population-based salt reduction strategies. In: Report of a joint technical meeting held by WHO and the Food Standards Agency, UK, WHO

Yap CH, Ihle BU (2003) Coagulopathy after snake envenomation. Neurology 61:1788

Taylor RS, Ashton KE, Moxham T, Hooper L, Ebrahim S (2011) Reduced dietary salt for the prevention [illegible]

[illegible]

Part IV
Sequels and Consequences

Infectious Causes of Epilepsy?

Charles R. Newton and Ryan G. Wagner

Abstract Epilepsy occurs throughout the world, and the prevalence appears to be higher in areas in which the incidence of infections of the central nervous system (CNS) is greater. However, establishing the causal relationship between infections and the development of epilepsy is difficult, since epilepsy occurs a variable period after an acute infection. In addition to this, not all people with chronic infestations of parasites develop epilepsy. We discuss possible CNS infections, particularly the neglected diseases associated with epilepsy, looking at possible mechanisms of epileptogenesis.

Keywords Seizures • Burden • Pathogenesis • Infections • Parasites • Neglected diseases

C.R. Newton (✉)
KEMRI/Wellcome Trust Collaborative Programme, Centre for Geographical Medicine (Coast), Kilifi, Kenya

Clinical Research Unit, London School of Hygiene and Tropical Medicine, London, UK

Neurosciences Unit, Institute of Child Health, University College London, London, UK

Department of Psychiatry, University of Oxford, Oxford, UK
e-mail: cnewton@kemri-wellcome.org

R.G. Wagner
MRC/Wits Rural Public Health and Health Transitions Research Unit (Agincourt), School of Public Health, University of the Witwatersrand, Johannesburg, South Africa

Division of Epidemiology and Global Health, Department of Public Health and Clinical Medicine, Umeå University, Umeå, Sweden

M. Bentivoglio et al. (eds.), *Neglected Tropical Diseases and Conditions of the Nervous System*,
DOI 10.1007/978-1-4614-8100-3_20,

1 Introduction

Recent estimates show that epilepsy is the twentieth leading contributor to global disability (Vos et al. 2012), affecting nearly 70 million people, of whom 80 % live in poor countries (Ngugi et al. 2010). The incidence and prevalence of epilepsy are thought to be much greater in the low- and middle- income countries than in high-income countries. In Africa the median prevalence is 5–74/1,000 (Preux and Druet-Cabanac 2005), whilst in South America the range is 6–43/1,000 (Burneo et al. 2005) and in Asia it is 2–14/1,000 (Mac et al. 2007) compared to 3–8/1,000 in Europe (Forsgren et al. 2005). The incidence is thought to be even higher, with 45.0 (Inter-quartile range 30.3–66.7) in high-income countries and 81.7 (IQR 28.0–239.5) for low- and middle-income countries (Ngugi et al. 2011).

The higher prevalence and incidence of epilepsy in poorer countries is attributed to a greater incidence of focal brain insults, in particular trauma and infections. The incidence of some infections of the brain associated with the development of epilepsy e.g. bacterial meningitis (Peltola 2001) are more common in these regions, whilst other infections e.g. onchocerciasis are only found in these regions.

Epilepsy has been defined as a condition characterized by recurrent (two or more) unprovoked seizures, occurring at least 24 h apart (ILAE 1993; Thurman et al. 2011). The seizures are caused by an abnormal electrical discharge, mainly within the cortex (gray matter) of the brain. All humans have the potential to develop seizures, but the threshold for seizures varies considerably between individuals, probably determined by the individual's genome. Seizures are often provoked by acute infections of the CNS, so called acute symptomatic seizures, but these types of seizures do not fulfill the international definitions of epilepsy (ILAE 1981). However, acute symptomatic seizures may lead to epilepsy at a later date, as a result of brain damage during the acute infections.

The seizures are classified as generalized or partial, depending on the area of the brain where the discharge occurs (ILAE 1981). The discharges can be detected by electroencephalography, but more often, the seizures are classified according to their clinical characteristics (semiology). The discharge may start in one part of the brain, spreading to other parts and thus become generalized (partial becoming generalized seizures). Most seizures, and nearly all generalized seizures, are characterized by a loss of consciousness. Epilepsy is often classified as syndromes (ILAE 1989), but few of these are associated with parasitic disease (Otero et al. 1989). Although brain insults such as granulomas cause localized discharges, these discharges are often not detected as focal seizures since the secondary generalization occurs rapidly, so that the history of the seizures does not provide sufficient information to identify these as partial seizures.

Epilepsy has many different causes, including specific genes, head trauma, tumors and infections. However in most patients with epilepsy, the cause of the seizures cannot be determined, and are classified as idiopathic (ILAE 1981). The proportion of patients with idiopathic epilepsy is determined, in part, by the tools available to investigate the causes, particularly neuroimaging facilities. In poor countries these facilities are often limited, and a greater proportion of epilepsy is labelled idiopathic than in high-income countries.

2 Infections Associated with Epilepsy

There are many infections and parasites associated with epilepsy, but determining the casual relationship is often problematic since (1) the acute infections may cause brain damage resulting in the development of epilepsy many years later; (2) many organisms cause chronic infections and it is not clear when and how these organisms cause epilepsy e.g. onchocerciasis; and (3) many people are exposed to infections and parasitic infestations that are associated with epilepsy and do not develop epilepsy. Most case-control studies that report an association between epilepsy and infection and parasitic infestations are retrospective studies using prevalent cases, and thus may have selection bias. Furthermore, since many infections are more prevalent in populations in which other causes of epilepsy are common e.g. head trauma, spurious associations may arise and reverse causality cannot be excluded. Thus, establishing the casual relationship is difficult and requires detailed epidemiological, immunological and neuro-radiological studies, but ultimately depends on measuring the change in the incidence of epilepsy after prevention of the infection.

3 Viral Infections

Over a hundred viruses have been associated with the seizures, and many are thought to lead to the development of epilepsy. Most viruses are only found in specific areas of the world, although some are ubiquitous e.g. herpes simplex. Viruses can cause meningitis, encephalitis or meningo-encephalilitis. They usually cause an acute illness, often associated with seizures (acute symptomatic seizures), with the epilepsy developing after meningo- and/or encephalitis as a result of brain damage caused by the acute illness.

Herpes simplex virus type 1 is the most common cause of sporadic encephalitis in the world, and is strongly associated with the development of epilepsy. It has a predilection for the mesial temporal lobe with epilepsy occurring in 44 % of children in one study (Elbers et al. 2007).

Japanese B encephalitis is transmitted by the culex mosquito. It affects mainly children and young adults, particularly in rural areas of South-East Asia. The development of epilepsy is less common possibly because of predominant involvement of deep grey matter with sparing of cortices (Singhi 2011).

Human immunodeficiency virus (HIV) is associated with seizures, but most seizures are probably caused by opportunistic infections. There is considerable debate as to whether HIV can cause seizures *per se*, and the relationship with epilepsy is even more tenuous. In a recent multi-site study of active convulsive epilepsy (ACE) in Africa, HIV was not associated with the development of ACE, despite differences in the prevalence of HIV between the sites (Ngugi et al. 2013).

4 Bacterial Infections

Bacteria cause meningitis and brain abscesses, both of which are associated with the development of epilepsy. In fact, bacterial abscess is the brain infection most strongly associated with the development of epilepsy.

The most common bacteria causing acute bacterial meningitis (ABM) in older children and adults, i.e. *Haemophilus influenza*, *Streptococcus pneumoniae* and *Neisseria meningitides*, are now being prevented by vaccination programmes. Neonatal meningitis is caused by other organisms, particularly Gram-negative bacteria.

The risk of developing epilepsy due to bacterial infections depends upon the age, organism, medical resources available, presence of seizures during the acute illness, low cerebrospinal fluid glucose on admission and persistent neurological and electroencephalographic (EEG) abnormalities (Singhi 2011). Younger ages, particularly neonates have the greatest risk of developing epilepsy. In this age group, Group B Streptococcal meningitis is associated with greater risk of epilepsy than *Escherichia coli* meningitis. In older children and adults, pneumococcal meningitis is associated with a greater risk of developing epilepsy. In addition, patients with ABM in low- and middle- income countries have a greater risk than those in high-income countries, probably because they present later in their illness and the treatment is often inadequate. In a systematic review of sequelae following ABM, 2.0 % (1.0–4.1 %) had epilepsy at follow-up (Edmond et al. 2010). However, this estimate was based on 27 studies in which the duration of the follow-up was variable.

Focal suppuration, such as brain abscesses or subdural or epidural empyema, may occur as a complication of meningitis or *de novo*. In a study of Polish patients with brain abscesses 34 % developed epilepsy (Koszewski 1991), whilst in a more recent study conducted in Taiwan only 6.4 % developed epilepsy, mostly within the first 3 years after the diagnosis of brain abscess (Chuang et al. 2010). Focal epilepsy is common.

Tuberculosis has increased in the last couple of decades, and is particularly associated with the increased incidence of HIV. It can affect the CNS be producing tuberculosis meningitis or lesions such as tuberculomas. Epilepsy is a common sequel of tuberculosis meningitis.

5 Neurocysticercosis

Neurocysticercosis is caused by people eating food contaminated with the larvae of the tape worm *Taenia solium* [see also Carpio (2014)]. This is thought to be the most common cause of acquired epilepsy is some regions of the world e.g. South America, and accounts for 20 million cases and 50,000 deaths each year.

The eggs of this helminth are shed by pigs, and hatch in the human intestine, with the larvae travelling via the blood to the CNS were they form cysts. In the brain, the cysts pass through four phases: vesicular stage with viable larvae contained within fluid, which is asymptomatic; colloidal stage in which the larvae degenerate and cause an inflammatory reaction; nodular stage in which the scolex is transformed into coarse mineralized granules; calcified stage in which the granulation tissue has collagen with calcification. Seizures associated with the first two stages are considered as acute symptomatic seizures, whilst those associated with the later stages are considered epilepsy.

The seizures are usually short and focal, although status epilepticus occurs in about one-third of affected individuals. Single, small, contrast-enhancing lesions with an eccentric nodule (scolex) surrounded by oedema are commonly seen on computed tomography scans. Multiple lesions in different stages ('starry sky' appearance) may be seen in some patients.

Most of the epidemiological evidence for the association between epilepsy and helminths comes from case-control studies. Thus, there is compelling evidence that cysticercosis is associated with epilepsy in Africa (Andriantsimahavandy et al. 1997; Nsengiyumva et al. 2003), Asia (Tran et al. 2007) and particularly South and Central America (Gracia et al. 1990; Jallon 1998; Cruz et al. 1999; Nicoletti et al. 2002; Del Brutto et al. 2005; Medina et al. 2005), where cysticercosis is thought to cause 25–30 % of cases of epilepsy in endemic areas (Montano et al. 2005). However, there is a lack of longitudinal studies to examine the incidence of epilepsy following infection (Carpio et al. 1998), or a reduction in epilepsy following control of cysticercosis. Furthermore, the variable response to treatment has complicated the interpretation of these data (Del Brutto et al. 2006).

6 Malaria

Plasmodium falciparum is the species of malaria parasite that causes neurological complications and is the only species of *Plasmodium* that has been associated with epilepsy. It is estimated that *P. falciparum* globally infects over two billion people, causing 500 million clinical episodes of malaria, with more than a million deaths per year (Snow et al. 2005). Over 70 % of the infections occur in young children living in malarial endemic areas of sub-Saharan Africa.

One of the unique features of *P. falciparum* is that the late stages of the erythrocytic cycle, i.e. schizonts, sequester within the vascular beds of the internal organs, particularly the brain. This feature is thought to be responsible for the complications, particularly cerebral malaria and possibly seizures.

The first study that reported an association was conducted in Kenyan children who had had cerebral malaria or malaria and complicated seizures (Carter et al. 2004). Of the children exposed to cerebral malaria or had complicated seizures during their acute illness, 9 % and 12 % had developed epilepsy 2–9 years later. Thus, the risk of developing epilepsy was in the order of four and six times that of the general population, respectively. This association was confirmed in studies in Mali, Gabon and Malawi.

In Mali, the incidence of epilepsy was 17.0 per 1,000 person-years following cerebral malaria, with a relative risk 14.3 [95 % confidence interval, CI, 1.6–132.0], after adjusting for age and duration of follow up (Ngoungou et al. 2006a). In a case-control study of Gabonese children with epilepsy, the odds ratio (OR) for exposure to cerebral malaria was 3.9 [95 % CI, 1.7–8.9] (Ngoungou et al. 2006b). In Malawi, 12 out of 132 children with retinopathy-positive cerebral malaria developed epilepsy (OR undefined; $p<0.0001$) (Birbeck et al. 2010). The risk factors for epilepsy were a higher maximum temperature (39.4 °C [SD 1.2] vs. 38.5 °C [1.1]; $p=0.01$) and acute seizures (11/12 vs. 76/120; OR 6.37, 95 % CI 1.02–141.2) (Birbeck et al. 2010).

The epilepsy documented in these epidemiological studies and from case series is characterized by generalised and focal seizures. In the seizures that occur following recovery from cerebral malaria, 40–66 % are generalised seizures, with the remainder focal or partial becoming generalised seizures. Complex partial seizures are infrequently reported from the African children, but this may be caused by difficulties in describing the semiology in this group of patients. Complex partial seizures have been reported in case reports of epilepsy following cerebral malaria in travellers (Schijns et al. 2008).

Epilepsy is reported to start from a month after the episode of cerebral malaria, but the range of time to onset is wide (Carter et al. 2004; Ngoungou et al. 2006a, b; Birbeck et al. 2010) and the cumulative incidence is likely to increase with longer follow-up. The frequency of seizures in the African studies in children with cerebral malaria varies considerably from one seizure per week to less than one in the last 2 years. Most of the children with epilepsy in these studies were not on treatment at the time of assessment (Carter et al. 2004; Ngoungou et al. 2006a, b). There are no comparable studies following adults exposed to cerebral malaria.

Electroencephalographic abnormalities occur in about a third of the African children with epilepsy following cerebral malaria (Carter et al. 2004; Ngoungou et al. 2006a, b; Birbeck et al. 2010). Focal features e.g. slowing is the most common feature, with only a quarter of children with epilepsy having epileptiform abnormalities (Carter et al. 2004). Discharges over the temporal lobe are recorded in only a few patients.

Several interacting mechanisms could be responsible. The most likely mechanisms are vascular or ischaemic damage, secondary to microvascular obstruction (Newton and Krishna 1998), since ischaemic lesions are detected in a few adults with severe malaria (Looareesuwan et al. 1995; Cordoliani et al. 1998). Family history of seizure disorders is more common in children who develop seizures during acute malaria than those who do not have seizures, suggesting that genetic predisposition may contribute to the epilepsy (Versteeg et al. 2003). An increase in excitotoxins, particularly quinolinic acid and glutamate have been measured in the cerebrospinal fluid of patients with malaria during the acute illness (Dobbie et al. 2000; Medana et al. 2002). Antibodies against voltage-gated channels are increased during the acute infection and this may lead to epilepsy (Lang et al. 2005).

7 Toxoplasmosis

Toxoplasma gondii is an ubiquitous organism transmitted by cats, infecting a third of the world's population. The oocysts are either ingested by humans via contaminated food or infected meat (pig or sheep), with later stages (bradyzoites) developing cysts in the brain.

A meta-analysis of three case-control studies conducted in Israel, Turkey and USA found an OR of 4.8 (95 % CI, 2.6–7.8), suggesting a strong relationship between Toxoplasmosis and the development of epilepsy (Palmer 2007). Furthermore, there was a significant relationship between the toxoplasmosis seroprevalence and prevalence of epilepsy (Palmer 2007). Recent studies in Africa have confirmed this association, although the association in a recent multi-site African study was 1.39 (95 % CI, 1.05–1.84) (Ngugi et al. 2013).

Toxoplasmosis cysts cause fibrotic gliosis, which may lead to epilepsy. However, the infection also interferes with the calcium transport, which may also contribute to the epileptogenesis.

8 Toxocara

Toxocara cati and *T. canis* are transmitted via the feces of cats and dogs, with infestation being more common in areas with poor sanitation. Studies examining the association between toxocara species and epilepsy are less compelling. A number of studies describe a significant association in Italy (Arpino et al. 2007; Nicoletti et al. 2008), USA (Glickman et al. 1979), Bolivia (Nicoletti et al. 2002), Burundi (Nicoletti et al. 2007) and Tanzania (Winkler et al. 2008); but a study in Turkey (Akyol et al. 2007), where the prevalence of toxocara is high (Yariktas et al. 2007), did not find an association, although there were only 100 cases and 50 controls. The type of epilepsy may be important, since the strongest associations are found with focal epilepsies (Nicoletti et al. 2007), and thus studies of generalised epilepsies may not detect an association.

The epilepsy is thought to result from granuloma formation caused by larvae entering the brain. The seizures may be precipitated by the granulomatous inflammation of the brain parenchyma and/or an allergic reaction, or are secondary to haemorrhages.

9 Onchocerciasis

The association between onchocerciasis and epilepsy is even more controversial [see also Njamnshi et al. (2014)]. A significant association was found in Burundi (Newell et al. 1997), Uganda (Ovuga et al. 1992; Kipp et al. 1994) and Tanzania

(Druet-Cabanac et al. 2004), but not in West or Central Africa (Druet-Cabanac et al. 2004). A meta-analysis found the association to be of borderline significance [relative risk 1.21(95 % CI, 0.99–1.47; p=0.06)], and was not able to find any relationship with endemicity (Druet-Cabanac et al. 2004). Differences in the strains of the *Onchocerca volvulus* between East Africa and other parts of Africa may explain the inconsistencies. Other factors such as differences in the host and the presence of mass treatment programs in some areas, may also affect the studies. A significant proportion of the subcutaneous nodules found in patients living in onchocerciasis endemic areas of Uganda were found to be caused by *Taenia solium*; questioning the attribution of epilepsy to onchocerciasis in areas where both conditions occur (Katabarwa et al. 2008). However, there is a report that the seizure frequency is reduced following mass ivermectin treatment for onchocerciasis in an endemic area in Uganda (Kipp et al. 1992), but it is unclear if this reduction is sustained, translates into a reduction in epilepsy, or is caused by possible anti-epileptic properties of ivermectin itself (Trailović and Varagić 2007).

The cause of epilepsy in onchocerciasis is unclear. Other clinical manifestations e.g. blindness of onchocerciasis are thought to be caused by the inflammatory responses to dead or dying microfilariae (>100,000 more microfilariae die every day in a heavily infected person) (Burnham 1998). The immune responses are predominantly antibody-mediated, but cellular components are also important. Other proposed mechanisms of epileptogenesis due to onchocerciasis include entry of the adult worm into the CNS, and the generation of autoantibodies. There are no pathological studies that have reported adult worms within the CNS, and there are no neuroimaging studies that have demonstrated lesions associated with adult worms. Microfilaria have been seen in the cerebrospinal fluid of untreated patients with onchocerciasis (Duke et al. 1976; Marin et al. 2006) and after the onset of treatment (Marin et al. 2006), but no studies have been reported that examined the association of microfilaria in the cerebrospinal fluid with epilepsy. Autoantibodies to the retinal photoreceptors have been found in the inner retina of patients with onchocerciasis (Chan et al. 1987), but the relationship with retinal damage is not clear. There have been no reports between the presence of autoantibodies and epilepsy in onchocerciasis.

10 Other Helminthic Infections

Of the other helminth infections, there is only one study reporting an association between exposure to spirometral infections (as detected by IgG antibodies) and epilepsy, in which 2.5 % of Korean patients with epilepsy had antibodies compared to 1.9 % controls (Kong et al. 1994). There are no case-control studies demonstrating an association between epilepsy and schistosomiasis, paragonimiasis, echinococcus or any other helminth. The seizures reported in these infections often occur during the acute infections of the brain (acute symptomatic seizures) and have not been reported to be associated with chronic epilepsy.

The larvae of schistosoma species usually infect humans via the skin, with the adult worms mating and producing eggs. The eggs can enter the brain, in which a granulomatous reaction develops, killing the eggs leaving a fibrotic scar which may be epileptogenic. *S. japonicum* is associated with cerebral lesions more than the other species.

Other organisms such as *Echinococcus sp*. produce cysts that cause granulomatous formation that may lead to epilepsy. Sparganosis and paragonimiasis are caused by *Spirometra mansoni* and *Paragonimus sp*., respectively, and are found in South East Asia. The definitive hosts are cats and dogs, with humans infected by drinking contaminated water or eating raw chicken, fish (sparganosis) or crabs (paragonimiasis). CNS involvement is relatively uncommon, but epilepsy is thought to develop in association with the chronic granulomatous lesions (Higashi et al. 1971).

The cause of epilepsy in most helminthic infections is not known. Parasites may produce substances that interfere with neuronal transmission. *P. westermani* and *S. mansoni* produce proteases which may cause brain damage (Lee et al. 2006). Helminths generate a marked immune response, including autoantibodies (Paganelli et al. 1980; Rahima et al. 1994; Macura-Biegun et al. 1998; Obwaller et al. 2004; Marin et al. 2006), and autoantibodies against neuronal elements may be responsible for some epilepsy (Peltola et al. 2000; Levite 2002). Specific antibodies to NMDA glutamate receptors are associated with an encephalopathy characterised by epilepsy. Antiphospholipid (Eriksson et al. 2001), anticardiolipin and antinuclear antibodies (Peltola et al. 2000; Eriksson et al. 2001) are more frequent in people with epilepsy than appropriate controls. Autoantibodies against neuronal elements such as glutamate receptors, voltage-gated calcium and potassium channels occur in patients with epilepsy (Palace and Lang 2000). Whether these immune-mediated mechanisms are responsible for the epilepsy in helminthic infections is unclear.

11 Conclusions

Infections are often associated with epilepsy, the causal relationship is often difficult to establish. Infections may be important causes of epilepsy in some parts of the world, particularly in low- and middle- income countries. The causal link may only be established by preventing the infection and monitoring decreases in the incidence of epilepsy.

Acknowledgements CRJCN is funded by The Wellcome Trust, UK.

References

Akyol A, Bicerol B, Ertug S, Ertabaklar H, Kiylioglu N (2007) Epilepsy and seropositivity rates of *Toxocara canis* and *Toxoplasma gondii*. Seizure 16:233–237

Andriantsimahavandy A, Lesbordes JL, Rasoaharimalala B, Peghini M, Rabarijaona L, Roux J, Boisier P (1997) Neurocysticercosis: a major aetiological factor of late-onset epilepsy in Madagascar. Trop Med Int Health 2:741–746

Arpino C, Gattinara GC, Piergili D, Curatolo P (2007) *Toxocara* infection and epilepsy in children: a case-control study. Epilepsia 31:33–36

Birbeck GL, Molyneux ME, Kaplan PW, Seydel KB, Chimalizeni YF et al (2010) Blantyre Malaria Project Epilepsy Study (BMPES) of neurological outcomes in retinopathy-positive paediatric cerebral malaria survivors: a prospective cohort study. Lancet Neurol 9:1173–1181

Burneo JG, Tellez-Zenteno J, Wiebe S (2005) Understanding the burden of epilepsy in Latin America: a systematic review of its prevalence and incidence. Epilepsy Res 66:63–74

Burnham G (1998) Onchocerciasis. Lancet 351:1341–1346

Carpio A (2014) Neurocysticercosis: neurology and neurobiology. In: Bentivoglio M, Cavalheiro EA, Kristensson K, Patel N (eds) Neglected tropical diseases and conditions of the nervous system. Springer, New York

Carpio A, Escobar A, Hauser W (1998) Cysticercosis and epilepsy: a critical review. Epilepsia 39:1025–1040

Carter J, Neville B, White S, Ross A, Otieno G et al (2004) Increased prevalence of epilepsy associated with severe falciparum malaria in children. Epilepsia 45:978–981

Chan C, Nussenblatt R, Kim M, Palestine A, Awadzi K et al (1987) Immunopathology of ocular onchocerciasis. 2. Anti-retinal autoantibodies in serum and ocular fluids. Ophthalmology 94:439–443

Chuang M, Chang W, Chang H, Lin W, Tsai N et al (2010) Predictors and long-term outcome of seizures after bacterial brain abscess. J Neurol Neurosurg Psychiatry 81:913–917

Commission on Epidemiology and Prognosis, I. L. A. E (1993) Guidelines for epidemiologic studies on epilepsy. Epilepsia 34:592–596

Cordoliani YS, Sarrazin JL, Felten D, Caumes E, Lévêque C, Fisch A (1998) MR of cerebral malaria. Am J Neuroradiol 19:871–874

Cruz M, Schantz P, Cruz I, Espinosa P, Preux P et al (1999) Epilepsy and neurocysticercosis in an Andean community. Int J Epidemiol 28:799–803

Del Brutto OH, Santibáñez R, Idrovo L, Rodrìguez S, Díaz-Calder6n E, Navas C, Gilman RH, Cuesta F, Mosquera A, Gonzalez AE, Tsang VC, García HH (2005) Epilepsy and neurocysticercosis in Atahualpa: a door-to-door survey in rural coastal Ecuador. Epilepsia 46:583–587

Del Brutto O, Roos K, Coffey C, Garcia H (2006) Meta-analysis: cysticidal drugs for neurocysticercosis: albendazole and praziquantel. Ann Intern Med 145:43–51

Dobbie M, Crawley J, Waruiru C, Marsh K, Surtees R (2000) Cerebrospinal fluid studies in children with cerebral malaria: an excitotoxic mechanism? Am J Trop Med Hyg 62:284–290

Druet-Cabanac M, Boussinesq M, Dongmo L, Farnarier G, Bouteille B et al (2004) Review of epidemiological studies searching for a relationship between onchocerciasis and epilepsy. Neuroepidemiology 23:144–149

Duke B, Vincelette J, Moore P (1976) The population dynamics of *Onchocerca volvulus* microfilariae during treatment with suramin and diethylcarbamazine. Tropenmed Parasitol 27: 133–144

Edmond K, Clark A, Korczak VS, Sanderson C, Griffiths UK, Rudan I (2010) Global and regional risk of disabling sequelae from bacterial meningitis: a systematic review and meta-analysis. Lancet infect Dis 10:317–328

Elbers JM, Bitnun A, Richardson SE, Ford-Jones EL, Tellier R, Wald RM, Petric M, Kolski H, Heurter H, MacGregor D (2007) A 12-year prospective study of childhood herpes simplex encephalitis: is there a broader spectrum of disease? Pediatrics 119:e399–e407

Eriksson K, Peltola J, Keränen T, Haapala A, Koivikko M (2001) High prevalence of antiphospholipid antibodies in children with epilepsy: a controlled study of 50 cases. Epilepsy Res 46:129–137

Forsgren L, Beghi E, Oun A, Sinnanpaa M (2005) The epidemiology of epilepsy in Europe—a systematic review. Eur J Neurol 12:245–253

Glickman L, Cypess R, Crumrine P, Gitlin D (1979) *Toxocara* infection and epilepsy in children. J Pediatr 94:75–78

Gracia F, de Lao SL, Castillo L, Larreategui M, Archbold C, Brenes MM, Reeves WC (1990) Epidemiology of epilepsy in Guaymi Indians from Bocas del Toro province, Republic of Panama. Epilepsia 31:718–723

Higashi K, Aoki H, Tatebayashi K, Morioka H, Sakata Y (1971) Cerebral paragonimiasis. J Neurosurg 34:515–527

ILAE (1981) Commission on Classification and Terminology of the International League Against Epilepsy. Proposal for revised clinical and electroencephalographic classification of epileptic seizures. Epilepsia 22:489–501

ILAE (1989) Proposal for revised classification of epilepsies and epileptic syndromes. Epilepsia 30:389–399

Jallon P (1998) L'épidémiologie des épilepsies en Amérique latine. Epilepsies 10:123–129

Katabarwa M, Lakwo T, Habumogisha P, Richards F, Eberhard M (2008) Short report: could neurocysticercosis be the cause of "onchocerciasis-associated" epileptic seizures? Am J Trop Hyg Med 78:400–401

Kipp W, Burnham G, Kamugisha J (1992) Improvement in seizures after ivermectin. Lancet 340:789–790

Kipp W, Kasoro S, Burnham G (1994) Onchocerciasis and epilepsy in Uganda. Lancet 343:183–184

Kong Y, Cho S, Kang W (1994) Sparganum infections in normal adult population and epileptic patients in Korea: a seroepidemiologic observation. Kisaengchunghak Chapchi 32:85–92

Koszewski W (1991) Epilepsy following brain abscess. The evaluation of possible risk factors with emphasis on new concept of epileptic focus formation. Acta Neurochir 113:110–117

Lang B, Newbold CI, Williams G, Peshu N, Marsh K, Newton CR (2005) Antibodies to voltage-gated calcium channels in children with falciparum malaria. J Infect Dis 191:117–121

Lee J, Cho G, Kwon J, Shin M, Lim J et al (2006) Macrophageal/microglial cell activation and cerebral injury induced by excretory-secretory products secreted by *Paragonimus westerman*. Neurosci Res 54:133–139

Levite M (2002) Autoimmune epilepsy. Nat Immunol 3:500

Looareesuwan S, Wilairatana P, Krishna S, Kendall B, Vannaphan S, Viravan C, White NJ (1995) Magnetic resonance imaging of the brain in patients with cerebral malaria. Clin infect Dis 21:300–309

Mac TL, Tran DS, Quet F, Odermatt P, Preux PM, Tan CT (2007) Epidemiology, aetiology, and clinical management of epilepsy in Asia: a systematic review. Lancet Neurol 6:533–543

Macura-Biegun A, Pituch-Noworolska A, Rewicka M, Mrozewia B, Noworolski J (1998) Antistriational antibodies during *Toxocara canis, Trichinella spiralis* infections. Comp Immunol Microbiol Infect Dis 21:101–106

Marin B, Boussinesq M, Druet-Cabanac M, Kamgno J, Bouteille B, Preux PM (2006) Onchocerciasis-related epilepsy? Prospects at a time of uncertainty. Trends Parasitol 22: 17–20

Medana I, Hien T, Day N, Nguyen H, Nguyen T et al (2002) The clinical significance of cerebrospinal fluid levels of kynurenine pathway metabolites and lactate in severe malaria. J Infect Dis 185:650–656

Medina MT, Durón RM, Martínez L, Osorio JR, Estrada AL, Zúniga C, Cartagena D, Collins JS, Holden KR (2005) Prevalence, incidence, and etiology of epilepsies in rural Honduras: the Salama Study. Epilepsia 46:124–131

Montano SM, Villaran MV, Ylquimiche L, Figueroa JJ, Rodriguez S, Bautista CT, Gonzalez AE, Tsang VC, Gilman RH, Garcia HH; Cysticercosis Working Group in Peru (2005)

Neurocysticercosis association between seizures, serology, and brain CT in rural Peru. Neurology 65:229–233
Newell E, Vyungimana F, Bradley J (1997) Epilepsy, retarded growth and onchocerciasis, in two areas of different endemicity of onchocerciasis in Burundi. Trans R Soc Trop Med Hyg 91:525–527
Newton C, Krishna S (1998) Severe falciparum malaria in children: current understanding of pathophysiology and supportive treatment. Pharmacol Ther 79:1–53
Ngoungou EB, Dulac O, Poudiougou B, Druet-Cabanac M, Dicko A, Mamadou Traore A, Coulibaly D, Farnarier G, Tuillas M, Keita MM, Kombila M, Doumbo OK, Preux PM (2006a) Epilepsy as a consequence of cerebral malaria in area in which malaria is endemic in Mali, West Africa. Epilepsia 47:873–879
Ngoungou EB, Koko J, Druet-Cabanac M, Assengone-Zeh-Nguema Y, Launay MN, Engohang E, Moubeka-Mounguengui M, Kouna-Ndouongo P, Loembe PM, Preux PM, Kombila M (2006b) Cerebral malaria and sequelar epilepsy: first matched case-control study in Gabon. Epilepsia 47:2147–2153
Ngugi A, Bottomley C, Kleinschmidt I, Sander J, Newton C (2010) Estimation of the burden of active and life-time epilepsy: a meta-analytic approach. Epilepsia 51:883–890
Ngugi AK, Kariuki SM, Bottomley C, Kleinschmidt I, Sander JW, Newton CR (2011) Incidence of epilepsy: a systematic review and meta-analysis. Neurology 77:1005–1012
Ngugi AK, Bottomley C, Kleinschmidt I, Wagner RG, Kakooza-Mwesige A, Ae-Ngibise K, Owusu-Agyei S, Masanja H, Kamuyu G, Odhiambo R, Chengo E, Sander JW, Newton CR; SEEDS group (2013) Prevalence of active convulsive epilepsy in sub-Saharan Africa and associated risk factors: cross-sectional and case-control studies. Lancet Neurol 12:253–263
Nicoletti A, Bartoloni A, Reggio A, Bartalesi F, Roselli M, Sofia V, Rosado Chavez J, Gamboa Barahona H, Paradisi F, Cancrini G, Tsang VC, Hall AJ (2002) Epilepsy, cysticercosis, and toxocariasis: a population-based case-control study in rural Bolivia. Neurology 58:1256–1261
Nicoletti A, Bartoloni A, Sofia V, Mantella A, Nsengiyumva G et al (2007) Epilepsy and toxocariasis: a case-control study in Burundi. Epilepsia 48:894–899
Nicoletti A, Sofia V, Mantella A, Vitale G, Contrafatto D et al (2008) Epilepsy and toxocariasis: a case-control study in Italy. Epilepsia 49:594–599
Njamnshi AK, Zoung-Kanyi Bissek A-C, Etya'ale D (2014) Onchocerciasis: neurological involvement. In: Bentivoglio M, Cavalheiro EA, Kristensson K, Patel N (eds) Neglected tropical diseases and conditions of the nervous system. Springer, New York
Nsengiyumva G, Druet-Cabanac M, Ramanankandrasana B, Bouteille B, Nsizabira L, Preux PM (2003) Cysticercosis as a major risk factor for epilepsy in Burundi, east Africa. Epilepsia 44: 950–955
Obwaller A, Duchêne M, Walochnik J, Wiedermann G, Auer H, Aspöck H (2004) Association of autoantibodies against small nuclear ribonucleoproteins (snRNPs) with symptomatic *Toxocara canis* infestation. Parasite Immunol 26:327–333
Otero E, Cordova S, Diaz F, Garcia-Teruel I, Brutto O (1989) Acquired epileptic aphasia (the Landau-Kleffner syndrome) due to neurocysticercosis. Epilepsia 30:569–572
Ovuga E, Kipp W, Mungherera M, Kasoro S (1992) Epilepsy and retarded growth in a hyperendemic focus of onchocerciasis in rural western Uganda. East Afr Med J 69:554–556
Paganelli R, Ngu J, Levinsky R (1980) Circulating immune complexes in onchocerciasis. Clin Exp Immunol 39:570–575
Palace J, Lang B (2000) Epilepsy: an autoimmune disease? J Neurol Neurosurg Psychiatry 69: 711–714
Palmer BS (2007) Meta-analysis of three case controlled studies and an ecological study into the link between cryptogenic epilepsy and chronic toxoplasmosis infection. Seizure 16:657–663
Peltola H (2001) Burden of meningitis and other severe bacterial infections of children in Africa: implications for prevention. Children 32:64–75
Peltola J, Haapala A, Isojärvi J (2000) Antiphospholipid and antinuclear antibodies in patients with epilepsy or new-onset seizure disorders. Am J Med 109:712–717

Preux P-M, Druet-Cabanac M (2005) Epidemiology and aetiology of epilepsy in sub-Saharan Africa. Lancet Neurol 4:21–31

Rahima D, Tarrab-Hazdai R, Blank M, Arnon R, Shoenfeld Y (1994) Anti-nuclear antibodies associated with schistosomiasis and anti-schistosomal antibodies associated with SLE. Autoimmunity 17:127–139

Schijns O, Visser-Vandewalle V, Lemmens E, Janssen A, Hoogland G (2008) Surgery for temporal lobe epilepsy after cerebral malaria. Seizure 17:731–734

Singhi P (2011) Infectious causes of seizures and epilepsy in the developing world. Dev Med Child Neurol 53:600–609

Snow R, Guerra C, Noor A, Myint H, Hay S (2005) The global distribution of clinical episodes of *Plasmodium falciparum* malaria. Nature 434:214–217

Thurman D, Beghi E, Begley C, Berg A, Buchhalter J et al (2011) Standards for epidemiologic studies and surveillance of epilepsy. Epilepsia 52:2–26

Trailović S, Varagić V (2007) The effect of ivermectin on convulsions in rats produced by lidocaine and strychnine. Vet Res Commun 31:863–872

Tran DS, Odermatt P, Le Oanh T, Huc P, Phoumindr N, Ito A, Druet-Cabanac M, Preux PM, Strobel M (2007) Risk factors for epilepsy in rural Lao PDR: a case-control study. Southeast Asian J Trop Med Public Health 38:537–542

Versteeg A, Carter J, Dzombo J, Neville B, Newton C (2003) Seizure disorders among relatives of Kenyan children with severe falciparum malaria. Tropical Med Int Health 8:12–16

Vos T, Flaxman AD, Naghavi M, Lozano R, Michaud C et al (2012) Years lived with disability (YLDs) for 1160 sequelae of 289 diseases and injuries 1990-2010: a systematic analysis for the Global Burden of Disease Study 2010. Lancet 380:2163–2196

Winkler AS, Blocher J, Auer H, Gotwald T, Matuja W, Schmutzhard E (2008) Anticysticercal and antitoxocaral antibodies in people with epilepsy in rural Tanzania. Trans R Soc Trop Med Hyg 102:1032–1038

Yariktas M, Demirci M, Aynali G, Kaya S, Doner F (2007) Relationship between *Toxocara* seropositivity and allergic rhinitis. Am J Rhinol 21:248–250

Preux PM, Druet-Cabanac M (2005) Epidemiology and aetiology of epilepsy in sub-Saharan Africa. Lancet Neurol 4:21–31

Rollinat [illegible] Bruck M, Arnold [illegible] Shoenfeld Y (1994) Antimicrobial antibodies [illegible] with [illegible] and [illegible] antibodies [illegible] Am [illegible]

[illegible]

Cognitive Impairment and Behavioural Disturbances Following Malaria or HIV Infection in Childhood

Michael Kihara, Amina Abubakar, and Charles R.J.C. Newton

Abstract While both falciparum malaria and human immunodeficiency virus infections are not classified under the neglected disease criteria, both have been shown to affect the central nervous system (CNS), which is of importance but neglected area of neuroscience research. The brunt of these two diseases is borne by children in sub-Saharan Africa, and unfortunately, the study of long-term effect cognitive deficits and disorders due the CNS infections in these children has been neglected. We review the evidence of the effect of falciparum malaria and HIV on the brain, describe the patterns of involvement and propose mechanisms by which these infections can alter the brain function. The results reveal that falciparum malaria results in different patterns of impairment, which may in part be explained by methodological and definition differences, however the cognitive impairment appear to cover all categories of cognition suggesting diffuse damage. HIV has been shown to impact on multiple developmental domains starting early in life and persisting into adolescence. Various biomedical and psychosocial factors have been observed to either exacerbate or ameliorate the negative effects of HIV.

M. Kihara (✉)
The Centre for Geographical Medicine Research (Coast), Kenya Medical Research Institute, Kilifi, Kenya

United States International University – Africa, Nairobi, Kenya
e-mail: Mkihara@kilifi.kemri-wellcome.org

A. Abubakar
The Centre for Geographical Medicine Research (Coast), Kenya Medical Research Institute, Kilifi, Kenya

Tilburg University, Tilburg, The Netherlands

Utrecht University, Utrecht, The Netherlands

C.R.J.C. Newton
The Centre for Geographical Medicine Research (Coast), Kenya Medical Research Institute, Kilifi, Kenya

Department of Psychiatry, University of Oxford, Oxford, UK

M. Bentivoglio et al. (eds.), *Neglected Tropical Diseases and Conditions of the Nervous System*, 369
DOI 10.1007/978-1-4614-8100-3_21, © Springer Science+Business Media New York 2014

Existing knowledge gap on impairment related to malaria and HIV shows significant gaps especially as it relates to elucidating pathways to poor outcome. Future research efforts need to focus on understanding these mechanisms so as to guide targeted intervention.

Keywords Neuro-developmental • Falciparum • Africa • Children • AIDS

1 Introduction

Children living in developing countries are exposed to a variety of infectious diseases, which not only threaten their survival but can also cause long-term cognitive and behaviour impairment in those that survive. Falciparum malaria and human immunodeficiency virus/acquired immunodeficiency syndrome (HIV/AIDS) are common infectious diseases, and while both are not classified as neglected diseases, they affect neuro-developmental outcomes, and the research and study on the long-term effect of these infections in children living in low-income countries is neglected.

In 2007, there were an estimated 451 million cases of malaria worldwide, with 60 % occurring in Africa (Hay et al. 2010). *P. falciparum* causes most of the severe cases and deaths in tropical regions such as sub-Saharan Africa and Southeast Asia. According to the World Malaria Report (2011), there were 655,000 malaria deaths worldwide in 2010, compared to 781,000 in 2009. It has been estimated that 91 % of deaths in 2010 were in the African Region. There are approximately 34 million people currently living with HIV (UNAIDS 2012a). Of these, 97 % reside in low- and middle-income countries (UNAIDS 2012b). Globally, there were 3.3 million children with HIV infection in 2011, with 330,000 new infections among children (UNAIDS 2012a), most being in sub-Saharan Africa. Neurocognitive involvement has been reported in those exposed to severe malaria and HIV (de Miranda et al. 2011; Kihara et al. 2006; Sherr et al. 2009). Despite the enormous numbers of children affected by these conditions, the impact of cognitive impairment and behaviour problems associated with these conditions is a neglected area of study.

2 Definition of Cognitive Impairment

Cognitive impairment is a generalized form of intellectual impairment which is usually defined as having an intelligence quotient (IQ) that is two standard deviations below the mean of the age-group i.e. IQ < 70. Cognitive impairment may be developmental in origin, usually linked to abnormalities in brain structure and function, present from birth or acquired through nervous system infection (BMJ-group 2011). Cognitive impairment may also arise due to exposure to environmental factors that may injure or damage the brain e.g. infectious pathogens. Behavioural disturbances refer to atypical behavioural patterns and can be categorized either as externalizing or internalizing. Externalizing behaviours are reactions that are directed toward others such as aggression or disruption, while internalizing behaviours, such as anxiety, withdrawal, and somatic complaints, are mainly directed toward the self.

3 Malaria

3.1 *Cognitive Impairment*

There are five species of malaria that infect humans, but only *Plasmodium falciparum* and *P. vivax* have been associated with effects on the central nervous system. *P. falciparum* has a particular propensity for the brain, since in the later stages of their erythrocytic cycle, there is preferentially adhere of the erythrocytes to the endothelium of the post-capillary venules in the brain, a process known as cytoadherence. The sequestration of these stages is thought to be responsible for the severe manifestations of falciparum malaria, i.e. cerebral malaria, although such sequestration occurs in children who are asymptomatic, albeit to a lesser degree.

Cognitive outcomes post malarial or parasitisation are of concern as a number of studies have documented persisting deficits that may affect the future cognitive development of children (Boivin 2002; Carter et al. 2003, 2005; Dugbartey et al. 1998; Holding and Kitsao-Wekulo 2004; Holding et al. 1999; Muntendam et al. 1996). In addition to the severity of the disease, various other risk factors have been identified that could affect cognitive function including multiple seizures, depth of coma, hypoglycaemia (Holding et al. 1999, 2004; Kihara et al. 2009), multiple infections (Fernando et al. 2003), multiple clinical complications at discharge (Holding et al. 2004), malnutrition, and social economic status (Boivin et al. 1993; Boivin 2002). The pathway that leads to cognitive impairment is obscure and is a result of a complex web of risk factors. We have, however, used available evidence to construct a simplified sketch of the probable pathways of cognitive impairment (Fig. 1).

The effects of falciparum malaria on cognition have been documented for wide range of exposures to the parasite (Fernando et al. 2010; Kihara et al. 2006). The studies used different methodologies of evaluating cognitive outcomes and looked at the five key cognitive functions: memory, language, attention, visual-spatial skills and executive functioning (Croisile 2004). Studies have also examined how exposure to the parasite, through repeated infections to uncomplicated malaria and severe malaria affects cognition (summarized in Table 1).

3.2 *Effect of Parasitaemia on Cognitive Function*

A few studies have examined the effect of asymptomatic parasitaemia on cognition (Al Serouri et al. 2000; Boivin et al. 1993; Vitor-Silva et al. 2009), whilst others have determined cognitive outcomes in children using malaria prophylaxis to

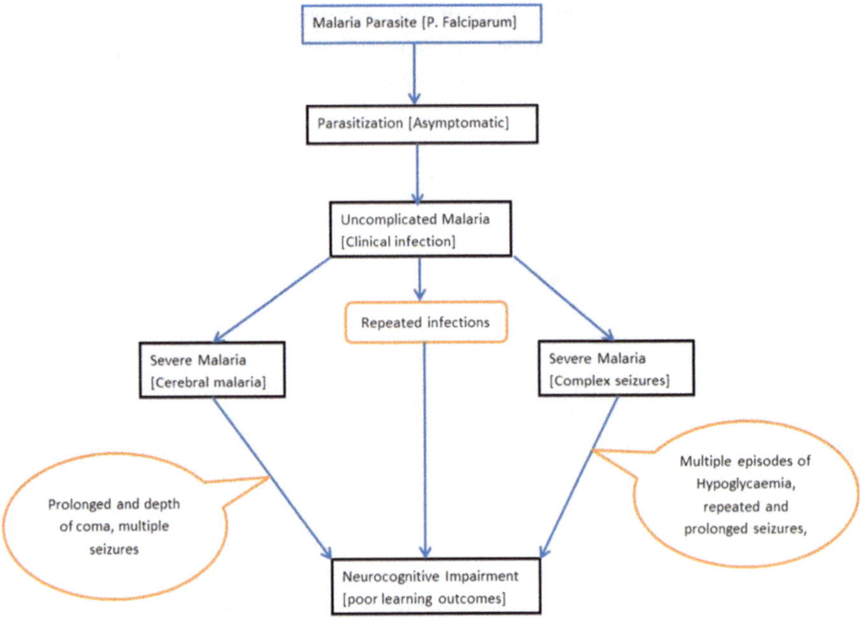

Fig. 1 Probable pathways of neurocognitive impairment in falciparum malaria

prevent the development of parasitaemia (Clarke et al. 2008; Fernando et al. 2006; Jukes et al. 2006; Thuilliez et al. 2010).

In the Yemeni study, 445 boys with asymptomatic parasitaemic boys were compared with 142 non-parasitaemic boys matched for age and schooling status (Al Serouri et al. 2000). Two weeks later, 150 children who remained parasitaemic and 150 children who were no longer parasitaemic were assessed using a battery of cognitive tests. The investigators found no significant differences in cognitive test scores between those who became non-parasitaemic and those who remained parasitaemic, although there was a difference in fine motor tasks.

Boivin et al. (1993) evaluated 97 Zairian schoolchildren for cognitive ability. Even though treating asymptomatic malaria did not improve cognitive ability, clearance of intestinal parasites showed improved performance on the Kaufman Assessment Battery for children.

A study of 198 Brazilian children aged 5–14 years sought to determine the cognitive outcomes of malaria parasite elimination for a period of 9 months (Vitor-Silva et al. 2009). Poor performance of language and mathematics was associated with at least one malarial attack. Overall, it remains unknown whether parasitaemia on its own results in poorer cognitive function or factors are involved.

Table 1 Summary of studies investigating neurocognitive outcomes among children exposed to falciparum malaria

	First author	Year	Country	Clinical description	No. cases	Controls	Follow-up length	Study design	Attention	Memory	Language	Visual spatial skills	Executive functions	Comment
1	Bangirana P et al.	2011	Uganda	CM	62	61	3 months	Case-control	Performed significantly poorer *($P<0.05$)*	Not significant	NT	Not significant	Not significant	Used Glasgow coma score of ≤14
2	Boivin M et al.	2011	Malawi	CM	83	95	1–40 months	Case-control	NT	NT	Performed significantly poorer *($p=0.003$)*	NT	NT	Retinopathy positive CM
3	Thuilliez J et al.	2010	Mali	Asymptomatic	227	0	N/A	Prospective	NR	NR	NR	NR	NT	N/A
4	John CC et al.	2008	Uganda	CM	44	87	2 years	Prospective	Performed significantly poorer *($P<0.05$)*	Not significant	Not significant	Not significant	Not significant	Used Blantyre coma score of ≤2 or Glasgow coma score ≤8
5	Clarke SE et al.	2008	Kenya	Asymptomatic	1,070	1,394	N/A	Cluster-randomised RCT	Performed significantly poorer *($P<0.05$)*	NT	NT	NT	Not significant	N/A
6	Boivin M et al.	2007	Uganda	CM	44	89	6 months	Cohort Study	Performed significantly poorer *($P<0.05$)*	Performed significantly poorer *($P<0.05$)*	NT	Not significant	Not significant	Used Blantyre coma score of ≤2 or Glasgow coma score ≤8
7	Jukes MCH et al.	2006	Gambia	Asymptomatic	291	288	>15 years	Cluster-randomised RCT	Not significant	Not significant	Not significant	Not significant	NT	N/A
8	Carter JA et al.	2005	Kenya	CM	152	179	>2 years	Cohort Study	Not significant	Not significant	Performed significantly poorer *($P<0.05$)*	NT	Performed significantly poorer *($P<0.05$)*	Used Blantyre coma score of ≤2

(continued)

Table 1 (continued)

	First author	Year	Country	Clinical description	No. cases	Controls	Follow-up length	Study design	Attention	Memory	Language	Visual spatial skills	Executive functions	Comment
9	Carter JA et al.	2005	Kenya	M/S	156	179	>2 years	Cohort Study	Not significant	Not significant	Performed significantly poorer *($P < 0.05$)*	NT	Not significant	Used Blantyre coma score of ≤2
10	Carter JA et al.	2003	Kenya	CM & Severe Malaria	25	27	>2 years	Cohort Study	NT	NT	Performed significantly poorer *($P < 0.004$)*	NT	NT	Used Blantyre coma score of <2 (W.H.O criteria)
11	Fernando SD et al.	2003	Sri Lanka	Malaria	171	154	1 year	Cross-sectional	NT	NT	Borderline performance *($P = 0.093$)*	Not significant	NT	N/A
12	Fernando SD et al.	2003	Sri Lanka	Malaria	385	213	1–6 years	Prospective	NT	NT	Performed significantly poorer ($P < 0.001$)	Performed significantly poorer ($P < 0.001$)	NT	N/A
13	Boivin MJ	2002	Senegal	CM	29	29	Avg 3.4 years	Matched case-control	Performed significantly poorer ($p < 0.05$)	Performed significantly poorer *($P < 0.05$)*	NT	Performed significantly poorer ($P < 0.05$)	Borderline significance *($P = 0.07$)*	Used W.H.O criteria for CM
14	Fernando SD	2001b	Sri Lanka	Malaria	295	292	N/A	Randomized control study	NT	NT	Performed significantly poorer *($P < 0.001$)*		NT	N/A

15	Fernando SD	2001a	Sri Lanka	Malaria	343	305	1–2 years	Prospective	NT	NT	Performed significantly poorer ($P<0.001$)	NT	NT	N/A
16	Al Serouri AW et al.	2000	Yemen	Asymptomatic	445	142	N/A	Matched case-control	Not significant	Not significant	NT	NT	Not significant	N/A
17	Holding PA et al.	1999	Kenya	Severe Malaria	87	87	3.5–6 years	Matched case control	Performed significantly poorer ($p<0.05$)	Not significant	Performed significantly poorer ($P=0.02$)	Not significant	Borderline performance ($P=0.06$)	Used Blantyre coma score of ≤4
18	Dugbartey AT et al.	1998	Ghana	CM	20	20	Avg 3.9 years	Matched case-control	Not significant	Performed significantly poorer ($p<0.01$)	Not significant	Not significant	Performed significantly poorer	Used W.H.O. criteria for CM
19	Mutendam AH et al.	1996	Gambia	CM	36	36	Avg 3.4 years	Matched case-control	NT	Not significant	Not significant	Not significant	Not significant	Used Blantyre coma score of ≤2
20	Dugbartey At et al.	1997	Ghana	CM	20	20	Avg 3.9 years	Matched case-control	NT	NT	NT	Performed significantly poorer ($P<0.01$)	NT	Used W.H.O criteria for CM
21	Boivin M et al.	1993	Zaire	Asymptomatic	97	0	N/A	Prospective	Not significant	Not significant	Not significant	Not significant	Not significant	N/A

NR not reported, *NT* not tested, *CM* cerebral malaria

3.3 Cognitive Function During Falciparum Malaria Illness

Fernando (2001) reported that the academic performance of 293 children experiencing uncomplicated malaria was poorer than that of 162 children experiencing non-malarial fever or that of 305 healthy controls. Although the performance of the children with malaria improved 2 weeks after treatment, it was still significantly lower than that of the healthy controls. This study was unable to examine whether poor performance was due to malaria per se, or the absenteeism from school caused by having an illness.

The same investigators also conducted a randomized double-blind placebo controlled study of antimalarial prophylaxis with school-age Sri Lankan children, in which they measured academic performance using school tests of language and mathematics (Fernando 2001; Fernando et al. 2006). Two hundred and ninety-five children received chloroquine while another 292 children received placebo. They found an improvement in academic performance in those on prophylaxis for a 9-month period. However, the placebo group also had significantly higher rates of absenteeism. A multivariate model identified absenteeism due to malaria and chloroquine prophylaxis as significant predictors of school performance.

Similar studies investigating the effect of intermittent preventive treatment (IPT) on education outcomes have been carried out. Clarke et al. (2008) and colleagues studied children aged 5–18 years who received three treatments at 4-month intervals and outcome was assessed through cross-sectional surveys 12 months post-intervention. Children in the intervention group performed significantly better on the sustained attention task but differences were shown in education achievement. Jukes et al. (2006) followed up children in the Gambia who were part of randomized trial done 15 years earlier. The study examined the long-term impact of malaria prophylaxis on cognitive abilities and found no overall intervention effect on cognitive abilities even though the intervention group had higher educational attainment. A prospective study of 227 Malian primary school children, in which each child had monthly clinical and lab examinations and tests for cognitive ability found poorer outcomes in the children with asymptomatic malaria, though which sub-tests showed significant differences was not reported (Thuilliez et al. 2010).

3.4 Cognitive Function Post-severe Malaria

There are a number of studies investigating the effects of severe malaria on cognitive function following exposure to severe malaria, (Bangirana et al. 2011; Boivin et al. 2007, 2011; Carter et al. 2003, 2005; Holding et al. 1999, 2004; John et al. 2008). Severe malaria manifests as impaired consciousness, prostration, seizures, severe anaemia, acidosis, hypoglycaemia or renal impairment. Impaired consciousness is defined as Blantyre Coma Score (BCS) of ≤ 4 for 4 or more hours. Cerebral malaria is the most severe neurological manifestation of falciparum malaria (Newton et al. 2000). The World Health Organization (2000) defines it as one's inability to localise a painful stimulus or a Blantyre coma score of ≤2 (Molyneux et al. 1989), a peripheral falciparum malaria parasitaemia and exclusion of other causes of encephalopathy. More recently features of retinal changes were associated with the sequestration of the parasites within the brain and define a more specific syndrome of cerebral malaria (White et al. 2009).

Most studies have revealed poorer performance in children exposed to severe malaria except one small Gambian study of children who survived without neurological deficits and showed no differences compared to community controls (Muntendam et al. 1996).

In Kilifi, Kenya, Holding et al. (1999) compared 87 children with a history of severe malaria with impaired consciousness (defined as a Blantyre score of ≤4) with community controls on tasks measuring information processing, language and behaviour based on the K-ABC (Kaufman and Kaufman 1983) and other locally-established tests. There was no significant impairment of information processing skills for the children, although a greater number of cases showed impaired performance. There were significant deficits in measures of language and attention/planning.

Holding et al. (2004) represented a re-analysis of their previously reported data, in which children with a history of severe malaria were divided into medium- and high-risk groups based on the severity of the disease. Seventeen children were considered high-risk and the rest, medium-risk. These children were compared to controls (low-risk group) matched for age, sex and socioeconomic status. The results revealed the adverse effects of schooling on the outcome of disease. Unschooled high-risk children performed significantly poorer than unschooled children in the other two groups.

Carter et al. (2003) compared 25 children previously admitted to hospital with cerebral malaria (CM) or severe malaria (defined as malaria prostration, multiple seizures or severe anaemia) and 27 children unexposed to either condition and found that language performance was poorer in the children exposed to either CM or severe malaria up to 6 years post-insult. Assessments of comprehension, syntax, lexical semantics, higher-level language abilities, pragmatics and phonology were administered to each child. Children with a history of CM or severe malaria performed significantly poorer on tests of comprehension, syntax and lexical semantics compared to the unexposed group. There was no evidence of a difference between the scores of children who had suffered CM and those who had severe malaria, although this may have been due to the small numbers.

In another study (Carter et al. 2005), 156 children exposed to malaria with complicated seizures (defined as >2 seizures within 24 h or focal or prolonged seizures for >30 min but without coma) showed significantly increased odds of impairment with respect to speech and language (pragmatics and phonology) compared to unexposed children. There was no evidence of a difference in performance on tests of memory, attention or other aspects of language.

3.5 Cognitive Function Post-cerebral Malaria

Many studies have reported cognitive deficits associated with severe malaria, particularly CM (Table 1). In a Ghanaian study, 20 children (aged 7–16 years) with a history of CM (defined using the WHO criteria) were compared with 20 age-, sex- and education-matched controls on a standardised neuropsychological battery (Dugbartey et al. 1998). Children with a history of CM performed significantly poorly relative to controls in bimanual tactile discrimination, accuracy of visual scanning, visual memory, perceptual abstraction and rule learning skill, right ear auditory information processing, and dominant-hand motor speed. The study found no significant differences between those with a history of CM and controls in non-verbal reasoning, visual-spatial processing, auditory attention and verbal fluency.

In Senegal, 29 children aged between 5 and 12 years with a history of CM (defined using WHO criteria but with coma duration adjusted to 12 h) were compared with 29 age- and education-matched controls (mild-malaria). Those with a history of CM performed significantly poorer on the simultaneous processing (spatial memory, photo series), mental processing, and sequential processing (hand movements, word order) tasks of the K-ABC and the attention capacity task from Test of Attention Variables (TOVA) than matched controls (Boivin 2002). The study found significant correlation between coma duration and attention capacity for the CM group emphasizing the importance of coma duration on outcome.

In Kenya, 152 children aged between 6 and 9 years previously exposed to CM were compared to 179 children unexposed to severe malaria (Carter et al. 2005). The performance of children previously exposed to CM was poorer than unexposed children on all the cognitive assessments administered: speech and language battery, attention, memory, and non-verbal functioning. Significant differences were found on tests of higher-level language abilities, lexical semantics, pragmatics and non-verbal functioning (construction task).

There was one study that did not detect any differences between CM and controls. This case-control study focused on the measurement of non-verbal functioning in 36 pairs of Gambian children who were discharged from hospital without any neurological deficits (Newton et al. 2000). The results showed no significant impairments in children after an average follow-up of 3.4 years (Muntendam et al. 1996). A curious finding from the study, though not statistically significant, was that the CM children performed better on all tests of intellectual development than the controls. A group difference of borderline significance was detected on a test of balance (sensori-motor development), with poorer performance among children exposed to CM.

A Ugandan study of children aged 5–12 years presenting with either cerebral malaria ($n=44$), or uncomplicated malaria ($n=54$) and compared with healthy community children ($n=89$) found a 3.67 increased risk of cognitive impairment in CM children compared to controls (Boivin et al. 2007; John et al. 2008). At 6 months post discharge, the deficits were in working memory and attention (Boivin et al. 2007). Deficits in these groups persisted 2 years post discharge especially in areas of attention (John et al. 2008).

Boivin et al. (2011) assessed Malawian children with retinopathy-positive CM for the domains of gross motor, fine motor, language and social skills. Eighty-three CM children were compared with 95 community controls. The results of cognitive performance showed poorer language development in those exposed to CM. A Stepwise regression demonstrated that coma duration and seizures while in hospital were predictive of poor outcome.

Finally, a prospective case-control study done in Uganda of 62 children with a history of malaria and neurological involvement and 61 community controls, assessed working memory, reasoning, learning, visual spatial skills and attention 3 months post-discharge (Bangirana et al. 2011). The results showed marginal non-significant evidence of lower attention scores and differences in the other cognitive abilities. The authors argued that lack of differences between the malaria and control groups could be as a result of gradual development of cognitive deficits such that the effect is less obvious in the short term.

3.6 Conclusion

The above sections have summarized the effects of falciparum malaria infection on cognitive function based on the severity of disease. The results show different patterns of impairment, although this may be due to methodological differences and also differences in definitions of the clinical conditions. The impairment appears to cover all categories of cognition: attention, memory, language, visual spatial skills and executive functioning (Kihara et al. 2006). Studies comparing cognitive functions before and after treatment for acute malarial illness show significant impairment in school performance and cognitive abilities even after recovery. Malaria prophylaxis is shown to improve cognitive function and school performance in clinical trials when compared to placebo groups.

4 Human Immunodeficiency Virus/Acquired Immunodeficiency Syndrome

4.1 Epidemiology

Worldwide an estimated 3.3 million children are infected with HIV (UNAIDS 2012a), with approximately 90 % of HIV-1 positive children living in sub-Saharan Africa (UNAIDS 2012b). Vertical transmission is the main mode of infection among young children (Wiktor et al. 1997). The mother-to-child transmission can occur *in utero*, at birth or through breast-feeding. In the absence of treatment, the rates of transmission from mother-to-child are 25–40 %. Of these less than 10 % of the infections are acquired *in utero*, about 10–20 % at delivery while breast-feeding accounts for 10–20 % of vertical infection (Dabis and Ekpini 2002). In the absence of treatment up to 50 % of HIV-infected children are expected to die before their second birth day; however with increased use of anti-retroviral agents (ARVs), HIV-infected children are living longer (Newell et al. 2004). With increased survival of the HIV-infected children, there is more focus on morbidity including the potential adverse effects of HIV on neurocognitive outcomes. The following section highlights the impact of HIV on various neurodevelopmental domains and some potentially moderating and mediating factors.

4.2 *Effects of HIV on Neuro-development*

HIV impacts on multiple neurodevelopmental domains at childhood.

4.2.1 Motor

Studies in different parts of the world indicate that motor functioning is consistently and severely affected by HIV infection. In the general population, motor impairments occurs in 2 out of 1,000 births (Stanley et al. 2000) compared with 22 % in HIV-infected children under the age of 3 years (Sherwood et al. 2008). Moreover, the extant literature reports that motor delay was correlated to neurological abnormalities (Blanchette et al. 2001). For instance, in one study it was observed that infants with vertically transmitted HIV scored significantly lower on motor subscales of the Bayley Scales of Infant Development (BSID) (Bayley 1993) compared to HIV-negative children born to HIV-positive mothers. Moreover, 36 % of the HIV-infected infants displayed abnormal computerized tomographic (CT) scans of the brain. The infants with abnormal CT scan also scored lower on the motor scale of the BSID compared to those with normal scans (Blanchette et al. 2002). These findings are consistent with a longitudinal study that reported a lower level of motor development for HIV-infected compared to uninfected infants (Gay et al. 1995). In this study, the infants' motor development continued to decline from the 6th to the 24th month. The lower rate of motor growth for these infants was found to positively correlate with neurological abnormalities. Studies among preschool and school aged children in high- and low- income countries indicate that older children infected with HIV experience various motor delays and disorders (Koekkeok et al. 2006; Ruel et al. 2012; Van Rie et al. 2008) an indication that motor delays and impairments presenting in infancy persist into childhood and adolescence.

4.2.2 Language

HIV-infected children present with language impairments as early as 18 months of age and these impairments persist into adolescence. In the only study so far identified investigating language functioning of vertically infected HIV-positive adolescents (Brackis-Cott et al. 2009), reported that HIV-positive adolescents performed significantly worse than controls in all measures of language functioning. Additionally, the published research seems to indicate that the patterns and degree of impairment may vary based on the domains assessed. Wolters et al. (1995, 1997) reported expressive language was significantly more affected than receptive language. Given the complexities involved in assessing language development, especially in low and middle income countries where there is a shortage of standardized tools, there has not been sufficient efforts to adequately disentangle the impact of HIV on various domains of language functioning. Further evaluation of differential impact of HIV on domains of language functioning is urgently needed.

4.2.3 Cognitive Function

The use of varied measures of cognition and difference in definition of control groups and in some cases small sample sizes makes it difficult to evaluate the true pattern, incidence and severity of neurocognitive impairments among HIV-infected children. However, the existing evidence indicates an early onset cognitive impairment that persists into childhood and adolescences. Among infants the Bayley Scales of Infant Development (BSID; Bayley 1993) has been the most widely used measure of mental development. Results from multiple studies indicate that HIV-infected children perform significantly worse than those who are not infected on the BSID mental scales (Gay et al. 1995; Drotar et al. 1997; Blanchette et al. 2001). In school going children and adolescents existing evidence indicates that an HIV infection contributes to impairments in multiple domains including attention, memory, visual spatial skills and executive function (Bisiacchi et al. 2000; Koekkoek et al. 2008) although a few the studies report a lack of effect or limited effects of HIV (Fishkin 2000; Bagenda et al. 2006). The adverse effects of HIV are often diffuse, infecting almost all brain regions, with observations indicating that HIV preferentially affects fronto-striato-thalamocortical loops (Thompson et al. 2005; Hult et al. 2008; Melrose et al. 2008) resulting in significant impairments in higher order cognitive abilities. For instance, Koekkoek et al. (2008) reports that '*Compared with age-appropriate norms, mean IQ of the HIV-infected children was in the average range. However, the HIV-infected children performed poorer on several neuropsychological tests compared with age-appropriate norms. Executive function (attentional flexibility, visuospatial and working memory) and processing speed emerged as the most sensitive cognitive measures in relation to HIV disease*'. However, given that there are few studies providing an in-depth evaluation of cognitive abilities in multiple domains, the conclusion on the influence of HIV on higher cognitive abilities remains preliminary and cannot be generalized.

4.2.4 Mental Health

In comparison to cognitive domain, relatively little is known of the impact of HIV on social, emotional and behavioural functioning of the infected children especially in the early years. Few studies have looked at the mental health outcomes of HIV-positive preschool children. The few that have done so highlight the possibility of adverse effects though the evidence is very scant. Among older children and adolescents there have been much more attention on the mental health outcomes of HIV-infected children. However, there have been inconsistent findings related to this, with certain studies reporting the existence of mental health problems among HIV-positive children while others fail to observe the same effects. For example, Menon et al. (2009) have reported one of the earliest studies on HIV-infected adolescent in sub-Saharan Africa. In this study, mental health outcome of HIV-infected adolescents (a significant majority of whom were vertically infected 73.2 %) were compared to that of normal school going children in Lusaka (HIV status and other risk factors unknown). They did not observe any statistically significant differences between HIV-positive adolescents and their peers in school in terms of mental health functioning. This is in contrast to what was reported in Uganda (Musisi and Kinyanda 2009), in which HIV-positive adolescents presented with significant mental health problems. In their study, 51 % of the subjects had significant psychological distress (SRQ-25 scores ≥ 6) and 17 % ($n = 14$) had attempted suicide within the last 12 months. Specific psychiatric disorders observed based on ICD-10 criteria (World Health Organization 2010) were anxiety 46 %, depression 41 %, somatisation 18 %, seizures 8 %, mania 1 % and HIV-associated progressive encephalopathy 5 %. Within the African context, various methodological shortfalls make it difficult to evaluate which of the two studies best presents the potential effects of HIV among the youth. Specific methodological limitations include the use of a non-validated measure in the context, not controlling for potentially confounding factors, e.g. highly active antiretroviral therapy (HAART), home environment, caregiver characteristics and cognitive functioning, and the lack of a well-defined controls. Studies investigating the aetiology, correlates and patterns of mental health outcomes in children who are HIV-infected are urgently needed especially in sub-Saharan Africa.

4.2.5 Potential Pathways

The effects of vertically transmitted HIV on a child's neuro-development can range from mild to severe impairment. The mother's status may impact upon the child's development in a number of ways. (1) Vertical transmission and subsequent HIV involvement in the central nervous system. Autopsy among HIV+ patients indicates that the brain is the second most frequently affected organ (Hardy and Vance 2009; Sullivan 2009; Woods et al. 2009). (2) Treatment—the use of antiretrovirals (ARVs) that penetrate the CNS is expected to contribute to better outcomes by controlling the effects of HIV on the CNS but results have been mixed. Mixed results on the impact of ARVs on child outcome may be due to either the drug used or the potential neurotoxicity of ARVs. (3) The prenatal environment leading to higher incidences of low-birth weight and prematurity. (4) Adverse postnatal environment through exposure to stressors such as suboptimal stimulation at home, orphan hood, and stigma amongst others. In the next section these different pathways are discussed divided into biomedical and psychosocial.

4.3 Biomedical Factors

4.3.1 Antiretroviral (ARVs)

The positive impact of ARVs on mortality, virological and immunological functioning in HIV-infected children is now well documented. In the pre-ARVs era the estimated mortality was high. For instance, in a pooled analysis of data from developing countries it was observed that by the first year of life, an estimated 35 % of HIV-infected children died and the percentage moved to 53 % by the second year. Access to ARVs has been associated with a significant reduction in mortality rates (Sutcliffe et al. 2008). Furthermore, studies indicate that the most severe forms of HIV related encephalopathy have been greatly reduced since the introduction of ARVs. Evidence on the effects of highly active anti-retroviral therapy (HAART) on neurocognitive outcomes among infected children is inconclusive. Some studies (Wolters et al. 1994; Shanbhag et al. 2005; Martin et al. 2006) have reported that initiation of HAART has resulted in improved performance on cognitive tasks while other studies found none while others a negative impact of HAART. Koekkeok et al. (2006) carried out a longitudinal study in Thailand to estimate the impact of ARVs on psychomotor function. In this study, four groups of children were studied, HIV-infected on ARVs for more than a year, HIV-infected newly treated, HIV-infected untreated and controls. Results indicated that at baseline the children performed similarly on all psychomotor tasks. Overall psychomotor performance did not change at the 4-month evaluation in all groups. At the 12-month evaluation, psychomotor performance had deteriorated substantially on all tasks in both the newly treated and the untreated children. Poor performance on psychomotor tests was observed even in the presence of immunological reconstitution. These results are in contrasts to what has been reported elsewhere e.g. Raskino et al. (1999) who observed that the mean scores of participants in their study improved by 11–13 points following the initiation of ARVs.

In a systematic review of literature it was concluded that even in the era of HAART, subtle impairments were still being consistently reported (Le Doare et al. 2012). Overall these impairments may arise from the inability of ARVs to remediate previous damage. There have been studies indicating that children placed on ARVs before the onset of AIDS-defining symptoms had less severe impairments compared to those introduced to ARVs when at the advanced stages of HIV. Other potential causes of continued neurocognitive disorders in the era of ARVs included potential neurotoxicity of ARVs and the use of ARVs that do not adequately penetrate the central nervous system.

4.3.2 Timing of Infection

Children infected prenatally are more likely to present with more severe forms of neurodevelopment delays compared to children vertically infected in the postnatal stage (i.e. during breastfeeding). These observations were first made in a cohort study in USA and were later replicated in Tanzania. In a longitudinal study involving more than 114 infants Smith et al. (2000) observed that early infected infants performed significantly worse than late infected infants in both motor and mental subscales as measured by BSID. Moreover, early-infected infants experienced a more rapid decrease in mental score (0.72 point per month) compared to the late infected infants (0.30 point per month). Similarly, McGrath et al. (2006) studied the relationship between the timing of mother-to-child transmission of HIV-1 and neurodevelopment among children born to HIV-1 infected mothers in Tanzania. They observed that children who acquire HIV in utero were at an elevated risk of neurodevelopmental delay compared to those who acquired HIV postnatally (i.e. through breastfeeding). According to McGrath et al. (2006) '*Testing HIV-1-positive at birth was associated with a 14.9 (95 % CI 5.0, 44.7) times higher rate of becoming developmentally delayed in mental function, while testing HIV-1-positive after birth was associated with a 3.2 (95 % CI 1.6, 6.4) times higher rate than in uninfected children*'.

4.3.3 Severity of HIV Disease

Two approaches can be used to evaluate the severity of HIV disease: CD4 counts and staging the clinical features. The WHO together with Centre for Disease Control, USA, produced an algorithm to define the severity of HIV infection. This algorithm starts from HIV-infected but asymptomatic to full blown AIDS. Studies in various parts of the world observed that the more severe the disease stage the more severe the neurodevelopmental or neurocognitive impairments. In a longitudinal study that compared HIV-infected infants born to HIV-negative mothers, it was similarly found that the stage of the HIV infection correlated with motor and cognitive impairments (Nozyce et al. 1994). Specifically, those infants who developed AIDS-defining conditions displayed major impairments in language, gross motor, fine motor, cognitive, and social development, suggesting that the decreasing immune functioning in children with HIV predicts impairments in neurodevelopment (Nozyce et al. 1994). Similar findings have been reported with older children. For instance, Ruel et al. (2012), in a study of HIV-positive 6–12 year old Ugandan children observed that those with CD4 count >350 cells/μL demonstrate significant cognitive and motor deficits that correlate with HIV plasma RNA levels. In a recent study of HIV-infected children and adolescents, it was observed that children who had a previous diagnosis of encephalopathy and a history of severe ill heath had poor performance in cognitive scores compared to their HIV-positive peers without the history of severe ill-health (Smith et al. 2012).

4.3.4 Nutrition

HIV-infected children are at a higher risk of experiencing growth failures. Studies such as those by Lepage et al. (1996) and Berhane et al. (1997) have shown that pre-ARVs HIV-infected children were at an increased risk of experiencing stunting and being underweight. Recent studies indicate that children on ARVs have better growth outcomes yet still show marked delay compared to those who are uninfected. Growth failure (i.e. being underweight and stunted) has been independently associated with neurodevelopmental delay in children. In the context of HIV, growth failure has been observed to exacerbate the adverse effects of HIV infection. For instance, Abubakar et al. (2009) in Kenya observed that HIV-infected children who were also underweight performed significantly more poorly than those who were infected but were not underweight. Moreover, impaired neurodevelopmental delay among HIV-infected children has also been associated with micronutrient deficiency both in the mother and the child. Intervention studies in Tanzania indicated that children of mothers who were HIV-positive performed significantly much better when the mothers received micronutrient supplementation. These results provide further evidence on the saliency of nutrition on developmental outcomes of children who are HIV-positive.

4.3.5 Psychosocial Risk Factors

Exposure to HIV among children often occurs in the context of environmental factors that pose equal or greater risks to a child's development, compared with the HIV infection, including unstable or multiple caregivers, maternal drug and/or alcohol use, poverty, and low maternal education (Brown et al. 2000). Though the study of the impact of psychosocial risk factors has received relatively less attention, research so far indicates the significant impact of these factors on the functioning of HIV-positive children. In a study, involving more than 41 children, Coscia et al. (2001) examined the relationship among home environment, socio-economic status (SES), cognitive functioning, and health status in a group of children with HIV-1 infection. They observed that home environment mediates the association between SES and child IQ, which provides support for the hypothesis that the home environment can either serve as a protective factor against or risk factor for the negative effects of poverty on cognitive functioning. Moreover, studies among HIV-infected children show that family structure and caregiver characteristics were correlated to performance. For instance, in a cohort study of children and adolescents in Thailand it was observed that HIV-infection alongside family structure, including living with caregivers other than biological parents, lower education of caregiver, and lower family income, were associated with higher risk of poor cognitive function (Puthanakit et al. 2010).

The impact of these psychosocial risk factors seem to be especially more salient when investigating mental health outcomes. For instance, in a study of psychiatric symptoms among HIV-infected and affected children in the USA, Mellins et al. (2003) report that while these children were at an elevated risk of experiencing emotional and behavioral problems, higher risk was more related to socio-demographic characteristics as opposed to HIV status.

4.3.6 The HIV Exposed but Uninfected Children

Children born of HIV-infected mothers though uninfected have been observed to also suffer different forms of developmental delay. As earlier alluded, the maternal HIV status may impact on the child in various ways including compromised prenatal environment, compromised postnatal environment and exposure to ARVs in utero. Studies in high income countries report a lack of a significant influence of HIV exposure on child cognitive, motor and language functioning, while studies from resource poor setting indicate a stronger impact of maternal HIV status on child outcomes (Le Doare et al. 2012). For instance, among preschool children in the Democratic Republic of Congo, Van Rie et al. (2008) observed that while HIV exposed uninfected children perform better than infected children, their performance was significantly worse that than of community controls on various parameters. They also observed that HIV exposed uninfected children were more likely to present delays and impairment in language expression (47.4 %, $n=9$) compared to community controls (12.9 %, $n=4$). Similar patterns of results were observed in a sample from Thailand, where HIV affected children were observed to have lower cognitive scores compared to controls, although they were no differences in behaviour and growth outcomes (Sanmaneechai et al. 2005). An evaluation of the correlations of these scores indicate that the poor performance of HIV exposed child may be related to psychosocial factors such as orphan hood, and low parental SES.

5 Conclusion

Current research indicates that HIV impacts on multiple developmental domains, its effects start early in life and tend to persists into adolescence. Various biomedical factors (e.g. severity of HIV illness, use of ARVs) and psychosocial (home environment) have been observed to either exacerbate or ameliorate the negative effects of HIV. However, there are still major knowledge gaps such as limited research evidence from the sub-Saharan Africa, home to 90 % of all HIV-positive children, and lack of sufficient knowledge on the impact of treatment on long-term survivors, which warrants further research. Moreover, little if anything has been done to understand the treatment and intervention needed at enhancing cognitive outcomes among HIV-infected children.

References

Abubakar A, Holding P, Newton C, Van Baar A, Van de Vijver F (2009) The role of weight for age and disease stage in poor psychomotor outcome of HIV-infected children in Kilifi, Kenya. Dev Med Child Neurol 51:968–973

Al Serouri AW, Grantham-McGregor SM, Greenwood B, Costello A (2000) Impact of asymptomatic malaria parasitaemia on cognitive function and school achievement of schoolchildren in the Yemen Republic. Parasitology 121:337–345

Bagenda D, Nassali A, Kalyesubula I, Sherman B, Drotar D, Boivin MJ, Olness K (2006) Health, neurologic, and cognitive status of HIV-infected, long-surviving, and antiretroviral-naive Ugandan children. Pediatrics 117:729–740

Bangirana P, Musisi S, Boivin MJ, Ehnvall A, John CC, Bergemann TL, Allebeck P (2011) Malaria with neurological involvement in Ugandan children: effect on cognitive ability, academic achievement and behaviour. Malar J 10:334

Bayley N (1993) The Bayley Scales of Infant Development, 2nd edn. Harcourt/Psychological Corporation, New York

Berhane R, Bagenda D, Marum L, Aceng E, Ndugwa C, Bosch RJ, Olness K (1997) Growth failure as a prognostic indicator of mortality in pediatric HIV infection. Pediatrics 100:1–4

Bisiacchi PS, Suppiej A, Laverda A (2000) Neuropsychological evaluation of neurologically asymptomatic HIV-infected children. Brain Cogn 43:49–52

Blanchette N, Smith ML, Fernandes-Penney A, King S, Read S (2001) Cognitive and motor development in children with vertically transmitted HIV infection. Brain Cogn 46:50–53

Blanchette N, Smith ML, King S, Fernandes-Penney A, Read S (2002) Cognitive development in school-age children with vertically transmitted HIV infection. Dev Neuropsychol 21: 223–241

BMJ-group (2011) Assessment of learning difficulty and cognitive delay. Retrieved from http://bestpractice.bmj.com/best-practice/monograph/884.html

Boivin MJ (2002) Effects of early cerebral malaria on cognitive ability in Senegalese children. J Dev Behav Pediatr 23:353–364

Boivin MJ, Giordani B, Ndanga K, Maky MM, Manzeki KM, Ngunu N, Muamba K (1993) Effects of treatment for intestinal parasites and malaria on the cognitive abilities of schoolchildren in Zaire, Africa. Health Psychol 12:220–226

Boivin MJ, Bangirana P, Byarugaba J, Opoka RO, Idro R, Jurek AM, John CC (2007) Cognitive impairment after cerebral malaria in children: a prospective study. Pediatrics 119:e360–e366

Boivin MJ, Gladstone MJ, Vokhiwa M, Birbeck GL, Magen JG, Page C, Taylor TE (2011) Developmental outcomes in Malawian children with retinopathy-positive cerebral malaria. Trop Med Int Health 16:263–271

Brackis-Cott E, Kang E, Dolezal C, Abrams EJ, Mellins CA (2009) The impact of perinatal HIV infection on older school-aged children's and adolescents' receptive language and word recognition skills. AIDS Patient Care STDS 23:415–421

Brown LK, Lourie KJ, Pao M (2000) Children and adolescents living with HIV and AIDS: a review. J Child Psychol Psychiatry 41:81–96

Carter JA, Murira GM, Ross AJ, Mung'ala-Odera V, Newton CR (2003) Speech and language sequelae of severe malaria in Kenyan children. Brain Inj 17:217–224

Carter JA, Mung'ala-Odera V, Neville BG, Murira G, Mturi N, Musumba C, Newton CR (2005) Persistent neurocognitive impairments associated with severe falciparum malaria in Kenyan children. J Neurol Neurosurg Psychiatry 76:476–481

Clarke SE, Jukes MC, Njagi JK, Khasakhala L, Cundill B, Otido J, Brooker S (2008) Effect of intermittent preventive treatment of malaria on health and education in schoolchildren: a cluster-randomised, double-blind, placebo-controlled trial. Lancet 372:127–138

Coscia JM, Christensten BK, Henry RR, Wallston K, Radcliffe J, Rustein R (2001) Effects of home environment, socioeconomic status, and health status on cognitive functioning in children with HIV-1 infection. J Pediatr Psychol 26:321–329

Croisile B (2004) Our cognitive functions: how they are organized, how they evolve with age, and how they are trained. *Brain Today (e-mag)*

Dabis F, Ekpini ER (2002) HIV-1/AIDS and maternal and child health in Africa. Lancet 359: 2097–2104

de Miranda AS, Brant F, Machado FS, Rachid MA, Teixeira AL (2011) Improving cognitive outcome in cerebral malaria: insights from clinical and experimental research. Cent Nerv Syst Agents Med Chem 11:285–295

Drotar D, Olness K, Wiznitzer M, Guay L, Marum L, Svilar G, Kiziri-Mayengo R (1997) Neurodevelopmental outcomes of Ugandan infants with human immunodeficiency virus type 1 infection. Pediatrics 100(1):E5

Dugbartey AT, Spellacy FJ, Dugbartey MT (1998) Somatosensory discrimination deficits following pediatric cerebral malaria. Am J Trop Med Hyg 59:393–396

Fernando SD (2001) The impact of malaria on the cognitive performance and physical development of school children in a malaria endemic area of Sri Lanka. Ph.D dissertation, University of Colombo, Sri Lanka

Fernando SD, Gunawardena DM, Bandara MR, De Silva D, Carter R, Mendis KN, Wickremasinghe AR (2003) The impact of repeated malaria attacks on the school performance of children. Am J Trop Med Hyg 69:582–588

Fernando SD, de Silva D, Carter R, Mendis KN, Wickremasinghe R (2006) A randomized, double-blind, placebo-controlled, clinical trial of the impact of malaria prevention on the educational attainment of school children. Am J Trop Med Hyg 74:386–393

Fernando SD, Rodrigo C, Rajapakse S (2010) The 'hidden' burden of malaria: cognitive impairment following infection. Malar J 9:366

Fishkin PE (2000) Brief report: relationship between HIV infection and WPPSI-R performance in preschool-age children. J Pediatr Psychol 25:347–351

Gay CL, Armstrong FD, Cohen D, Lai S, Hardy MD, Swales TP, Scott GB (1995) The effects of HIV on cognitive and motor development in children born to HIV-seropositive women with no reported drug use: birth to 24 months. Pediatrics 96:1078–1082

Hardy DJ, Vance DE (2009) The neuropsychology of HIV/AIDS in older adults. Neuropsychol Rev 19:263–272

Hay SI, Okiro EA, Gething PW, Patil AP, Tatem AJ, Guerra CA, Snow RW (2010) Estimating the global clinical burden of *Plasmodium falciparum* malaria in 2007. PLoS Med 7(6):e1000290

Holding PA, Kitsao-Wekulo PK (2004) Describing the burden of malaria on child development: what should we be measuring and how should we be measuring it? Am J Trop Med Hyg 71(2 Suppl):71–79

Holding PA, Stevenson J, Peshu N, Marsh K (1999) Cognitive sequelae of severe malaria with impaired consciousness. Trans R Soc Trop Med Hyg 93:529–534

Holding PA, Taylor HG, Kazungu SD, Mkala T, Gona J, Mwamuye B, Stevenson J (2004) Assessing cognitive outcomes in a rural African population: development of a neuropsychological battery in Kilifi District, Kenya. J Int Neuropsychol Soc 10:246–260

Hult B, Chana G, Masliah E, Everall I (2008) Neurobiology of HIV. Int Rev Psychiatry 20: 3–13

John CC, Bangirana P, Byarugaba J, Opoka RO, Idro R, Jurek AM, Boivin MJ (2008) Cerebral malaria in children is associated with long-term cognitive impairment. Pediatrics 122:e92–e99

Jukes MCH, Pinder M, Grigorenko EL, Smith HB, Walraven G, Bariau EM, Bundy DAP (2006) Long-term impact of malaria chemoprophylaxis on cognitive abilities and educational attainment: follow-up of a controlled trial. PLoS Clin Trials 1(4):e19

Kaufman AS, Kaufman NL (1983) Kaufman Assessment Battery for Children. American Guidance Service, Circle Pines

Kihara M, Carter JA, Newton CR (2006) The effect of *Plasmodium falciparum* on cognition: a systematic review. Trop Med Int Health 11:386–397

Kihara M, Carter JA, Holding PA, Vargha-Khadem F, Scott RC, Idro R, Newton CR (2009) Impaired everyday memory associated with encephalopathy of severe malaria: the role of seizures and hippocampal damage. Malar J 8:273

Koekkeok S, Eggermont L, De Sonneville L, Jupimal T, Wicharuk S, Apateerapong W, Ananworanich J (2006) Effects of highly active antiretroviral therapy (HAART) on psychomotor performance in children with HIV disease. J Neurol 253:1615–1624

Koekkoek S, de Sonneville LM, Wolfs TF, Licht R, Geelen SP (2008) Neurocognitive function profile in HIV-infected school-age children. Eur J Paediatr Neurol 12:290–297

Le Doare K, Bland R, Newell ML (2012) Neurodevelopment in children born to HIV-infected mothers by infection and treatment status. Pediatrics 130(5):e1326–e1344

Lepage P, Msellati P, Hitimana DG, Bazubagira A, Van Goethem C, Simonon A, Dabis F (1996) Growth of human immunodeficiency type 1-infected and uninfected children: a prospective cohort study in Kigali, Rwanda, 1988 to 1993. Pediatr Infect Dis J 15:479–485

Martin SC, Wolters PL, Toledo-Tamula MA, Zeichner SL, Hazra R, Civitello L (2006) Cognitive functioning in school-aged children with vertically acquired HIV infection being treated with highly active antiretroviral therapy (HAART). Dev Neuropsychol 30:633–657

McGrath N, Fawzi WW, Bellinger D, Robins J, Msamanga GI, Manji K, Tronick E (2006) The timing of mother-to-child transmission of human immunodeficiency virus infection and the neurodevelopment of children in Tanzania. Pediatr Infect Dis J 25:47–52

Mellins C, Smith R, O'Driscoll P, Madger L, Brouwers P, Chase C, Matzen E (2003) High rates of behavioral problems in perinatally HIV-infected children are not linked to HIV disease. Pediatrics 111:384–393

Melrose RJ, Tinaz S, Castello JMB, Courtney MG, Stern CG (2008) Compromise to fronto-stratial functioning in HIV: an FMRI investigation of the semantic event sequencing. Behav Brain Res 188:33–347

Menon A, Glazebrook C, Ngoma MS (2009) Mental health of HIV positive adolescents in Zambia. Med J Zambia 36:151–156

Molyneux ME, Taylor TE, Wirima JJ, Borgstein A (1989) Clinical features and prognostic indicators in paediatric cerebral malaria: a Study of 131 comatose Malawian children. Q J Med 71:441–459

Muntendam AH, Jaffar S, Bleichrodt N, Hensbroek MB (1996) Absence of neuropsychological sequelae following cerebral malaria in Gambian children. Trans R Soc Trop Med Hyg 90: 391–394

Musisi S, Kinyanda E (2009) Emotional and behavioral disorders in HIV seropositive adolescents in urban Uganda. East Afr Med J 86:16–24

Newell M, Coovadia H, Cortina-Borja M, Rollins N, Gaillard P, Dabis F (2004) Mortality of infected and uninfected infants born to HIV-infected mothers in Africa: a pooled analysis. Lancet 364:1236–1243

Newton CR, Hien TT, White N (2000) Cerebral malaria. J Neurol Neurosurg Psychiatry 69: 433–441

Nozyce M, Hittelman J, Muenz L, Durako SJ, Fischer ML, Willoughby A (1994) Effect of perinatally acquired human immunodeficiency virus infection on neurodevelopment in children during the first two years of life. Pediatrics 94:883–891

Puthanakit T, Aurpibul L, Louthrenoo O, Tapanya P, Nadsasarn R, Insee-ard S, Sirisanthana V (2010) Poor cognitive functioning of school-aged children in Thailand with perinatally acquired HIV infection taking antiretroviral therapy. AIDS Patient Care STDS 24(3):141–146

Raskino C, Pearson DA, Baker CJ, Lifschitz MH, O'Donnell K, Mintz M, Englund JA (1999) Neurologic, neurocognitive, and brain growth outcomes in human immunodeficiency virus-infected children receiving different nucleoside antiretroviral regimens. Pediatric AIDS Clinical Trials Group 152 Study Team. Pediatrics 104(3):e32

Ruel TD, Boivin MJ, Boal HE, Bangirana P, Charlebois E, Havlir DV, Wong JK (2012) Neurocognitive and motor deficits in HIV-infected Ugandan children with high CD4 cell counts. Clin Infect Dis 54(7):1001–1009

Sanmaneechai O, Puthanakit T, Louthrenoo O, Sirisanthana V (2005) Growth, developmental, and behavioral outcomes of HIV-affected preschool children in Thailand. J Med Assoc Thai 88(12):1873–1879

Shanbhag MC, Rutstein RM, Zaoutis T, Zhao H, Chao D, Radcliffe J (2005) Neurocognitive functioning in pediatric human immunodeficiency virus infection: effects of combined therapy. Arch Pediatr Adolesc Med 159(7):651–656

Sherr L, Mueller J, Varrall R (2009) A systematic review of cognitive development and child human immunodeficiency virus infection. Psychol Health Med 14:387–404

Sherwood B, Maria H, Peggy J (2008) A review of pediatric HIV effects on neurocognitive development. Issues Compr Pediatr Nurs 31:107–121

Smith R, Malee K, Charurat M, Magder L, Mellins C, Macmillan C, Moye J (2000) Timing of perinatal human immunodeficiency virus type 1 infection and rate of neurodevelopment. The Women and Infant Transmission Study Group. Pediatr Infect Dis J 19:862–871

Smith R, Chernoff M, Williams PL, Malee KM, Sirois PA, Kammerer B, Rutstein R (2012) Impact of HIV severity on cognitive and adaptive functioning during childhood and adolescence. Pediatr Infect Dis J 31:592–598

Stanley FJ, Blair E, Alberman E (2000) Cerebral palsies: epidemiology and causal pathways. Mac Keith, London

Sullivan EV (2009) Special section of neuropsychology review on HIV/neuroAIDS. Neuropsychol Rev 19(2):143

Sutcliffe CG, van Dijk JH, Bolton C, Persaud D, Moss WJ (2008) Effectiveness of antiretroviral therapy among HIV-infected children in sub-Saharan Africa. Lancet Infect Dis 8:477–489

Thompson PM, Dutton RA, Hayashi KM, Toga AW, Lopez OL, Aizenstein HT, Becker JT (2005) Thinning of the cerebral cortex visualized in HIV/AIDS reflects CD4 and T lymphocyte decline. Proc Natl Acad Sci U S A 102:15647–15652

Thuilliez J, Sissoko MS, Toure OB, Kamate P, Berthelemy JC, Doumbo OK (2010) Malaria and primary education in Mali: a longitudinal study in the village of Doneguebougou. Soc Sci Med 71(2):324–334

UNAIDS (2012a) Report on the global AIDS epidemic. Retrieved from http://www.unaids.org/en/media/unaids/contentassets/documents/epidemiology/2012/gr2012/20121120_UNAIDS_Global_Report_2012_en.pdf

UNAIDS (2012b) Core slides: global summary of the AIDS epidemic. Retrieved from http://www.unaids.org/en/media/unaids/contentassets/documents/epidemiology/2012/201207_epi_core_en.pdf

Van Rie A, Mupuala A, Dow A (2008) Impact of the HIV/AIDS epidemic on the neurodevelopment of preschool-aged children in Kinshasa, Democratic Republic of the Congo. Pediatrics 122(1):e123–e128

Vitor-Silva S, Reyes-Lecca RC, Pinheiro TR, Lacerda MV (2009) Malaria is associated with poor school performance in an endemic area of the Brazilian Amazon. Malar J 8:230

White VA, Lewallen S, Beare NAV, Molyneux ME, Taylor TE (2009) Retinal pathology of pediatric cerebral malaria in Malawi. PLoS One 4(1):e4317

Wiktor SZ, Ekpini E, Nduati R (1997) Prevention of mother-to-child transmission of HIV1 in Africa. AIDS 11(Suppl B):S7987

Wolters PL, Brouwers P, Moss HA, Pizzo PA (1994) Adaptive behavior of children with symptomatic HIV infection before and after zidovudine therapy. Pediatr Psychol 19:47–61

Wolters PL, Brouwers P, Moss HA, Pizzo PA (1995) Differential receptive and expressive language functioning of children with symptomatic HIV disease and relation to CT scan brain abnormalities. Pediatrics 95:112–119

Wolters PL, Brouwers P, Civitello L, Moss HA (1997) Receptive and expressive language function of children with symptomatic HIV infection and relationship with disease parameters: a longitudinal 24-month follow-up study. AIDS 11:1135–1144

Woods SP, Moore DJ, Weber E, Grant I (2009) Cognitive neuropsychology of HIV-associated neurocognitive disorders. Neuropsychol Rev 19:152–168

World Health Organization (2000) Severe falciparum malaria. World Health Organization, Communicable Diseases Cluster. Trans R Soc Trop Med Hyg 94(Suppl 1):S1–S90

World Health Organization (2010) International classification of diseases, 10th revision. Retrieved from http://apps.who.int/classifications/icd10/browse/2010/en

World Malaria Report (2011) World Health Organization. Retrieved from http://www.who.int/entity/malaria/world_malaria_report_2011/9789241564403_eng.pdf

Index

A

Acanthamoeba, 93–105
Acquired immunodeficiency syndrome (AIDS)
  biomedical factors
    ARVs, 383
    HIV expose to children, 386
    nutrition, 384–385
    psychosocial risk factors, 385
    severity of HIV disease, 384
    timing of infection, 383
  effects on neuro-development
    cognitive function, 381
    language, 380–381
    mental health, 381–382
    motor, 380
    potential pathways, 382
  epidemiology, 379–380
Active convulsive epilepsy (ACE), 357
Acute bacterial meningitis (ABM), 358
Acute disseminated encephalomyelitis (ADEM), 83, 115
Acute schistosomal encephalopathy (ASE), 115
  diagnosis, 118
  pathogenesis, 116
  treatment, 122
Acute schistosomiasis, 37, 114
Acute symptomatic schistosomiasis (ASS), 115, 116, 118
Acute symptomatic seizures, 356, 359
ADEM. *See* Acute disseminated encephalomyelitis (ADEM)
Aedes, 25, 270, 279
African eyeworm, 36
African trypanosomiasis, 31–33
  perspectives and neuroscience role, 195–196
  trypanosome
    CNS invasion, 186–192
    growth in CNS, 192–194
    and host immune system, 185–186
    infections, 194–195
    parasites, 184–185
AHD. *See* Antihelminthic drugs (AHD)
AIDS. *See* Acquired immunodeficiency syndrome (AIDS)
Albendazole (ALB), 75, 82, 140
Alzheimer's dementia, 56
American trypanosomiasis. *See* Chagas disease (CD)
Amoebiasis, 93–94
  clinical diagnosis, 102–103
  infectious agents, 94–97
  pathogenesis, 100–101
  pathophysiology, 97–98
  predisposing factors, 98–99
  treatment, 103–106
Amphetamine, 308
  behavioral studies, 312
  cathinone and, 314–315
  pharmacological effects, 311
Angiostrongyliasis, 73–75
  brain CT finding, 79
  ocular, 74, 75
Angiostrongylus cantonensis, 68, 73–75
ANS. *See* Autonomic nervous system (ANS)
Antibody dependent enhancement (ADE), 281
Anticonvulsants, 122, 285
Antihelminthic drugs (AHD), 139–141
Anti-helmintic treatment, 34, 82
Anti-microbial molecules, 15
Anti-retroviral agents (ARVs), 380, 382, 383

M. Bentivoglio et al. (eds.), *Neglected Tropical Diseases and Conditions of the Nervous System*,
DOI 10.1007/978-1-4614-8100-3,

Antivenin
development, 346
snake venom and, 345–346
Apoptosis, 254, 255, 260, 261
Arbovirus, 269
Arcuate nucleus-median eminence complex (AMC), 195
Arrhythmias, 209, 210, 212, 342
ARVs. *See* Anti-retroviral agents (ARVs)
Ascariasis treatment, 71
Ascaris suum, 71
ASS. *See* Acute symptomatic schistosomiasis (ASS)
Atopic myelitis, 71
Atrioventricular (AV) block, 207, 209, 210
Autochthonous trichinellosis, 81
Autonomic nervous system (ANS), 203
chronic cardiac form, 209–210
chronic digestive form, 210–212
chronic indeterminate form, 208–209
disorder, 11

B
Bacteria
infections, 358
NTD cause by, 11
propagation in cooler tissues, 225–226
Balamuthia mandrillaris, 93–105
Bat rabies virus, 235–236
Batson's venous plexus, 115, 116, 119
Bayley Scales of Infant Development (BSID), 380, 381, 384
Baylisascaris procyonis, 71–72
BBB. *See* Blood-brain barrier (BBB)
B7-H1, 258, 260–262
Black stone, snake bite, 345
Blantyre Coma Score (BCS), 376
Blood-brain barrier (BBB), 16, 100, 101, 103, 186
Blood-CNS barrier dysfunction, 73
Blood-CSF barrier, 187
Brain
astrocytes, 257
dysfunctions mechanisms, 194–195
immunocytochemistry, 260
lesions, 207
parenchyma, 188–192
RABV, 254–255
Brugia malayi, 82
BSID. *See* Bayley Scales of Infant Development (BSID)

C
Calcified neurocysticercal lesions (CNL), 133
Calloselasma rhodostoma. *See* Malayan pit viper
Canine rabies, 14, 235, 243
Card agglutination test for trypanosomiasis (CATT), 172
Cardiomegaly, 30
Cardiovascular disease, 337–339
Cassava
bitter varieties, 326
cyanogens, 328, 329
toxicity, 322
Catha edulis. *See* Khat
Cathinone, 308, 310
addictive and abuse potential, 313
and amphetamine, 315
dopamine D1 receptor, 313
drugs substitute, 314
hyperthermic response, 312
khat *vs*., 315–316
pharmacological effects, 311
psychostimulants, 314
CD. *See* Chagas disease (CD)
Central nervous system (CNS), 25, 29–30
acute infections, 356
complications, 8
gnathostomiasis, 78
granulomatous lesions in, 101
helminthic infections of, 75
HIV on, 382
involvement
in Chagas disease, 207–208
in epilepsy, 363
in strongyloidiasis, 81
mediated immune-unresponsiveness, 261–262
pathogens, 68
rabies virus in, 238
trypanosome growth in, 192–194
trypanosome invasion of, 186–192
Cerebral gnathostomiasis, 78, 79
Cerebral haemorrhage, 344
Cerebral infarction, 343–344
Cerebral malaria
children with, 378–379
epilepsy, 360
in Kenyan children, 359
Cerebrospinal fluid (CSF), 73–75, 78, 83
analysis in rabies, 239
barriers, 186
eosinophils in, 70
Chagas disease (CD), 10, 11, 28–31, 202–203

chronic indeterminate form, 208–209
clinical phases and forms, 203–204
CNS involvement in, 207–208
etiological agent, 206
nervous form, 203
nervous system involvement in, 203–204, 208–212
pathogenesis and evolution, 204–206
strains and organ tropism, 206
Child Attitude Toward Illness Scale (CATIS), 56
Children
with cerebral malaria, 378–379
cognitive outcomes in, 371
exposion to falciparum malaria, 373–377
HIV (*see* Human immunodeficiency virus (HIV))
Ugandan study, 378
Choroid plexus, 186–188
Chronic neuroinflammation, 196
Chronic schistosomiasis, 38
Circumventricular organs (CVOs), trypanosomes into, 186–188, 195
CNS. *See* Central nervous system (CNS)
Cognitive impairment
AIDS (*see* Acquired immunodeficiency syndrome (AIDS))
definition, 370–371
during falciparum malaria illness, 376
HIV (*see* Human immunodeficiency virus (HIV))
malaria, 371
parasitaemia effect, 371–375
post-cerebral malaria, 378
post-severe malaria, 376–377
Computed tomography (CT) scan
contrast-enhanced, 136
HIV, 380
Japanese encephalitis, 275
in rabies, 239
Conditioned place preference (CPP), 313
Conventional DCs (cDCs), 259
Cooler tissues, bacterial propagation in, 225–226
Corticosteroids, 71, 140, 277
Coyote rabies, 236
CSF. *See* Cerebrospinal fluid (CSF)
Cutaneous larva migrans (CLM), 83
Cyanogens, cassava, 328, 329
Cysticerci, 33, 34
colloidal, 130
developmental phases, 136
genetic diversity, 130
infection of humans by, 128
larval phase, 131
life cycle, 129
MRI, 138
T. solium, 135
vesicular, 130
Cysticercosis, 9, 33–35, 128, 131, 141
Cytoadherence, 371

D

Demyelination
in leprosy, 227–228
Schwann cells, 221–222
Denervation, 205, 208, 211
Dengue, 25. *See also* Dengue virus (DENV)
associated immune mediated syndromes, 283
associated transient muscle dysfunction, 283–284
diagnosis, 284–285
encephalitis, 283
encephalopathy, 282–283
epidemiology, 278
neurological complications, 282
vaccines, 285–286
Dengue fever (DF), 25, 278, 281, 285
Dengue hemorrhagic fever (DHF), 278, 281
Dengue shock syndrome (DSS), 278, 281
Dengue virus (DENV), 278–279. *See also* Dengue
clinical features, 281–282
diagnosis, 284–285
differential diagnosis, 285
management, 285
neurological manifestations, 282–284
outcome, 285
coagulopathy, 280
entry and replication, 280
life cycle, 279–280
pathogenesis, 280–281
prevention, 285–286
vaccines, 285–286
Depigmentations, 54
Diet, 337–338, 346
Dirofilaria immitis, 83
Dirofilaria repens, 83
Dopamine, 314
Catha edulis and, 316
pharmacological effects, 311–312
Drugs of abuse, 308, 310, 311, 316
Dystonia, 275

E

Echinococcosis, 25–26
Echinococcus, 363
  E. granulosus, 25, 26
  E. multilocularis, 25, 26
Electroencephalography (EEG), 50, 276, 356
Electromanometry, 211
Elephantiasis, 83
ELISA. *See* Enzyme-linked immunosorbent assay (ELISA)
Encephalitic rabies, 234, 238, 260
Encephalopathy
  dengue, 25, 282–283
  development, 150
  and filariasis, 83
  HIV, 383
  Lagochilascaris minor, 72
  neurobiology of *Loa loa*, 158
  post-treatment, 157–158
Enzootic infection, 202–203
Enzymelinked immunoelectrotransfer blot (EITB) assay, 138
Enzyme-linked immunosorbent assay (ELISA), 69, 79, 138, 276
Eosinophilic encephalitis, 73
Eosinophilic meningitis, 73, 75
Eosinophilia, 37, 69
  CSF, 74, 78
  peripheral blood, 78–79, 119
Eosinophilic leukocyte-mediated toxicity, 115
Epilepsy, 8, 16, 52–53
  ACE, 357
  bacterial infections, 358
  definition, 356
  helminthic infections, 362–363
  incidence, 356
  infections associated with, 357
  malaria, 359–360
  neurocysticercosis and, 132–133, 358–359
  onchocerciasis, 361–362
  prevalence, 356
  stigmatizing burn deformities in, 54
  toxocara, 361
  toxoplasmosis, 361
  viral infections, 357
Epileptogenesis, 133
Ergosterol biosynthesis pathway, 104, 105
Etiopathogenesis, of optic nerve disease, 155
Evasive strategy, in RABV, 260
Explanatory Model Interview Catalogue (EMIC), 56
Extraparenchymal neurocysticercosis, 134, 138

F

Falciparum malaria, 370, 379
  children exposion to, 373–377
  cognitive function during illness, 376
  effects, 371
  neurocognitive impairment, 372–375
Febrile disease, 25
Filarial nematodes, 36, 148, 150
Filariases, 35–36
Filariasis, 82–83
Flavivirus, 25, 269, 271, 273
Fluid attenuated inversion recovery (FLAIR), 136, 137
5-fluorocytosine (flucytosine), 104, 106
Food preservation, 336–339
Fox rabies, 236

G

GAE. *See* Granulomatous amoebic encephalitis (GAE)
Gambiense, HAT, 32, 33
Gastrin-releasing peptide receptor (GRPR), 152
Germ theory of disease, 5
Glaucoma, 154, 155
Glial cells, of PNS, 219–221
Glossina morsitans, 32, 166
Glossina palpalis, 32, 166, 168
Glycerol, 140
Gnathostoma, 76–79
  G. spinigerum, 77
Gnathostomiasis, 76, 78
G protein-coupled receptors (GPCRs), 152
Granuloma, 98
  formation, 70, 121, 222, 361
Granulomatous amoebic encephalitis (GAE), 93–94
  challenges in therapy, 107
  clinical diagnosis, 102–103
  infectious agents, 95–97
  model of, 100
  pathogenesis, 100–101
  pathophysiology, 97–98
  predisposing factors in, 98–99
  risk factors associated with, 99
  treatment, 103–106
Granulomatous lesions
  chronic, 363
  in CNS, 101

H

Hansen's disease. *See* Leprosy
HAT. *See* Human African trypanosomiasis (HAT)
Heart block, 207, 209, 210

Helminthic infections, 75, 362–363
Helminths, 363
  soil-transmitted, 39
Hepatomegaly, 114, 119, 177
Herpes simplex virus (HSV), 277
  type 1, 357
Highly active antiretroviral therapy (HAART), 107, 383
HIV. *See* Human immunodeficiency virus (HIV)
Human African trypanosomiasis (HAT), 10, 32, 55, 166–167
  clinical features, 171–177
  epidemiology, 167
  gambiense, 32, 33
  history, 167–169
  prevalence and geographical distribution, 169–171
  rhodesiense, 32, 33
  sleep-wake disruption in, 176
  transmission, 167, 169
  in travelers and migrants, 177
Human immunodeficiency virus (HIV), 357
  biomedical factors
    ARVs, 383
    exposion to children, 386
    nutrition, 384–385
    psychosocial risk factors, 385
    severity of disease, 384
    timing of infection, 383
  effects on neuro-development
    cognitive function, 381
    language, 380–381
    mental health, 381–382
    motor, 380
    potential pathways, 382
  epidemiology, 379–380
  infected adolescent in Africa, 382
Human rabies, 237–239
  due to lyssaviruses, 240
  management, 242–245
  prognosis, 241–242
Human T-cell lymphotropic virus type I (HTLV-I), 325, 326
Hydrophobia, 234, 238
Hyper reactive Malarial Splenomegaly (HMS), 31
Hypersensitivity reactions, 114, 122, 223, 346
Hypertension
  portal, 114, 115
  salt intake *vs.*, 337–339
  stroke risk factors, 335, 337
  in urban areas, 339
Hypotrophy, 35

I
Immune cells
  destruction, 261
  infiltration, 252
  localization, 101
Immune response molecules, 15, 184, 190, 191, 194
Immunization
  Japanese encephalitis, 277
  rabies, 236
Immunoblotting, 69, 71, 75
Indeterminate leprosy, 228
Indirect immunofluorescence (IIF) assays, 102
Infantile paralysis, 7
Infections. *See also Specific types*
  associated with epilepsy, 357
  nematode (*see* Nematode, infections)
  of nervous system, 5
  trypanosome, 194–195
Inflammatory responses, 362
  in lepromatous patients, 228
Innate immune response, 252, 253
  in nervous system, 256–258
  in periphery, 255–256
Intensity theory, 151
Interferon (IFN), response in neurons, 256–257
Internalized Stigma Mental Illness Scale (ISMI), 56
International Federation of Anti-Leprosy Associations (ILEP), 46
Intracutaneous itch mediators, 152
Itch-dedicated nociceptor neurons, 152
Itching
  mediators, 152
  neuroscience of, 151–152
  onchocercal (*see* Onchocercal itch)

J
Jad/chaad. *See* Khat
Japanese B encephalitis, 270, 357
Japanese encephalitis (JE)
  anterior horn cell involvement, 274
  epidemiology, 270–271
  movement disorders, 274–275
  prevention, 277–278
Japanese encephalitis virus (JEV), 14, 270–272
  clinical features
    acute flaccid paralysis, 274
    diagnosis, 276
    differential diagnosis, 277
    electrophysiology, 276
    imaging, 275–276
    laboratory findings, 275
    management, 277

Japanese encephalitis virus (JEV) (*cont.*)
movement disorders, 274–275
outcome, 277
seizures, 274
cytokines, 273
epidemiology, 271
immunology, 273–274
life cycle, 272
pathogenesis and pathology, 272–273

K
Katayama fever. *See* Acute schistosomiasis
Khat
abuse potentials, 312–315
addiction, 296–298, 312–315
behavioral studies, 312
vs. cathinone, 315–316
conditioned place preference, 313
dependence, 296–298
development from niche to cash crop, 294
drug discrimination experimental studies, 313–315
in Ethiopia, 294–295
health problems, 296
in Kenya, 295
mental disorders, 299–300
pharmacological effects, 311–312
physiological disorders, 300
prevalence of use, 294–296
psychosis and psychiatric disorders, 298–300
psychostimulants, 308, 312–314, 316
self-administration study, 312–313
self-reinforcing study, 312–313
in Somalia, 295
tolerance, 315
use, 308–310
in Yemen, 294
Konzo
ancillary investigations, 324–325
clinical electrophysiology, 325
differential diagnosis, 325–326
epidemiology, 322–323
linamarin metabolism, 328
neuroepidemiology, 325
neurological examination, 324
neurological signs, 323–324
OCN, 329
pathogenesis, 326–329
SCN, 329
spastic/tetra paraparesis, 327
treatment and prevention, 329

L
Lagochilascaris minor, 72–73
Late-onset effects, 16
Leishmania, 25
Leishmaniasis, 25, 103
Lepromatous leprosy (LL), 52, 53, 228
Leprosy, 11, 12, 51–52
autonomic nerve involvement in, 227
bacillus, 51
causative organism, 218–219
clawed and amputated fingers, 53
clinical manifestations
neuropathies and skin lesions, 224–226
phenotypes, classification, 224
demyelination in, 227–228
indeterminate, 228
inflammation with neurodegeneration, 223
laboratory findings, 227–228
lepromatous, 228
nerve injury in, 227
neural infection and bacterial propagation, 219
neuritic, 226
as neurodegenerative diseases, 229
pathogenesis, 219–223
pathological features, 228
PNS involvement in, 217, 219
stigma, 57
tuberculoid, 228
Leptomeninges, trypanosomes into, 186–188
Lethargic encephalitis, 8
LL. *See* Lepromatous leprosy (LL)
Loa loa, 83, 150
encephalopathy, neurobiology, 158
infection, 35, 36
microfilaremias, 36
Loiasis, 36
Low cord syndrome, 120
Lychenification, of skin, 54
Lymphatic filariasis, 82, 83
Lyssavirus
human rabies due to, 240
rabies and, 234

M
Macrophages, 98, 185, 186, 192, 280
Magnetic resonance imaging (MRI)
brain image analyses using, 102
Japanese encephalitis, 275–276
rabies, 239
Malaria, 6, 359–360
in Africa, 370
behavioural disturbances, 370

cognitive impairment, 371
effect of parasitaemia on cognitive function, 371–375
falciparum (*see* Falciparum malaria)
post-cerebral, 378
post-severe, 376–377
Malayan pit viper, 341
Mannitol, 122, 140
Marungi. *See* Khat
Matrix metalloproteases (MMPs), 188
Mebendazole, 82
Megaesophagus, 205, 210–212
Memory T cells, DENV, 281
Meningoencephalitis, 73, 76, 203
Mental disorders, khat, 299–300
Mental health, 381–382
Microbe-host interactions, 15–17
Microbe-host symbiotic behavior, 15–16
Microbes, 6
for nervous system, 16
pathogenic, 185
Microfilaria, 54, 158
in cerebrospinal fluid, 362
detection, 83
Miltefosine, 103
Milwaukee Protocol, for rabies, 243–245
Miraa. *See* Khat
Mood disorders, 174
Mosquito-borne Japanese encephalitis virus, 14
Mosquito borne viruses, 270
Motor conduction velocity (MCV), 227
Motor dysfunction, 226
Motor neuron disorder, 322–329
Movement disorders, in Japanese encephalitis, 274–275
MRI. *See* Magnetic resonance imaging (MRI)
Multiple-drug therapy (MDT), 229
Muscle atrophy, 53
Mycobacterium leprae (ML), 12, 13, 51, 218, 219
Myenteric plexuses, denervation, 205

N

NC. *See* Neurocysticercosis (NC)
Necrotic-exudative granulomas, 113, 116
Neglected tropical diseases (NTD), 4, 8
snake bite and, 339
stigma and, 49–50
travel, migration and, 21–25
African trypanosomiasis, 31–33
Chagas disease, 28–31
cysticercosis, 33–35
dengue, 25
echinococcosis, 25–26
filariases, 35–36
leishmaniasis, 25
onchocerciasis, 26
rabies, 26–28
schistosomiasis, 36–39
strongyloidiasis, 39–41
Neglected tropical diseases and conditions (NTDC)
challenges and problems, 15–16
lethargic encephalitis, 8
neglected nervous system in, 4, 16–17
neurosyphilis, 5–7
poliomyelitis, 7
Negri bodies, 13, 14
in apical dendrite, 240
in rabies, 240
Nematode, 16
filarial, 36, 148, 150
infections, 8
angiostrongyliasis, 73–75
Ascaris suum, 71
Baylisascaris procyonis, 71–72
filaria, 82–83
Gnathostoma species, 76–79
Lagochilascaris minor, 72–73
Strongyloides stercoralis, 80–81
Toxocara species, 69–70
Trichinella species, 81–82
Neonatal meningitis, 358
Nerve conduction
leprosy, 227
peripheral and central, 324
Nervous system (NS)
Chagas disease and, 203–204
inflammatory response, 257–258
innate immune response in, 256–258
involvement in Chagas disease, 208–212
RABV in (*see* Rabies virus (RABV), in NS)
Neural-immune interactions, 15
Neural larva migrans (NLM), 68–70
Neural lesions, 112
Neurocysticercosis (NC), 34, 358–359
causative agent, 129–131
clinical manifestations, 132–134
diagnosis
immunodiagnosis, 137–139
neuroimaging, 136–137
and epilepsy, 132–133
epileptogenesis and, 133
eradication, 141
as etiology of epilepsy, 133
extraparenchymal, 134, 138
host and parasite factors, 134–135

Neurocysticercosis (NC) (*cont.*)
immunological response, 131–132
life cycle, transmission, 128–129
parenchymal, 132–134, 136
risk of seizure recurrence, 134
treatment, 139–140
Neurodegeneration, 329
in leprosy, 223
Schwann cells infection to, 222
Neurodegenerative diseases, leprosy as model for, 229
Neurodermatitis. *See* Leprosy
Neuro-development, HIV effects
cognitive function, 381
language, 380–381
mental health, 381–382
motor, 380
potential pathways, 382
Neurognathostomiasis, 68, 76, 79
causative agent of, 77
clinical manifestations of, 78
Neuroinflammation
chronic, 196
limitation of, 258
Neurological diseases, stigma
epilepsy, 52–54
human African trypanosomiasis, 55
leprosy, 51–53
onchocerciasis, 53–55
schistosomiasis, 55
Neuronematodiasis
causative pathogens, 68
clinical features, 68
diagnosis, 69
Neurons
IFN response in, 256–257
preserving integrity of, 254–255
Neuropathy
clinical signs and symptoms, 224–225
optic, 152–154
Neuroschistosomiasis (NS), 112–113
clinical features, laboratory investigations and diagnosis, 118–120
CNS involvement, 112
experimental investigations on, 117–118
pathogenesis, 115–117
prognosis, 122
treatment, 121–122
Neuroscience
itching, 151–152
sleeping sickness, 195–196
stigma behaviour, 50–51
Neurosyphilis, 5–7
Neurotropic virus, 14, 16, 253, 258, 282
Nicotinic acetylcholine receptor, 237
NLM. *See* Neural larva migrans (NLM)
Nociceptor neurons, 152
Non-rapid eye movement (NREM) sleep, 176, 194
Norepinephrine systems, pharmacological effects, 311–312
NS. *See* Neuroschistosomiasis (NS)
NTD. *See* Neglected tropical diseases (NTD)
NTDC. *See* Neglected tropical diseases and conditions (NTDC)
Numerical rating scale (NRS), 150

O
Ocular angiostrongyliasis, 74, 75
Ocular larva migrans (OLM), 69, 70, 72, 83
Ocular onchocerciasis, 153, 154, 159
Onchocercal eye disease, 149, 154
Onchocercal itch
clinical importance, 150–151
definition and clinical assessment, 150
peripheral nervous system, 150–152
Onchocercal skin disease, 54, 149, 150
Onchocerca volvulus, 53, 83, 148
Onchocerciasis, 26, 53–55, 361–362
blind spot of, 152–155
distribution and clinical forms, 148–150
epidemiology, 152–154
and clinical manifestations, 157–158
epilepsy, 155–156
ocular, 153
on optic nerve disease, 154
optic neuropathy, 152–155
PNS involvement, 150–152
post-treatment encephalopathy, 157–158
seizures and evolution, 157
Onchodermatitis, 54, 149
Opium, 309, 310
Optic atrophy, 153, 154
Optic nerve disease (OND)
etiopathogenesis, 155
onchocerciasis on, 154
pathogenesis, 155
reduction in, 154
Optic neuropathy, 152–154, 324

P
Panstrongylus megistus, 11
Paralysis, 77, 226
acute flaccid, 274
infantile, 7
lower limb, 173

Paralytic rabies, 234, 238
Parenchymal neurocysticercosis, 132–134, 136
Parkinsonian features movement disorders, 275
Parkinson's disease, 56
PD-1 (programmed death protein-1), 258, 260
Pentamidine, 33, 104
People with epilepsy (PWE), 57
Peripheral nerve ganglia, 186–188
Peripheral nervous system (PNS), 203
  chronic infection attacks, 11
  glial cells of, 219
  onchocercal itch, 150–152
Peripheral neuropathy, 83, 224
Periphery
  adaptive immune response in, 259
  innate immune response in, 255–256
PES. *See* Pseudotumoral encephalic schistosomiasis (PES)
Plasmacytoid dendritic cells (pDCs), 259
Plasmodium falciparum, 359, 370, 371
Plasmodium vivax, 6
Pleocytosis, 78, 102
  CSF, 239
  lymphocytic, 131
PNS. *See* Peripheral nervous system (PNS)
Poliomyelitis, 7
Polysomnographic syndrome, 175, 176
Polysomnography (PSG), 175
Post-exposure prophylaxis (PEP), rabies, 26, 27
Poverty-promoting chronic infectious diseases, 22
Praziquantel (PZQ), 38, 121, 122, 140
Procyon lotor, 71
Prostaglandin D_2 (PGD_2), 194
Pruritus
  pathogenesis, 151
  peripheral nervous system, 150–152
Pseudotumoral encephalic schistosomiasis (PES), 115, 118
PZQ. *See* Praziquantel (PZQ)

Q

Qat. *See* Khat
Quantitative buffy coat (QBC) test, 29, 31

R

Rabies, 13–15, 26–28
  bat, 235–236
  canine, 235
  cardiac complications, 239
  coyote, 236
  diagnostic investigations, 239–240
  encephalitic, 234, 238
  encephalomyelitis, 234
  endocrine complications and, 239
  epidemiology, 235–237
  excitotoxicity, 243
  fox, 236
  human, 237–239
    due to lyssaviruses, 240
    management, 242–245
    prognosis, 241–242
  lyssavirus genotypes, 234
  Milwaukee Protocol for, 243–245
  Negri bodies in, 240
  paralytic, 234, 238
  post-exposure prophylaxis, 26
    WHO guidelines, 27
  prodromal symptoms, 238
  raccoon, 236
  salivary glands and, 238
  skunk, 236
  sterilisation of infection, 260
  woodchucks, 236
Rabies virus (RABV), 234, 240
  antigen, 239
  bat, 243
  epidemiology, 235–236
  in human, 237–238
  RNA, 242
  vaccine, 13
Rabies virus (RABV), in NS, 251–254
  adaptive immune response, 258–259
    in periphery, 259
    T cells, 260–261
  CNS mediated immune-unresponsiveness, 261–262
  IFN response in neurons, 256–257
  inflammatory response, 257–258
  innate immune response
    in NS, 256–258
    in periphery, 255–256
  morbidity and mortality, 261
  preserving integrity of neurons, 254–255
  virus replication, 252
Raccoon rabies, 236
Radiculoencephalomyelitis, 76
Radiculomyelitis, 76
Rapid eye movement (REM) sleep, 176
Regeneration
  of peripheral nerve, 220
  Schwann cells, 221–222
Respiratory complications, and rabies, 239
Reverse transcription polymerase chain reaction (RT–PCR), 239
Rhodesiense, HAT, 32, 33
River blindness. *See* Onchocerciasis

S

Salivary glands, and rabies, 238
Salt, 336, 346
  intake, 337–339
  restriction, 338
Scambler's hidden distress model of stigma, 48, 58
Schistosoma haematobium, 37
Schistosoma japonicum, 37
Schistosoma mansoni, 37, 117, 121
Schistosomiasis, 9, 36–39, 112
  acute, 37
  chronic, 38
  history of, 113–114
  stigma in, 55
Schizophrenia, 299–300
Schwann cells, 13, 219
  demyelination and regeneration, 221–222
  infection, 221–222
    to neurodegeneration, 222
  signaling pathways in, 222–223
SCS. *See* Spinal cord schistosomiasis (SCS)
Seizures, 356. *See also* Epilepsy
  acute symptomatic, 356, 359
  classification, 356
  complex partial, 360
  frequency, 362
    in African studies, 360
  HIV with, 357
Sensory loss, 225–226
Sensory-motor deficit, 35
Serological tests, 75, 79, 284
Serotonin, 314
  depletion of, 316
  pharmacological effects, 311–312
Skin
  lychenification of, 54
  nodules, 54
  ulcerations, 100
Skin lesions, 52, 172
  clinical signs and symptoms of, 224–225
Skin rash (trypanides), 173
Skunk rabies, 236
Sleeping sickness. *See* African trypanosomiasis; Human African trypanosomiasis (HAT)
Sleep onset rapid eye movement (REM) sleep periods (SOREMP), 175–177
Sleep-wake disorder, 175–177
Smallpox, 6
Snake bite, 347. *See also* Snake venom
  black stone use, 345
  cerebral haemorrhage, 344
  cerebral infarction, 343–344
  cerebrovascular accidents, 343
  cerebrovascular treatment complications, 345
  local symptoms and signs of, 342
  neglected tropical disease and, 339
  neurological complications, 342–343
  systemic manifestations, 342
Snake venom
  antivenin, 345–346
  components, 341
  composition
    cytotoxic, 340, 341
    neurotoxins, 341
  stroke and (*see* Stroke)
  toxins, 341
  types, 340
Sodium chloride. *See* Salt
Soil-transmitted helminths, 39
Spinal bombesin-recognized neurons, 152
Spinal cord schistosomiasis (SCS), 115
  diagnosis, 119–121
  histopathological findings, 116–117
  treatment, 122
Spinal gnathostomiasis, 80
Stigma, 46–49
  accepted, 48
  anticipated, 48
  assessment, impact, 55
  behaviour, neuroscience of, 50–51
  consequences of, 57–58
  determinants of, 57
  enacted, 47
  endorsed, 48
  fair and legitimate discrimination, 47
  felt, 47–48
  internalized, 48
  measurement, 55–57
  model, 47, 48
  neurological diseases associated with, 51–55
  and NTDs, 49–50
  reduction
    approaches to, 58–59
    challenges to, 59
    domain of, 58
  Scambler's hidden distress model of, 48, 58
  in schistosomiasis, 55
Stigma experience scale (SES), 56
Stigma impact scale (SIS), 56
Stigma Scale for Chronic Illnesses:8-Item Version (SSCI-8), 56
Stroke, 335
  salt intake and, 339
  snake bite and
    cerebral infarction, 343–344

cerebrovascular accidents, 343
cerebrovascular treatment
complications, 345
local symptoms and signs, 342
neurological complications, 342–343
systemic manifestations, 342
snake venom and
antivenin, 345–346
composition, 340–341
Strongyloides stercoralis, 39, 40, 80–81
Strongyloidiasis, 39–41, 80
hyperinfection syndrome, 81
Subarachnoid hemorrhage (SAH), 76
Subcutaneous filariasis, 82–83
Suppressors of cytokine signaling (SOCS), 258
Suprachiasmatic nucleus (SCN), 195
Syphilis, 5, 6
Systemic hypersensitivity reaction, 114

T
Taenia multiceps, 26
Taeniasis, 33, 131
Taenia solium, 33, 128, 131, 362
T cells
invasion of brain parenchyma, 260–261
memory, 281
penetration of brain parenchyma, 191
and RABV, 260–261
Terrifying disease, 13
Thiabendazole, 73
Thromboembolia, 28
Toxicity, eosinophilic leukocyte-mediated, 115
Toxocara, 361
NLM, 70
OLM, 71
species, 69–70
VLM, 71
Toxocara canis, 69, 70
Toxocara cati, 69, 70
Toxocariasis, 68–71
Toxoplasma gondii, 10, 361
Toxoplasmosis, 4, 361
Treponema pallidum, 5
Triatomines, 10, 11
Trichinella species, 81–82
Trichinellosis, 81, 82
Trophozoites, 95–97
Trypanosoma brucei, 10, 166
gambiense, 32, 166, 167, 170, 177
molecules/factors involve in, 190
rhodesiense, 32, 166, 167, 177
Trypanosoma cruzi, 10, 11, 28, 29, 202, 204, 206
Trypanosoma fever, 168
Trypanosome
BBB crossing
host-derived molecules, 189–192
parasite-derived molecules, 189
control of, 192–194
growth in CNS, 192–194
and host immune system, interactions between, 185–186
infections, 194–195
invasion of CNS, 186–192
parasites, 184–185
Trypomastigote, 29
Tsetse fly, 10, 166–169, 184
Tuberculoid leprosy (TT), 228

U
Ulnar nerve, 226, 227

V
Vaccines
against JEV infection, 278
RABV, 13
Variant surface glycoproteins (VSGs), 185
Vasoactive intestinal peptide (VIP), 193
Vector-borne disease, 82
Vector-borne transmission, 202
Venomous snakes, 340
Verbal rating scale (VRS), 150
VFR ("visiting friends and relatives") travellers, 25
Viral infections, epilepsy, 357
Viral metagenomic analyses, 16
Virulence factors, 101
Visceral larva migrans (VLM), 69, 71, 83
Visual analogue scale (VAS), 150

W
Wolbachia, 158
Woodchucks rabies, 236
Wuchereria bancrofti, 36, 82, 83

Z
Zoonosis, 28
Zoonotic nematodiasis, 68
Zoonotic pathogens, 68

MIX
Papier aus verantwortungsvollen Quellen
Paper from responsible sources
FSC® C105338

FSC
www.fsc.org

If you have any concerns about our products,
you can contact us on
ProductSafety@springernature.com

In case Publisher is established outside the EU,
the EU authorized representative is:
Springer Nature Customer Service Center GmbH
Europaplatz 3, 69115 Heidelberg, Germany

Printed by Libri Plureos GmbH
in Hamburg, Germany